Chemical Geology, 34 (1981) 179—180
Elsevier Scientific Publishing Company, Amsterdam — Printed in The Netherlands

Book Review

Kerogen — Insoluble Organic Matter from Sedimentary Rocks. B. Durand (Editor). Éditions Technip, Paris, 1980, xxviii + 519 pp., Ffr. 490.00.

It is safe to say that there are few books on the market that fill an obvious gap as does this volume on kerogen, the finely dispersed carbonaceous material in sedimentary rocks commonly defined as the acid-insoluble, polycondensed end-product of diagenetic alteration of organic substances derived from living organisms. Being by far the most abundant form of organic matter on Earth, kerogen represents an important intermediate in the geochemical transformation of organic carbon. Moreover, it is the ultimate source of most petroleum, and the widespread occurrence of kerogenous substances (and their metamorphosed derivatives) in rocks dating back as far as 3.8 Ga is likely to constitute the oldest evidence of life. In spite of these crucial implications, kerogen studies have been neglected for a long time, and it was basically during the last decade that substantial progress has been made towards an elucidation of the chemical nature of this intriguing constituent of sedimentary rocks. Accordingly, a review of the field as a whole seemed to be called for at this stage, and the editor of this volume has rendered an invaluable service to the community at large for submitting an updated summary covering the principal facets of the kerogen story.

Introduced by an editorial keynote chapter by B. Durand ("Sedimentary Organic Matter and Kerogen — Definition and Quantitative Importance of Kerogen"), the book contains 15 papers (3 of which are written in French) dealing with selected aspects of the kerogen problem. Among the topics reviewed by specialists in the respective fields are the microscopic (B. Alpern, A. Combaz, P. Robert) and electron microscopic (A. Oberlin, J.L. Boulmier, M. Villey) methods of kerogen investigation as well as the wide field of analytical techniques currently in use for chemically characterizing kerogenous substances (thermogravimetric analysis, C. Durand-Souron; infrared spectroscopy, P.G. Rouxhet, P.L. Robin, G. Nicaise; electron paramagnetic resonance, A. Marchand, J. Conard; degradational and related chemical techniques, D. Vitorović). Exhaustive overviews are, furthermore, devoted to the elemental composition (C, H, N, O, S, Fe) of kerogens and its change during the maturation pathway (B. Durand and J.C. Monin), and to the carbon isotopic composition of sedimentary organics (E.M. Galimov). The rest of the contributions deal with procedures for kerogen isolation (B. Durand and G. Nicaise), the soluble extracts (M. Vandenbroucke), the incorporation of organic matter in recent sediments and the formation of proto-kerogen (A.Y. Huc), and the geochemical evolution of organic matter in general (R. Pelet).

Although not an encyclopedia in the proper sense, the volume is nonetheless of unrivalled comprehensiveness in its field, providing a rich source of data and general information for the specialist and student alike. Minor flaws are related to the fact that non-Anglo-Saxon authors do their very best to write English texts; however, the few French chapters interspersed throughout the pages do not fail to teach the average reader which to regard as the lesser evil. Inspecting a current frontier, the volume is certainly a "must" for anybody dealing with the impact of life on Earth in the widest sense — whether from an economic or purely academic standpoint. Indeed, here is one of the rare cases in which the reviewer is tempted to quote the German naturalist—philosopher G.C. Lichtenberg:

> "He who owns two pairs of trousers, let him sell one and buy this book."

In view of the formidable price of this well-printed and lavishly illustrated volume, this dictum may be taken almost literally.

M. SCHIDLOWSKI (Mainz)

KEROGEN

INSOLUBLE
ORGANIC MATTER
FROM
SEDIMENTARY ROCKS

AMONG OUR BOOKS

- Formation des gisements de pétrole
  Étude des phénomènes géologiques fondamentaux.
  C. SALLÉ, J. DEBYSER

- Éléments de pétrologie dynamique des systèmes calcaires
  Tome 1 : Description macroscopique et microscopique. Diagenèse. Applications.
  Tome 2 : Atlas photographique.
  L. HUMBERT

- Les grès du paléozoïque inférieur au Sahara
  Sédimentation et discontinuités. Evolution structurale d'un craton.
  S. BEUF, B. BIJU-DUVAL, O. de CHARPAL,
  P. ROGNON, O. GARIEL, A. BENNACEF

- Dépôts évaporitiques
  Illustration et interprétation de quelques séquences.

Publication in English Language

- Advances in Organic Geochemistry 1973
  Proceedings of the 6th international Meeting on Organic Geochemistry.

- Evaporite Deposits
  Illustration and Interpretation of some Environmental Sequences.

KEROGEN

INSOLUBLE ORGANIC MATTER FROM SEDIMENTARY ROCKS

Edited by
Bernard Durand
Institut Français du Pétrole

1980

ÉDITIONS TECHNIP 27 RUE GINOUX 75737 PARIS CEDEX 15 technip

ISBN 2-7108-0371-2

Printed in France
by Imprimerie Bayeusaine, 14401 Bayeux

Foreword

This book devoted to kerogen – the insoluble organic matter of sedimentary rocks and the source for most petroleum – is a very timely publication. Originally, kerogen had been studied in oil shales and was merely considered as a raw material which yielded shale oil when the rock was retorted at high temperature. Later, petroleum geochemists resumed studies on kerogen from source rocks, and this field has been very active over the last ten years. The composition and properties of kerogen are now an essential aspect in understanding the origin of oil and gas. This study also constitutes a practical and important tool for petroleum exploration.

No one other than Bernard Durand was more qualified to edit such a book on kerogen. Dr. Bernard Durand graduated from the *Ecole Nationale Supérieure des Mines* in 1961 and started research in mining geology and inorganic geochemistry. After completing his thesis in that field, he joined the *Institut Français du Pétrole (IFP)* in 1966 and has worked continuously in petroleum geochemistry since that time. For many years kerogen has been his main subject of interest and activity. Besides his personal work at *IFP,* Dr. Durand has made incessant efforts to promote cooperation among well-known specialists in various fields from several universities and industrial research centers and to have them focus their activity on kerogen research. The result of this cooperation has been a considerable advance in the understanding of kerogen and its significance in the formation of petroleum. Many international meetings and research conferences have confirmed the prominent role played by Dr. Durand and his co-workers in the main achievements of kerogen studies.

The principal results of this collective enterprise are presented in this book, mainly by the persons who where individually responsible for each specialized field. Furthermore, the book is enriched by the participation of various European scientists who have carried out independent research in their respective countries. Kerogen is considered here not only as a specific constituent of sedimentary rocks, but also as a particular step in the cycle of organic carbon on Earth. Its composition and chemical properties are explained by the succession of events which finally resulted in the formation of kerogen: photosynthesis, food chain, sedimentation, biochemical and chemical actions. In turn kerogen composition, together with subsequent geological history, will be responsible for the compositions of the oil and gas to be generated. From that point of view, this book should be of great interest to all scientists concerned with earth sciences and to all professionals interested in prospecting, producing and transforming fossil fuels.

B. TISSOT

Director, Science and Advanced Technology
Institut Français du Pétrole

Introduction

The initial aim of this book was to describe the research that has been done on kerogen for about 10 years by a group of French scientists under the initiative of the *Institut Français du Pétrole (IFP)* and *Compagnie Française des Pétroles (CFP)*. However, with the help of a few additional contributions, it has appeared possible to deal with kerogen if not exhaustively at least relatively thoroughly. I am pleased that professors B. Alpern, E.M. Galimov and D. Vitorović as well as engineer P. Robert from *Société Nationale Elf Aquitaine — Production (SNEA (P)* agreed to join us in seeing that the initial project received ample development.

It has not been possible to deal with some subjects as fully as I would have wished. This is mainly true of the structure of kerogens, the simulation of their evolution during burial and their sedimentology, all of which are subjects which are partially dealt with at times but not given individual treatment. These subjects are difficult ones in which knowledge is somewhat spotty and which should be developed in the future.

Readers must not look upon this book as an encyclopedia of kerogen but rather as a succession of images which the different authors formulate for the time being through their specialized outlooks. One must not be surprised to find points of disagreement. The divergences highlight the difficulty which still exists in obtaining a synthetic understanding of organic sedimentary matter. Indeed, whereas this has been possible in the more restricted field of coal — and the famous book *Coal* by D.W. Van Krevelen is one of the best examples — the generalization of such understanding to all types of organic sedimentary matter comes up against three main difficulties:

(a) The great diversity of sedimentation media.
(b) The need to isolate organic matter from its mineral context.
(c) The insoluble nature of most of this organic matter (kerogen).

These difficulties have delayed advances in knowledge but have not prevented them, and I hope that this book will convince readers that an important step forward has been made.

The study of kerogen has been helped along enormously by the many research projects previously done on coal. I hope that our colleagues who are coal specialists can in turn find results here that are of use to them at a time when there is evidence of renewed interest in their discipline. I also hope that this book will likewise be read by those concerned with the upgrading of heavy oil cuts. Many methods developed for kerogen can indeed be transposed for the analysis of resins and asphaltenes which make up a large portion of such cuts.

This book includes three Chapters in French. Everybody should read them carefully given the importance of these Chapters and the quality of their authors.

We owe a great deal to our colleagues in the organic geochemistry department of the *Institut Français du Pétrole* thanks to whom an inestimable total of results concerning kerogen have been gathered. We also owe a great deal to our colleagues from oil companies, universities and government agencies in different countries who have helped us and listened to us in the last few years. Among them I am thinking in particular of those who have provided sediment samples, the indispensable basis for research, i.e. Mrs. M. Teichmüller, Miss Y. Sommers and Messrs. R. Cane, G. Claypool, J. Claret, J. Connan, G. Demaison, G. Dunoyer de Segonzac, H. Hagemann, A. Hood, B. Housse, J. Hunt, J.L. Oudin, R.G. McCrossan, P. Montmessin, W. Robinson and D. Welte.

I would like to extend my personal thanks to my wife, Nicole, as well as to Miss C. Lefèvre, Mrs. E. Deroze, Mrs. M. Vandenbroucke and Messrs. G. Brace and J.C. Marcolin for the considerable help they have given me in assembling and compiling the final version of the book, and to my colleagues Mrs. Y. Debyser, Mrs. J. Roucaché and Messrs. G. Deroo, J. Espitalié, J. Herbin and M. Madec for fruitful discussions in all the last years.

Lastly, I owe special thanks to A. Combaz, R. Pelet and B. Tissot who have given initial impulse to much of the research described and have provided precious advice, as well as to A. Douglas for his critical review and his translation of some of the Chapters. I am also grateful to *Compagnie Française des Pétroles, Société Nationale Elf Aquitaine* and *Institut Français du Pétrole,* and more particularly to Mr. Cl. Sallé, Director of Exploration, and Mrs. J. Funck, Director of Information and Documentation, for help and encouragment for many years now.

B. DURAND
Institut Français du Pétrole

Contents

1

Sedimentary organic matter and kerogen. Definition and quantitative importance of kerogen

B. DURAND

2
Procedures for kerogen isolation

B. DURAND and G. NICAISE

3
Les kérogènes vus au microscope

A. COMBAZ

4
Elemental analysis of kerogens (C, H, O, N, S, Fe)

B. DURAND and J.C. MONIN

5

Thermogravimetric analysis and associated techniques applied to kerogens

C. DURAND-SOURON

6

Characterization of kerogens and of their evolution by infrared spectroscopy

P.G. ROUXHET, P.L. ROBIN and G. NICAISE

7

Electron microscopic study of kerogen microtexture. Selected criteria for determining the evolution path and evolution stage of kerogen

A. OBERLIN, J.L. BOULMIER and M. VILLEY

PART ONE
KEROGENS AMONG CHARS. HOW TO DETERMINE MICROTEXTURE (MICROSTRUCTURE)

PART TWO
EXPERIMENTAL RESULTS AND DISCUSSION

PART THREE
PYROLYSIS AS A MODEL OF CATAGENESIS

GENERAL CONCLUSIONS

8

Electron paramagnetic resonance in kerogen studies

A. MARCHAND and J. CONARD

9

C^{13}/C^{12} in kerogen

E.M. GALIMOV

10

Structure elucidation of kerogen by chemical methods

D. VITOROVIĆ

11

Pétrographie du kérogène

B. ALPERN

12
The optical evolution of kerogen and geothermal histories applied to oil and gas exploration

P. ROBERT

13

Structure of kerogens as seen by investigations on soluble extracts

M. VANDENBROUCKE

14
Origin and formation of organic matter in recent sediments and its relation to kerogen

A.Y. HUC

15
Évolution géochimique de la matière organique

R. PELET

Abstracts of the Chapters

(p. 13)

CHAPTER 1

Sedimentary organic matter and kerogen. Definition and quantitative importance of kerogen

B. DURAND

Less than 1% of the organic matter (OM) synthesized by photosynthesis each year succeeds in escaping from being recycled or biochemically or physicochemically degradated so as to be incorporated in sediments and then to evolve through burial toward becoming a residue containing carbon and simple molecules such as CO_2, H_2O, CH_4, etc. During this incorporation, the portion which preserves the structural features of the living OM is termed **inherited,** as opposed to the so-called **neoformed** portion. In sedimentary OM, a distinction is also made between an **autochthonous** fraction, which is derived from organisms living in the sedimentation environment, as opposed to the **allochthonous** fraction as well as a **reworked** fraction resulting from the erosion of ancient sedimentary strata or crystalline and volcanic rocks. Sedimentary OM is analyzed by **petrographic** methods to create categories by optical techniques as well as by **chemical** methods to create categories by chemical fractionation or degradation techniques. One of these chemical categories is **kerogen.** This term has had and still has various meanings which are passed in review. It is defined in this book as the insoluble (in relatively unpolar solvents) fraction of sedimentary OM of any kind, i.e. ancient or recent sediments etc., and in a dispersed or concentrated form. Its analysis mainly corresponds to an examination of the relations between the hydrocarbons which are solvent extracted from it and the remainder of the sedimentary OM, and therefore it has mainly been done as part of research on the origin of petroleum. But it is also needed as part of the effort to understand all types of sedimentary OM on account of its quantitative importance. Indeed, kerogen represents an average of more than 80% weight of such matter.

The amounts of kerogen present in sediments are somewhere in the vicinity of 10^{16} t, i.e. approximately 1 000 times the ultimate resources of fossil fuels.

(p. 35)

CHAPTER 2

Procedures for kerogen isolation

B. DURAND and G. NICAISE

At the present time procedures for kerogen isolation must be adapted either for physico-chemical analysis or for petrographic analysis. Physico-chemical analysis require good recovery with little chemical alteration of the organic matter (OM), while petrographers need satisfactory preservation of the morphology of macerals and microfossils. In each case the sediment must be ground differently. The former require a fine grain size which results in the breaking of the microfossils, while the latter need a coarse particle size with the result that recovery of OM is not satisfactory. Both use physical and chemical methods. In general physical methods are better suited to petrographic analysis, while chemical methods are more appropriate for physico-chemical analysis. The physical methods used are mainly separation by flotation, centrifuging and dense liquids. The chemical methods are generally the destruction of most of the mineral phase with HCl and HF, sometimes followed by the destruction of pyrite by $LiAlH_4$, NO_3H or ferric sulfate and/or redissolving of neoformed fluorides with hot HCl. The use of HCl and HF obviously results in the hydrolysis of little evolved OM (recent sediments, soils, peat, shallow sediments) and the organic residue that is no longer kerogen. Evolved OM is altered only to a slight degree. $LiAlH_4$ and ferric sulfate only partly eliminate pyrite, and OM is not very much altered. Use of NO_3H results in good pyrite destruction but also in a high degree of oxidation and nitration of the kerogen.

(p. 55)

CHAPITRE 3

Les kérogènes vus au microscope

A. COMBAZ

This Chapter presents an overview and a typology of organic components in sedimentary rocks as seen through the microcospe.

During the past decade, there has been a remarkable convergence of the research being conducted in fields of study as different as paleobotany, palynology, coal petrography, organic geochemistry, pedology and geomicrobiology on the organic components of rocks, and hence of kerogen. The naturalist's procedure, which consists essentially of direct observation of thin sections of rocks, is complementary and indispensable to any geological research. When isolated from their mineral context and viewed under the microscope, the organic components are seen to be of highly varied composition and origin, and given a suitable sampling, this microscopic observation can lead to the reconstruction of the sedimentary and paleogeographic environments.

Every sedimentary rock is able to have carbon content. Some of them are almost exclusively formed by organic matter in more or less thick beds.(Plate 3.1):

(a) The (ligno) humic coals, formed from higher-form vegetation fragments of varying sizes; humic gel (vitrinite), fusinization remains (inertinite), and secondary components, i.e., spores and pollens (exinite). The predominant elements of these rocks were generated under aerobic conditions.

(b) The sapropels, on the other hand, are essentially formed from carbonaceous material of aquatic origin (algal material). To this category belong the bogheads (kerosene shale, torbanite, balkachite, kukersite, coorongite, etc.) formed almost exclusively from a microscopic alga *Botryococcus brauni;* the tasmanites formed from *Tasmanites;* the shungites found in Precambrian terrain and also formed from microscopic algae and/or bitumens formed from the latter; the diatomites formed from the accumulation of the siliceous frustules of diatoms and their organic components.

Sedimentary rocks with low organic content include:

(a) the laminites (or oil shales) with a varved structure in which argilaceous-carbonate and carbonaceous microbeds alternate (the Silurian shales of the Sahara, the «Hot shales» of the North Sea, the «schistes-carton» of the Paris area, the «Green River shales» of Colorado, etc.).

(b) Most frequently the amount of organic carbon is small or very small: the grey and black shales, (Plate 3.2) differ from the preceding shales. On the other hand we can't observe distinct laminations. But they can be excellent source-rocks for oil.

(c) the carbonates which also can contain significant levels of organic matter if they were formed in confined sedimentary environments;

The types of organic components are either form-preserving or amorphous.

In the form-preserving fraction, there are three families:

1. The terrestrial microfossils and vegetation fragments (Plate 3.3) which include **the spores and pollens** whose integument (exine) is a biochemical complex composed of lignin nucleii and carotenoid polymers: sporopollenine; and vegetation fragments, or organic dust derived from decaying plants of all sizes (degraded cellulosic and lignitic compounds).

2. The algal microfossils and fragments (Plate 3.4) from phytoplancton and phytobenthos. Several groups have been distinguished: the **Acritarchs** *(incertae sedis),* the **Dinoflagellates,** various unicellular, cenobitic and colonial **Algae.**
Identifiable fragments of benthonic pluricellular algae are rare.

3. The microfossils and animal remains most commonly found in Paleozoic rocks (Plate 3.5): the **Chitinozoa,** the **Graptolite** secula, the **Scolecodonts,** the epidermis fragments of **Gigantostraca. Microforaminifera** are frequently found in Mesozoic and Cenozoic systems.

The amorphous —and by far the largest— fraction (Plate 3.6) cannot be biologically defined because of lack of observable structure.

All categories of origin are to be found in this fraction: microfossil membranes, various macrofossil tissues rendered amorphous by early diagenesis and catagenesis, primary organic substances, catabolic organic material. One can distinguish, without specific interpretation, organic matter which is aggregate, granular, sub-colloidal, pellicular, or gelled. It is important not to confuse these descriptive terms with explanatory terms such as sapropelic, humic, etc.

Aside from the occasionally observed soluble fraction (Plate 3.6, photo 14) one other fraction should be mentioned: the mineral fraction included within the organic fraction, usually pyrite (Plates 3.8 and 3.5, photo 1).

The palynofacies families are logically grouped according to the three principal types of kerogen as represented in the Van Krevelen diagram (Plate 3.10). The upper type represents algal kerogen (sapropelic), the lower type is (ligno) humic kerogen and the middle type is liptinic kerogen, or a facies made up of the other two.

With microscopy, it is possible to distinguish the facies which characterize kerogen in the various stages of its formation: early or diagenetic changes, and late or catagenetic changes (Fig. 3.7). It is thus possible, with microscopic observation, to master the complexity of organic matter in rocks. Of great value in organic geochemistry, optical research on kerogens is a remarkable tool for geologic retrodiction and for oil prediction.

(p. 113)

CHAPTER 4

Elemental analysis of kerogens (C, H, O, N, S, Fe)

B. DURAND and J.C. MONIN

Elemental analysis of kerogen involves the usual micro-analytic techniques for organic substances. Some preparation is necessary, i.e. grinding and drying, because of the heterogeneity and hygroscopic characteristics. Organic sulphur is obtained by measurement of the difference between total sulphur and pyrite sulphur which is often abundant. The latter is determined by measurement of iron contents. Oxygen contents is often not carried out but is indispensable. In the field there are considerable differences between the various laboratories because of inadequate analytic techniques. The weight shares for C, H, O, S, and N in kerogen vary respectively within the following approximate limits: 60-90% (average ≈ 76%), 1-11% (average ≈ 6%), 1-30% (average ≈ 11%), 0-12% (average ≈ 3.6%) and 0-4% (average ≈ 2%). In spite of careful preparation pyrite contents can exceed 70% and other residual mineral contents may be more than 30%.

The interpretation is mainly based on C, H and O diagrams. The Van Krevelen diagram (H/C, O/C, atomic ratios) is the most suitable. By means of elemental analysis the various types of organic matter

(OM) and evolution paths can be distinguished. The types are defined on the basis of a series of reference kerogens isolated from sediments in which the OM was of the same nature at the time of deposit. Geochemical techniques show that the characteristics are identical. Three types I, II and III are proposed as references.

Evolution paths are the succession of physicochemical transformations which the type has undergone during evolution under the given geological conditions. This succession can be determined by elemental analysis. There are several kinds of evolution paths depending on the geological phenomena. Special attention is given to evolution paths due to burial and their relation to oil formation.

It is not possible yet to situate the minor elements, sulphur and nitrogen, within a coherent theoretical framework. It is first necessary that analytical techniques be improved.

(p. 143)

CHAPTER 5

Thermogravimetric analysis and associated techniques applied to kerogens

C. DURAND-SOURON

Thermogravimetric analysis (TGA) and differential thermal analysis (DTA) are two efficient and fast techniques which are not yet very widely used for analyzing the thermal degradation of kerogens. In order to successfully compare results from different origins, the experimental conditions must be carefully defined, i.e. test equipment, size of sample, ambient atmosphere and heating rate. For series of samples, standard conditions are thus defined. The products generated are analyzed by mean of techniques described in other Chapters of this book for residue, as well as by mass spectrometry (MS) for volatile products.

In a non oxidizing atmosphere, between 25 and 600° C, thermogravimetric curves show three stages. The first one, to about 350° C, corresponds to the release of oxygenated products. The second one, to about 500° C, corresponds to the release of hydrocarbons. And the third one, above 500° C, corresponds to a structural reorganization without any great weight loss. The relative amount of the weight losses during the first two stages enables a classification of kerogens to be made.

The natural evolution of different types of kerogens results in different rates and temperatures of decomposition, and also in a different ratio of the two first stages. Besides, thermal degradation is on the whole a rather good simulation of the natural evolution of organic matter and a way of producing hydrocarbons.

Results obtained in a hydrogen atmosphere and in an oxidizing atmosphere, which have not yet been thoroughly investigated, are briefly recorded.

As a conclusion, these rather simple and fast techniques result in a good characterization of kerogens. They are used as a preparation for other analytical techniques and they can be used to simulate hydrocarbon production for scientific or industrial purposes.

(p. 163)

CHAPTER 6

Characterization of kerogens and of their evolution by infrared spectroscopy

P.G. ROUXHET, P.L. ROBIN and G. NICAISE

The infrared (IR) spectra of kerogens present a limited number of moderately broad bands; as a consequence, they do not offer a fingerprint of the complex organic solid but the measurement of absorption coefficient provides a sort of functional analysis of the material, which can be analyzed as a whole. The assignment of the different bands is presented and a systematic comparison is made with the spectra of coals; spectra of asphaltenes, of humic acids and of a model kerogen precursor are also presented. The experimental aspects involved in the quantitative use of the IR spectra are discussed in details; this concerns sample preparation, spectrograph setting, spectrum handling and combination of IR analysis with chemical treatment.

The comparison of kerogens of different origins and the examination of the effect of catagenesis and pyrolysis provides an illustration of the possibilities of infrared spectroscopy (IR) for characterization of kerogens and of other carbonaceous solids, of interest to the industry.

The spectra of shallow kerogens differ markedly according to the nature of parent material, for instance algae or higher plants. The differences concern the concentration of saturated hydrocarbons, the hydroxyl content, the concentration of carbonyl groups, esters and carboxylic acids, the intensity of the bands containing an important contribution of aromatic rings.

The sequence of spectra obtained for samples of increasing burial depth reflects the main chemical modifications taking place upon catagenesis, i.e. the removal of the various chemical groups held by the polyaromatic sheets, and the formation and subsequent removal of aromatic CH groups which are closely parallel to the evolution of the free radicals concentration. The IR spectra may thus help the geologist in identifying the type of parent material of a kerogen and evaluating its degree of maturation.

Data obtained on pyrolyzed kerogens indicate a strong influence of oxygenated functions on the chemical behaviour of carbonaceous matter at high temperature.

(p. 191)

CHAPTER 7

Electron microscopic study of kerogen microtexture. Selected criteria for determining the evolution path and evolution stage of kerogen

A. OBERLIN, J.L. BOULMIER and M. VILLEY

Various kerogens (I, II and III series) belonging to three paths on the Van Krevelen diagram were studied by high resolution electron microscopy (EM). Specific criteria have been defined which characterize their evolution stage, then each evolution path; this helps to evaluate kerogen oil potential.

(a) Kerogen evolution is comparable to carbonization. During both processes, the same elementary domains appear. They are made of 2 to 3 piled-up aromatic molecules, each of them containing 20 to 30 carbon atoms, i.e. 5 to 12 aromatic rings. These stacks, less than 10 Å in size, are distributed at random in the kerogen. They are connected by non aromatic groups.

As catagenesis progresses, molecule stacks do not grow. As the oil production zone gets close to the end, the stacks suddenly get a local parallel preferred orientation (molecular orientation), i.e. they suddenly gather into small oriented clusters.

(b) When thermal simulation is applied to kerogens, heat-treated and natural samples are similar when their representative points are neighbours in the Van Krevelen diagram.

(c) Molecular orientation extension (clusters size) is primarily affected by the chemical composition of the less evoluted kerogen. In I′ series (heat-treated I series) which is hydrogen rich and oxygen poor, orientation extends over 1 000 Å; cluster size is less than 100 Å for III series samples which are rich in oxygenated functions and poor in hydrogenated functions; in II series, it averages between these two values (> 150 Å).

For a better understanding of both carbonization and carbonification, it appeared necessary to pyrolyse simple substances representative of kerogen precursors. Homogeneous materials with a well defined chemical composition were selected: sporopollenin, extracted from lycopodium spores and lignite. On a Van Krevelen diagram, sporopollenin (H/C = 1.56, O/C = 0.34) is situated on the exinite evolution path, close to the kerogen II series. Lignite (H/C = 1.12, O/C = 0.44) is situated near the III series. Heat-treated products have been characterized by such methods as EM, differential thermal analysis (DTA), thermogravimetry analysis (TGA), infrared (IR) spectrometry and electron spin resonance (ESR).

Sporopollenin melts with decomposition at 436° C. Molecular orientation (200 Å in extent) appears at 460° C after the weight loss rate maximum. It coincides with the aromatic CH maximum, the OH and carbonyl-carboxyl minima and comes before the aliphatic CH minimum and before the maximum spin concentration as well. Experimental conditions variations (quantity, pressure, action of oxygen) noticeably modify molecular orientation extension (from 50 Å to more than 2 000 Å).

During pyrolysis, lignite (H/C = 1.12, O/C = 0.44) does not melt and shows a small molecular orientation extension which unlike sporopollenin is not sensitive to experimental conditions. This behaviour is due to persistent oxygenated functions, some of them bonding the aromatic layer stacks.

Sporopollenin preheat-treated in air (reticulation) appears comparable to lignite, which confirms the major part played by oxygen as far as melting and molecular orientation are concerned.

In conclusion molecular orientation extension integrates the entire history of a kerogen (chemical composition of the parent organic material, oil retention or migration and oxidation if any). Since it is the result of oil departure it is also absolute criterion for evaluating the oil potential of a sediment. When it is smaller than expected, the tar yield is also smaller. To sum up, if the orientation extension of a heat-treated kerogen is 500-1 000 Å, the oil yield is high; if it is 100 Å or smaller, the oil potential is low; for 200 Å it is medium.

(p. 243)

CHAPTER 8

Electron paramagnetic resonance in kerogen studies

A. MARCHAND and J. CONARD

From previous work on coals and carbons, the presence of free radicals was to be expected in kerogens (Part. I). They are conveniently detected and studied by electron paramagnetic resonance (EPR -also called electron spin resonance: ESR).

Part. II of this Chapter gives a brief description of the paramagnetic resonance phenomenon and of the apparatus used for EPR measurements. The various kinds of information (signal intensity, signal shape or width, *g* factor value) yielded by an EPR experiment are discussed in relation to the concentration of free-radicals and their interactions with the surrounding matter (relaxation processes and various couplings).

The general results of the EPR study of natural kerogens are presented in Part. III. The importance of purification is stressed upon. The process of formation of the free radicals is outlined, and gives an understanding of the behaviour of the various characteristics of the EPR signal (intensity, line-width, *g*-value) as a function of the kerogen degradation ("catagenesis"). These results are also examined in relation to those of other physical measurements (O/C and H/C ratios, vitrinite reflectance, infrared absorption) and to the nature of the parent biological material. Some significant examples are given: kerogens from the Douala Basin (Logbaba) from the lower Toarcian and from algal materials.

The studies of natural kerogens have been completed by similar studies on artificially degraded samples (Part. IV) and it is shown that thermal treatments in the laboratory allow a fair simulation of the evolution of the paramagnetism during the natural catagenesis of kerogens.

Part. V is devoted to the possible applications of EPR to oil exploration. A discussion is conducted about the value of the signal intensity as a "paleotemperature index". It is concluded that such an index would be ambiguous in many cases, because temperature is only one among several other factors which determine the paramagnetism of kerogens. The ambiguities could be solved by using simultaneously EPR intensity measurements and one other significant parameter (H/C or O/C ratios, vitrinite reflectance, infrared absorption, *g*-value), but an estimation of the potential oil production of a kerogen directly from its paramagnetism is easy enough, and there is no real need to determine its "paleotemperature".

Finally the possible future developments of EPR in kerogen studies, such as can be foreseen in not too distant a future, are outlined in Part. VI.

(p. 271)

CHAPTER 9

C^{13} C^{12} in kerogen

E.M. GALIMOV

Correlations between the experimentally-measured carbon isotope compositions of biochemical compounds and theoretically estimated thermodynamic values (β-factors) of the same biomolecules is considered as evidence for a thermodynamically ordered carbon isotope distribution in biological systems. A number of regularities in the carbon isotope distribution in living organisms, as well as the behaviour of carbon isotopes during the diagenetic transformation of organic matter (OM) may be predicted and recognized on the basis of the principles stated.

The carbon isotope composition of kerogen is determined both by the character of its biological precursors and the isotope fractionation occuring during formation of the kerogen. The fractionation seems to be due to the elimination of functional groups, which have distinctive isotope compositions, and an isotope effect during polymerization. The latter has been investigated experimentally and is shown to result in an enrichment of about 3‰ in the C^{12}-isotope of the polymer carbon.

The hypothesis that the melanoidin type of protein-carbohydrate reaction serves as a basis for study of the humification of OM and kerogen formation has been examined by studying the carbon isotope composition of melanoidin reaction products. It has been found that the relative carbon isotope distributions in "fulvic" acid, "humic" acid and "kerogen"-like materials obtained in the experiment are essentially the same as in the corresponding natural humic substances. This fact suggests that the melanoidin process in an adequate model of geopolymer formation.

A review of C^{13}-kerogen data of sedimentary rocks allows one to conclude that no essential change in the carbon isotope composition of kerogen occurs during catagenesis and that the C^{13}/C^{12} balance in kerogen is derived at an early stage of diagenesis when all of the humic acid enters the kerogen structure.

A correlation between the carbon isotope composition of a kerogen and the bitumen from the same deposit indicates their genetic relationship. Such a correlation between kerogen and oil may, in principle, be used to recognize the process of oil migration. Also, the relationship between the carbon isotope composition of methane and the rank of kerogen may be used to reconstruct pathways of gas migration.

Variations in the C^{13}-content of kerogens through Phanerozoic and Precambrian times are characterized by peculiarities which are possibly due to some turning-points in the evolution of the biosphere.

(p. 301)

CHAPTER 10

Structure elucidation of kerogen by chemical methods

D. VITOROVIĆ

Knowledge of the chemical nature of kerogen, of its composition and structure, may help in a better understanding of the source of the precursor material, the environmental conditions of deposition, the type of diagenetic, catagenetic and maturation processes, the gas and oil potential of kerogen and the source rock potential of the sediment.

However, the determination of the chemical structure of kerogen represents one of the most complex problems. Chemical methods generally appear to be most useful for the examination of kerogen structure, although physical and other methods are also very helpful for the same purpose.

In this Chapter a review of chemical methods in structural studies of kerogen such as oxidative degradation, treatment with hydrogen iodide and other reagents, alkaline hydrolysis, and functional group analysis, is given. Since oxidative degradation seems to be one of the most useful chemical methods for the investigation of kerogens, the largest part of the text is concerned with this method (potassium permanganate, chromic acid, nitric acid, ozone, air and oxygen, and other oxidants).

Furthermore, examples of oxidative degradation studies, and of oxygen functional group determination, of kerogens from various shales, particularly from Green River Formation shale, Estonian kukersite, and Aleksinac (Yugoslavia) shale, are presented.

It was shown that chemical methods, especially oxidative degradation, have been successful in kerogen structural studies, although they have not yet given as good results as might have been expected. Problems have arisen from the insolubility, inhomogeneity, complexity and diversity of kerogens and even from the nonuniformity of kerogen from one formation.

Therefore, much work still remains to be done using chemical methods. More success may be expected by using a combination of either several chemical methods or a carefully chosen combination of chemical and other methods.

(p. 339)

CHAPITRE 11

Pétrographie du kérogène

B. ALPERN

The concept of **kerogen** is critically examined from a naturalistic and genetic viewpoint. The **insolubility** which defines kerogen develops during its maturation and can thus be used to measure it, but this insolubility also depends on early aerobic influences such as fusinization and oxidation. Therefore, this concept is ambiguous because it cannot be used to define the **rank** (coalification) and the natural **type** (sapropelic, humic, etc.). For chemists, the binomial **bitumoid-kerogen** (soluble-insoluble) is proposed together with the term **organoclast** for petrologists. The latter term includes all soluble and insoluble bitumens as well as all other amorphous or figured microcomponents. A project for classifying organoclasts is proposed in parallel to the maceral classification of coals, although much vaster in that it covers all the organic constituents (phyto and zoological) of sedimentary rocks of all ages, whether visible in normal or fluorescent microscopy.

An effort is also made to clarify and quantify the **diagenesis-catagenesis** terms on the basis of organic matter (OM) reflectometry.

The limitations and uncertainties of organic petrology —a new discipline— are inventoried. Emphasis is placed on the difficulty of diagnosing microconstituents separated from their lithological context and which are often amorphous. Particular importance is given to vitrinite and to all substances which may resemble it and are not of lignocellulosic origin, and sometimes not even of botanical origin. In oil shales and petroleum source rocks (of the Paris Basin type), true vitrinite is shown to be in a small minority and usually detrital and reworked. Its use as a paleothermometer is thus particularly delicate for such rocks, and so it becomes preferable to use the optical properties of algal components whose syngenetism is not in doubt.

After reflectance which is redefined and illustrated by way of its modern applications, fluorescence is dealt with at length, and the example of the Paris Basin is chosen for comparing the two techniques and their results. There is a striking parallel between the evolutionary curve of quotient Q (red/green) of the fluorescence spectrum and the yield in soluble extracts. Fluorescence is very useful in helping out reflectance for detecting the entrance to the petroleum window. In rocks where true vitrinite is in the minority, a great many causes of distorsion can alter the relationship between optical evidence and thermal history.

The concept of a petroleum window becomes richer and more refined when the variety of the petrographic types capable of generating oil and gas is taken into consideration. A successive series of maturation zones is usually observed, corresponding to the successive cracking of different components which have been redefined by geochemists as **kerogens I, II and III** (Tissot *et al.).* The strength and originality of organic petrography stem from the fact that it can be used to distinguish and to separately measure the **rank** (coalification) and the original petrographic **type.**

(p. 385)

CHAPTER 12

The optical evolution of kerogen and geothermal histories applied to oil and gas exploration

P. ROBERT

Optical methods for studying kerogens and their catagenetic evolution increasingly contribute to petroleum exploration. They provide a thermal range for the generation and conservation of hydrocarbons which parallels the generation of coals during coalification. These maturation phenomena are governed by the rate and duration of burial of sediments, and, even more, by variations in their thermal histories of which geothermal gradients reflect only the present phase.

This paper covers the three main aspects of the optical study of kerogens:

(a) **The application of analytical procedures** initially used for coals to the exploration by borehole analyses of whole sedimentary series. This includes physical concentration procedures, vitrinite and bitumen reflectance methods (R_o and R_oB), difficulties inherent in statistical measurement diagrams and the need to use various optical methods (incident, transmitted light, fluorescence).

(b) **An overall view of the thermal generation of hydrocarbons,** as noted in the literature of the last 20 years, with regard to its statistical and economic aspect, as well as by direct observation under the microscope.

The role of optical methods is developed in two ways — understanding thermal generation of hydrocarbons and recognizing source rocks — The theoretical chemical surveys of hydrocarbon generation are resumed and discussed.

(c) **Analysis of thermal histories of sediments** as performed with the help of reflectance data: starting from a "normal" statistical increase, the reflectance curves against depth suggest paleo-thermal anomalies which can be referred to various causes (magnetic, tectonic, rock conductivities).

An attempt to classify the deep thermal flows results in distinguishing two kinds of basin:

(a) Basins with a normal geothermy.
(b) Hyperthermal provinces related with the active parts of the earth's crust.

(p. 415)

CHAPTER 13

Structure of kerogens as seen by investigations on soluble extracts

M. VANDENBROUCKE

Petroleum products extracted from rocks originate from the natural evolution of sedimentary organic matter (OM) by way of physico-chemical processes linked to burial. Since extracts are soluble and made up of compounds with fairly low molecular weights, a great many analytical procedures can be applied to them, leading to their accurate structural determination. Their study thus enables at least one part of the chemical structure of kerogens to be specified, whereas such procedures cannot be applied to insoluble materials. Moreover it is possible, at least in part, to go back to possible biological precursors of the different types of OM, and hence to the condensation and insolubilization processes which could take place in the formation of kerogens.

Structural characteristics of extracts are examined in the three main reference types of OM, and a possible structure of the corresponding kerogens is deduced for each type, at the beginning of the principal zone of oil formation:

(a) **Type I kerogen** (example of the Green River Shales formation, Uinta Basin, U.S.A.) appears to be made up mainly of linear structures with long chains (about forty carbon atoms) linked to one another by oxygenated groups such as ester or ether. Insolubility could be caused by the steric intricateness of these long chains.

(b) **Type II kerogen** (example of the Toarcian Shales, Paris Basin, France) appears to be built from polycyclic units generally comprising four of five cycles, one of them being aromatized, with frequent substitutions by heteroatoms (mainly oxygen and sulfur). These cycles seem to have side chains, more or less branched, with an average of less than twenty carbon atoms. Insolubility could be caused by interaction between aromatic nuclei and by bonding between the polycyclic units, mainly by oxygenated functions.

(c) A structure for **Type III kerogens** cannot be inferred from studies of corresponding extracts. These extracts are never obtained in sufficient amounts to be representative of the total OM. However some specific features, such as the length of carbon chains or the possible nature of some chemical bonds, may be determined by analyzing extracts.

(p. 445)

CHAPTER 14

Origin and formation of organic matter in recent sediments and its relation to kerogen

A.Y. HUC

This Chapter aims to investigate the origin, the formation and the early evolution of kerogen. It is partly based on a bibliographic review of biological, oceanographic and ecologic phenomena which may be considered to have caused the qualitative and quantitative diversity of organic deposits on the surface of the earth, and partly on more specific investigations into the nature of the diagenesis of organic matter (OM) in recent sediments. The following points are stressed:

(a) The chemical composition of living organisms, notably the presence or absence of lignin.

(b) Phenomena of decomposition (more or less complete destruction of biological structures) and neogenesis (synthesis of complex compounds from decomposition products at various stages of decomposition), linked to environmental factors (protection of OM in an anaerobic medium, the way the synthesis of humic substances takes place in different types of superficial formations).

(c) The entrainment and transportation processes which for an allochthonous supply depend directly on the geological context and on the climate of the bordering continent, processes which can be viewed as such in the light of the theory of biorhexistasy.

(d) The mode of sedimentation (aggregates, fecal pellets, etc.) which is generally closely related to the behaviour of the finest-grained mineral fraction (hydrodynamic equivalences, organo-mineral associations) and to the entire sedimentary environment (importance of low-energy depocenters).

(e) The action of early evolution (diagenesis), resulting in OM becoming functionless (nitrogen and oxygen compounds), accompanied by an increase in resistance to hydrolysis and, even more so, by a conservation of the relative importance of the saturated CH groups. This character is customarily used to define the physico-chemical classification of kerogen, and it thus seems to be determined as soon as OM begins being deposited.

Studying OM in existing sedimentary environments is therefore of utmost importance because it acquires characteristics which are preserved later on and which can be traced back to an early stage of its evolutionary history (start of catagenesis).

(p. 475)

CHAPITRE 15

Évolution géochimique de la matière organique

R. PELET

1. Organic chemistry, which was initially concerned solely with analyzing the components of living organisms, quickly became generalized. It is now the science of the compounds of covalent carbon. An equivalent effort must be made to define the organic matter (OM) of rocks and OM in general by getting free of any preconceived biological assumption. OM is what is studied by organic chemistry. A fundamental consequence of this definition, as contemporary astrophysics shows us, is that OM seen in this way is a normal constituent of the universe and even an important quantitative constituent. Whereas simple molecules that are easy to detect were the first to be discovered, the existence in interstellar dust of solid OM which is highly polymerized no longer raises any doubt. Nonetheless, in the heart of stars as in planetary depth, carbon is found in elemental form. Therefore, OM has a cosmic cycle of which the earth's-geochemical-cycle is a part or a variant.

2. Besides astrophysics, the study of meteorites brings us new information on the origin of OM. Although metallic meteorites are very poor in OM (but they do contain elemental carbon and carbide), the amounts increase in carbonaceous chondrites which are evidence of planetary crustal formations. Qualitatively, this OM appears to be very complex. Essentially it is formed of solid material which is highly polymerized and difficult to distinguish from terrestrial kerogen. The main difference is that, up to now, neither the simple molecules accompanying this kerogen nor the simple products of its experimental degradation can be assimilated with terrestrial molecules of biological origin. Therefore, it is of abiotic origin. This comment is supported by the fact that, during synthesis tests of prebiological molecules from paleoplanetary atmospheres, in addition to simple molecules, solid polymeric by-products are always obtained in more or less large amounts. From the very beginning there has been a duality between living matter and kerogen. The same constituent groups are used, but in a rigorous order for the biological molecule and in total disorder for kerogen. Therefore, there is no doubt about the fact that, prior to the appearance of life, kerogen had been created and accumulated on earth. The amounts must never have been very great, and present evidence must be rare because it has been reabsorbed into the cycle and furthermore, in the present state of our analytical methods, it is difficult to distinguish. The massive appearance of terrestrial organic matter is thus linked to the emergence of life. It is difficult to evaluate the amount of biomass at a given time by the OM content of the corresponding rocks. The most comprehensive and best smoothed compilations give a slowly increasing curve in time, which may have several plausible explanations. From the qualitative standpoint on the other hand, the appearance then explosion in the Paleozoic of lignin-rich terrestrial vegetable matter marks a cutoff with prior time when all living matter was oceanic, as well as it sets up the familiar differentiation between autochthonous continental and marine influxes in the sedimentation.

3. Whereas geography —taken in its widest sense, including climatic conditions— determines both continental and oceanic biological productivity, it also controls the different stages of sedimentation, i.e. erosion and continental transport, chemical precipitations and subsequent deposition. Therefore, it is geography which is the immediate cause of the composition of sediments. In the world at present, this control leads to the predominance of detrital sedimentation and, with regard to organic material, the predominance of continental influxes. This has not always been true in the past. The history of the opening up of the Atlantic Ocean shows, in the Lower Cretaceous, the existence of vast anoxic marine basins in which a large proportion of the planktonic OM has been preserved. Before the subsequent depositional phase, continental and marine OM undergo quite different destinies. Continental OM in subaerial soils generally undergoes an initial geochemical evolution before even being affected by erosion and aquatic transport which in turn bring about other changes. On the other hand, marine OM that has escaped biological recycling undergoes little physicochemical changes in the marine environment which is much less aggressive than the subaerial environment. The direction of this presedimentary evolution is nonetheless the same in both cases. The most labile or assimilable fractions preferentially disappear, and kerogen, a most inert fraction, increases in relative proportion. When it arrives at the bottom of a sedimentation basin, OM is thus largely made up of kerogen, with a few lipides which are not very reactive, together with reduced amounts of carbohydrates and proteins which are labile components and active metabolites. For continental OM these latter are more abundant in the fraction which is pseudo-soluble in water, transported quickly and protected by organomineral complexes, and they are rarer in the solid fraction, which has often undergone long pedogenetic evolution before erosion, followed by long transportation, and which contains most of the products derived from lignin. In the geological history of the sedimentary column, sedimentation is succeeded by early diagenesis which is the time when the influence of living beings is still felt, first of the benthos and then of microorganisms. Quantitatively, this biological influence even further reinforces the relative share of kerogen. Qualitatively, combined with a series of abiotic reactions, it leads to the elimination of peptidic bonds in kerogen and hence to a considerable decrease in its nitrogen content. In geochemical evolution as a whole, early diagenesis is the time of nitrogen.

4. At the beginning of the next stage, i.e. diagenesis itself, sedimentary OM comprises, apart from kerogen, only a very small amount of lipides. In addition to carbon and hydrogen, this kerogen mainly contains oxygen which is principally involved in functional groups such as $\geqslant C-O-$, $>C=O$, $-OH$. Diagenesis will see the disappearance of these functional groups which are mainly eliminated in the form of CO_2, accessorily H_2O and CH_4, all of which moreover is available for mineral diagenesis. At the end of these reactions, kerogen is impoverished in oxygen and almost completely defunctionalized. It ceases to be hydrolyzable, and its humic-products content drops to zero. In general geochemical evolution, diagenesis is the time of oxygen.

5. Catagenesis which comes next inherits a quantitatively impoverished kerogen which is essentially reduced to a hydrocarbon-containing structure accompanied by a few free lipides also evolving toward hydrocarbons. The beginning of catagenesis is generally recognized by a vitrinite reflectance value of 0.5%. During catagenesis, kerogen sets free hydrocarbons having a low to medium molecular mass, thus losing its hydrogen and increasing its degree of condensation, i.e. its aromaticity, with the carbon lattice corresponding to the temperatures and pressures encountered being a hexagonal lattice. Catagenesis can thus be seen to be the time of hydrogen.

6. The OM which undergoes metagenesis contains a quantitatively highly impoverished kerogen with only very minor amounts of elements other than carbon accompanied by large amounts of hydrocarbons having various molecular masses. Metagenesis brings about a generalized dismutation of these hydrocarbons into methane and kerogen (residual). Kerogen as a whole expulses these latter heteroatoms and evolves toward a crystalline state which is not graphite, the crystallites of the elemental layers being slightly offset with regard to each other. It is only during general metamorphism that true graphite appears. Much farther down in the terrestrial depths, even diamonds will appear in a continental plate undergoing subduction. Therefore metagenesis is the time of carbon.

7. The geochemical evolution of OM thus appears as a specific cycle, i.e. the sedimentary cycle which is half-way between the short biological cycle and the long cosmic cycle. It depends on the existence of the biological cycle which supplies it with its raw material in sufficient quantity and quality, while it is a stage that is not even necessary in cosmic evolution. It is the fire of the supernovae that will restore mobility to carbon which is fast asleep in the crystalline perfection of planetary cores.

1

Sedimentary organic matter and kerogen. Definition and quantitative importance of kerogen

B. DURAND*

I. INTRODUCTION

The object of this Chapter is to define the term kerogen as used in this work, to show the importance of the study of this subject and to indicate a quantitative evaluation.

Since the term kerogen is now used in many different ways, it seemed of interest to inventory these uses and thus to supply the historical background for the concept of kerogen before giving the definition we use here. In order to do so, it is first necessary to describe the main features of the formation and evolution processes for sedimentary organic matter (OM) and the analysis procedures used. This is the only way to bring out the links between various uses of the word kerogen and to clearly show the meaning of the term.

This presentation will also contribute to the continuity between the various Chapters of this work since they introduce several subjects which are later further developed. At the same time a general conception of sedimentary OM is brought out which is essentially that of the co-authors of this work.

II. ORIGIN AND FORMATION OF SEDIMENTARY ORGANIC MATTER

Sedimentary OM is derived from living organic matter (and the products of its metabolism).

After the death of organisms the component substances such as carbohydrates, proteins, lipids, lignin, etc. are subjected to decomposition to various degrees depending on the sedimentation medium (especially its redox properties). Part of the products of this decomposition is recycled by other organisms which use them as energy sources. Simple molecules such

*Institut Français du Pétrole. Rueil-Malmaison, France.

as CO_2, H_2O, CH_4, NH_3, N_2, SH_2, etc. are formed as the products of metabolic processes. Another part is transformed into simple molecules (CO_2, H_2O, etc.) by physico-chemical processes (e.g. oxidation). The rest, which in most cases represents only a very small fraction of the initial quantity of living mater, escapes complete biological recycling or physico-chemical decomposition and is incorporated into sediments. This fraction is therefore the primary source of sedimentary OM.

This material is preserved in various ways (Chapter 14). As examples we can mention such possibilities as that the toxicity of the products or their incapacity for use as energy sources keeps them from being reused by organisms or that rapid structural modifications, polycondensation ("polymerization"), associations with minerals make enzymatic deterioration impossible or that the sedimentation medium contains neither organisms nor oxygen, etc.

Some of the preserved products have clearly and sometimes completely preserved the chemical structures which they had in living matter. These are, for example, substances with low chemical reactivity which in the organisms often had a protective role against modifications of the external medium (some saturated hydrocarbons, vegetable waxes, resins, sporopollenin, chitin, etc.) or substances derived from pigments (isoprenoids, porphyrins, etc.) or other metabolites (steroids, terpenoids, etc.).

These molecules, whose structural filiation with living matter is evident, are said to be **inherited** or may be called geochemical **fossils** or **geochemical markers.**

Other complex molecules (not inherited) are said to be **neoformations.** This signifies that they theoretically have no equivalent in living matter. The idea of neoformation is less clear than that of inheritance since:

(a) Neoformed structures can incorporate inherited elements.
(b) There are still many unknown structures in living matter.
(c) Many structures in living matter as well as in preserved products are much too complex to be analyzed by existing methods.

It is thus difficult to evaluate the share of inheritance and that of neoformation. However the very complexity of a large part of the preserved products supports the importance of the role of neoformation.

Organic products incorporated in sediment can be derived from organisms living in the sedimentation basin and are then called **authochtonous.** The material may also come from organisms living outside of the sedimentation medium such as that, for example, which is introduced by rivers, wind action, etc. This material is then called **allochthonous.** To these are added materials from organisms contemporary with more ancient sedimentary strata and organic products from metamorphic and crystalline rock or from the leaching of soils which have been transported by erosion, transgressions or other less important geological modifications. These products are called **reworked** and are generally allochtonous.

All these processes, decomposition of living material and incorporation into sediments of the fraction preserved, introduction of the various other organic materials, etc. are part of what is more generally called the process of formation of sedimentary OM.

III. GENERAL PROCEDURES FOR THE ANALYSIS OF SEDIMENTARY ORGANIC MATTER

The organic content of a sediment is therefore a complex and heterogenous group which is also, as will be seen below, in continuous evolution. The analysis generally proposes to define criteria for origin (progenitors, physico-chemistry of the initial materials, etc.); homogeneity (distribution of the various materials: autochthonous, allochthonous,, contemporary, reworked etc., physico-chemical variety of these materials, etc.); distribution (concentration, relations between organic and mineral phases) and evolution (nature of evolutionary processes, stage reached by these processes, etc.) It is then possible to begin the study of the main geological and geochemical problems (for example, the origin of oil, paleoenvironment of sedimentary deposits, etc.) and to make classifications.

At the present time there are two main approaches:

(a) Firstly there is the approach wich essentially uses optical analysis methods for the study of the **petrography of OM.**

(b) Secondly there is the approach using chemical analysis methods for **coals physico-chemistry, petroleum geochemistry, pedology** (soil study), **environmental geochemistry**, etc.

A. Optical methods.

Petrographers work on thin slices and polished sections made from sediments or concentrates of OM which are generally obtained by physical dressing (grinding and dense fluid separation, flotation, etc. — *cf.* Chapters 3, 11, and 12). On the basis of experience obtained in coal petrography, they identify and classify debris from organisms and microfossils, but their main task is to define the petrographic objects called macerals (the term maceral was created by Mary Stopes [67], by analogy with mineral), i.e. groups which are characterized by the same optical properties (color, reflectance, fluorescence, etc.) so that they can be identified and classified.

In coal petrography three main families of macerals are usually distinguished ([63], Chapter 11) which are differentiated within the same sample by different reflectances. Thus in increasing order of reflectivity we have the following :

(a) Macerals of the exinite family (also called liptinite) generally containing spore cases, leaf cuticles, resin globules, single-cell algae, etc. and also, more generally, debris derived from lipid — rich parts of higher plants.

(b) Macerals of the vitrinite family. They are gels derived especially from the lignin-cellulose walls of cells from higher plants.

(c) Macerals of the inertinite family. These elements are generally of the same origin as macerals of the vitrinite family but have been altered or oxidized for various reasons such as reworking, changes in redox conditions, action of microorganisms, forest fires, etc.

The optical properties of macerals also make it possible to study coal evolution processes. It is conventional to use the reflectance of macerals of the vitrinite family to classify coals according to their **rank** in the natural process of coalification. This process progressively transforms peat into anthracite during burial mainly, it is believed, because of the effects of increased temperatures accompanying burial (geothermal gradient) with a resulting increase in the reflectance of vitrinite.

The study of coals on the basis of maceral analysis has made it possible to utilize coal petrography as an instrument for the analysis of coal genesis and evolution processes. There is, in effect, a good correlation between the petrographic and morphological characteristics of macerals, their origin and physico-chemistry and the stage which they have reached in the evolutionary processes [78]. The application of maceral analysis principles to the study of dispersed OM has not yet, however, made it possible to etablish this coherence (Chapter 11). This is true because of the variety of deposit media and progenitors so that the relations between petrographic characteristics, genesis and physico-chemistry are more difficult to establish.

Petrographers also work with transmitted light on OM collected after the destruction of minerals by hydrochloric and hydrofluoric acid, according to methods inherited from palynology.

They then distinguish between recognizable and amorphous OM. Recognizable organic matter is "the sum of the fragments of identifiable biological origin" [52]. In them can be observed fragments of vegetable tissue, ligneous fibers, epidermises, cuticles, resins, etc. and microfossils especially pollens, which are one of the first objects of study for palynology, but also spores and algae: *Dinoflagellata, Tasmanaceae, Botryococcaceae,* etc. Important groups such as *Diatomaceae, Foraminifera, Radiolaria* and *Coccolithophoridae,* whose tests are destroyed by acids, cannot be identified.

Amorphous OM consists of flakes, powder and agglomerates without definite shapes. The proportion of the various amorphous and figured constituents makes it possible to define the palynofacies ([12], Chapter 3).

By means of these techniques it is possible to follow the evolution of OM, especially that which takes place during burial. During burial, spores, pollen, algae and in most cases amorphous OM as well, which are transparent and light yellow to light brown in color in the case of slightly or moderately evolved samples, become increasingly black and opaque [15, 64, 30].

Amorphous OM and at the same time OM which is very rich in amorphous fractions are often called colloidal, algal, sapropelic or marine. For many people the term sapropelic is a synonym for being derived from algae, although according to the definition of R. Potonie (in [63]) it designates a gel derived from organic debris which has putrified in an anaerobic medium.

The terms sapropelic, algal and marine should be used prudently when qualifying this amorphous material since the algal origin cannot be demonstrated by the amorphous characteristic alone even if much algal debris is observed at the same time, nor can the marine origin (i.e. the amorphous OM originates in marine organisms).

OM which is rich in opaque debris and especially in ligneous debris is often called ligneous, coaly, humic or terrestrial (i.e. originating in terrestrial organisms). To some extent the same criticisms which are made about the terms sapropelic, algal and marine can also be made about the use of these terms. In theory they mean that the OM is derived from coals, soils or at least from a continental sedimentation environment. This is not always clear on the basis of optical examination alone. In this case, however, the origin of the OM can be shown best through this examination.

B. Chemical methods.

In a relatively arbitrary way we will distinguish between the chemical methods intended to isolate fractions of OM without damaging the structures and those whose object is to degrade these structures in a controlled way.

The first consist of solvent extraction, gas entrainment techniques or gas phase extraction for the volatile compounds.

At the present time the most used of these methods are the following:

(a) Extraction with organic solvents at moderate temperatures (< 80°): chloroform, benzene, methanol-benzene mixture, etc. This method extracts hydrocarbons and more complex products which are heteroatomic and/or high atomic weight. The latter products will be called resins and asphaltenes hereunder. The organic fraction extracted in this way is often called **bitumen,** but this bitumen should not be confused with road bitumen or with the bitumens of petrographers. The latter are organic substances filling rock pores and under the microscope appear as homogeneous. These bitumens may or may not be soluble in organic solvents. Therefore the term bitumoid has been proposed (Alpern, *in:* Chapter 11) for calling the part of OM which is soluble in organic solvents. The insoluble fraction left in the rock is generally called **kerogen.** This is the conventional fractionation in petroleum geochemistry.

(b) Cold diluted base solutions possibly associated with dechelating agents (for example a NaOH 1*N*, 1% sodium pyrophosphate mixture) for the extraction of the products called fulvic and humic acids. The fraction remaining in the rock is called **humin.** This fractionation is classic in pedology (soil science).

The second group consist of the use of specific reagents (Chapter 10) for characterizing and quantifying chemical functions and bonds; or controlled hydrolysis, oxidation and hydrogenation methods or thermal degradation (pyrolysis), etc. The latter are less specific but can modify or fractionate structures which are highly complex into simpler elements.

Among these methods one of the most used is hydrolysis with strong hot concentrated mineral acids (for example HCl 6*N* at boiling point) which liberate amino acids and carbohydrates and other substances. The fraction remaining in the sediment is called stable residue. This fractionation is widely used for the study of recent sediments.

In all cases, the degradation or fractionation products are then refractionated and/or analyzed in detail by means of current techniques of physico-chemical analysis, i.e. elemental analysis, various spectroscopic techniques [visible , infrared (IR), ultraviolet (UV), nuclear magnetic resonance (NMR), electron spin resonance (ESR)], chromatography (liquid and gas phases), mass spectrometry (MS), etc.

Study of the organic compounds left in the sediment presents special difficulties, and for a long time this was not done. If the analysis is made directly on the rock, interferences with minerals hamper correct determination of the properties of OM, since rocks in which OM predominates are rare and the most frequent contents are about 1% (see below).

Modifications, often of a considerable extent, occur when these fractions are to be isolated from their mineral matrix (*cf.* Chapter 2).

C. Complementary nature of optical and chemical methods.

Each of these methods gives a partial picture of sedimentary OM. Theoretically optical methods give more comprehensive results. Used alone, however, these methods have the following disadvantages:

(a) It is not possible to see objects with a size below the power of resolution of the microscope, so that most of the organic content of the sediment cannot be seen. This tendancy is often further increased by the preparation methods, which favor elements of large size. The use of fluorescence techiques, which are becoming wide-spread at the present time, overcomes this drawback to a certain extent.
(b) Some specialists have a tendancy to systematize on the basis of morphological criteria from transmitted light study alone. Thus abundance of algae is considered a single chemism while amorphous matter is called algal, sapropelic or marine.
(c) The usual preparation techniques result in the loss of much organic matter.

The geochemical method used alone has, among other, the following disadvantages:

(a) Too often OM is considered a single entity which is to be fractionated, while in fact it is an assemblage of many entities. Thus this method does not give direct access to the different constituents (allochthonous, autochthonous, reworked).
(b) Some specialists have a tendancy to systematize on the basis of the compounds which they know how to analyze best, for example non volatile saturated hydrocarbons, while these are often not very representative of the OM as a whole.
(c) The method does not sufficiently take into account the fact that the same chemism can originate in rather different circumstances.

It is becoming increasingly clear that these two methods are both necessary for good comprehensive understanding of sedimentary OM and more and more laboratories are using them jointly.

IV. EVOLUTION AT DEPTH

A. General

The biochemical and physico-chemical processes mentioned above are replaced here by processes which we will call geochemical and geodynamic:

(a) Weathering phenomena due to readjustements or changes in the redox conditions. (These phenomena are often difficult to distinguish from formation processes). These phenomena are of relatively great importance in soils which are not subjected to subsidence.
(b) Phenomena due to temperature and pressure increases accompanying burial and possibly to the action of mineral catalysts and are called diagenetic and catagenetic phenomena.
(c) Phenomena due to the displacement of a part of the sedimentary organic content called migration (for example migration of hydrocarbons from source rocks to reservoir rocks).

These phenomena finally lead to a further deterioration of the organic matter which was already relatively degraded at the formation stage, i.e. to produce simple, thermodynamically stable molecules (CO_2, CH_4, H_2, H_2O, etc.) and a residue very rich in aromatic carbon from complex structures inherited from living or neoformed matter. This stage is generally reached before the zone of metamorphism where the sediments will be transformed into crystalline rocks and where the carbon residue may be transformed into graphite.

Although there is a tendency to move towards thermodynamically stable compounds, experience shows that these processes can be described only in kinetic terms since the carbon compounds formed are mainly metastable under subsurface conditions even on the geological time scale.

In addition, the existence of migration phenomena makes the sedimentary system an open system in which equilibria, even if they had the time to establish themselves, would be continually displaced. No doubt exceptions exist for closed systems such as fluid inclusions in which partial equilibria have the time to establish themselves between compounds with low molecular weights. In addition, since reaction rates increase as temperatures rise, it is possible that the OM system and its degradation products may be increasingly well described in thermodynamic terms as the zone of metamorphism are approached. However the zones of metamorphism have not been studied from this point of view.

These phenomena are indicated by progressive modifications in the properties of organic matter which can be brought out by many methods. We have already spoken of the increased reflectance of vitrinite which accompanies coalification and the variations in spore and pollen color and transparency. There are, therefore, a large number of properties whose evolution can be followed. Further details on this subject will be given in other Chapters of the present work.

It should be noted that one of the general effects of this evolution is the convergence of the properties of OM of different origins since the result is always the same relatively simple system. In particular, there is convergence of optical properties.

B. Evolution of soluble, insoluble and volatile fractions of organic matter (Fig. 1.1.).

Study of many sedimentary basins has little made by little made it possible to bring out the following main trends:

Recently sedimented OM (recent sediments, peat, soils) is not very soluble in organic solvents. In addition, this soluble fraction contains few hydrocarbons. On the other hand, it is partly soluble in bases (humic and fulvic acids) and in acids (hydrolyzable fractions).

There may be a volatile fraction which is the result of physico-chemical and/or biochemical degradation, and in certain cases there may be so-called biogenic methane (marsh gas) and hydrogen sulphide.

The beginnings of burial (Zone A in Fig. 1.1.) are marked by:

(a) The disappearance of soluble fractions in acids and bases.
(b) The formation of CO_2 and H_2O.
(c) A slight increase in the size of the fraction soluble in organic solvents which consists essentially of resins and asphaltenes and a small quantity of hydrocarbons.

(d) A relative decrease in the amount of oxygen in the fraction insoluble in organic solvents which is accompanied by various modifications in its physico-chemical characteristics, in particular, the oxygenated functions disappear (Chapter 6).

Beyond a certain depth (transition A > B in Fig. 1.1), which is generally about 1 km, there is an increase in the relative size of the fraction soluble in organic solvents and the volatile fraction. This phenomenon is principally due to the formation of hydrocarbons. The distribution of these hydrocarbons changes with depth through the progressive reduction in the average molecular weight. Thus in successive stages we can observe an oil zone, a wet gas zone and a methane zone. Thus less and less resins and asphaltenes are recovered. The frac-

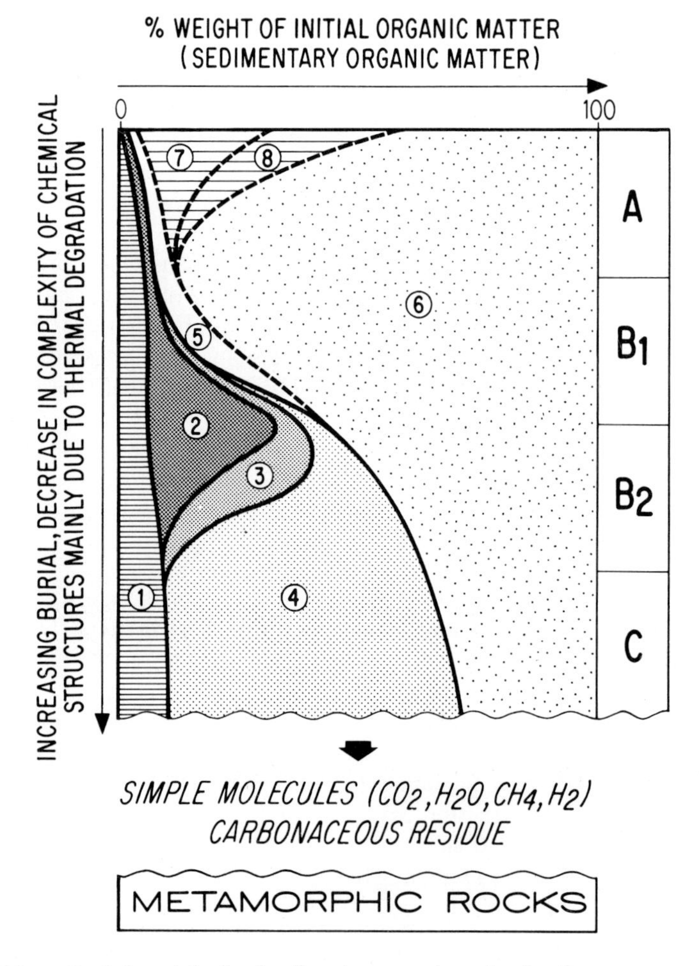

Fig. 1.1. — Evolution at depth of sedimentary organic matter fractions:

1. CO_2 + H_2O. **2.** Oil. **3.** Wet gas. **4.** Methane. **5.** Resins + Asphaltenes. **6.** Insoluble matter in organic solvents (kerogen). **7.** Soluble matter in alkalis. **8.** Soluble matter in acids.

Dotted lines between **5**, **6**, **7** and **8** mean possible overlapping of these fractions.

tion soluble in organic solvents, and which can be recovered after evaporation of the solvent, thus ends by disappearing. Zones B (B_1 + B_2) and C in Fig. 1.1 have been located in reference to what Vassoyevitch [79] called the main oil formation zone. B_1 and B_2 correspond to the presence of oil with the boundary between the two corresponding to the maximum formation, while C indicates the absence of oil and wet gas and the presence of dry gas (methane).

The fraction insoluble in organic solvents which had lost oxygen during the preceding stage (A) now becomes poor in hydrogen. At the same time many of its physico-chemical properties are modified (disappearance of saturated CH functions, *cf.* Chapter 6), increase in reflectance (Chapters 11 and 12), variation in the number of free radicals (Chapter 8), etc.).

Transformation of organic material into lighter molecules is never complete. Light molecules such as CH_4, CO_2, H_2O, H_2, etc. coexist with an insoluble residue very rich in aromatic carbon (carbonaceous residue) which could produce graphite.

It is supposed that the evolution which we have just described is principally due to increased temperatures [42, 43, 44, 51, 79] and the observed phenomena are called diagenetic and/or catagenetic. There is a great deal of variation in the definitions and names of the characteristic zones depending upon the authors (Chapter 4), and the zones which we have called A, B, C are not necessarily those used. However the disagreement concerns the criteria to be used and the position of the zone boundaries rather than the description of the entire group of phenomena. In any case it should be noted that the location of the zones which may be used changes a great deal with the nature of the initial OM [73] and that exact boundaries cannot be assigned unless the organic content in the entire sedimentary column is very well known.

It is possible to introduce a quantitative element into the preceding sheme. As a first approach, the relative extent of hydrocarbon formation rises to the extent to which the OM (at the recent sedimentary stage) is rich in hydrogen and therefore poor in oxygen. Inversely, the extent of CO_2 and H_2O formation will be greater if the OM is initially highly oxygenated.

If we refer to the petrographic patterns, the first case (high hydrogen contents) will result from OM rich in macerals of the exinite family and/or with a significant amorphous fraction, while the second case (high initial oxygen content) will result from OM rich in macerals of the vitrinite family and/or with a large coaly fraction. However this correspondence between maceral analysis or the palynofacies and the capacity to form hydrocarbons has only a statistical character, and there are many exceptions. Indeed we shall see later on (Chapter 4) that optical properties are only an approximate indicator of hydrogen content.

C. Comparison between evolution at depth and pyrolysis.

The evolutionary process which has just been described is analogous in general to that which is observed during laboratory pyrolysis of sedimentary OM. This analogy was noted as early as 1915 by D. White [84, 85]. Indeed coalification and carbonization are both carbon enrichment processes of the solid phase of OM where fractions are progressively formed which are volatile or soluble in organic solvents and rich in hydrocarbons, while the insoluble residue loses first oxygen and then hydrogen and thus becomes richer in carbon. In both cases the final result is the formation of light molecules including methane and a residue which is very rich in carbon. The hydrocarbon yield of pyrolysis depends, as in the case of natural evolution, on the chemical structure of the original substance and especially on the H/C ratio.

Pyrolysis is used industrially to make coke from coal and graphite from organic substances, so-called graphitizable substances (for instance tars from coals or petroleum). It has also been used and may possibly be used again for producing petroleum products called shale oil from sediments called oil shales. From what has been said it can be understood that oil shale is rock which, under present technical and economic conditions, has the following favourable characteristics:

(a) Sufficient OM.
(b) The OM has a sufficiently high hydrogen content.

It should be noted that effects analogous to those obtained by burial of sediments at temperatures of about 20 to 200° C can be obtained in the laboratory only at significantly higher temperatures ranging from about 200 to 600° C, depending on the stage of natural degradation which it is intended to reproduce.

However the natural modifications observed in sedimentary OM are also due to the temperatures, and relatively low temperatures are compensated for by the very long geological periods involved.

It is not necessary to invent the existence of catalytic effects in order to explain temperature differentials between the artifical and natural processes. A calculation explains it more simply on the basis of a description of the phenomena in kinetic terms. By means of the same kinetic formulation based on first-order kinetics and the same range of activation energies (20-80 kcal/mol) [71, 72, 73], it is possible to correctly simulate both the laboratory phenomena (time scale = 1 hr, temperature range = 200 to 600° C) and those observed underground (time scale = millions of years, temperature range = 20 to 200° C).

There are however important differences between pyrolysis and natural evolution (Chapter 4). In general, the lower the activation energy of the phenomena, the more difficult it is to simulate them on the laboratory scale by increasing the temperature. This is true because secondary reactions are also activated at the same time which would be much less extensive at low temperatures.

V. DEFINITION OF KEROGEN

A. History and different meanings of the word kerogen.

According to Steuart [66], the word was proposed by A. Crum-Brown to describe the organic content of the oil shale in the Lothians (Scotland) which produce, by distillation, oils with a waxy (paraffinic) consistency (from Greek Keros = wax). However the term kerogen covered and still covers a variety of concepts which historically originated with the exploitation of oil shale and are closely linked to research on the origin of petroleum and especially with the development of theories on its organic origin. When we inventory the principal meanings of the word kerogen we find the following:

1. Organic matter in oil shale.

Historically, this definition is the result of the first extension of the definition of Crum-Brown. Since the OM in a variety of oil shales showed, when examined with the methods of the time, the same general characteristics as the Lothian shales, the conception was formed that this was a relatively well defined substance. For Cunningham Craig [16] it was the result of the inspissating of petroleum through argillaceous beds (and was consequently an asphaltoid substance). For most observers [2, 3, 53, 75] however, it was a syngenetic product. Under the microscope it consisted of a light yellow to brownish gel containing sometimes abundant "yellow-bodies" which were finally recognized as being derived from algae. This kerogen was therefore formed from algae and from products probably resulting from their decomposition, so that the terms algal and sapropelic were frequently used.

The very concept of OM in oil shale now seems very approximate since the notion of oil shale is essentially economic. Capacity for distillation under economically profitable conditions at a given time can be attained for a relatively diverse range of petrographic and chemical compositions and generating media [2, 3, 22, 31, 53]. In itself, therefore, it is not a sufficient basis for defining a certain type of OM.

2. *Sedimentary organic matter with a distillation yield of at least 50%.*

Historically, this concept was formulated by Himus and his coworkers [22, 31]. It defines a type of OM with a high capacity for distillation. At the same time, Himus describes another type of OM with little capacity for producing oil by distillation (10% at the most) called coaly OM. He supposes that intermediate situations can be obtained mainly by the mixture of these two types. Rocks whose organic content is composed essentially of kerogen are called kerogen rocks. When the concentration of OM is very high, the substance is called kerogen coal. Coal sensu stricto, also called humic coal, is the concentrated form of the type of organic matter with little capacity for distillation.

Under the microscope, this kerogen is characterized by the great abundance of macerals of the exinite family and/or of the amorphous fraction. It is often called sapropelic. The fraction of the OM which is not kerogen is characterized by the great abundance of macerals of the vitrinite family and/or of ligneous debris. This fraction is often called coaly or humic. We thus find here the mixture which we have mentioned above of optical and genetic criteria.

3. *Sedimentary organic matter capable of producing oil through artificial distillation (pyrolysis) or natural distillation (evolution during burial).*

This concept is historically derived from the parallel already indicated in 1915 by White [84, 85], between pyrolysis and evolution due to burial. It was then formulated by Trager [75] and applied concretely by Takahashi [69] to research on petroleum source rocks in the oil-bearing sedimentary basins of Japan.

This concept indicates a more comprehensive view than the preceding ones concerning the group of sedimentary organic substances since according to it the OM which produces oil through distillation is of the same nature as that producing oil during burial.

In addition, the characteristics of this sapropelic kerogen are those of the preceding definition. It is thus here again opposed to a coaly OM whith little capacity for producing oil by natural or artificial distillation.

We note that, in the three preceding concepts, kerogen loses its distinctive properties during burial. For (see above) as hydrocarbon formation proceeds, the remaining OM loses hydrogen and therefore its capacity for the distillation and formation of hydrocarbons. The optical characteristics also change, and there is a convergence of the optical properties of kerogen and of coaly OM.

4. Insoluble organic matter resulting from the condensation of lipids.

This concept was defined by Breger [6, 7, 8, 9] and is based on the following considerations:

Of the main chemical families consituting living OM, i.e. lipids, carbohydrates, proteins, lignin, etc., only lipids and lignin would be preserved to a significant degree during the formation of sedimentary OM. Part of the lipids would produce oil during burial. Another part would be condensed (polymerization) into an insoluble product, kerogen, which is also capable of producing oil but only under exceptional conditions (for example extensive burial). For structural reasons lignin derivatives are not able to form oil.

Erdman [26] gives a definition of kerogen which is close to that of Breger but in more general terms. In recently formed sedimentary OM he distinguishes two main classes of constituents. Only one of these will produce oil and kerogen during burial. The opposition mentioned above between a type of OM capable of generating hydrocarbons and a type unable to do so is thus found again in Breger and Erdman, this time at the level of the chemical structures. Kerogen, however, which elsewhere was a potential source of hydrocarbons, is here rather a by-product of their formation.

5. The fraction of organic matter dispersed in sediments which is insoluble in organic solvents.

Historically, the creators of this concept seem to have been Forsman and Hunt [28] who defined it on the basis of the OM disseminated in ancient sediments. Today this is the most widespread meaning of the word kerogen.

Originally this was a petroleum geochemistry concept since organic solvent extraction is used for examining the relations between hydrocarbons and the rest of the OM. In this sense a synonym of the word kerogen is **kerabitumen** which was recommended by the nomenclature committee of the *4th World Petroleum Congress.* However the term kerabitumen is still not in general use.

The definition depends on the nature and the conditions of use of the solvent. In practice (*cf.* paragr. III.B) the solvents are not very polar (chloroform, benzene, methanol-benzene) and are used at temperatures below 80° C. Under these conditions (Chapter 2), the quantities of OM dissolved (bitumen) still vary considerably, depending upon the solvent and the extraction conditions. However there is little variation in the quantities of hydrocarbons dissolved.

6. Miscellaneous.

The following meanings can also be mentioned:

(a) Total sedimentary OM [46].
(b) The insoluble OM in any kind of rock [61].
(c) The sedimentary OM insoluble in organic solvents and alkalis.
(d) The sedimentary OM non hydrolyzable by hot mineral acids.

This variety of concepts reflects the complexity of the sedimentary OM, and a clearer idea of this complexity can be formed on the basis of attempts at classification [74, 8, 14, 68]. This also indicates the diversity of the interpretations of the transformation processes in sediments, especially as concerns petroleum formation.

B. The definition used here: sedimentary organic matter insoluble in the usual organic solvents. Why this fraction should be studied.

In the present book we **call kerogen sensu stricto the fraction of sedimentary OM which is insoluble in the usual organic solvents** (under the conditions indicated in paragr. V.A.5) as opposed to the soluble fraction called **bitumen.** However we do not limit our use of the word to organic matter disseminated in ancient sediments as is the case of current uses of the term, and we apply it to **all sedimentary OM.** Thus we include in sedimentary OM the humic coals of various ranks (peat, lignite, bituminous coal, anthracite), boghead coal, cannel coal, asphaltoid substances (asphalts, bitumens, tar in tar sands), OM in recent sediments and soils.

The use of a single term for designating the insoluble fraction of sedimentary OM in general is justified by the progressive convergence now observed between the methods and philosophies of the various disciplines dealing with sedimentary OM. To an increasing degree a general conception of sedimentary OM is being developed which we are presenting here.

According to this view, the organic materials found in soils, recent and ancient sediments, coals, etc. are considered as special cases. Formation and evolutionary processes account for the transitions from one special case to the next. Thus the types of OM generating hydrocarbons are thus less and less contrasted with those which are not capable of doing so. There does not seem to be any criterion whether of a genetic or physico-chemical kind which seems to be sufficient to establish such sharply defined categories in spite of technical developments, and there are transitions between the various evolutionary stages of the OM and between the types of OM. (We will, however, see that some types are better represented, and this phenomenon is the basis for classifications which are too sharply defined).

The term used for designating this insoluble fraction is finally of secondary importance. It is more important to see the significance of its study.

The main reason is that examination of the relations between the soluble and insoluble fractions in standard organic solvents is a necessary stage in the understanding of the origin and formation of petroleum in sediments.

The soluble fraction contains "free" hydrocarbons which are formed during geological processes, whereas the insoluble fraction is where the largest proportion of compounds can be found that are liable to form hydrocarbons (resins and asphaltenes recovered from the soluble fraction are also capable of such formation).

Secondly, let us note that the insoluble nature is linked to a condensation state of OM, denoting a certain degree of stability under geological conditions (and even cosmic, *cf.* Chapter 15) which may justify, as is done in several of the texts presented, the use of the word **kerogen** to designate, in a more general way, a "polycondensed" or "polymerized" state of OM not belonging to the living realm (geopolymers).

From a more practical standpoint, it must be pointed out that what we call kerogen must be analyzed to achieve a physico-chemical description of sedimentary OM. Indeed, if the structures are not to be degraded, the only procedure now possible for as complete a physico-chemical analysis as possible is extraction by a solvent not having any solvolysis effect and the analysis of the products thus extracted, followed by an analysis of the insoluble residue. Yet this residue is by far the largest quantitative proportion, and so we cannot pretend to describe the sedimentary OM without having examined it.

Because of this quantitative importance, the physico-chemical description of kerogen is thus a good approximation of that of the sedimentary OM. This is why its analysis, as will be seen later on, serves to go well beyond an examination of oil-forming phenomena. (This quantitative importance also justifies the procedure that was used to deal with the optical analysis of kerogen (Chapters 3, 11 and 12). The litterature contains almost no optical description of kerogen, sensu stricto, because petrographers only very rarely practice solvent extraction before analysis. Their observations and conclusions are however applicable to kerogen).

Insolubility requires the development and use of methods which are not generally used for analyzing soluble fractions. It is thus a federating element from the analytical standpoint which by itself justifies the assembling of most of the texts presented in this book.

There are thus technical, theoretical and finally historical reasons for analyzing the insoluble fraction of sedimentary organic matter per se. However, the disadvantages of making it into a separate category must be looked at from the methodological standpoint.

This fraction is a heterogeneous mixture in which a priori neither the origin of the constituents nor the stages they have attained in the processes of oil formation are distinguished. As a result, kerogen contains:

(a) OM incapable of forming oil because, for example, it has undergone an oxidation process at the same time as it was incorporated into the sediment or before it was thus incorporated, such as fusinite, or again because it has undergone a previous sedimentary cycle, such as various reworked products.

(b) OM at varying stages of the oil formation process, ranging from little evolved OM (peat, insoluble organic contents of recent sediments) to insoluble residues that are very rich in carbon and are created by this formation;

(c) The insoluble fraction of "bitumens" as defined by petrographers, which are probably themselves the result of the migration of part of the hydrocarbons formed.

The fact of creating this category "kerogen" thus has, a priori, no genetic significance from either the standpoint of the origin of the constituents of the sedimentary OM or even from the standpoint of the origin of oil (hence we are a long way off from the etymological standpoint). It is only by a more thorough analysis, with emphasis being placed on optical methods for evaluating the heterogeneity of the organic content, that affiliations can be established between the different constituents.

The creation of this category thus has a highly pragmatic nature and so cannot satisfy naturalists (*cf.* Chapter 11) and even less philosophers. Unfortunately this is the case of most categories that we now know how to distinguish in sedimentary OM, and a great deal of progresses will have to be made in our knowledge of such matter before entirely satisfactory classifications are established.

VI. QUANTITATIVE IMPORTANCE

A. Quantities of kerogen imbedded in sediments.

Sediments almost always contain OM. Sandstone contains the least while some rocks (coal, boghead coal and veins of bitumen or asphaltite, etc.) consist almost entirely of it.

Kerogen, defined as the fraction of this sedimentary OM (including coal, asphalt and bitumen, the OM in soils and recent sediments, etc.) which is insoluble in organic solvents, makes up the largest part. It generally represents more than 95% of the weight in the case of recent OM. This proportion progressively falls during burial because of the formation of soluble and/or volatile products, especially hydrocarbons. However the extent of this formation varies considerably depending on the nature of the origin of the OM (10-80% of the initial OM depending on the case). In view of the variety of the situations, it is hardly possible to estimate the average proportion of kerogen in the total sedimentary OM.

The content by weight of kerogen for a sediment is generally estimated on the basis of its organic carbon content. This method leads to excessive underestimates for slightly evolved sediments (recent or shallow buried). This is true because the carbon content of slightly evolved kerogens is about 60-70% and, as we have seen (paragr. IV.B), at this stage kerogen constitutes almost all the OM. In order to obtain the kerogen content it is therefore necessary to multiply the organic carbon content by 1.5 or 1.6. During evolution due to burial, the organic carbon content of kerogens increases and can exceed 90% for very evolved sediments. At the same time the proportion of kerogen in the OM falls, but part of the products formed is no longer in the rock at the time of analysis, either because these products have left the rock in situ (migration) or because they have been lost during handling. Finally, at these stages, the remaining soluble and/or volatile products rarely form more than 20% of the total OM, and the organic carbon content in the end gives an estimate which is in the case quite close of kerogen content.

For all sediments, the organic carbon content of a sediment is an underestimate of its kerogen content, and it is generally admitted that the result must be multiplied by a coefficient of 1.2 to 1.3. However in view of the approximate nature of the calculations below, we will evaluate kerogen content on the basis of the organic carbon content.

The distribution of carbon in the earth's crust has recently been estimated [36, 49, 80]. We shall use here the estimate by Hunt [36], based on the work of Ronov *et al.* [56]. According to him, the total mass of organic carbon contained is sediments is 1.2×10^{16} t. Other studies, most of which are older [20, 34, 45, 81, 83] indicate quantities of between 10^{15} and 3.5×10^{16} t. The calculations of Hunt are partly based on a more recent evaluation of the total mass of sediments existing in the oceanic domain now thought to constitute about 8% of the total sediment mass, and also on a new estimate of the average organic carbon contents about which more data is now available. These calculations are still, however, largely subject to revision, especially as concerns the estimates for average organic carbon content. The latter figure is till distorted by the sampling since there is more data for the present continental domain than for the present subsea area, and there is more for shallow sediments than for deep ones. The figure of 10^{16} t will therefore be considered as indicating only an order of magnitude for the quantities of organic carbon and therefore of kerogen which are present in sediments. This figure represents [36] 10-15% of the total carbon (organic and inorganic) contained in the earth's crust (carbonates in sedimentary rock 50-60%, carbonates in non-sedimentary rock 25-30%), while the rest is found in the form of elementary carbon (graphite, diamonds, carbonaceous particles, etc.) in non-sedimentary rocks, i.e. eruptive, crystalline and metamorphic rocks. The quantities of carbon contained in the biomass, atmospheric CO_2 and that dissolved in the oceans are smaller by several orders of magnitude.

Table 1.1 compares quantities for the different forms of carbon. Most of the figures given are from Hunt [36] but have been rounded off. For the biomass, a range has been given on the basis of various estimates [82, 36, 13]. It is difficult to attain a high degree of accuracy for this figure.

TABLE 1.1
QUANTITIES OF THE DIFFERENT FORMS OF CARBON IN THE EARTH'S CRUST
(from Hunt [36])

Forms of carbon	Quantities (t)
Kerogen	1.10^{16}
Carbonates	
in sedimentary rocks	6.10^{16}
in non sedimentary rocks	1.10^{16}
dissolved in oceans	5.10^{13}
«Elemental» carbon (mainly in basaltic, granitic and metamorphic rocks)	1.10^{16}
Dissolved in oceans (CO_2, organics)	1.10^{12}
Atmosphere (CO_2)	1.10^{12}
Biomass (living organisms)	$0.3\text{-}3.10^{12}$

According to various sources [1, 4, 18, 21, 24, 41, 58, 62, 65, 86, 87, 88] the mass of carbon entering synthesized organic products each year through photosynthesis processes (primary organic productivity) is 1.5 to 7.10^{10} t for the oceans and 1.5 to 8.10^{10} t for the continental areas. A small fraction estimated at 0.01% to 10% of this quantity, depending on the sedimentation media [38, 48, 54, 82], escapes the biological cycle or surface alterations and is incorporated in sediments. The ratio between the quantities of kerogen present in sediments and the quantities incorporated annually through the water-sediment interface might possibly give an idea of the length in years of the "long" organic carbon cycle (as opposed to the "short" biological cycle). This estimate is, however, very doubtful since the estimate for the rate of incorporation by sediments is very inaccurate while this rate varies a great deal depending on the sedimentation media. It is also probable that this rate has varied at various geological epochs. The calculation gives about 10^8 years for the length of this long cycle.

The elementary carbon in crystalline rocks or the carbon in carbonates is finally partly of organic origin. The former is partially the result of the degradation of kerogen during burial ("carbonaceous residue", *cf.* Chapter 4 and above) while the second is often a by-product of biological activity. Theoretically, the reincorporation of these forms of carbon into the biological cycle takes place through their transformation into CO_2 or CO_3^{--} by the action of surface alteration, degradation in sediments and magmatic and volcanic phenomena. The broadest estimate for this long carbon cycle should therefore take into account the other forms of sedimentary carbon, so that the preceding length of time must be multiplied almost by 10.

The length of time obtained in this way is so great that the validity of this kind of calculation becomes very doubtful.

Figure 1.2 situates the total quantity of kerogen in relation to the ultimate resources for the main categories of fossil fuels. By ultimate resources we mean all tonnage recoverable from deposits, whether these are proven or to be discovered and whether or not this tonnage is economically workable under present conditions but which might become workable in the future according to reasonable projection. The above tonnage is opposed to reserves which constitute a small part of the resources which are both proven and recoverable at the present time. For the ultimate resources for oil, gas and asphalt we have selected the figures of 4×10^{11} t, 2×10^{11} t and 3×10^{11} t, taking into account the fact that in this field the margin of error is large [5].

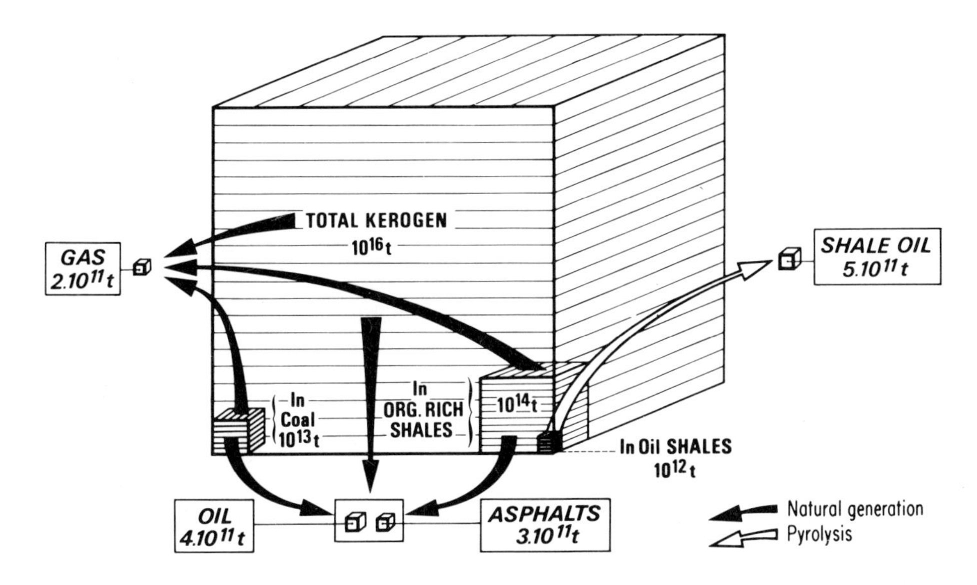

Fig. 1.2. — Comparison between the quantity of kerogen and the ultimate resources of fossil fuels.

Resources for asphalts may appear very high in the light of current figures. This is based on Demaison's recent estimate [19] which gives the figure of about 1.5×10^{11} t for the in situ asphalt in the three major asphaltic provinces: Orinoco (Venezuela), Athabasca (Canada) and Melekess (USSR).

If we suppose that most of the hydrocarbons found in oil, gas or asphaltic fields come from the transformation of kerogen, a comparison of the quantities of kerogen and quantities of oil, gas and asphalt resources expresses the average final yield of natural transformations (petroleum and gas formation, migration) and them of recovery, leading from kerogen to recoverable hydrocarbons. This yield is considered as between 10^{-5} and 10^{-4}.

Hunt [36] estimates the quantities of oil (liquid hydrocarbons) disseminated in sediments (the micropetroleum of Russian authors) at about 2×10^{14} t. The quantities of petroleum *in situ* in exploitable fields which are either proven or to be discovered would amount to 10^{12} t (if we multiply ultimate oil resources by 2.5, i.e. fixing at 40% the average rate of recovery which can finally be expected). If we consider on the other hand that asphalt also derives from this micropetroleum the average yield of the concentration processes leading from disseminated oil to exploitable fields would then be about 1%.

A similar calculation gives an average yield of about 2% for the transformation of kerogen into oil (liquid hydrocarbons). This is a mean value for the whole sedimentary column; therefore the mean maximum rate of this transformation is higher.

However these figures are very questionable. Firstly they do not reflect the diversity of geological situations, and secondly they are based on rough estimates.

The coal ressources indicated of 10^{13} t are from estimates from *World Coal* 1965 [89]. According to Feys [27] this estimate varies from 7 to 17 $\times$ 10^{12} t. The present reserves are about 0.5 $\times$ 10^{12} t [89]. The estimate of oil shale resources is very uncertain. We have therefore selected two values, one for oil shale in the broad sense, i.e. shale whose organic carbon content is more than 5%. According to Duncan *et al.* [23] their organic content amounts to 10^{15} t. This estimate seems very high, and based on the experience gained at *Institut Français du Pétrole (IFP)* we would be inclined to reduce this value by an order of magnitude. This figure would also be more in agreement with the figure by Vassoyevitch *et al.* [80], who gives 2.10^{14} t for the organic content of shales having an organic carbon content of between 4 and 10% and that by Nesterov *et al.* [49], which gives about 10^{13} t for shales having an organic content of more than 3%. We would thus have 10^{14} t of kerogen which could give 3 $\times$ 10^{13} t of oil by pyrolysis, i.e. 75 times the petroleum resources. The other value was obtained by multiplying by 2 an estimate of oil quantities obtainable through the pyrolysis of the presently inventoried oil shales on the basis of current technology (oil shale reserves), i.e. by fixing at 50% the average rate of transformation of kerogen from these shales into oil. This estimate [5, 10, 23] is about 0.5 $\times$ 10^{12} t which therefore gives 10^{12} t for the kerogen in these shales.

B. Distribution of kerogen in sediments.

Our knowledge of this distribution is based principally on the work of Trask *et al.* [76], who have compiled organic carbon contents in sediments in the US on a large scale, and on the work of Ronov [55] and Ronov *et al.* [56] who have done the same thing for the Russian platform, and also, for some aspects, on the work of Hunt [34, 35, 36] and of Gersanovic *et al.* [*in:* 50], Vassoyevitch *et al.* [80], Nesterov *et al.* [49].

The distribution of organic carbon, depending upon geological age since the Cambrian, shows two minimums. The first was in the Silurian and the second in the Triassic (Fig. 1.3). These classic results can be interpreted [25] as a variation in primary organic productivity which was due to variations in the climate or in the quantities of atmospheric CO_2. It is also possible to suppose a relative reduction, due to regressions of great amplitude which seem to have marked these epochs, in the surface of the continental margins which appear to be especially favorable for the accumulation and preservation of organic matter [25, 50].

The distribution of organic carbon, depending upon the main types of sediments, has been the subject of statistics for the present continental and oceanic domains (Table 1.2). The continental domain includes not only the continents but also the continental shelves and slopes. Average organic carbon contents are low. On the average clays and shales are richer than carbonates which are in turn richer than sandstone. Within each of these categories, the finest grained sediments are usually richest in organic carbon. The published literature does not

TABLE 1.2
ORGANIC CARBON IN SEDIMENTS
(from Hunt [36])

Location	Mean Values (Wt. %)	Mass (10^{16} t)
Continents, Shelf and Slope		
Clays and Shales	0.99	0.82
Carbonates	0.33	0.08
Sands	0.28	0.09
Oceanic		
Clays and Shales	0.22	0.07
Carbonates	0.28	0.10
Siliceous	0.26	0.04

give distributions of values for each sedimentary category. It is, however, possible to deduce on the basis of what has been said above that more than 95% of the kerogen is found in sediments with less than 5% organic carbon.

Kerogen seems to be linked to the finest grain size fractions of the sediments. For example, Hunt [35] found the following distribution in Viking Shale (Alberta) in terms of grain size:

Grain Size	Average OM Wt. %
Siltstone	1.79
Clay 2-4 μ	2.08
Clay less than 2 μ	6.50

However it should be observed that the method by which the fractions are recovered, generally centrifuging or decantation, can lead to the enrichment of fine fractions with kerogen although there may not be a real bond.

Examination of the distribution between the oceanic and continental domains shows that the largest quantities of kerogen are to be found in the clay and shale of the continental margins. This is confirmed by the work of Gersanovic *et al.* [*in:* 50] according to which more than 85% of the OM sedimented in the Holocene is to be found on the continental slopes.

Observations on the distribution of kerogen in terms of geological age, type of sedimentation and sedimentation domain are important for petroleum exploration since there is a clear correlation between the capacity of a sedimentary medium to fossilize OM and its petroleum possibilities. Thus Ronov [55] observes that the average organic carbon content of sediments is three times higher in petroliferous basins than in non-petroliferous basins. There is thus a correlation between the distribution of petroleum reserves and that of the average carbon content of sediments in relation to the geological age (Fig. 1.3).

While these observations are very interesting, much more fruitful observations could no doubt be made within the framework of studies on sedimentary models. In previous research, a very small number of sedimentary categories have been chosen and they are therefore badly defined. Only average organic carbon content values have actually been used, whereas it would also be interesting to study the distributions.

The nature of kerogen has not been taken into consideration. Moreover, when more detailed studies are available, they concern anomalous accumulations such as coals, oil shale, etc. which, as seen above, represent a very small portion of sedimentary OM. Therefore a true sedimentology of OM has still not been created. The work will be difficult since the phenomena determining sedimentation of OM have still not yet been well understood and weighted (Chapter 14).

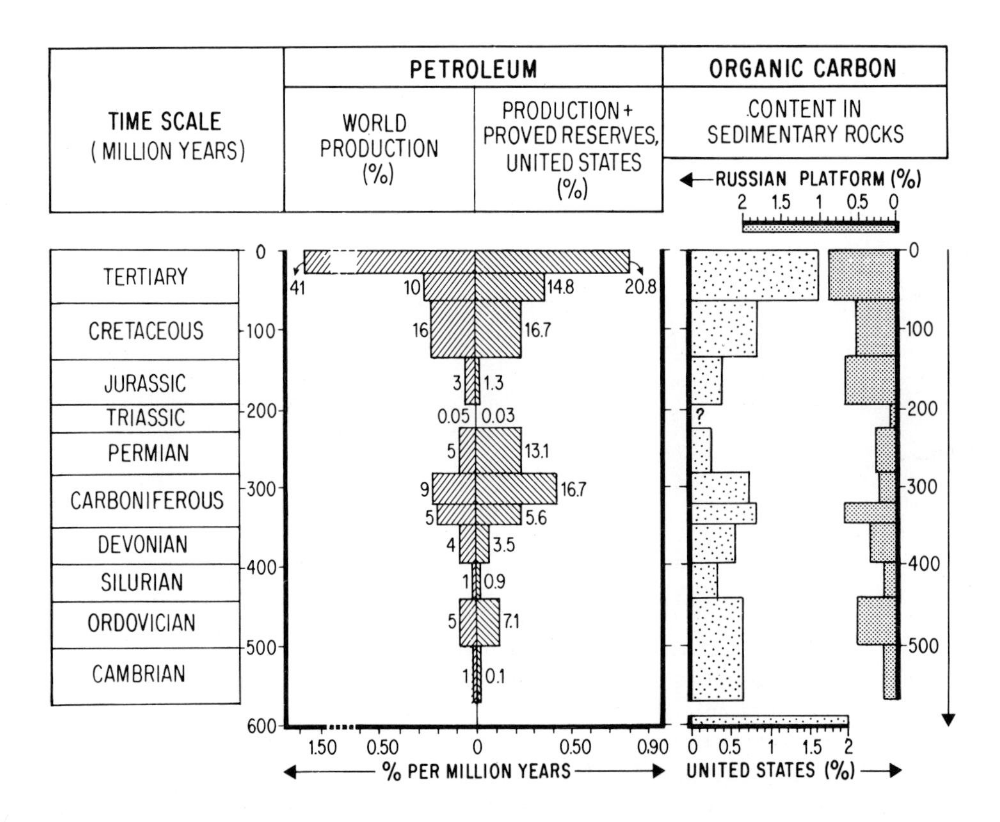

Fig. 1.3. — Geological distribution of oil and organic carbon (from J. Debyser and G. Deroo, (1969), *Rev. Inst. Franç. du Pétrole,* **XXIV**, 1, 21).

Little work in this direction has been published. As an example we can cite an interesting study by Cazes *et al.* [11] on the distribution of kerogen in the Oxfordian in the E of the Paris Basin which is a marly-carbonate medium with reefal limestone. This statistical study takes into consideration a large number of sedimentological and petrological variables. The fossilization of kerogen in this medium finally appears to be linked to a very calm type of sedimentation, in most cases accompanied by an argillaceous element facilitating the preservation of the OM. We can also cite the publications of the *Orgon* Mission in which *Centre National pour l'Exploitation des Océans (CNEXO), IFP* and French oil companies are associated [17].

REFERENCES

1. Basilevich, N.I., Rodin, L.E. and Rozov, N.N. (1971), "How Much Does the Living Matter of Our Planet Weight?" (in Russian), *Prinoda,* No. 1.
2. Bertrand, C.E., (1893), *Bull. Soc. Belge de Géologie,* **7,** 45.
3. Bertrand, C.E., (1897), *2nd Lecture to the Belgian Society of Geology, Paleontology and Hydrology,* 19 Oct. 1897.
4. Bogorov, V.G., (1971), *in: Organic Matter of Recent and Ancient Sediments,* (in Russian), N.B. Vassoyevitch ed. Akad. Nauk SSSR, Moscow.
5. Boy de la Tour, X., Brasseur, R. and Le Duigou, J., (1976), *Rev. Inst. Franç. du Pétrole,* **XXXI,** 741.
6. Breger, I.A., (1960), *Geochim. et Cosmochim. Acta,* **19,** 297.
7. Breger, I.A., (1961), "Kerogen", *in: McGraw Hill Encyclopedia of Science and Technology,* **9,** 337.
8. Breger, I.A., (1963), *in: Organic Geochemistry,* Pergamon Press, Oxford, **3,** 50.
9. Breger, I.A., (1976), 2nd IIASA Conference, *Future Supply of Nature-made Oil and Gas,* **51,** 913.
10. Burger J., (1973), *Rev. Inst. Franç. du Pétrole,* **XXVIII,** 315.
11. Cazes, P. and Reyre, Y., (1976), *Bull. BRGM* (2), IV, **2**-1976, 85.
12. Combaz, A., (1964), *Rev. Micropaléont.,* **7,** 4, 205.
13. Combaz, A., (1973), *in: Advances in Organic Geochemistry 1973,* B. Tissot and F. Bienner ed., Editions Technip, Paris, 1974, 423.
14. Combaz, A., (1973), *in: Pétrographie de la Matière Organique des Sédiments — Relations avec la Paléotempérature et le Potentiel Pétrolier,* B. Alpern ed., CNRS, Paris, 1975, 93.
15. Correia, M., (1967), *Rev. Inst. Franç. du Pétrole,* **XXIV, 12,** 1417.
16. Cunningham. Craig E. H., (1916), *Journ. Inst. Petr. Tech.,* 1916, **2,** 246.
17. CEPM-CNEXO, (1974), *Géochimie Organique des Sédiments Marins Profonds — Orgon I — Mer de Norvège,* CNRS, Paris, 1977.
18. Dalton, L.V., (1909), *Econ. Geol.,* **4,** 603.
19. Demaison, G.J., (1977), *AAPG Bull.,* **61, 11,** 1950.
20. Dietrich, G., (1963), *General Oceanography — An Introduction,* Interscience Pub., New York, 229.
21. Douce, R. and Joyard, J., (1977), *La Recherche,* **79,** 527.
22. Down, A. L. and Himus, G. W., (1940), *J. Inst. Petroleum,* **26,** 329.
23. Duncan, D. C. and Swanson, V. E., (1965), *Geological Survey Circular,* **523.**
24. Duvigneaud, P., *in:* Dajoz, (1970), *Précis d'Ecologie,* Dunod, Paris.
25. Debyser, J., (1975), *Rev. Inst. Franç. du Pétrole,* **XXIV,** 22.
26. Erdman, J.G., (1975), *Proceedings of the 9th World Petroleum Congress,* **2,** 139.
27. Feys, R., (1976), *Annales des Mines,* January 1976, 9.
28. Forsmann, J.P., and Hunt, J.M., (1958), *in: Habitat of Oil,* L.G. Weeks ed., AAPG, 747.
29. Forsmann, J.P., (1963), "Geochemistry of kerogen," *in: Organic Geochemistry,* I.A. Breger ed., Pergamon Press, 1963, 148.
30. Gutjahr, C.C.M., (1966), *Leidse Geol. Meded.,* **38,** 1.
31. Himus, H.W. (1941), *in: Oil Shale and Cannel Coal,* **2,** Institute of Petroleum, London, 1951.
32. Hunt, J.M., Steward, F., and Dickey, P.A., (1954), *AAPG Bull.,* **38,8,** 1671.
33. Hunt, J.M., and Jamieson, G.W., (1958), *in: Habitat of Oil,* L.G. Weeks ed., AAPG, 735.
34. Hunt, J.M., (1961), *Geochim. et Cosmochim. Acta,* **22,** 37.
35. Hunt, J.M., (1963), *in:* V. Bese, *Vortrage der III int. viss. Konf. für Geochemie, Microbiologie und Erdölchemie,* Budapest, 8-13/10/1962, Budapest, 1963.
36. Hunt, J.M., (1972), *AAPG Bull.,* **38,** 377.
37. Hunt. J.M., (1974), *in: Advances in Organic Geochemistry 1973,* B. Tissot and F. Bienner ed., Editions Technip, Paris, 1974, 593.
38. Ivanenkov, V.N., (1977), "Oxidation Rate of Organic Matter in the World Ocean". *Proceedings of the 8th International Meeting of Organic Geochemistry,* Moscow, 1977.
39. Kinney, J.W., (1923), PhD Thesis, McGill University.
40. Kinney, C.R., and Schwartz, D., (1957), *Ind. Eng. Chem.,* **49,** 1125.
41. Koblenz-Mishke, O.M., Volkovinsky, V.V., and Kabanova, J.G., (1968), *in: Symp. Sci. Explos. South Pacific,* N.S. Wooster ed., Scripps Institute of Oceanography, La Jolla, California, 183.

42. Larskaya, Y.S., and Zhabrev, V.D., (1964), *Dokl. Akad. Nauk SSR, 157,* N° **4** 897 (in Russian).
43. Louis, M.C., (1964), *in: Advances in Organic Geochemistry 1964,* G.D. Hobson and M.C. Louis ed., Pergamon Press, Oxford, 85.
44. Louis, A., and Tissot, B., (1967), *Proceedings of the 7th World Petroleum Congress,* **2,** 47.
45. Mason, B., (1966), *Principles of Geochemistry,* 3rd ed., John Wiley, New York.
46. Manskaya, S.M., Kodina, L.A., and Generalova, V.N., (1976), *Geokhimiya,* **1,** 3.
47. Menzel, D.W. and Ryther, J.H., (1970), *in: Organic Matter in Natural Waters,* Alaska Univ. Marine Sci. Inst. Publ, **1,** 31.
48. Menzel, D.W., (1974), *in: The Sea,* E. Goldberg ed., John Wiley, New York **5,** 659.
49. Nesterov, I.I., and Salmanov, F.K., (1976), 2nd IIASA Conference, *Future Supply of Nature-Made Oil and Gas,* **51,** 913.
50. Pelet, R., Cauwet G. and Combaz, A., (1975), *Rev. Inst. Franç. du Pétrole,* **XXX,** 18.
51. Philippi, G.T., (1965), *Geochim. et Cosmochim. Acta,* **29,** 1021.
52. Raynaud, J.F., and Robert, P., (1976) Bull. Centre Rech. Pau, SNPA, **10,** 1, 109.
53. Renault, B., (1899), "Sur Quelques Microorganismes des Combustibles Fossiles,", *Bull. Soc. Industries Min.,* 3e Ser. **13,** 865; 1900, 3[e] Ser. **14,** 5.
54. Romankevitch, E.A., Artemiev, V.E. and Belyaeva, A.N., (1977), "The Geochemistry of Organic Matter in Bottom Sediments of Seas and Oceans," *Proceedings of the 9th International Meeting on Organic Geochemistry,* Moscow, 1977.
55. Ronov, A.B., (1958), *Geochemistry,* **5,** 510.
56. Ronov, A.B., and Yaroshevkiy, A.A., (1969), *in: The Earth's Crust and Upper Mantle,* Am. Geophys. Union Geophys. Mon. Ser., **13,** 37.
57. Rubey, W.W., (1951), *Geol. Soc. America Bull.,* **62,** 1111.
58. Ryther, J.H., (1963), *in: The Sea,* **2,** M.N. Hill ed., Interscience Pub., New York.
59. Ryther, J.H., (1969), *Science,* **166,** 72.
60. Ryther, J.H., (1970), *Nature,* **227, 5276,** 374.
61. Saxby, J.D., (1976), *in: Oil Shale,* T.F. Yen and G.V. Chilingarian ed., Elsevier, 104.
62. Skopintsev, B.A., (1961), *in: Recent Sediments of Seas and Oceans,* Akad. Nauk. SSR. Moscow, 285 (in Russian).
63. *Stach's Textbook of Coal Petrology,* 2nd ed., Borntraeger, 1975.
64. Staplin, F., (1969), *Bull. Canad. Petrol. Geol.,* **17,** 1, 47.
65. Steeman, Nielsen E. and Jensen, J.A., (1957), *Galathea Rep.,* **1,** 49.
66. Steuart, D.R., (1912), *in: The Oil Shales of the Lothians — Part III: The Chemistry of the Oil Shales,* 2nd ed., Memoirs of the Geological Survey, Scotland, 143.
67. Stopes, M., (1935), *Fuel,* **14,** 4.
68. Subbota, M.I., Khodjakuliev, I.A., and Romaniuk, A.F., (1976), *in: Research on the Organic Matter in Contemporary and Fossil Deposits* (in Russian), Nauk ed., Moscow, **65.**
69. Takahashi, J., (1935), *Sci. Rpts. Tohoku Univ., 1,* 3, 63.
70. Teichmüller, M., (1958), *Rev. Industrie Minérale,* spec. issue, 15 July 1958, 99.
71. Tissot, B., (1969), *Rev. Inst. Franç. du Pétrole,* **XXIV,** 4, 470.
72. Tissot, B., and Pelet, R., (1971), *Proceedings of the 8th World Petroleum Congress,* **2,** 35.
73. Tissot, B., and Espitalié, J., (1975), *Rev. Inst. Franç. du Pétrole ,* **XXX,** 5, 743.
74. Tomkeieff, S.I., (1954), *Coals and Bitumens,* Pergamon Press London.
75. Trager, E.A., (1924), *AAPG Bull.,* **8,** 301.
76. Trask, P.B., and Patnode, H.W., (1942), *Source Beds of Petroleum,* AAPG, Tulsa.
77. Trask, P.B., (1932), *Origin and Environment of Source Sediments of Petroleum* Gulf Publ. Col, Houston, Texas.
78. Van Krevelen, D.M., (1961), *Coal (Typology, Chemistry, Physics, Constitution),* Elsevier.
79. Vassoyevitch, N.B., Korghagina, Y.I., Lopatin, N.V. and Chernyshev, V.V., (1969), *Vestnik Mosk. Univ. 1969,* **6,** 3 (in Russian).
80. Vassoyevitch, N.B., Koniukov, A.I. and Lopatin, N.V., (1976), International Geological Congress, 25th Session, *Report of Soviet Geologists on Combustible Mineral Resources,* 7.
81. Weeks, L.G., (1958), *in: Habitat of Oil,* AAPG 1958, Tulsa, 1.
82. Welte, D.H., (1970), *Naturwissenschaften,* **57,** 17.
83. Wickman, F.E., (1956), *Geochim. et Cosmochim. Acta,* **9,** 136.
84. White, D., (1915), *J. Wash. Acad. Sc.,* **6,** 189.
85. White, D., (1916), *Geol. Soc. America Bull.,* **28,** 1917.
86. Whittle, K.J., (1976), *in: concepts in Marine Organic Chemistry,* Symposium, Edinburgh.
87. Williams, P.J., (1975), *in: Chemical Oceanography,* 2nd ed., J.P. Riley and G. Skinow ed., **2,** 301.
88. Winberg, G.G., (1960), *Primary Productivity of Aquatoria,* Akad, Nauk. SSSR. Minsk.
89. *World Coal,* (1975), Miller Freeman Publication, San Francisco.

2

Procedures for kerogen isolation

B. DURAND* and G. NICAISE*

I. INTRODUCTION

Several reviews have been made of research on the separation of kerogen from its mineral matrix [13, 32, 33, 34]. They lead to the conclusion that the ideal method for isolating kerogens does not yet exist and that the isolated kerogen, i.e. the object of the analysis, does not always coincide perfectly with the concept of kerogen.

Under these conditions, it is necessary to make the best possible adaptation of the separation technique to the analysis methods which are to be used afterwards. This separation is not necessary in some cases, i.e. analysis using pyrolysis (Chapter 5), chemical degradation (Chapter 10) etc. However if physico-chemical analysis are to be made, kerogen must be isolated in such a way that the isolated fraction is as representative as possible of in situ kerogen. In particular, recovery must be sufficient and chemical alteration must be avoided. If a petrographic analysis is to be made the principal objective is to bring out identifiable organic debris and macerals in large areas and to avoid altering the morphology.

One of the basic obstacles to be overcome in isolating kerogen from its mineral matrix is in the diversity of the spatial relations and the chemical bonds between the mineral and organic phases. Kerogen contents in sedimentary rock are most often about 1% (*cf.* Chapter 1) and can reach any value between ϵ and 100%. The physico-chemical properties depend on the nature of the initial organic matter (OM) and are modified by the evolution of the processes affecting the sediments. The same is true for the relations between the mineral phase and the organic phase. The grain size of the various constituents, whether they are organic or mineral matter, is in most cases extremely small and also varies a great deal from one sample to the next, and even within the same sample.

An example which clearly illustrates the difficulties encountered is that of the relations between pyrite and OM. The close association between OM and pyrite in sediments is a well known fact [3, 24, 38]. It is illustrated by photos of Chapter 3.

---

*Institut Français du Pétrole. Rueil-Malmaison, France.

II. ISOLATION METHODS FOR PHYSICO-CHEMICAL ANALYSIS

Among the methods used, we will distinguish physical methods from chemical methods.

A. Physical methods.

Almost all ore dressing processes were used, i.e. flotation, ultrasonic methods, electromagnetic and electrostatic methods, processes based on the wetting properties of OM and minerals, etc. [32, 33, 34].

These processes are applied after fine grinding of the sediment. Organic fractions are thus obtained which should not, in theory, be chemically altered. However the rate of recovery for OM containing small amounts of minerals is low, and the chemical composition of the recovered fraction is rarely representative of the whole. In order to improve recovery it is necessary to carry out successive grindings and to repeat the operations several times, which is long and costly.

We will however note that, according to Robinson [32], use of a method derived from that of Quass [29, 37] has made it possible to considerably reduce (24-99%) the proportion of the mineral fraction from oil shales of various origins. This method is based on wettability differences between kerogen and the mineral matrix. It consists in incorporating water and an oil phase (mineral oil or *n*-hexadecane in the cases mentioned) during attrition grinding. The kerogen tends to form progressively a separate phase with the oil.

Good results were also obtained with Green River Shales by centrifuging in a $CaCl_2$ solution [32, 36] with a method based on that of Luts [25].

It is easy to understand the lack of effectiveness and general applicability of these methods in view of the diversity of the situations which occur. Success cannot therefore be expected except in particular cases, and the methods must be modified in terms of the characteristics of the samples.

B. Chemical methods.

Theoretically the objective of these methods is to isolate the kerogen by making the mineral phase soluble while keeping modifications in the chemical composition of the OM as small as possible.

They do in fact make possible good recovery of the OM which is little altered except for recent OM (recent sediments, soil, peat, etc.).

Following the work of Down [7], Giedroyc [14] and Dancy [4] the methods now used are all based upon the dissolution of carbonates, sulfides, sulfates, oxides and hydroxides by HCl and the dissolution of silicates by HF, after fine grinding of the rock ($\sim$ 100 μ) in order to facilitate the reactions. It is usually recommended that operations be carried out at a temperature which is sufficient to dissolve all carbonates but not too high to prevent possible oxidation of the organic matter [5, 14, 36] i.e. about 70° C.

1. *Effects of acid attack on minerals.*

If we consider only the destruction of the mineral phase, the principal limitations encountered are as follows:

(a) All the minerals cannot be dissolved. The most frequent of these residual minerals is iron sulfide FeS_2 (pyrite, cubic form, and less frequently marcassite, orthorhombic form). Heavy oxides and sulfates and some silicates are also found. At *Institut Français du Pétrole (IFP)* the following were identified: zircon, rutile, anatase, brookite, tourmaline and barite. The latter mineral in most cases probably comes from pollution, for example from drilling mud [1]. From time to time tungsten carbide is found in the residual minerals and is probably due to the drilling bits and grinders.

(b) During HF attack there is frequently neoformation of complex fluorides [12] and especially ralstonite [16]: $Na_x\ Mg_x\ Al_{2-x}\ (F.OH)_6\ H_2O$.

(c) The minerals are sometimes protected from chemical attack by the OM.

2. *Elimination of residual minerals and neoformed minerals.*

The elimination of pyrite is by far the principal problem. Most of the other residual minerals have little quantitative importance, and their chemical stability is such that they have hardly any effect during the later analyses of kerogen other than as diluents. However the barite and neoformed fluorides are very undesirable because they interfere with the OM, especially during element analyses (*cf.* Chapter 4).

a. Elimination of pyrite.

The principal methods proposed for the elimination of pyrite are:

(a) Separation in dense liquors.

(b) Destruction by oxidizing agents: dilute nitric acid [5, 7, 13, 22, 23, 33, 34], aqueous solution of soda [30] or ferric iron salts [15].

(c) Destruction by reducing agents: nascent hydrogen (formed by the action of hydrochloric acid on zinc) [13, 2, 12, 22, 5], sodium boron hydride $NaBH_4$ [34, 26] or lithium aluminium hydride $LiAlH_4$ [22, 33].

The temperatures for chemical processing are around 100° C.

At the present time none of these methods gives results with general applications. Separation in dense liquors leads in most cases to only fractionation into a light supernatant fraction and a heavy pyrite-enriched fraction.

In many cases the supernatant fraction still contains an excessive quantity of pyrite due to the extreme fineness of the constituents and the close bond between this pyrite and the OM (*cf.* paragr. 1). In addition, the properties of the OM in each fraction may be different. Destruction of the pyrite by chemical agents is very unequal. According to Saxby [33] the most effective agent is NO_3H followed by $LiAlH_4$, while other reagents are less effective.

---

[1] It has been verified that the most frequent sulfates in the sediments, i.e. gypsum and anhydrite, were destroyed by acid attack although the kinetics of their destruction is slow.

TABLE 2.1
EFFECT OF FERRIC SULFATE TREATMENT
ON ELEMENTAL ANALYSIS OF KEROGEN

Sample, IFP No.		H/C*	O/C* (× 100)	Organic S/C* (× 100)	N/C* (× 100)	Fe (weight %)	C (weight %)	C recovery (%)
Green River Shales, USA, Eocene	1	1.64	7.8	0.6	1.64	0.1	78.0	95.9
15760	2	1.55	8.1	0.4	1.65	0.7	76.0	
Green River Shales, USA, Eocene	1	1.16	3.3	1.7	2.5	1.9	75.1	101.4
15790	2	1.12	3.6	0.9	2.1	1.1	77.8	
Torbanite, Australia, Permian	1	1.54	3.5	0.6	1.2	0	83.1	97.1
15844	2	1.53	3.7	0.6	1.1	0	80.8	
Messel Shales, Germany, Eocene	1	1.39	21.7	1.7	2.7	1.0	65.3	94.6
15132	2	1.33	19.6	0.9	2.3	0.6	67.1	
Lower Toarcian Shales, France	1	1.27	14.9	2.5	2.3	6.2	58.8	98.6
10696	2	1.24	14.2	2.9	2.2	1.2	68.1	
Siluro-Devonian Shales, Algeria	1	0.81	5.0	0.7	3.8	8.2	66.9	101.8
14085	2	0.74	8.6	2.3	3.8	0.8	70.3	
Jurassic, Mexico	1	1.27	13.8	14.5	2.9	15.6	24.8	101.9
25838	2	1.08	6.5	7.2	2.3	10.7	53.7	
Springhill Shales, Chile, Jurassic	1	0.60	11.1	0	0.8	31.9	20.9	95
11463	2	0.70	9.9	3.9	1.8	4.7	50.7	
Peat, Holland, recent	1	1.00	54.5	0.8	1.4	0	53.7	< 80
15557	2	0.85	50.9	2.3	1.1	0	50.3	
Lignite, Germany, Eocene	1	0.82	35.6	2.7	1.3	0	59.8	88.3
15288	2	0.88	40.8	3.4	1.6	0	55.7	
Green River Shales, USA, Eocene	1	1.65	17.9	4.4	2.8	10.7	38.8	100.8
21566	2	1.39	5.7	1.8	2.7	4.0	69.2	
Green River Shales, USA, Eocene	1	1.47	7.0	1.6	2.7	0.6	68.8	—
21565	2	1.40	6.1	1.8	2.6	2.9	69.4	

1. Before treatment
2. After treatment
* Atomic ratios

The effectiveness in eliminating pyrite at the same time produces an alteration in the OM. It appears that the later forms a protective coating for part of the pyrite, and this makes it inaccessable to the reagents. It is therefore necessary to degrade this coating in order to destroy the pyrite. The critical study by Saxby takes into account all the reagents which we have mentioned except for weak oxidizing agents. At IFP, we have investigated the effects of ferric sulfate (*cf.* paragr. II.B.3b) showing (*cf.* Table 2.1) that this reagent also makes it possible to eliminate a large part of the pyrite at the cost of slight modifications of the kerogen.

b. *Elimination of other residual minerals.*

The elimination of other minerals which cannot be destroyed by acid attack does not generally encounter the problem of protection by OM. In this case we have rarely observed a close link with the OM. We will however mention the case of kerogen isolated from Black Sea sediments in which anatase crystals recovered with carbon were observed (Chapter 7). However the stability of these minerals makes it impossible to consider chemical elimination. It is therefore necessary to make use of physical separation (heavy liquids, centrifuging, etc.) with the risk of fractionating which is implied.

Minerals resulting from pollution, especially barite, can be eliminated by sorting and washing before grinding. The introduction of tungsten carbide can be avoided in the same way by utilizing the appropriate grinding methods or at least by avoiding overgrinding. In any event, sorting and washing before grinding are recommended because organic pollution is also possible.

c. *Elimination of neoformed fluorides.*

For the elimination of neoformed fluorides it was proposed to use sodium or ammonium carbonate [13, 22, 5] or boric acid [32]. It was shown [31] that treatment with refluxed concentrate HCl was an effective process. Infrared (IR) spectroscopic control [31] clearly shows the disappearance in the spectra (Fig. 2.1a) of bands in the 500-700 cm^{-1}, 1 650 $cm^{-1}$and 3 600 cm^{-1} ranges which are characteristic of the product, as shown on Fig. 2.1b which is an IR spectrum of ralstonite. In any case it seems preferable to keep fluorides from forming by using only HF mixed with HCl [20, 21] and by following a strict operating procedure, i.e. effective washings between destruction of carbonate and that of the silicate, and limitation on the quantities of rock attacked. In some cases however it seems impossible to avoid the formation of fluorides at the temperatures of 60-70° C which are usually used for the destruction of the carbonate and silicate. In the cases studied by *IFP* it seems that the access of the HCl to the carbonate has been slowed down by the presence of coatings of very hydrophobic OM. It is therefore assumed that the elimination of the Ca^{++}, Mg^{++}, K^{+} and Na^{++}ions had been insufficient during later attacks by HF and that this favored the formation of complex fluorides.

The addition of a wetting agent (methanol or acetone) in these cases effectively accelerates acid attack. However the formal proof that neoformation is due to a wetting problem and that the addition of a wetting agent is sufficient to eliminate it is not yet available. In addition, methanol or acetone should not be generally used because of the risk that these products will be adsorbed by the kerogen (especially the very oxygenated kerogens which have hydrophilic functions).

In the end the attitude towards the problem of mineral impurities in the kerogen isolated by chemical means should be pragmatic. It has been observed at *IFP* that statistically it is possible to tolerate a pyrite content as high as 40% by weight and/or a content of other minerals as high as 15% by weight (Chapter 4) without excessive difficulties for later analyses.

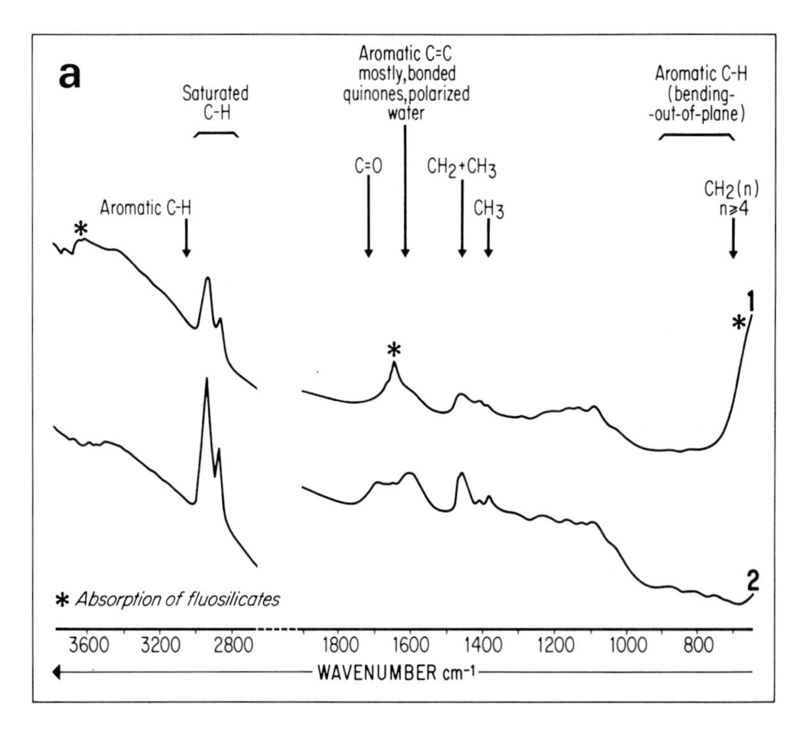

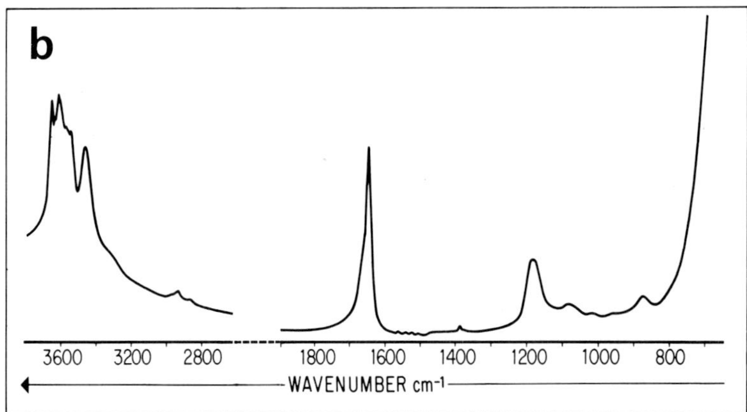

Fig. 2.1.a. — IR spectra of a kerogen sample from Mexico (IFP No 25838):
1. Before. **2.** After boiling HCl treatment, showing the disappearance of absorption bands of neoformed fluorides.

Fig. 2.1.b. — IR Spectrum of a ralstonite sample.

The problem of their elimination does not in general arise except above these thresholds, i.e. for about one kerogen in three. If it is desired to separate the pyrite from these kerogens, the best methods at the present time seem to be the use of $LiAlH_4$ or ferric sulfate, for they do not lead to any fractionating and seem to be relatively efficient and have little altering effect. Nevertheless the use $LiAlH_4$ is a relatively long method involving dangerous products. When the residues of the acid attacks contain a large proportion of residual minerals other than pyrite (this can be checked by element analysis including iron analysis), these residual minerals probably largely consist in neoformed fluorides which can be substantially eliminated by reflux HCl treatment for several hours.

It is possible to prefer separation by decantation or centrifuging although there is a risk of fractionation.

TABLE 2.2
EFFECTS OF HCl and HF TREATMENTS ON ELEMENTAL ANALYSIS OF KEROGENS

Sample, IFP No.		Atomic H/C	Atomic O/C (× 100)	Atomic Organic S/C (× 100)	Atomic N/C (× 100)	C (recovery weight %)
Peat, Holland, recent	1	1.21	53	0	2	1 → 3 81.6
	2	0.97	54	0.8	4	
15557	3	1.00	53	0.4	2	
Lignite, Germany, Miocene	1	0.97	36	0.3	2	1 → 3 96.7
	2	0.86	39	0.4	1	
15287	3	0.81	39	—	2	
Kukersite, Estonia, Ordovician	1	—	—	—	—	1 → 3 97.9
	2	1.43	16	1	1	
15326	3	1.36	16	1	0.5	
Torbanite, Australia, Permian	1	1.57	4	0.3	1	1 → 3 97.2
	2	—	—	—	—	
15844	3	1.49	3	0.8	—	
Green River Shales, USA, Eocene	1	—	—	—	—	1 → 3 98.6
	2	1.58	10	0.5	3	
15409	3	1.57	10	1	1	

1. Original sample
2. After HCl treatment
3. After HCl and HF treatment

3. *Effects of reagents on organic matter.*

Surveys of this question have been made by Saxby [33, 34, 35].

a. *HCl and HF.*

HCl and to a lesser extent HF have a hydrolyzing effect on the OM of sediments. The lesser evolved the OM, the stronger is the effect. Reactions other than hydrolysis are possible [33, 34]. (According to the experience acquired at *IFP,* the addition of chlorine and fluorine is probable). Oxidation of the OM can be feared if the temperature is too high ($> 70°$ C), especially if the preparations run dry [5, 14, 36]. Therefore performing the attacks under an inert atmosphere is advisable.

The extent of hydrolysis can be evaluated by an organic carbon balance. Durand *et al.* [10] have shown that for a series of coals the carbon loss was about 10 to 15% at the peat stage and was negligible beyond the lignite stage. For an equal degree of evolution, this carbon loss is greater for so-called "marine" or "lacustrine" OM (as opposed to so-called "continental" which like coal is largely derived from the constituents of higher vegetation and especially lignin). It is probable that this figure can in certain cases exceed 50% at the recent sediment stage [1, 6, 18]. Measurements made at *IFP* indicate however that for ancient sediments the carbon loss due to hydrolysis during the preparation of the kerogen is generally low. Statistics for 45 samples of very varied origin show an average carbon loss of 5.8%. Most of this carbon loss moreover is very probably due to handling.

The effect of HCl and HF on the chemical composition can be studied with rock which is very rich in OM. Table 2.2. shows the results of elemental analysis made before and after the treatment of such rock. The corresponding OM represents a quite broad range of the chemical structures to be found in kerogen. With the exception of slightly evolved rocks, elemental analysis are not modified to a great extent. This result was also obtained for a series of coals [10]. However, fixation of Cl and F has been experienced in some cases. It increases with evolution of OM due to burial, owing probably to the increase of microporosity and aromaticity, and may reach several percents weight; this fixation could occur by absorption of HCl and HF in the microporosity or by addition and substitution reactions.

Other methods of investigation such as IR spectroscopy, dark-field electron microscopy (EM), electron spin resonance (ESR), spectroscopy and thermogravimetry do not show important modifications either ([10], Fig. 2.2.). Once again immature OM is the most affected. For example we can indicate the very marked increase in the function C = O at 1 700 cm^{-1} in the IR spectra of peat and lignite ([10], Fig. 2.3.)

Since all these observations were made with overall study methods, it is obviously not possible to state that there were no modifications in detail. In addition, in the case of sediments with less OM, the chemical bonds between the organic phase and the mineral phase would necessarily be destroyed.

The effect of hydrolysis on OM which is not greatly evolved and especially on recent sediments is such that in this case the term kerogen should not be used for the residue of acid attack. The analytical object (the residue of the attacks) is very different here from the initial concept. In the case of so-called "marine" or "lacustrine" OM where the effect is greater, a large part of the organic residue of the acid attack probably consists of the organic fraction which is not contemporary with sedimentation (reworked products) which are more resistant

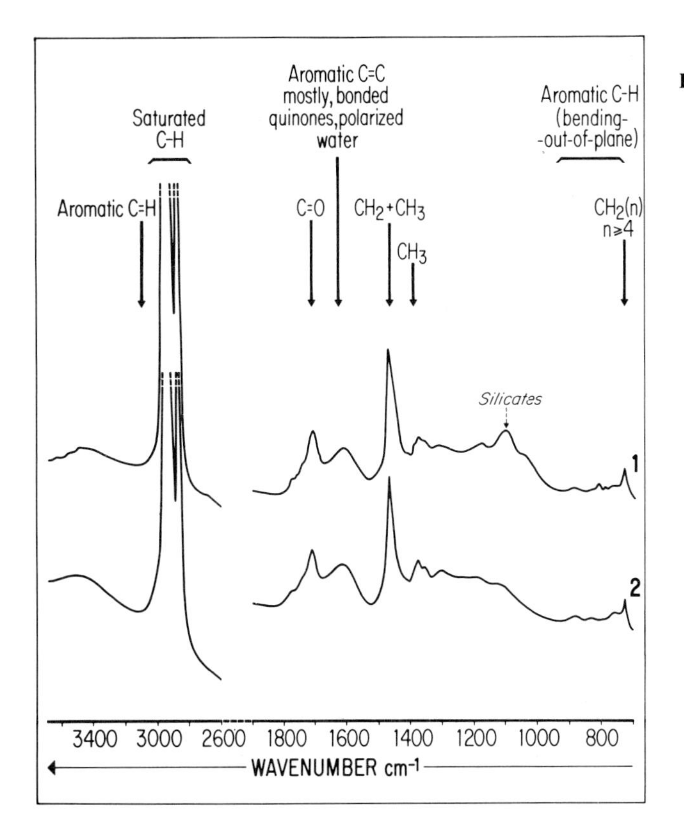

Fig. 2.2. — IR spectra of a torbanite sample from Australia (IFP No 15844):

1. Before. **2.** After HCl and HF treatments. The only change is the disappearance of silicate absorption near 1 100 cm^{-1}.

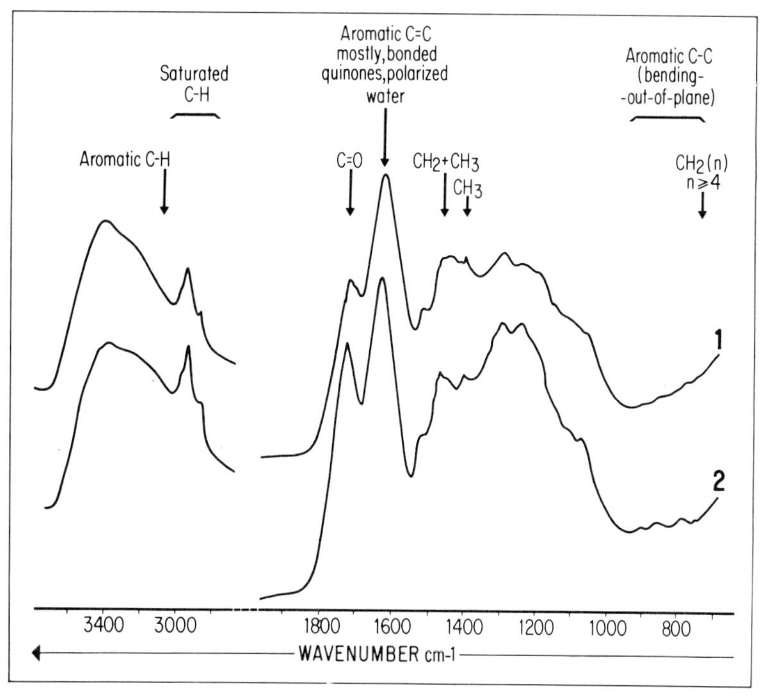

Fig. 2.3. — IR spectra of a lignite sample from Germany (Reinisches Revier, IFP No 15287):

1. Before. **2.** After HCl and HF treatments. On this poorly evolved sample, an increase of C = O groups absorption near 1 700 cm^{-1} can be noticed.

to hydrolysis. On the other hand for ancient sediments this organic residue seems to be very close to kerogen.

It has been proposed [17, 39] to use acetic acid or carbonic acid for reducing the possible effects of hydrochloric acid. The first has the disadvantage of being an organic acid. There is the risk that it will not be possible to eliminate it completely after the attack. The second is effective only at high pressures. In any case the fact that it is necessary to use HF for the destruction or silicates does away with the improvement which was intended.

b. *Agents for destroying pyrite.*

Reducing agents.

$LiAlH_4$: this is the only reducing agent whose effects on kerogen have been systematically studied. It is used in solution in tetrahydrofuran [22, 33], or dioxane mixed with diethyl ether

TABLE 2.3
EFFECT OF NO_3H TREATMENT ON ELEMENTAL ANALYSIS OF KEROGEN
(NO_3H 40%, 70° C)

Sample, IFP No.		H/C*	O/C* (× 100)	Organic S/C* (× 100)	N/C* (× 100)	Fe (weight %)	C (weight %)	C (recovery %)
Lignite, Germany, Eocene	1	0.77	35.7	—	1.0	—	59.7	—
15288	2	0.55	57.8	1.8	6.9	0.4	49.5	
Green River Shales, USA, Eocene	1	1.64	4.6	0.4	1.8	0.5	80.0	68.8
15760	2	1.65	13.5	0.6	4.1	0.0	72.3	
Green River Shales, USA, Eocene	1	1.56	10.1	1.3	0.7	2.0	72.4	61.9
15409	2	1.48	21.9	0.5	7.5	0.2	65.2	
Lower Toarcian, Shales, France	1	1.26	13.1	3.1	2.7	6.2	61.5	45.0
10696	2	1.19	35.5	2.0	8.2	0.2	57.0	
Douala Basin, Cameroon, Cretaceous	2	0.54	6.5	0.0	2.4	4.8	77.4	61.3
13356	2	0.45	27.4	0.5	7.6	0.1	64.1	
Green River Shales, USA, Eocene	1	1.48	20.7	14.3	0.0	36.0	7.8	—
15789 (Heavy fraction)	2	1.01	22.5	0.6	7.5	0.2	61.4	

1. Before NO_3H treatment
2. After NO_3H treatment
* Atomic ratios

[31]. Reflux boiling treatment is used. After cooling, the solution is poured in water in order to eliminate excess $LiAlH_4$ and then acidified with hydrochloric acid and brought to the boiling point. The residue is then washed until neutral. Many organic functions are reduced at the same time as the pyrite, especially groups containing a C=O bond (carboxylic acids, esters, ketones etc.) which are transformed into alcohol. These modifications can be indicated by IR spectroscopy [22, 31] and have been used to characterize these functional groups in kerogens. The bonds C=N and C=S are also affected [31]. According to Saxby [34] the general effect on the element analysis is a slight increase in the H/C ratio. Carbon losses are also very low. These losses could be partly due to the fact that part of the kerogen or the products of its reduction have been made soluble, especially by tetrahydrofuran [33]. No dissolution phenomena have been noted when dioxane [31] is used.

Other reducing agents.

$NaBH_4$ and nascent H are less powerful reducing agents than $LiAlH_4$. The groups C=O, C=N and C=S are in principle not reduced. The effects on the OM are even less [33] than in the case of $LiAlH_4$.

Oxidizing agents.

(a) **NO_3H:** The effects of NO_3H on OM are considerable. It has been shown [7, 13, 33, 34] that oxidation, nitration and a decrease in the hydrogen-content are produced. The identical results have been obtained at *IFP* (Table 2.3). This table also shows that there is a considerable carbon loss. The IR spectra of the OM (Figs. 2.4 and 2.5) show a considerable increase in the C=O functions at 1 700 cm^1 and the appearance of vibrations at 1 330 and 1 540 cm^1 assigned to R-NO_2 [33].

These effects can be minimized by the use of conditions which are as mild as possible, but for good destruction of the pyrite temperatures higher than 50° C and concentrations over 1N are required. Saxby [33] suggests that NO_3H 2N be used at 50° C.

(b) **Ferric sulfate:** Ferric salts were successfully used to eliminate pyrite in coal [15]. The use of ferric sulfate seems preferable to that of ferric chloride. Up to the present time they have not been used to eliminate pyrite from kerogen. At the suggestion of R. Pelet we have investigated the effects of ferric sulfate in a very acid medium at *IFP*. The action on the pyrite is to selectively oxidize the sulfur which it contains by forming elementary sulfur and a sulfate ion. At *IFP*, we have operated with reflux boiling treatment with a solution of 150 g of hydrated ferric sulfate ($(SO_4)^3$ Fe^2, 9 H_2O) for 1 litre of HCl 4N. A few drops of methanol were added for good dispersion and wetting of the kerogen. Washings were performed by filtration.

The study was carried out for 12 kerogen samples, 6 of which contain a considerable proportion of pyrite while the others have practically none (Table 2.1).

Results now available show that:

(a) The elimination of pyrite is unequal but there is generally a substantial improvement. It is possible to improve this elimination by further grinding (under non-oxidizing conditions).

(b) Evolved OM is little affected. Organic carbon recovery is generally excellent, taking into account the inevitable losses due to the handling. The element analyses are not affected to a significant degree nor are the IR spectra. It should however be noted that there is sometimes a slight fall in H/C and a slight reduction in the intensities of the CH bands at 2 900 cm^1.

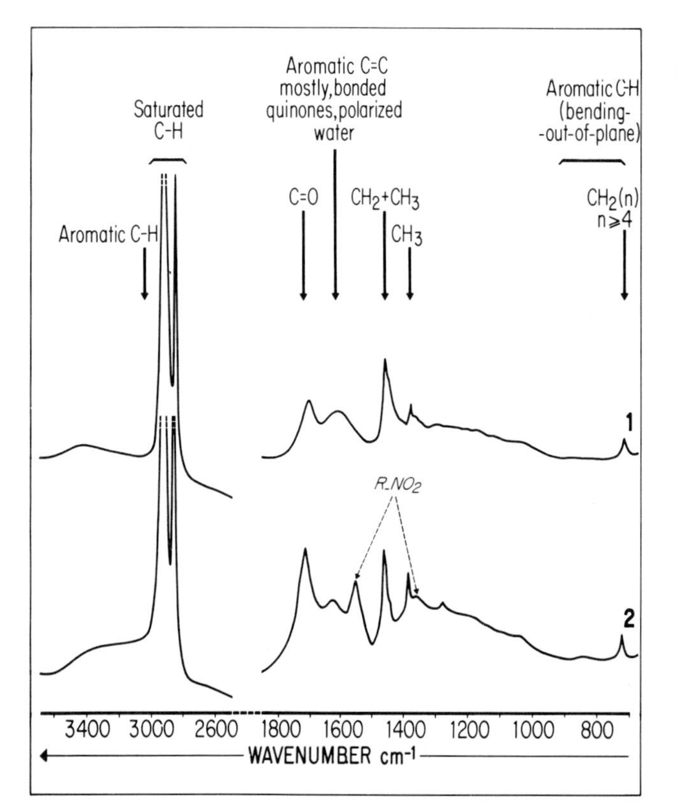

Fig. 2.4. — IR spectra of a Green River Shales kerogen sample (IFP No 15760):

1. Before. **2.** After NO_3H treatment. Increase of $C=O$ groups absorption at 1 700 cm^{-1} and appearance of $R-NO_2$ groups absorption at 1 540 and 1 300 cm^{-1} can be noticed.

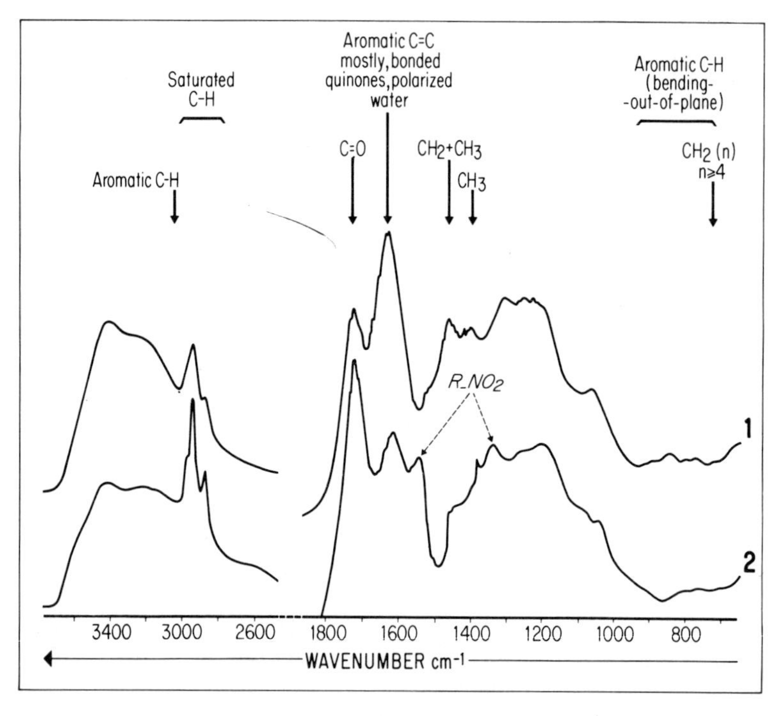

Fig. 2.5. — IR spectra of a lignite sample (IFP No 15288):
1. Before. **2.** After NO_3H treatment.

In some cases sulfur and iron are clearly incorporated. The sulfur is probably elementary sulfur liberated by the oxidation of pyrite. At the present time the form in which the iron is found is unknown. Samples of little evolved OM (Nos 15.557 and 15.288) undergo considerable modifications. In this case it is observed that carbon recovery is poor. The IR spectra show a reduction in the intensity of the saturated CH bands and a modification in the distribution of the oxygenated functions, i.e. an increase in the intensity of the C=O groups at 1 700 cm^{-1}, and a reduction in the intensity of the OH functions between 2 700 and 3 700 cm^{-1} and of C–O bonds between 930 and 1 800 cm^{-1}. In the case of recent OM, it therefore seems that the OM is also being oxidized. As a result of the hot acid treatment, a part of the neoformed fluorides is eliminated.

Use of this method thus leads to a perceptible reduction in the residual minerals without altering to a notable extent the OM, except for recent OM. Moreover it is easy to use; it is therefore to be recommended for the recovery of kerogen which cannot be used because the pyrite content is too high. There is the disadvantage that elementary sulfur is liberated. A large part is deposited on the walls of the refrigerator of the reflux apparatus and can therefore be easily eliminated. However if exact analysis of sulfur are desired, it is necessary to consider extraction with an organic solvent (for example chloroform) or vacuum distillation.

C. Extraction with organic solvents.

This extraction is an important operation because it is part of the definition of kerogen. Few critical studies have however been made on this subject. The solvent to be used, the procedure and the stage of operation must be specified.

There is a consensus as to the fact that solvents should be used with little or no solvolysis effect. The objective of the extraction is in fact to eliminate products, especially hydrocarbons, which exist in the free state in the sediment as opposed to "fixed" products (kerogen) which can in turn liberate "free" products during the later evolution of the sediment. Even if solvents without solvolysis effects are used, the line of division between the two groups remains quite vague and the complexity of the sedimentary OM is such that all the transitions exist between total liberty and a strong chemical bond (covalent).

This problem is complicated by the effects of adsorption, diffusion or extraction kinetics. In practice, benzene, methanol-benzene mixtures in varying proportions and chloroform are the most used. Extraction takes place in a Soxhlet extractor, a mixer, an ultrasonic apparatus or simply in a beaker with heating magnetic agitation at temperatures from ambient temperature to the boiling points of the solvents used.

In theory extraction can be performed at all stages of kerogen preparation. In practice it is performed immediately after grinding since in most organic geochemical studies the extractible fraction is studied rather that the kerogen. This extractible fraction can be altered and/or partially lost if extraction takes place at another stage. The rock is finely ground (grain size with a mode of around 100 μ) before extraction so that the extraction kinetics will not be too slow. The grain size is also very well suited for chemical methods of kerogen isolation.

Under the somewhat restrictive conditions which we have just described, the quantities of the soluble fraction and their composition vary considerably. According to Monin *et al.* [27] for a given solvent the quantities of extract are practically independent of the apparatus (Soxhlet extractor, ultrasonic device, etc.) and reach the limiting value after a period of time

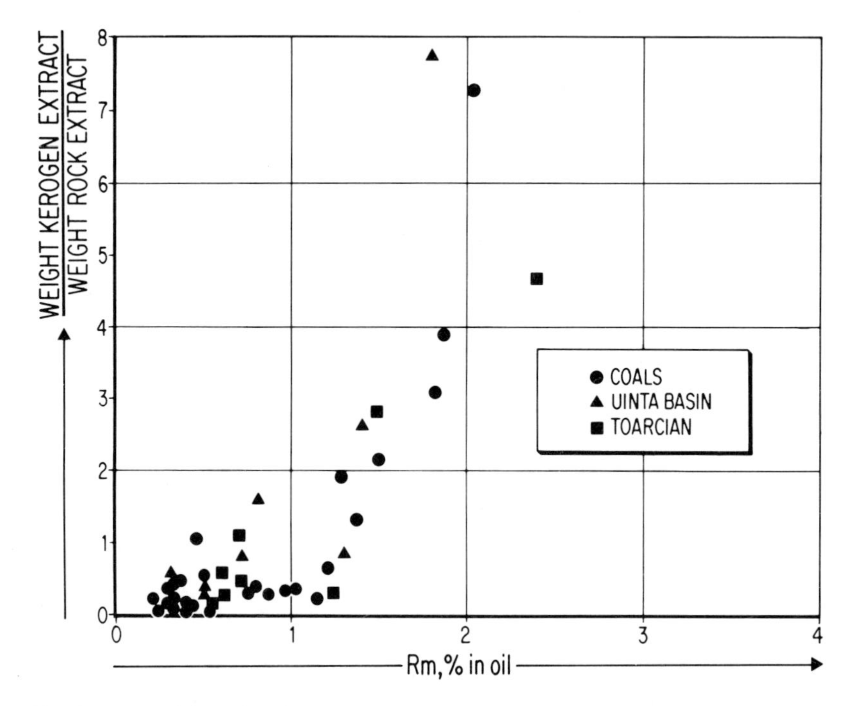

Fig. 2.6. — Weight ratio of kerogen extract to rock extract as a function of vitrinite reflectance for:

● Coals. ▲ Uinta Basin series. ■ Lower Toarcian series (Paris Basin and Posidonien Schiefer from Germany).
Kerogen is prepared from the $CHCl_3$ extracted sediment, then extracted.

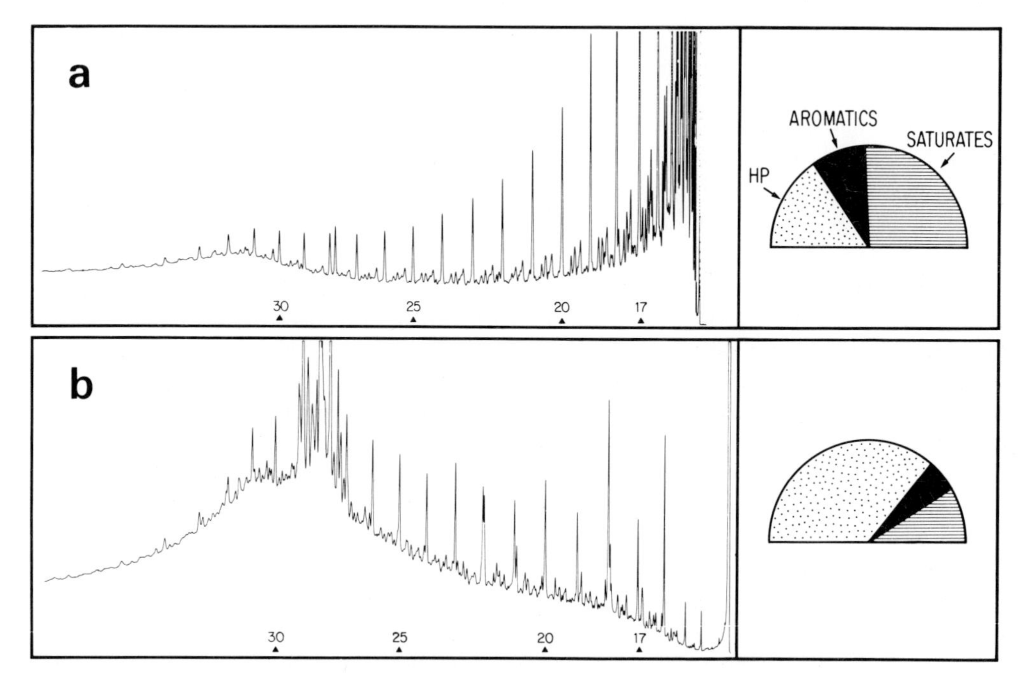

Fig. 2.7. — Gas chromatograms of saturates from:

1. $CHCl_3$ extract of the sediment. **2.** $CHCl_3$ extract of the kerogen prepared from the $CHCl_3$ extracted sediment. Lower Toarcian sample from Germany (Posidonien Schiefer, IFP, No 20508).

which varies depending on the apparatus. A temperature increase within a range of 20-60° C does not significantly change this limit but make it possible to reach the limit more quickly. The quantities of extract depend largely on the solvent used. The differences are however small as concerns the hydrocarbons sensu stricto, and it is the quantities of resins and asphaltenes which vary. Thus under identical conditions, chloroform extracts twice as much of these products as benzene, and the methanol/benzene mixture extracts three times more.

Since critical studies and standardisation, which should apparently begin with the choice of a single solvent, are not yet available, the choice of a method should be determined principally on the basis of rapidity and simplicity.

Some carry out a second extraction after the destruction of the minerals by acid attack. A study was made at *IFP,* comparing a sediment extract and the extract of the residue from the acid attack of the extracted sediment. The extractions were made under identical conditions (1 hr at 50° C by chloroform in a beaker with magnetic agitation).

Figure 2.6 shows the results obtained for a series of coals and a series of Lower Toarcian sediments from France and Germany and Green River Shales from the U.S.A. In this case a comparison was therefore made of the quantities of extract obtained from ground rock and that obtained from rock extracted and then treated with HCl and HF. The results are classified in terms of the stage of evolution evaluated according to the reflectance power of vitrinite. For the coal it is observed that the ratio of the quantities extracted before and after acid treatment is generally under 0.5 for the reflectance power values under about 1% and then increases rapidly. Similar but more scattered results are obtained for other sediments. This effect does not seem to be due to hydrolysis since only slightly evolved OM can be hydrolyzed to a considerable extent, while the effect is greater for the more evolved kerogen. This cannot be interpreted as uncertainty in the measurements since the quantities of extract are sufficient to obtain a good degree of precision.

Kerogen extracts are generally richer in resins and asphaltenes and to a lesser extent in aromatic hydrocarbons than the corresponding rock extracts. In addition saturated hydrocarbons are enriched with high-molecular-weight hydrocarbons. In Fig. 2.7, for example, it is possible to note the relative enrichment in saturated polycyclic hydrocarbons of the family of steranes and triterpanes (region of chromatogram between n-C_{27} and n-C_{35}).

III. ISOLATION METHODS FOR PETROGRAPHIC ANALYSIS

The principles of the methods used by the petrographers are strictly identical to those which we have just described. Their principal object, however, is to recover large sized elements whose morphology will have been changed as little as possible. The chemist on the other hand considers it important to carry out recovery and purification of OM which will be as complete as possible. These requirements are to a certain extent contradictory. This is true because, as we have seen, good recovery and purification require fine grinding which can destroy the shapes.

So as not to risk altering the morphology, petrographers carry out relatively coarse grinding (grain size around 1 mm). Under these conditions correct extraction of the rock by solvents cannot be expected, and if the recovered organic fraction is to correspond to the definition of kerogen this extraction must be carried out after recovery. In addition, the OM recovered will be richer in minerals, either original (physical and chemical methods) or neoformations (chemical methods). Petrographers generally prefer physical methods to chemical methods. Physical methods have a tendency to favor the recovery of large-size elements, especially microfossils, organic debris and large areas of macerals which are the special objects of study. Separation methods by centrifuging, heavy liquids or flotation (Chapters 3 and 12) are the most frequently used. In addition, chemical methods are thought to modify the optical properties which are used for analysis and especially for fluorescence (Chapter 12).

Among petrographers, palynologists are perhaps the most exigent as to the preservation of the morphology since the ornementation of pollen are often indispensable for the determinations. The problem of grinding is therefore even more important for them. They prefer chemical methods to physical methods since they work principally with transmitted light methods. In this case chemical methods make it possible to obtain cleaner preparations. For them, pyrite, because of its opaqueness, is an obstacle to observation, and they often eliminate it by nitric acid treatment which obviously has the result of making the recovered fraction of OM still less representative of the whole (*cf.* paragr. II.B.3b). This treatment does not change the morphology and has some effect on the colors (the preparation becomes brighter or darker). Pyrite and other residual minerals may also be eliminated by centrifuging.

It is probable that the contradictory requirements of chemists, petrographers and palynologists could be reconciled by means of a systematic study which would define, more closely than has been done up to now, the consequences of grinding on morphology, the quality of extractions and chemical attacks as well as the consequences of chemical attacks on optical properties.

A unification of operating procedures if possible would greatly facilitate the synthesis of petrographic, palynologic and geochemical analysis for the same samples.

IV. OPERATING PROCEDURE USED AT IFP

Operating procedures are very varied in spite of the fact that there are many restrictive conditions. As an example we will describe the procedure used at *IFP*.

This procedure was designed for physico-chemical rather than petrographic analysis. However many petrographic analysis with reflected light have been made with kerogen prepared in this way. On the other hand it seems to be difficult at the present time to use these preparations in palynology, mainly because drying and grinding.

The sediment is first ground in an Aurec ring mill for 30 s. In most cases the grain-size mode produced in this way is 80 to 100 μ.

There is then chloroform extraction in a beaker placed on a magnetic heating agitator covered with a cap filled with water. The operation lasts for an hour and the temperature of the chloroform is around 50° C. The extracted rock is then collected by filtration with fiberglass paper.

The rock is then placed in an apparatus in which acid attacks and washings are carried out followed by filtration with an 0.6 μ filter [9], (Fig. 2.8). All operations take place between 60 and 70° C in a nitrogen atmosphere so as to avoid possible oxidation of the kerogen during preparation.

The successive operations are then:

(a) Two attacks with HCl 6*N*, the first of which is short (2 hrs) while the second is long (1 night) separated by washing with distilled water.

(b) Three washings with distilled water.

(c) Two attacks with a mixture of HCl 6*N*, HF 40% (1/3; 2/3), the first of which is short (2 hrs) and the second long (1 night) separated by a washing with distilled water.

(d) An attack by HCl 6*N* for 2 hrs.

(e) Two washings with distilled water.

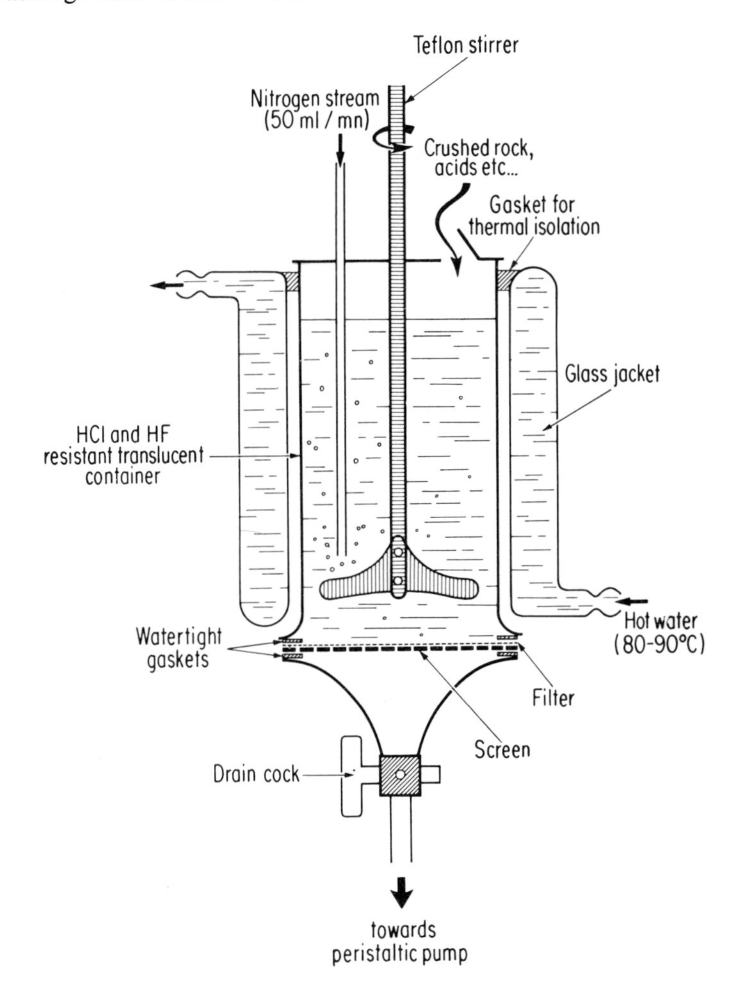

Fig. 2.8. — IFP apparatus for kerogen isolation. French patent No 74/37.072.

All these operations are carried out with stirring. It is a primary importance to maintain a temperature of at least 65° C for the effectiveness of the operations. Except for the last two, the washings are carried out in acidulated (HCl) distilled water so as to avoid deflocculation of the clays, which would clog the filter, and minimize the formation of fluorides. The fluids are removed from the reactor by means of a peristaltic pump. The proper temperature is maintained by means of hot water circulation in heating jackets. The reactors have a capacity of 0.750 litre and are made of a transparent fluorinated polymer. The maximum amounts of rock subjected to attack are about 50 g for highly silicate rock and 100 to 200 g for very carbonate rock.

The residue of the acid attacks is recovered by means of the 0.6 μ filter. At this stage an aliquot part may also be recovered for microscopic study and is preserved in acidulated water. Indeed kerogens are more difficult to study under the microscope if they have been dried.

There is no extraction from the residue with organic solvents. It is dried in a vacuum at 100° C (80° C for slightly evolved OM). Aliquot parts for various types of analysis are placed in sealed small containers (1-10 ml). It has been verified that they can be preserved in these containers without significant alteration by means of elementary analyses carried out at regular intervals on kerogen which has been frequently handled and therefore exposed to the light and air.

At the present time residual minerals are not eliminated.

Is has been observed empirically that physico-chemical analysis were generally possible if the iron content was under 18% (corresponding to a pyrite content of around 40%) and/or if the contents for minerals other than pyrite were under 15%.

Most of the physico-chemical analysis mentioned in this study were carried out with kerogen prepared in this way.

It is planned to make the method for eliminating pyrite with iron sulfate in an acid medium a routine operation. For this purpose the acid attack residues will be given a treatment for two hours with a ferric sulfate solution (150 g/l) in HCl 4*N*. Only the kerogen which, according to element analysis contains iron contents of more than 10% by weight will be treated in this way. Under this value, the risk of oxidizing the kerogen cancels out the advantage of working on cleaner kerogen. The operation is followed by treatment with HCl 4*N* for 1 hr and then by washings in order to eliminate the residual iron sulfate.

It is also planned to retreat the kerogen with low pyrite contents (Fe $<$ 10% by weight) but which is rich in other residual minerals ($>$ 7-8%) by hot HCl 6*N* for 2 hrs in order to eliminate possible neoformed fluorides.

All this can been made in the apparatus shown on Fig. 2.8 by replacing water by polyethyleneglycol in the heating jackets, which makes it possible to increase the treatment temperature up to 95° C.

It is worthwhile to note that this apparatus can be used for other kinds of preparation: It is for example well suited to palynological preparation by modification of the procedure (coarse grinding, no stirring, shorter attack times).

REFERENCES

1. Bordowskiy, O.K., (1965), *Marine Geology,* **3**, 3.
2. Burlingame, A.L., Haug, P.A., Schnoes, H.K. and Simoneit, B.R., (1969), *in: Advances in Organic Geochemistry* 1968, P.A. Schenck and I. Havenaar ed., Pergamon Press, Oxford, 85.
3. Combaz, A., (1970), C.R.A.S. **270 D**, 2240.
4. Dancy, T.E., (1948), Ph. D. Thesis, London University.
5. Dancy, T.E. and Giedroyc V., (1950), *J. Inst. Petroleum,* **36**, 593.
6. Debyser, Y., Dastillung, M. and Gadel, F., (1977), Proceedings of the 9th International Meeting of Organic Geochemistry, Moscow, 1977.

7. Down, A.L., (1939), *J. Inst. Petroleum,* **25,** 813.

8. Durand, B., Espitalié, J., Combaz, A. and Nicaise, G., (1972), *Rev. Inst. Franç. du Pétrole,* **XXVII,** 865.

9. Durand, B., (1974), French Patent No 74/37.072.

10. Durand, B., Nicaise, G., Roucaché, J., Vandenbroucke, M. and Hagemann, H.W., (1977), *in: Advances in Organic Geochemistry,* 1975, R. Campos and J. Goni Ed., Enadimsa, Madrid, 601.

11. Fedina, I.P. u Danyushevskaya, A.I., (1971), *Gazoobrazovaniu,* Usloviy, 37.

12. Forsman, J.P. and Hunt, J.M., (1958), *in: Habitat of Oil,* L.G. Weeks Ed., *AAPG,* 747.

13. Forsman, J.P., (1963), *in: Organic Geochemistry,* I.A. Breger ed., Pergamon Press, Oxford, **5,** 148.

14. Giedroyc, V., (1948), Ph. D. Thesis, London University.

15. Hamersma, J.W., Kraft, M.L., Koutsoukos, E.P. and Meyers, R.A., (1972), *Abstr. 164th Nat. Meet. Am. Chem. Soc.,* **Fuel,** 16.

16. Hitchon, B., Holloway, L.R. and Bayliss, P., (1976), *Canadian Mineralogist,* **14,** 391.

17. Hubbard, A.B., Smith, H.N., Heady, H.H. and Robinson, W.E., (1952), U.S. Bureau of Mines, *Rept. Invest. 4872.*

18. Huc, A.Y., Durand, B. and Monin, J.C., (1977), *Initial Report of the Deep Sea Drilling Project,* **42 B.** XLII-2, 737, U.S. Government Printing Office. Washington D.C.

19. Jones, D.G. and Dickert, J.J., (1965), *Chem. Eng. Progr. Symp. Ser.,* **61,** 33.

20. Langmyhr, F.J. and Sween, S., (1965), *Anal. Chim. Acta,* **32,** 1.

21. Langmyhr, F.J. and Kringstad, K., (1966), Anal. Chim. Acta, **35,** 131.

22. Lawlor, D.L., Fester, J.I. and Robinson, W.E., (1963), *Fuel,* **42,** 239.

23. Long, G., Neglia, S. and Favretto, L., (1968), *Geochim. et Cosmochim. Acta,* **32,** 647.

24. Love, L.G. and Amstutz, G.C., (1966), *Fortschr. Mineral.,* **43,** 273.

25. Luts, K., (1928), *Brennstoff. Chem.* **9,** 217.

26. McIver, R.D., (1967), *Proceedings of the 7th World Petroleum Congress,* **2,** 25.

27. Monin, J.C., Pelet, R. and Février, A. (1978), *Rev. Inst. Franç. du Pétrole,* **XXXIII,** 233.

28. Moore, L.R., (1969), *in: Organic Geochemistry,* G. Eglinton and M.T.J. Murphy Ed., Springer Berlin, **11,** 265.

29. Quass, F.W., (1939), *J. Inst. Petroleum,* **25,** 813.

30. Reggel, L., Raymond, R., Wender, I. and Blaustein, B.D., (1972), *Abstr. 164th Nat. Meet. Am. Chem. Soc., Fuel,* 10.

31. Robin, P., (1975), Thesis, Université Catholique de Louvain.

32. Robinson, W.E., (1969), *in: Organic Geochemistry,* G. Eglinton and M.T.Y. Murphy ed., Pergamon Press 6, 181.

33. Saxby, J.D., (1970), *Chem. Geol.* **6,** 173.

34. Saxby, J.D., (1976), *in: Oil Shale,* T.F. Yen and G.V. Chilingarian ed., Elsevier, 1976, 104.

35. Saxby, J.D., (1970), *Geochim. et Cosmochim. Acta,* **34,** 1317.

36. Smith J.W., (1961), U.S. Bureau of Mines *Rept. Invest. 5725.*

37. Smith, J.W. and Higby, L.W., (1960), *Anal. Chim. 32,* 17.

38. Sweeney, R.E., (1972), Ph. D. Thesis, Univ. of California, Los Angeles.

39. Thomas, R.D., (1969), *Fuel,* **48,** 75.

40. Thomas, R.D. and Lorentz, P.B., (1970), U.S. Bureau of Mines, *Rept. Invest. 7378.*

3

Les kérogènes vus au microscope

A. COMBAZ*

On trouvera dans ce chapitre une vue globale des constituants organiques des roches sédimentaires dans leur approche microscopique : un bref rappel historique du développement de celle-ci en relation avec l'approche chimique, et une perspective générale de la typologie desdits constituants et de la diversité des méthodes d'observation.

I. GÉNÉRALITÉS SUR LA MATIÈRE ORGANIQUE DES ROCHES

A. Rétrospective.

La présence de matière organique dans les roches argilo-carbonatées est connue depuis longtemps, que ce soit en couches individualisées importantes (charbon), en lits microscopiques (laminites) et, dans le cas le plus fréquent à l'état d'inclusions dans la matière minérale pour les faibles ou les très faibles teneurs.

L'étude systématique des charbons a été pratiquée à la suite des travaux de C.E. Bertrand [7, 8] et de A. Duparque [39, 40], par de nombreux pétrographes des charbons (*cf.* chap. 11).

L'observation fréquente, en lumière transmise, de corps translucides jaunâtres sur sections minces, conduisit certains d'entre eux : H. Witham [89], P.F. Reinsh [74], Fayol [45], B. Renault [75], etc., à tenter de dégager ce qu'ils soupçonnaient être des spores végétales. Par « macération » dans divers réactifs chimiques capables de détruire la fraction la plus labile du charbon (vitrain), ils ouvrirent un vaste champ de connaissances nouvelles avec l'étude microscopique des microfossiles à test organique. Cette discipline fut baptisée « sporologie », ou étude des spores et pollens fossilisés dans les charbons de tous rangs. Cependant, depuis le 18e siècle déjà, avec Geoffroy Le Jeune [48], les botanistes étudiaient les pollens de l'Actuel. Et Von Post [86] cherchait à préciser la stratigraphie du Quaternaire en concentrant, pour l'observer, le matériel pollinique des niveaux tourbeux.

Plus tard, R. Potonie [72] pour les charbons, et G. Erdtmann [42] pour le Quaternaire, forgeaient ainsi un remarquable instrument paléontologique à l'usage des stratigraphes. La « sporologie » étendait son champ d'action à tous les terrains stériles jusqu'aux plus anciens,

---

*Compagnie Française des Pétroles. Paris, France.

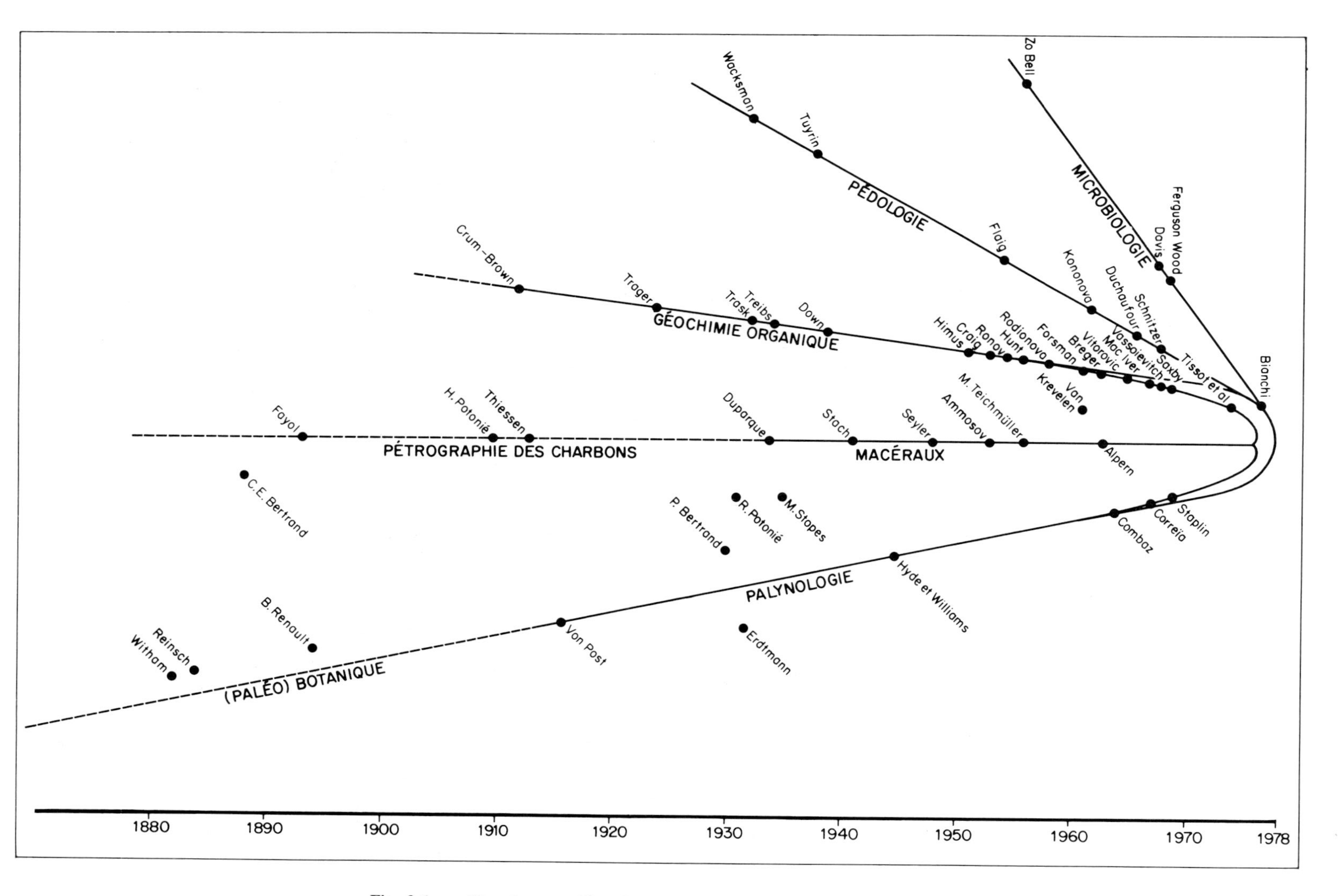

Fig. 3.1. — Cheminement historique de l'étude des constituants organiques.

quand H.A. Hyde et D.A Williams [52], créèrent le terme de « palynologie », appliqué à l'étude des pollens aériens (du grec παλνηο, je sème, je saupoudre).

Par erreur, et par contagion, cette présumée étude des « poussières aériennes », notamment des pollens anémophiles, s'étendit bientôt à toute la « sporologie » de l'actuel et de l'ancien, puis à tous les microfossiles organiques retrouvés parmi les constituants organiques des roches sédimentaires. C'est donc par le biais de l'observation en **lumière transmise,** de la matière organique dégagée par la préparation palynologique qu'ont été classés tous ces constituants organiques rassemblés par les hasards de la sédimentation : débris organiques divers, microfossiles organiques et fraction non immédiatement identifiable car amorphe.

Leur ensemble et leurs proportions relatives déterminent ce que les palynologistes appellent le **palynofacies** [21].

Parallèllement se développait une technique d'étude sur sections minces et surfaces polies, adaptée des techniques d'étude des pétrographes des charbons. Par extension à tous les types de roches sédimentaires en particulier sous l'impulsion des géologues pétroliers, elle devenait **pétrographie organique** [2].

Elle se distingue des méthodes palynologiques parce qu'elle privilégie l'analyse en **lumière réfléchie.**

En partie également sous l'impulsion des pétroliers, cette voie de recherche connut aussi un développement rapide tant à partir de l'étude des hydrocarbures naturels : pétrole, bitumes, extraits chloroformiques de roches, qu'à partir de celle de la matière organique insoluble des roches sédimentaires. L'ensemble de ces travaux qui sont du domaine de la **géochimie organique,** complète l'approche naturaliste. Dans ce domaine le terme de **kérogène** recouvre un concept très évolutif depuis sa création par Crum-Brown en 1912 (voir chap. 1). A l'instar du mot « palynologie », il devait s'affirmer de plus en plus, concurremment à ceux de « macéral », « résidu palynologique », ou « résidu organique ». Le schéma proposé ci-dessous essaie de rendre compte des principales étapes historiques de l'étude des constituants organiques des roches sédimentaires et de l'évolution des concepts qui leurs sont liés (fig. 3.1).

B. Les constituants organiques des roches.

En règle générale, les sédiments les plus riches sont issus de milieux alliant une forte productivité primaire (milieu de genèse), au caractère confiné du milieu de dépôt à l'interface eau/sédiment, ce qui suppose presque toujours un transport court. Ce sont les roches organiques proprement dites : tourbes, lignites, charbons, sapropélites, laminites (schistes bitumineux), dont la teneur en carbone organique varie de 10 à 80 %.

Ces roches ne représentent cependant qu'une faible partie du stock de carbone organique de la lithosphère. La part prédominante revient à l'énorme masse des roches argileuses, argilo-détritiques et argilo-carbonatées qui, malgré leur teneur généralement inférieure à 1 % de carbone organique, ou atteignant parfois quelques %, représentent plus de 90 % du stock organique fossile (voir chap. 1).

Comme le bilan en matériel organique est désormais susceptible d'être pratiqué sur tous les types de roches moyennant une adaptation technologique adéquate, nous en proposons ci-dessous un examen aussi complet que possible tout en restant concis. L'objet de ce chapitre vise essentiellement l'observation en lumière transmise. Le chapitre 11 traite des méthodes d'étude en lumière réfléchie et également de la fluorescence UV (transmise et réfléchie).

1. L'observation directe sur section mince.

Ce stade de l'observation des roches doit être considéré comme la base indispensable à toute étude à finalité géologique. Elle représente une démarche naturaliste authentique dans laquelle la roche est regardée comme **un tout construit**, reflétant le phénomène sédimentologique d'un bassin donné, en un point précis, et à un moment donné de son histoire. C'est une « microséquence » sédimentaire enregistrant les influences reçues du continent, du milieu de dépôt, du contexte climatique, des phénomènes précoces et tardifs liés à la sédimentation, et des diagenèses subies au cours des temps.

La texture, la structure, et la nature de la roche, exprimées par tous ses composants organiques et minéraux doivent donc, pour tout chercheur, être l'image de référence sans laquelle aucune interprétation finale, au terme d'analyses optiques et/ou géochimiques nombreuses et parfois complexes, ne saurait être satisfaisante (voir pl. 3.1 et 3.2), puisque celles-ci supposent la concentration artificielle de la phase organique qui a pour conséquence la destruction du contexte minéral, et donc également des tests et débris d'organismes minéralisés. A moins que le concentré organique ne soit obtenu par séparation purement physique par liqueurs denses après broyage de la roche.

2. Le résidu palynologique.

C'est ainsi que l'on désigne, au terme d'attaques chimiques destinées à éliminer la phase minérale, la matière organique résiduelle qui fait l'objet des études palynologiques permettant non seulement la définition d'une stratigraphie souvent précise, mais aussi la reconstitution des environnements sédimentaires et de la paléogéographie.

II. ROCHES ORGANIQUES ET CONSTITUANTS ORGANIQUES

A. Méthodes de préparation.

• Sections minces : méthode classique, consistant à coller sur la lame porte-objet, une esquille de roche surfacée, puis à user l'autre face par moyens automatiques et finition à la main jusqu'à l'épaisseur de 4 à 6 centième de mm. Le montage est fait au baume du Canada, avec pose d'une lame couvre-objet, ou polissage de la surface libre.

• La préparation du résidu palynologique vise à concentrer au mieux le matériel d'étude en limitant les pertes au maximum, et à offrir les microfossiles dans l'état de « propreté » maximum pour permettre une bonne observation. Diverses techniques ont été mises en œuvre et de nombreuses méthodes plus ou moins élaborées ont été conçues par combinaison de procédures, au point que chaque laboratoire a plus ou moins « sa » méthode de préparation.

On en distinguera donc deux types :

1. Les méthodes de type « standard » et
2. Les méthodes adaptées.

1. Méthode standard [54].

La roche étant préalablement broyée à une granulométrie de 0,2 à 3 mm, selon l'intérêt que l'on porte aux microfossiles de grande taille (jusqu'à 1 mm), on lui fait subir une première attaque à l'acide chlorydrique concentré pendant 2 heures, avec un ou deux brassages en cours d'attaque. Après lavage et décantation, on a éliminé les carbonates et des sulfures. Une attaque à l'acide fluorhydrique, dans les mêmes conditions, permet l'élimination de la silice et des silicates. Le résidu lavé à l'eau est alors affiné grâce à une certaine centrifugation dans une liqueur dense telle qu'une solution aqueuse de chlorure de zinc.

L'élimination des vestiges minéraux permet de recueillir le « résidu palynologique » prêt au montage pour étude microscopique.

Cette méthode convient généralement aux roches de type argilo-carbonaté. Elle reste toutefois un peu sommaire, et peut donner une image souvent imparfaite du véritable bilan organique de la roche car une partie de la matière organique (souvent sous forme organo-minérale complexe) n'est pas désagrégée.

2. Les méthodes adaptées.

Elles sont adaptées à la nature de la matrice minérale : grès et silstones, argiles, calcaires et dolomies, anhydrite et sel.

Les acides utilisés restent ClH et HF à froid ou à chaud. Le sel est dissout à l'eau chaude.

Dans le cas des charbons notamment, on utilise les agents oxydants, en surveillant leur action de façon à préserver le contenu sporopollinique.

Quant aux sédiments meubles sub-actuels et actuels, on s'efforce d'en extraire les microfossiles organiques par simple lavage et centrifugation, de façon à éviter l'intervention d'agents chimiques éventuellement corrosifs pour le matériel récent.

Les agents oxydants (NO_3H, chlorates, H_2O_2) sont les plus agressifs et sont capables de détruire rapidement les microfossiles organiques à test mince, et d'endommager les plus résistants.

Les autres difficultés techniques concernent l'élimination plus ou moins sélective des microfossiles au cours des préparations, du fait d'un mauvais contrôle de la solution acide entraînant la floculation des aluminosilicates.

Au moment du lavage et de la décantation de la solution surnageante, une partie plus ou moins importante du contenu organique mal décanté peut être éliminée.

Il faut ajouter qu'en matière de préparation palynologique des actions concertées dans le cadre de la *Commission Internationale de Microflore de Paléozoïque* [1], puis la *Commission Internationale de Pétrographie des Charbons* ont montré que chaque laboratoire a sa méthode propre, éprouvée par l'expérience. La comparaison des résultats de préparation d'une même série d'échantillons homogènes a cependant montré, à côté de certaines disparités importantes, que la plupart des laboratoires concernés obtenaient des résultats très sensiblement équivalents [2].

3. Les autres préparations, destinées aux autres méthodes d'études optiques.

Bien entendu le produit obtenu peut faire l'objet d'un examen optique plus ou moins approfondi. Si tel est le cas, il convient alors de prévoir au stade de broyage de la roche, une granulométrie de l'ordre du mm de façon à préserver autant que possible la fraction figurée du contenu organique. Dans ces conditions, le protocole préparatoire prévu au chapitre 2 est le mieux adapté à l'étude des palynofaciès, dans la mesure aussi où un lavage chloroformique élimine toute fraction hydrocarbonée et lipidique susceptible d'interférer avec la phase insoluble amorphe.

Les préparations destinées à **l'étude réflectométrique** de la vitrinite nécessitent généralement une concentration de ce matériel. Elles sont abordées aux chapitres 11 et 12. S'il s'agit de **l'étude en fluorescence UV**, le sujet est également traité par ailleurs (chap. 11). Il faut noter ici que le lavage au chloroforme est préjudiciable à ces observations, ce qui indique bien à quel point le phénomène de fluorescence est lié à l'existence d'hydrocarbures dans l'intimité de la roche.

L'étude microscopique du kérogène **en lumière infrarouge,** (IR) ou au moyen des **rayons X** (RX) ne nécessite pas de préparations particulières. Dans l'un et l'autre cas, il s'agit d'explorer les constituants les plus sombres (du brun au noir) pour tenter d'y discerner les traits structuraux (IR) ou des inclusions minéralisées, surtout pyriteuses (RX) (voir pl. 3.8 et 3.9).

Enfin, les **examens en électromicroscopie** nécessitent les méthodes de préparation normalement requises par ces études : sections microtomiques et minéralisation pour la transmission; surfaces de roches brutes (ou polies), ou kérogène en inclusion — ou déposé sur le porte-objet — et métallisation à l'or — ou tout autre élément — selon le type d'investigation dont l'échantillon doit être l'objet, pour la microscopie électronique à balayage.

B. Types de roches carbonées (pl. 3.1).

Nous entendons par là, les roches dont le contenu en carbone organique est élevé (≥ quelques %), c'est-à-dire la famille des charbons, *sensu stricto* de type de (ligno) humique, celle des sapropélites : bogheads, et autres, dans la série des alginites, et enfin celle des laminites (souvent dénommées « schistes bitumineux »), appartenant à la lignée des alginites ou des exinites [84].

1. Les charbons humiques (50 à 90 % de carbone organique).

a. Gisements.

Diversement répandus dans toutes les parties du monde, ils se présentent sous forme de couches dont l'épaisseur est généralement inférieure au mètre, mais qui peuvent aussi atteindre 2 à 6 m voire dépasser 10 m. Les couches les plus anciennes sont siluro-dévoniennes, époque où commence le peuplement des terres par les végétaux supérieurs : Cryptogames vasculaires, puis Gymnospermes dès le Carbonifère supérieur, et Angiospermes au Jurassique supérieur.

b. Fréquence.

Elle varie en fonction des époques. Les formations du Carbonifère recèlent environ 60 Gt de réserves exploitables, celles du Permien 50 Gt, du Jurassique 14 Gt, du Tertiaire 76 Gt, du Quaternaire 70 Gt [3]. Ajoutons que 3 pays : Etats-Unis, URSS et Chine, disposent de 90 % des ressources de la planète.

c. *Nature.*

Observées tant en lumière réfléchie qu'en lumière transmise ces roches apparaissent essentiellement constituées de débris de taille variable de végétaux supérieurs. Les matériaux d'origine : lignine, subérine, cellulose ayant subi de profondes transformations biochimiques sont réduits à l'état de gel humique **(vitrinite)** et de vestiges fusinisés **(inertinite).** Au sein de cette masse sombre, l'observation microscopique révèle une autre famille de constituants qui ont conservé une couleur jaune, et une structure moléculaire proche de celle qu'ils possédaient au moment du dépôt, c'est le groupe de l'« **exinite** ». Les fossiles ou fragments de fossiles qui le composent sont surtout les spores végétales et les pollens, les cuticules de feuilles. Il s'y ajoute les résines et autres sécrétions végétales.

d. *Faciès.*

Les conditions de sédimentation (milieu ± oxygéné) entraînent de très sensibles modifications de faciès, par le type de transformations biochimiques (bactéries, champignons, algues) dont elles favorisent le développement. Ces considérations ont conduit l'école russe [82] à établir une classification à la fois génétique et morphographique des charbons. Si le milieu est très réducteur **(anaérobiose),** la gélification l'emporte largement et la vitrinite est le constituant majoritaire. Si par contre le milieu connaît un régime d'**aérobiose** (plus ou moins oxydant), la fusinisation l'emporte largement et il en résulte la présente dominante d'inertinite. Par ailleurs, une autre variable vient diversifier ces faciès fondamentaux : c'est l'apport plus ou moins important de constituants secondaires, spores, cuticules, résines, algues, qui enrichit en hydrogène la composition élémentaire moyenne du charbon considéré. Dans les **cutinites** dominent les cuticules de feuilles, dans les **sporinites** (*cannel coal*) dominent les spores végétales : mégaspores et miospores. Alors que les teneurs en spores, observées après « macération », sont couramment de quelques dizaines ou centaines de milliers par gramme de charbon, dans le cas des sporinites, les teneurs sont de plusieurs millions par gramme de roche.

D'autres types de charbons se distinguent par l'abondance de sclérotes de champignons, ou d'algues microscopiques, liés aux conditions de dépôt. Cependant, nous restons bien ici dans le cortège sédimentologique des charbons.

Dans un tel complexe sédimentaire, pour une séquence donnée, certains sites favorisaient le dépôt de ces types particuliers de composants biologiques à la faveur d'un véritable classement dû aux conditions écologiques (milieu de génèse), et/ou aérodynamiques (pluies de pollens et de spores, chutes et accumulation de feuilles) et enfin hydrodynamiques dans le milieu de dépôt (suspension et vannage par les courants). En tout état de cause, les constituants qui dominent ici sont d'origine terrestre et ont donc été générés en milieu aérien; il s'y ajoute dans le milieu de dépôt divers accompagnateurs d'origine aquatique.

2. ***Les sapropélites*** *(30 à 90 % de carbone organique).*

A l'inverse de la précédente famille, nous sommes ici en présence de roches dont les constituants organiques dominants ont été générés en milieu aquatique. Il s'agit dans tous les cas de matériel algaire, où les algues microscopiques sont souvent reconnaissables.

L'importance de la fraction minérale dans la roche ne change rien à cet aspect déterminant.

a. Les bogheads.

Ils représentent une sous-famille très caractéristique de cet ensemble. Leur constituant algaire dominant est une Chlorophycée très commune dans l'Actuel, et dont les accumulations sont connues depuis le Cambrien. Il s'agit de *Botryococcus brauni,* dont l'abondance est telle que ces petites colonies algaires de quelques centièmes de mm de diamètre sont serrées les unes contre les autres et atteignent parfois 250 000/cm^3 [7]. Une « gelée fondamentale » brune cimente ces micro-organismes.

Ainsi **kerosene shales** d'Australie, **torbanite** d'Ecosse, **n'hangellite** du Mozambique, **balkachite** du Kazakhstan et **coorongite** d'Australie, représentent un même type de roche sapropélique.

Bertrand et Renault (1892), ont observé 166 lits de *Botryococcus* dans le boghead d'Autun sur une épaisseur de 24 mm [8]. Mais certains lits peuvent atteindre une épaisseur de l'ordre de 25 cm. Quant aux vases actuelles à *Botryococcus* du lac Beloë (Russie), elles atteignent 9 m d'épaisseur. La **kukersite** de l'Ordovicien supérieur d'Estonie est une sapropélite carbonatée à rattacher directement aux bogheads. Les *Gloeocapsomorpha prisca* qui en sont le composant organique principal, ont en effet été assimilés aux Botryococcacées [50].

Dans une taxinomie cohérente, toutes ces roches « particulières » aux noms d'inspiration géographique, toutes formées par le même type d'algue mériteraient l'appellation de « botryococcites » de préférence au terme traditionnel de bogheads.

b. La tasmanite.

Elle doit également son origine à la géographie : la Tasmanie, mais aussi aux *Tasmanites punctatus* Newton, qui la composent [69].

Il s'agirait encore d'Algues chlorophycées [87] accumulées ici en quantités énormes, au point de former des couches compactes dans les formations du Permien inférieur de Tasmanie, du Jurassique d'Alaska, du Dévonien du Canada, etc. Bien que les Tasmanacées soient très répandues depuis le Silurien dans les argiles de type sapropélique, les concentrations assez fortes pour justifier l'appellation de « tasmanite », restent relativement exceptionnelles. Ces accumulations doivent en conséquence résulter de phénomènes de prolifération locale et d'un classement sédimentaire, en plus de conditions physico-chimiques favorables à leur préservation, et ceci pendant des durées appréciables, puisque les couches atteignent parfois plusieurs mètres d'épaisseur.

c. La chounguite (1) (20 à 96 % de carbone organique).

C'est une roche carbonée compacte, à cassure conchoïdale, de couleur noire à reflets d'acier, que l'on observe dans certains terrains précambriens notamment de Russie septentrionale et de Sibérie, sous forme de lits peu épais (quelques mm, exceptionnellement quelques cm) [57].

Elle est généralement associée aux formations lagunaires plus ou moins métamorphisées graphitifères. Selon toutes probabilités, il s'agit des premières sapropélites formées par l'accumulation de très anciennes algues microscopiques (1,8 à 3 milliards d'années). Ayant subi une sévère catagenèse [pouvoir réflecteur (PR) de 5 à 6 %] les structures biologiques en sont très estompées. Certaines se présentent en filonnets (type I) d'autres (types II à IV) sont

(1) du village de Chounga (Karélie).

plus minéralisées. Dans le premier cas, il s'agit du bitume transformé issu des sapropélites, lesquelles sont souvent enrichies en éléments métalliques : U, V, Re, etc.;

d. La diatomite (ou tripoli ou kiselguhr) (10 à 30 % de carbone organique).

Il s'agit d'une sapropélite de type différent, beaucoup moins riche en matière organique puisque les frustules des Diatomées (*Bacillariophycées*) qui forment la masse principale de cette roche sont de nature siliceuse [35].

Ce phytoplancton dont la taille de chaque individu ne dépasse pas quelques 1/100e de mm, prolifère aussi bien en milieu marin que lacustre et les sédiments qui en sont issus rassemblent 4 à 5 millions d'individus/cm^3. Le carbone se présente surtout sous forme de matière organique amorphe probablement issue de la composante organique de ces organismes : gouttelettes lipidiques, caroténoïdes, mucilages, et en particulier de la membrane pectosique imprégnée de silice qui constitue le frustule.

D'autres organismes, dont l'abondance est parfois comparable à celle des Diatomées, sont observables dans ces roches. Ce sont les Silicoflagellés, dont la taille est de l'ordre de quelques microns.

Ces roches très répandues depuis le Jurassique, constituent parfois des dépôts épais de quelques dizaines, voire quelques centaines de mètres. On les connaît notamment dans le Tertiaire sur le pourtour méditerranéen, en Californie, en URSS, Autriche, Allemagne, etc.

*3. **Les laminites ou « schistes bitumineux »** (15 à 30 % de carbone organique).*

Le sens donné ici au terme laminite n'est pas strictement celui que lui a donné son créateur (A. Lombard 1963) puisque ce dernier ne faisait pas référence à une teneur élevée en matière organique.

L'appellation ancienne de « schistes bitumineux » est tout à fait inadéquate, puisque les roches ainsi désignées ne sont ni **schisteuses** ni **bitumineuses,** mais seulement **laminées** et enrichies en matière organique **oléogène** (ou si l'on veut **bitumogène**). Les laminations visibles à l'œil nu sur des cassures perpendiculaires à la stratification de la roche, ne sont autres que l'accumulation de films sédimentaires de nature alternativement organique et minérale, déposés au gré des apports saisonniers **(structure varvée)** [16]. De tels sédiments impliquent un milieu calme, protégé, peu profond (?) et confiné au niveau des eaux de fond. Ces conditions sont réalisées dans les mers épicontinentales, les lacs et les lagunes.

Certaines laminites présentent une grande richesse en microfossiles organiques surtout marins, parfois terrestres. Citons à titre d'exemple :

— **les argiles radioactives** du Silurien du Sahara qui contiennent des sicules de Graptolites, des Chitinozoaires, des Acritarches, des Tasmanacées, des Gigantostracés;

— **les argiles radioactives** du Frasnien du Sahara, du Brésil et de Libye, qui contiennent en abondance des Acritarches, des Chitinozoaires, des Tasmanacées, des spores, etc. Mais le plus souvent, ces roches contiennent une matière organique amorphe très abondante;

— **les schistes carton du Bassin Parisien** (Toarcien inférieur) comportent occasionnellement des niveaux à Tasmanacées dans une matière organique amorphe dominante et peu ou pas de microfossiles d'origine terrestre;

— **les « schistes » de l'Irati du Parana** (Brésil) (Permien inférieur), d'origine probablement lacustre. On y rencontre des niveaux à poissons, crustacés et reptiles, mais la matière organique est presque exclusivement de type amorphe, avec cependant quelques pollens;

— **les « Green River Shales »** du Colorado (Eocène), sont pratiquement « azoïques ». A l'exclusion de quelques spores de champignons, le reste du matériel organique est amorphe.

C. Les roches sédimentaires communes.

Les autres roches à faible ou très faible teneur en matière organique ne peuvent être qualifiées de « carbonées ». Nous devons cependant les examiner dans la lignée de ce qui précède au titre de leur contenu organique.

1. Les argiles grises et noires (pl. 3.2) (1 à 5 % de carbone organique).

Si certaines formations argileuses ne contiennent que des traces de matière organique malgré une couleur gris sombre à noir, alors due aux inclusions de pyrite et autres sulfures (comme des « Terres Noires » du Jurassique supérieur des Alpes du Sud), dans la plupart des cas, la couleur sombre est l'indicateur d'une appréciable teneur en matière organique. Elles ne sont pas pour autant des roches-mères (de pétrole), qualité qui résulte du type de matière organique sapropélique par opposition à la matière organique détritique (« charbonneuse ») qui n'est capable de produire que du gaz.

La frontière entre les roches du type « schistes noirs » (« black shales ») et les laminites est relativement arbitraire, l'appartenance à cette catégorie étant fondée à la fois sur le caractère plus ou moins apparent des laminations et sur la teneur en carbone organique. Une même formation argileuse de ce type peut donc comporter les horizons de laminites typiques. Comme exemple de roches-mères de ce type, citons :

— **les argiles noires** du Kimmeridgien d'Angleterre et de la mer du Nord : leur teneur en carbone organique atteint 8 %. Les microfossiles organiques sont abondants : Dinoflagellés, Spores et pollens, Tasmanacées; leur radioactivité, surtout due à l'uranium, est sensiblement plus élevée que celle des autres formations argileuses;

— **les « schistes noirs »** de la marge occidentale africaine (Crétacé inférieur) : ce sont des marnes brun foncé à intercalations de calcaire sparitique riche en Foraminifères avec 2 à 10 % de carbone organique. Certains de ces « black shales » contenant une matière organique de type sapropélique, d'autres une matière organique de type détritique;

— **les « marl slates »** du Permien supérieur de la mer du Nord : ce sont des argiles dolomitiques feuilletées, parfois gréseuses et ferrugineuses. La microflore y est abondante : pollens de Gymnospermes, débris végétaux, Acritarches, Microforaminifères;

— **les argiles radioactives siluriennes** du Sahara et de Lybie, etc.

De telles formations argileuses grises ou noires, à faible potentiel organique, sont très répandues dans le monde et dans toutes les séries. Cependant certaines époques ont été particulièrement marquées par ces dépôts dont l'extension atteint une très grande ampleur. On constate alors le développement de mers épicontinentales généralement peu profondes, à la faveur d'affaissements continentaux qui entraînent des transgressions marines. C'est le cas, par exemple, au Silurien inférieur : ses argiles noires, radioactives, sont connues dans tout le nord de l'Afrique, le Moyen-Orient, le Brésil, l'Espagne, la France, la Grande-Bretagne, la Baltique; au Frasnien qui se présente comme un épisode récurrent du précédent, sensiblement dans les mêmes régions; au Mississipien, au Permien supérieur, au Jurassique inférieur

(Toarcien) et supérieur (Kimmeridgien), au Crétacé inférieur : Turonien-Cénomanien notamment dans l'Atlantique sud, et encore au Tertiaire inférieur.

D'autres régions, d'extension limitée, sont aussi le lieu d'une sédimentation organique par le jeu d'invasions marines récurrentes et de retraits, au fil de très longues périodes. C'est le cas du Moyen-Orient dont l'actuel Golfe Persique représente un stade résiduaire, et aussi du Golfe de Venezuela-Maracaïbo.

2. Les carbonates (pl. 3.2).

L'environnement carbonaté comporte des faciès très divers. Leur teneur moyenne en carbone organique est très faible.

En revanche, certains calcaires souvent argileux, à structure laminée, ont un taux élevé en carbone organique. Ils sont issus de milieux marins confinés [17] très calmes, où se déposent en alternances saisonnières des lits de quelques dixièmes de millimètres, de couleur sombre et claire. Les premiers sont pétris de micro-organismes, d'argile et de matière organique, les seconds sont formés d'agrégats de fins cristaux de calcite [16]. Les environnements les plus typiques correspondant au « piégeage » de matière organique correspondent aux :

— faciès lagon des complexes récifaux;
— faciès lagunaire hypersalé à évaporitique;
— faciès organogènes noirs en sédimentation vaseuse.

On connaît des thanatocénoses à Ptéropodes, à Coccolithophoridés, à Globigérines, Foraminifères, à Dinoflagellés, etc., par exemple : les ptéropodites du Dévonien de l'Ouest canadien, les foraminiférites de l'Eocène de l'offshore sénégalais, les calcaires laguno-lacustres de Limagne, les calcaires bitumineux kimmeridgiens d'Armailles (Ain), les calcaires bitumineux couviniens de l'ouest canadien, etc.

3. Les silstones, sables et grès.

Ils sont a priori défavorables à la concentration et à la conservation de la matière organique. Les constituants organiques qu'ils peuvent contenir occasionnellement, sont de nature biodétritique, d'origine continentale. Beaucoup plus rarement cependant on observe des vestiges planctoniques dans les silstones (Crétacé inférieur de Hollande). La sédimentation des éléments biodétritiques avec les clastiques tient à leur origine, au mode de transport fluviatile puis par courants marins, et à leur masse (taille, densité) qui permet un tel classement hydrodynamique. Cependant, le plus fort contingent de matériel organique biodétritique est réduit en fines particules, de faible densité, qui se sédimentent au sein de la fraction argileuse.

4. Les évaporites.

Ils sont a priori défavorables à la matière organique. Liés au faciès de transition et d'oscillation entre le milieu marin normal, voire limnique normal, et les lagunes sursalées, ils peuvent contenir une matière organique allochtone. Des thanatocénoses répetées, entraînent souvent un enrichissement local des sédiments en matière organique. Les conditions de dépôt étant très favorables à la préservation ultérieure de ce potentiel carboné. Exemple : argiles en lits ou inclusions dans le sel triasique du Sahara septentrional, à teneur conséquente en matière organique.

D. Productivité et sédimentation organique (voir chap. 14).

D'une manière générale, les zones de productivité en matière organique sont très largement répandues dans les mers et sur les continents. La zonéographie de la productivité primaire fait ressortir sur les continents : la ceinture tropico-équatoriale (forêt et savane), les zones tempérées (forêts et cultures) et sub-boréales (forêts et steppes), et dans les mers : les régions antarctiques et arctiques, et les zones aval des courants froids (côtes occidentales d'Afrique et d'Amérique du Sud).

En revanche, dans la nature actuelle comme au cours de la plupart des périodes passées, les zones de forte concentration en matière organique dans les sédiments restent des faits d'ampleur limitée dans l'espace et dans le temps (à l'exclusion des époques particulières dont il est fait mention au paragraphe II.C.4 ci-dessus). En se reportant au chapitre 1, figure 3, il apparaît que le stock de matière organique dispersée dans les sédiments est donc mille fois plus important que celui concentré dans les charbons, pétroles et autres roches combustibles. Il ne représente cependant que 4 % du stock de carbone organique total de la planète. Ce carbone non recyclé par la biosphère permet l'existence de 20 % d'oxygène libre dans l'atmosphère. Ce phénomène a été d'une importance capitale dans le déroulement de l'évolution biologique.

III. TYPES DE CONSTITUANTS ORGANIQUES

Après le tour d'horizon des familles de roches à contenu organique concentré et dilué, des milieux de sédimentation qui leur donnent naissance, et de leur répartition dans le temps et l'espace, il convient d'aborder l'examen systématique des constituants microscopiques organiques du kérogène de ces roches. Nous en distinguerons trois fractions principales : figurée, amorphe et remaniée, la première comportant elle-même trois classes de constituants selon leur milieu de génèse et/ou leur origine biologique (fig. 3.2.) :

1. Microfossiles et fragments végétaux terrestres.
2. Microfossiles et fragments algaires.
3. Microfossiles et vestiges animaux.

A. Fraction figurée.

C'est ainsi que micropaléontologistes et palynologistes désignent les éléments ayant une structure morphologique définie, permettant de leur attribuer une origine biologique plus ou moins précise. Ainsi microfossiles proprement dits, altérés ou non, et débris de tissus, de membranes, d'organismes divers, moulages cellulaires, sont considérés comme des éléments « figurés ».

1. Microfossiles et fragments végétaux terrestres (pl. 3.3).

Ils appartiennent à la classe de constituants dont les éléments sont produits en milieu aérien, sur les continents. Ils sont issus des végétaux terrestres primitifs ou supérieurs, depuis

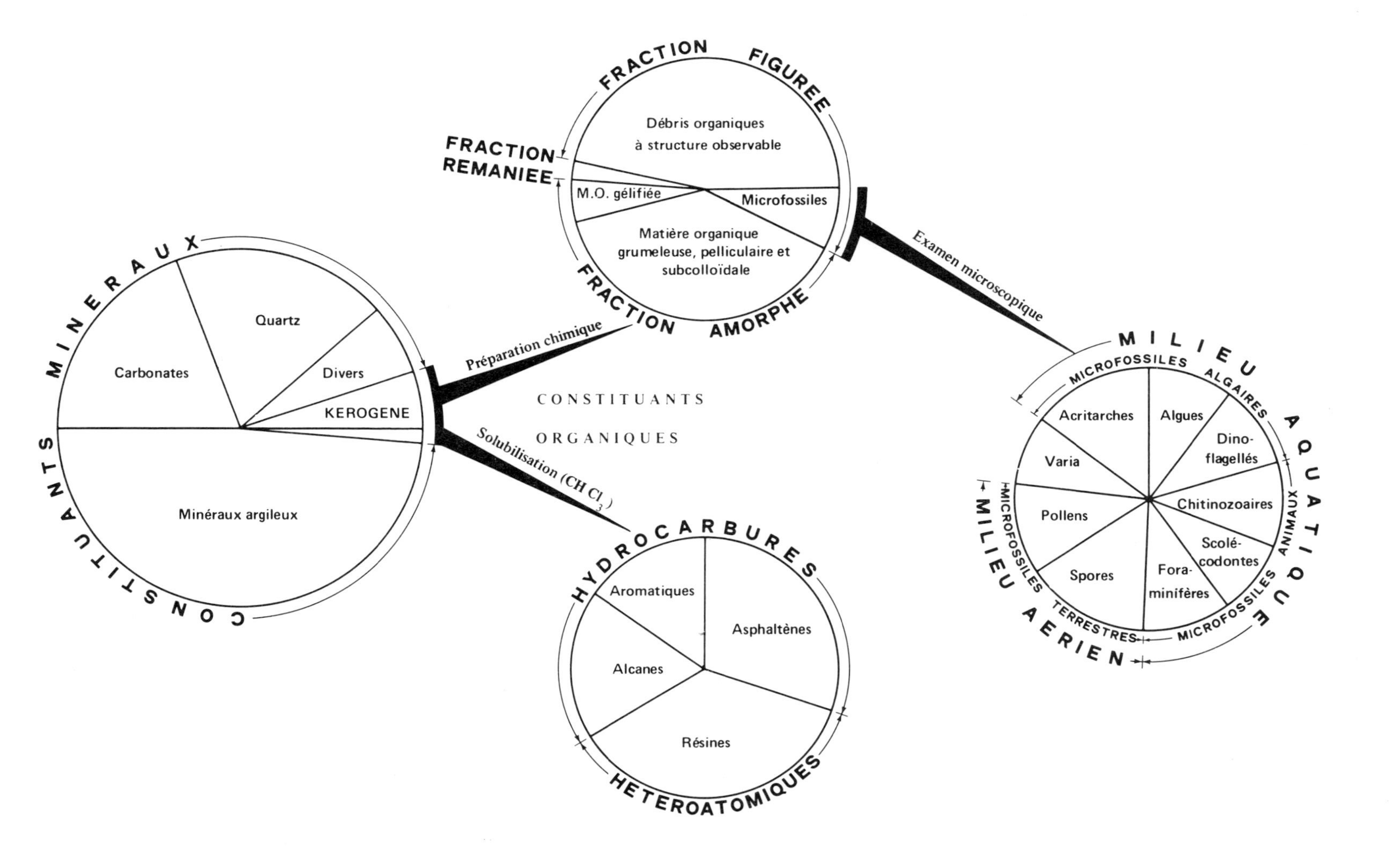

Fig. 3.2. — Les constituants des roches sédimentaires [26].

les mousses jusqu'aux plus grands arbres, tous représentés, en l'occurrence, par des particules microscopiques : grains de pollens, spores, fragments de bois, d'écorce, de cuticules de feuilles, de téguments, de fructifications et de graines, de vaisseaux, de pilosités, de membranes, etc.

a. Spores et grains de pollen.

On les désigne souvent par l'expression « microflore fossile », alors qu'ils ne sont que les téguments de cellules reproductrices microscopiques d'une macroflore fossile détruite. Cette « microflore » est donc intimement liée à l'évolution de la flore proprement dite.

Les spores (ou mi(cr)o spores, ou tétraspores) dont la taille varie de 5 à 200 μ, sont les organes de dissémination et de reproduction des Cryptogames.

Chez les moins évolués, Algues et Champignons, elles peuvent être des formes de résistance, tout comme chez les Bactéries, ou de multiplication végétative ou asexuée, mais elles peuvent aussi avoir la même signification que chez les Mousses et les Cryptogames vasculaires. Chez ces végétaux, elles sont toujours issues de la méïose, double partition cellulaire au cours de laquelle intervient la recombinaison génétique. Leur germination donne naissance à un organisme autonome : tige, feuille de Mousse, ou prothalle de Cryptogame vasculaire, qui sera producteur des éléments sexuels participant à la fécondation. Les spores peuvent elles-mêmes êtres, ou non, sexuées.

Chez les Phanérogames les spores sont toujours sexuées, et la spore femelle demeure incluse dans l'organe qui l'a produite au sein de la fleur (Phanérogames monoïques). Seule la spore mâle est disséminée : c'est le grain de pollen qui donnera presque immédiatement l'élément fécondant mâle. Transportés par le vent (anémophilie), l'eau (hydrophilie), les insectes (entomophilie), ils vont féconder les fleurs femelles par le développement de leur tube pollinique qui achemine l'élément fécondant mâle jusqu'à l'ovule.

Précisons que chez certaines Phanérogames dioïques, les fleurs femelles produisent une **mégaspore** fonctionnelle. On les trouve en relative abondance dans les charbons où elles ont été bien étudiées.

Les pollens existent en fait dans les roches sédimentaires depuis le Carbonifère supérieur qui a vu l'apparition des premières Gymnospermes, et sont extrêmement répandus dans toutes les séries sédimentaires depuis cette époque, en particulier dans le cortège des roches argileuses.

Les pollens d'Angiospermes sont très répandus dans les sédiments, depuis les débuts du Cénophytique (Crétacé inférieur). Ces plantes essentiellement terrestres, comptent aussi de nombreuses variétés aquatiques.

L'abondance des pollens est extrême : une inflorescence de pin peut produire 2.10^7, de bouleau : 6.10^6, d'oseille : 4.10^8 grains de pollen. Chaque plante peut donc en produire des milliards, et G. Erdtman [43] a pu évaluer à 75 000 t la production annuelle de pollens de pin de la Suède méridionale. L'exine, ou tégument externe, qui est seul apte à la fossilisation, est formé par l'un des matériaux les plus résistants du monde organique : la sporopollénine, dont l'analyse élémentaire à l'état frais varie dans la mesure suivante [77] :

$$C_{90}\,H_{138}\,O_{20}$$
$$\text{à } C_{90}\,H_{158}\,O_{44}$$

La surface externe des pollens et spores est souvent très ornementée, et de façon assez typique pour que l'on puisse caractériser chaque espèce végétale à partir de ces critères morphologiques qui font précisément l'objet de la palynologie (fig. 3.3).

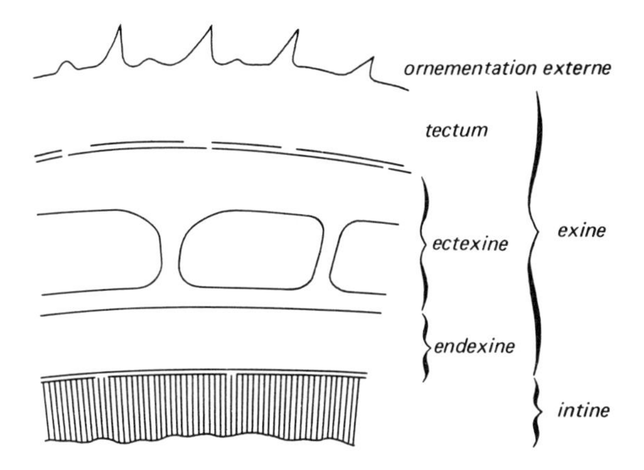

Fig. 3.3. — Section du tégument d'un grain de pollen.

A l'état frais, cette surface est tapissée de substances cireuses, tandis qu'à l'intérieur, une autre membrane : l'intine, de nature cellulosique, aisément périssable, entoure le contenu cytoplasmique. Les agents capables de détruire le tégument sporopollinique sont essentiellement : certaines diastases de micro-organismes, les actions oxydatives prolongées, et surtout les actions thermiques.

Dans la sporopollénine fraîche, et donc en amont de la sporopollénine fossile, on trouve les composants suivants [58] :

— lignine;
— albumines;
— lipides;
— phénols solubles
— hydrates de carbone;
— aminoacides;
— polysaccharides;
— polymères de caroténoïdes.

Le tégument des spores et pollens fossiles est donc le résultat d'une condensation de ce complexe biochimique sous forme d'un assemblage de noyaux de lignine et de polymères de caroténoïdes. L'effet thermique postérieur lié à l'enfouissement progressif de ces substances dans les roches sédimentaires (catagenèse), va accuser encore la condensation moléculaire avec élimination progressive des chaînes latérales liées aux cycles aromatiques. Ces modifications de la structure chimique se traduisent à la fois par une coloration de plus en plus sombre, et une réduction de la taille du microfossile.

En résumé, spores et pollens fossiles héritent d'une double complexité : la première est biologique, c'est celle du tégument frais dont la structure et la substance varient selon les groupes végétaux, auxquels il appartient; l'autre, diagénétique, est due aux aléas des transformations biochimiques et chimiques subies depuis le stade de la sédimentation jusqu'aux effets thermiques liés à la profondeur.

L'appellation « sporopollénine » est donc une notion simplicatrice suggérant une substance cohérente, de structure moléculaire définie, alors que, selon toutes probabilités, cette cohérence n'existe ni dans la substance fraîche ni dans la substance fossile.

b. Fragments végétaux.

Alors que les organes de reproduction et de dispersion des espèces (gamétophytes), simples cellules microscopiques, sont très protégés à la fois par leur test, chimiquement très résistant, et par leur petite taille, la plante proprement dite (sporophyte) de la mousse à l'arbre, est très vulnérable aux actions destructrices. Comme tous les métaphytes, elle est rapidement anéantie en débris plus ou moins dispersés, et les édifices moléculaires complexes du stade vivant subissent le retour à l'état de molécules plus ou moins simples.

Selon les hasards des milieux de sédimentation et de diagenèse cependant, certains tissus formés de substances hautement polymérisées sont susceptibles d'être conservés avec les structures cellulaires d'origine, ou leurs empreintes. Ce sont évidemment les substances ayant une fonction de protection dans la plante : cuticules, cires, résines qui présentent les meilleures qualités de résistance à la dégradation chimique et biochimique, tandis que la masse des substances constituant les tissus de soutien : cellulose, hémicellulose, subérine, lignine, est avec le temps, et selon les vicissitudes de l'histoire sédimentologique et géologique du milieu de dépôt, plus ou moins complètement transformée par hydrolyse en produits gélifiés (fulviques et humiques) ou fusinisés.

Aux stades de la tourbe et du lignite, les plantes d'origine peuvent être facilement identifiables (exemple : diverses essences sylvatiques aisément déterminées par les paléobotanistes ayant étudiés les lignites oligocènes de Rhénanie). L'altération en milieu oxydant (sols aériens) conduit à une dégradadation beaucoup plus profonde et rapide des vestiges végétaux tels que la lignine. Leur dispersion ultérieure dans les sédiments par abrasion des sols et transport fluviatile, puis par les courants marins, complète l'anéantissement du sporophyte de départ. Ces débris noirs anonymes correspondent à l'inertinite des charbonniers (paragr. II.B.1). C'est sous cette forme ou à un moindre degré de fusinisation, qu'est répandue et stockée dans les sédiments marins la plus grande partie de la masse des constituants carbonés fossiles. (chap. 15).

Cependant, une grande diversité de structures végétales conservées a déjà été étudiée en lumière transmise, parmi les constituants organiques des roches (phytoclastes) (pl. 3.3).

Toutes sortes de tissus ont été identifiés : cuticules de feuilles, vaisseaux divers (spiralés, aréolés, etc.), tissus ligneux, sporanges, fructifications, etc. Certaines préparations microscopiques issues de roches peu diagénisées, livrent à l'observation une véritable « bouillie » végétale où dominent ces éléments biodétritiques. (pl. 3.10).

2. *Microfossiles et fragments algaires (pl. 3.4).*

On regroupe ici, dans la fraction figurée, une grande variété de microfossiles organiques à position systématique bien définie, aussi bien que des *incertae sedis*. Les fragments d'algues benthiques, homologues des débris végétaux terrestres, sont beaucoup plus rares et difficiles à identifier comme tels. L'ensemble appartient au monde aquatique qui n'est plus seulement le milieu de dépôt, mais aussi le milieu de genèse; phyto et zooplanctons sont représentés par divers groupes de micro-organismes à test entièrement organique ou à la fois organique et minéralisé. Les algues benthiques comme les autres métaphytes terrestres ou aquatiques, de par leur taille même, leur différenciation histologique, la nature de leurs constituants biochimiques étant beaucoup plus vulnérable, sont beaucoup moins aptes à la fossilisation, et les structures caractéristiques conservées à l'état de substance carbonées restent exceptionnelles [25].

a. *Acritarches [44].*

Ce terme recouvre une classe de microfossiles *incertae sedis* appartenant au milieu aquatique, notamment marin. Probablement polyphylétiques, elles appartiennent en tous cas au monde des Algues dont elles représentent les formes de dispersion, en particulier dans le groupe des **Prasinophycées** ([1]).

Leur taille varie de quelques microns à quelques centaines de microns. On connaît de telles formes en abondance dans le Cambrien, voire l'Antécambrien. Elles sont très abondantes et diversifiées à travers tout le Paléozoïque (surtout inférieur), un peu moins durant les Méso-et Cénozoïque. Certains horizons dans les roches argileuses peuvent être littéralement « farcis » de spécimens d'Acritarches appartenant à la même espèce ou au moins au même genre Micrhystridium, Veryhachium, etc. qui forment parfois de véritables amas.

La composition élémentaire des membranes d'Acritarche [41] serait la suivante :

C	H	O + S	N
71,8 %	7,84 %	18,12 %	2,16 %

b. *Les Leiosphaeridacées, [41] et Tasmanacées.*

Ce sous-groupe défini morphologiquement (formes lisses à tégument fin ou épais) n'a pu être situé de façon certaine dans la systématique et est encore souvent maintenu parmi les Acritarches. Cependant les Leiosphaeridacées (tégument fin) sont connues depuis le Cambrien, les Tasmanacées (tégument épais), depuis le Silurien, et dans l'Actuel les genres *Halosphaera* et *Pachysphaera* (Chlorophycées monadoïdes) seraient leurs homologues fossiles [86].

Le genre *Tasmanites* (*cf. supra,* paragr. II.B.2) est extrêmement répandu à certaines époques : Silurien, Dévonien, Permien, Jurassique... Il semble que ce foisonnement soit lié à un milieu marin hyposalin [22]. Leur test jaune translucide, plus ou moins perforé, est formé d'une substance voisine de la sporopollénine fossile, dénommée tasmanine, dont l'analyse élémentaire conduit aux résultats suivants :

C	H	O + S	N
76,78	9,56	12,10	1,56

c. *Les Dinoflagellés (ou Péridiniens ou Dinophycées).*

Organismes végétaux aquatiques unicellulaires planctoniques, bien représentés dans les mers actuelles (Océan Indien, Mer Rouge) où ils sont à l'origine du fameux phénomène des « marées rouges » quand leur extrême prolifération (quelques millions par litre d'eau) parvient à teinter la mer et à intoxiquer le milieu. Leur taille varie de 25 à 500 μ. On les rencontre encore dans les milieux saumâtres et dans les eaux douces — les Dinophycées actuelles montrent une alternance de générations entre stade motile et stade enkysté.

Seuls les kystes de Dinoflagellés sont conservés dans les sédiments — depuis le Norien — car le stade motile est protégé par une thèque cellulosique fragile [44]. L'origine de ce groupe remonterait au Silurien [18], mais son expansion commence vraiment au Jurassique. Certains horizons dans les roches argileuses d'origine marine témoignent de très nombreuses populations au cours du Mésozoïque ou du Tertiaire, évoquant des proliférations comparables aux « marées rouges » actuelles, sans toutefois pouvoir établir de parenté certaine entre les deux observations.

([1]) Jacques EMBERGER. Communication orale.

d. *Les Algues coloniales.*

Par opposition aux formes unicellulaires ou monadoïdes, il existe deux types d'Algues distinctes des véritables algues benthiques :

— les formes cénobiales du type *Deflandrastrum,* possibles ancêtres de *Pediastrum* [20] que l'on ne rencontre jamais en concentrations importantes;

— les formes coloniales, qui comptent divers genres parmi lesquels *Botryococcus* tient une place particulière (paragr. II.B.2). Il s'agit d'une algue chlorophycée planctonique d'eau douce, mais pouvant également se développer en eaux saumâtres ou salées, calmes [19]. La cellule élémentaire ne dépasse pas une dizaine de microns mais les colonies forment des pelotes pouvant atteindre 1 mm de diamètre (fig. 3.4).

Les *Botryococcus* constituent une mousse parfois abondante à la surface des étangs et des lacs. Ce phénomène donne naissance à la **coorongite** (ou « caoutchouc australien »), dans les Nouvelles Galles du sud (région du Coorong).

La « cuticule » ou matrice générale d'une colonie, est constituée d'hydrocarbures représentant 76 % du poids de l'algue sèche. Il s'agit d'hydrocarbures très condensés insaturés ($C_{34}H_{58}$) [65] sous les formes botryococcène et isobotryococcène dans le rapport 9/1. La forme saturée dérivée est le botryococcane ($C_{34}H_{70}$).

$$\begin{array}{c} CH_3 \\ | \\ C_6H_{13} - C_7H_{14} - C - C_{11}H_{22} - C_6H_{13} \\ | \\ CH_2 \\ | \\ CH_3 \end{array}$$

La polymérisation des botryococcènes conduit à la formation de la coorongite et à la forme fossilisée ancienne : le boghead. Aussi la pyrolyse qui provoque la dépolymérisation de ces substances, produit-elle des alcanes en abondance. Les *Botryococcus* témoignent d'une

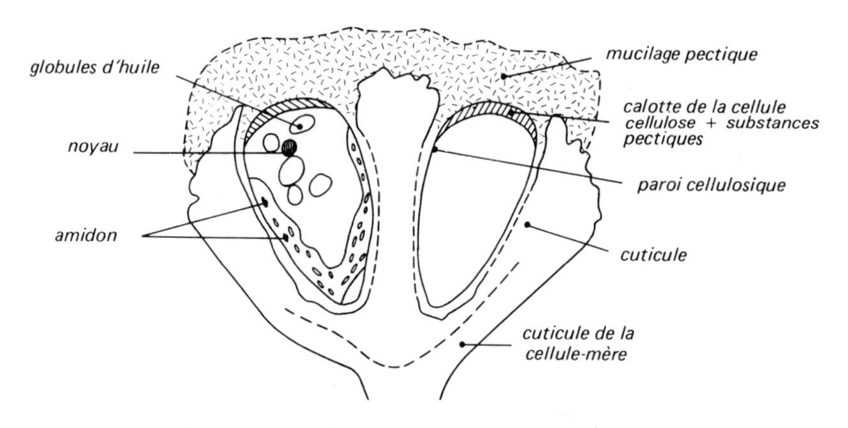

Fig. 3.4. — Section longitudinale de *Botryococcus* [11].

grande ubiquité dans l'espace, sous tous les climats, et dans le temps, depuis l'Antécambrien [59] jusqu'à l'Actuel [19].

e. Les Algues benthiques.

Leurs vestiges fossiles sont rares, du moins sous la forme de structures organiques conservées, mais leur intérêt est grand car elles constituent une source potentielle de matière organique considérable. Dans certains cas privilégiés, la structure algaire peut être identifiée. C'est d'abord le cas des Algues calcaires, lesquelles conservent souvent une trame organique filamenteuse. Mais parfois aussi certaines Algues molles ont imprimé leurs traces dans les schistes tels que ceux du Carbonifère — dits à « fucoïdes » — de Pensylvanie, où dès 1878, Lesquereux [62] voyait l'origine des champs de pétrole de cette région. Dans les Carpathes, d'autres schistes à fucoïdes se sont vus attribuer [64] les mêmes aptitudes pétroligènes.

Dans l'Asghill du Sahara septentrional, Arbey [6] a découvert les traces d'Algues géantes et soupçonne, à cette époque, l'existence d'immenses prairies d'Algues benthiques. Les caractères particuliers de cette formation d'argiles dites « charbonneuses », fortement radioactives et très riches en microfossiles organiques (notamment les *Tasmanites*) l'ont, depuis longtemps déjà, fait reconnaître comme une excellente roche-mère de pétrole [25]. L'extension maximale du peuplement algaire côtier se développe sur les substratums à faible déclivité. La transgression silurienne sur le continent africain arasé par les glaciers, peut donc développer de véritables « tourbières marines » à l'échelle continentale [22].

Parmi les sédiments actuels et récents, certains témoignent à l'évidence d'accumulations d'Algues ou de débris algaires conduisant à des concentrations élevées en carbone organique. Des bancs de goémon fossile sont observés dans plusieurs anses de Bretagne [46]. Ils ont l'aspect de tourbières plus ou moins submergées de 20 à 40 cm d'épaisseur que l'on peut suivre parfois sur 1 500 m de rivage. Ils sont formés d'une masse amorphe brune à noire, d'aspect charbonneux, avec des algues comprimées dont certaines parties sont reconnaissables (flotteurs, crampons).

Dans le sud-ouest du Golfe Persique, de véritables tourbières d'Algues se constituent actuellement en bordure d'une lagune salée : le Khor el Bazam (Abu Dhabi). Les mattes algaires peuvent atteindre 40 km de long sur une largeur moyenne de 2 km et une épaisseur de l'ordre de 30 cm [25].

La grande richesse des Algues en β-carotène, divers caroténoïdes, et en lipides, peut déjà dans une large mesure expliquer l'aptitude pétroligène de leurs vestiges plus ou moins reconnaissables, et du kérogène qui en est issu.

3. Microfossiles et vestiges animaux (pl.3.5).

D'une manière générale, ils ne représentent qu'une source négligeable de matière organique fossile. Ils ne sont fréquents que dans le Paléozoïque. Dès l'Ordovicien et surtout le Silurien, on rencontre en effet une relative abondance de fragments d'épidermes de Gigantostracés (Scorpions marins), de sicules et de débris de rhabdosomes de Graptolithes, de Chitinozoaires dont le développement se poursuit au Dévonien, enfin de restes de poissons notamment dans les schistes du Permien. Au cours du Mésozoïque et du Cénozoïque, on n'observe plus guère que des membranes internes de Microforaminifères, et exceptionnellement des fragments de carapaces de Crustacés, voire d'Insectes. Quant aux Scolécodontes (mâchoires d'Annélides), ils sont relativement répandus depuis le Paléozoïque jusqu'à l'Actuel.

a. Les Chitinozaires.

Ce groupe de micro-organismes est remarquable à bien des égards. Leur forme à symétrie axiale est celle de bouteilles, de manchons, de bourses...

Leur tégument, généralement lisse, s'orne parfois d'attributs ou d'ornements divers : carène ou appendices périaboraux, épines, branchioles, striations à la surface externe du tégument. Certains spécimens paraissent avoir eu une existence indépendante, d'autres forment des colonies généralement linéaires. Toutefois, Kozlowski [60] a décrit des groupements circulaires et même de véritables grappes de Chitinozoaires dépassant 5 mm et entourées d'une membrane formant une manière de cocon.

Leur taille, qui atteignait quelques dixièmes de mm, et même plus d'un millimètre, au moment de leur apparition au Trémadocien, a sensiblement diminué par la suite : 150 μ en moyenne à leur disparition au Dévonien terminal.

Malgré la découverte d'une structure interne particulière, soulignant le monophylétisme de ce groupe, et de nombreuses autres observations tendant à établir sa paléobiologie, l'énigme de son appartenance systématique n'est toujours pas résolue [28].

Découverts par A. Eisenack [41], ils doivent leur nom à la nature chitineuse de leur tégument qui fut établie par leur inventeur (fig. 3.5).

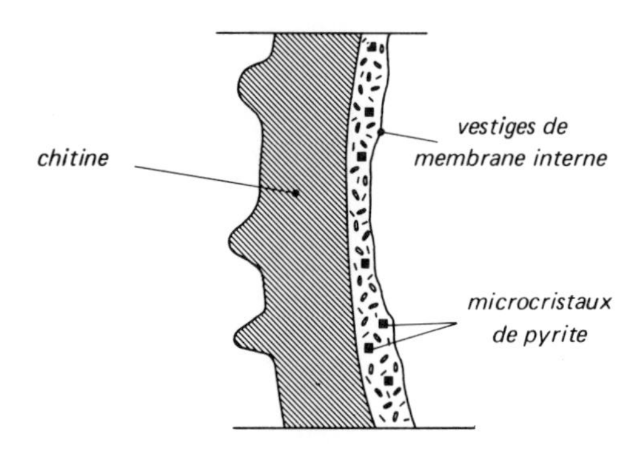

Fig. 3.5. — Section du tégument d'un Chitinozoaire.

Selon les plus récents travaux cependant [63], les Chitinozoaires ou Chitinomycètes occuperaient une position charnière en systématique, entre le monde végétal des *Cellobiontes* comme « ancêtres » des champignons, et le monde animal des *Chitinobiontes.*

b. Les Gigantostracés ou scorpions de mer.

Illustrés par les genres *Eurypterus, Pterygotus, Stylonorus,* ils sont abondamment représentés au Silurien sous forme de débris de carapaces chitineuses annelées, esquilleuses et/ou pileuses.

c. Les Graptolithes.

Les débris de rhabdososomes et les sicules (stade juvénile) de Graptolithes, sont relativement fréquents dans les sédiments marins depuis l'Ordovicien supérieur jusqu'au Dévonien inférieur. Leur tégument est également formé de chitine.

d. Les Scolécodontes, ou armatures buccales de vers Annélides.

Ils sont communs dans les formations marines depuis le Siluro-Dévonien; ils sont aussi de nature chitineuse.

e. Les Microforaminifères.

On les rencontre communément dans le Mésozoïque et le Tertiaire (rarement dans le Paléozoïque). Leur taille est beaucoup plus petite que celle des Foraminifères : 1 à quelques dixièmes de mm. Ce sont les vestiges de formes minéralisées dont ne subsiste que la membrane chitineuse interne.

La figure suivante résume les fluctuations dans le temps des principaux groupes d'organismes cités (fig. 3.6).

B. Fraction amorphe (pl. 3.6).

Ce qui reste, à ce point de notre inventaire, ce sont les constituants organiques biologiquement indéfinis faute de forme, de structure, c'est-à-dire la fraction organique pondéralement la plus importante. Toutes les catégories d'origine s'y retrouvent :

— les membranes de microfossiles, et tissus divers de macrofossiles très dégradés et amorphisés par diagenèse précoce (sol, sédiments, etc.) et catagenèse, au point de ne plus être identifiables optiquement;
— les substances organiques primaires, héritées directement du contenu cellulaire ou des sécrétions des êtres vivants : pigments, lipides, cires, résines, etc.;
— la matière organique catabolique, c'est-à-dire les déchets issus du métabolisme des êtres vivants;
— la matière organique néoformée, issue :
  . des recombinaisons chimiques et de la floculation de la fraction organique solubilisée dans l'eau interstitielle, et éventuellement combinée à des fractions de nature chimique différente;
  . des transformations cataboliques : effluents lourds, insolubles au chloroforme;
  . des transformations catagénétiques : hydrocarbures, bitumes.

Mais ces catégories ne sont pas aisément reconnaissables par l'observation microscopique. Sous réserve d'une élimination préalable, par lavage chloroformique de la plus grande partie des hydrocarbures présents, on peut envisager le classement pratique suivant, indépendamment de toute interprétation :

1. Matière organique grumeleuse.
2. Matière organique granuleuse.
3. Matière organique sub-colloïdale.
4. Matière organique pelliculaire.
5. Matière organique gélifiée.

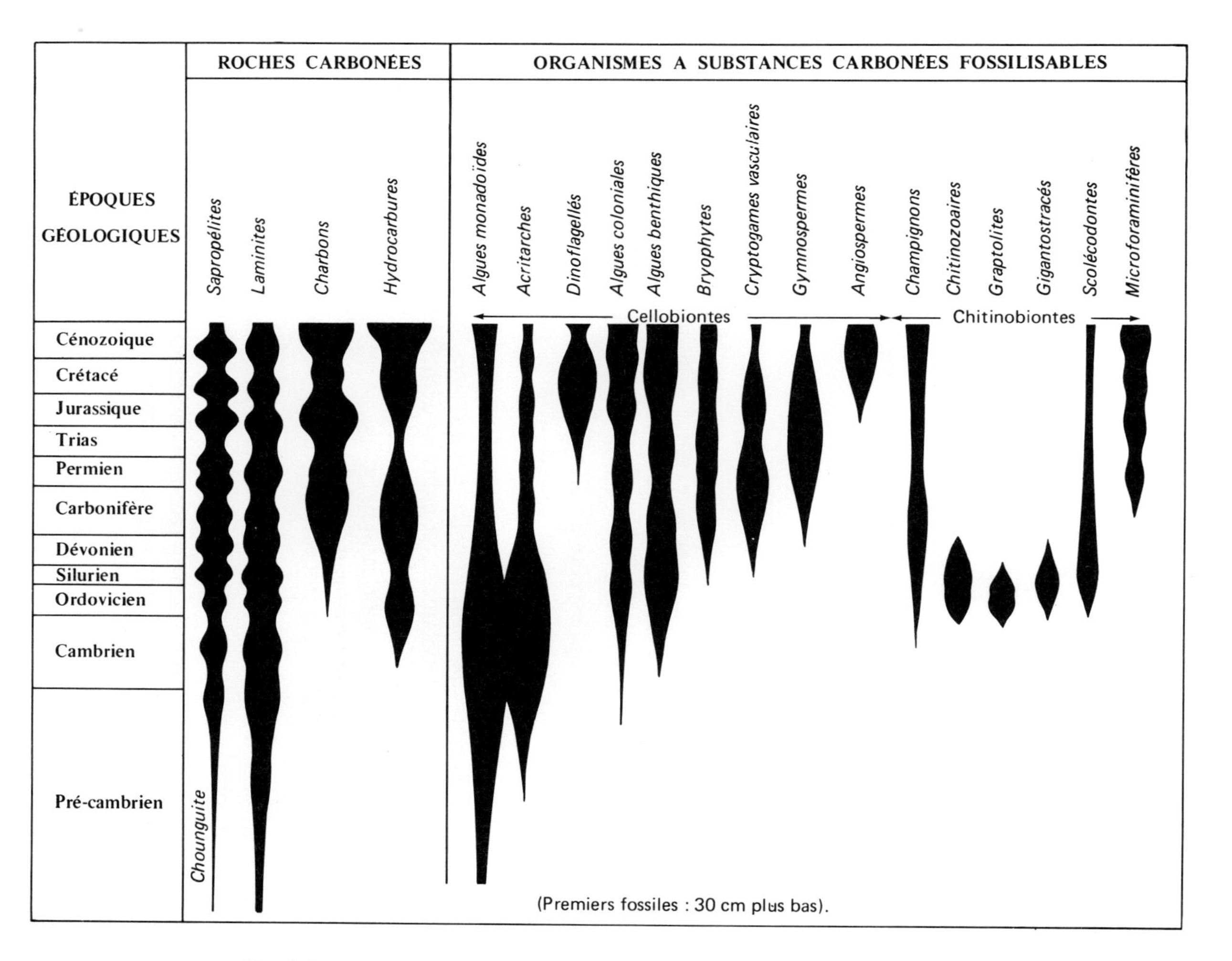

Fig. 3.6. — Extension stratigraphique des groupes de (micro) fossiles organiques.

1. La matière organique grumeleuse.

Elle apparaît en effet au microscope dans le kérogène extrait sous forme de **grumeaux** de tailles variables.

Rapportées aux roches d'origine — en se référant aux plus typiques — elle constitue la masse principale des laminations brunes ou noires (délits organiques) dans les laminites (schistes bitumineux), des veinules et pseudolaminations des argiles et marnes noires, et c'est aussi la « gelée fondamentale » des bogheads et autres sapropélites où elle enrobe les microfossiles présents.

Une fois extraite, elle se présente en flocons ou grumeaux plus ou moins compacts et étendus, selon le mode de préparation mis en œuvre, et selon la teneur de la roche extraite. La couleur est brun clair à jaunâtre si l'épaisseur n'est pas trop grande (flocons ténus ou frange des îlots compacts), et si la catagenèse n'a pas fait sentir ses effets.

2. La matière organique granuleuse.

Cette variété concerne les roches riches en matière organique : sapropélites, laminites, argiles noires plus ou moins carbonatées. On peut la regarder comme une variante de la catégorie précédente, dont elle se distingue par l'existence dans sa masse d'innombrables granules, ou glomérules, d'une taille moyenne de l'ordre du micron. Les moyens d'observation habituels (microscope photonique) ne permettent pas d'y distinguer de structure.

Lorsqu'elles ne sont pas opaques, ces granulations sont caractérisées par une certaine réfringence les faisant apparaître comme hétérogènes à la masse organique ambiante.

3. La matière sub-colloïdale.

Dans des conditions de préparation standard, elle peut être issue de roches très pauvres en matière organique et peut être suspectée d'être un artéfact de préparation ou lié à la boue de forage dans le cas de « *cuttings* ». Elle peut aussi être issue de sédiments frais (actuels ou subactuels). En préparations microscopiques, elle se présente sous un aspect pulvérulent dispersé, de teinte brune à jaunâtre très clair, selon son origine.

4. La matière organique pelliculaire.

On ne l'observe que dans le cas des roches carbonatées = dolomies ou calcaires à grains fins, de couleur sombre. Son aspect semble dû au moulage compressif de minéraux de calcite, aragonite ou dolomite. La matière organique comprimée dans l'espace intergranulaire au moment de la formation des cristaux, subsiste seule après l'attaque acide.

5. La matière organique gélifiée.

Il s'agit ici de fragments d'une substance cohérente, compacte, consolidée, se présentant sous forme de grains de taille très variable (de 1 à quelques centaines de μ) et aux contours plus ou moins francs ou émoussés, de couleur brune à noire.

C'est l'examen approfondi de cette matière organique amorphe qui permet à l'observateur expérimenté de proposer une interprétation de ces catégories primaires quant à leur nature et leur origine. Chacune de ces catégories « optiques » ne pouvant être automatiquement attribuée à une origine définie, il importe de ne pas confondre les termes **descriptifs** et **interprétatifs.** Dans la fraction grumeleuse en effet, on peut trouver des substances dégradées d'origine algaire (sapropélique), d'origine terrestre (produits humiques); des substances primaires (sécrétions); des substances cataboliques (déjections); des substances néoformées (floculation) ou des substances catagénétiques (bitumes, résidus de carbone fixe).

Aussi leur interprétation demande-t-elle une observation approfondie, alliée au témoignage des éléments figurés éventuellement présents (palynofaciès), avec un éventuel recours à d'autres méthodes d'étude, en particulier aux moyens de l'analyse chimique. Chacune des catégories évoquées peut alors être interprétée comme suit, en fonction de critères complémentaires qui constituent une bonne indication d'origine :

— **le type sapropélique** présente souvent un aspect scoriacé; il s'illustre aussi très souvent par la pullulation de « granules » de l'ordre de grandeur du micron. Ceux-ci généralement sphéroïdes ou oblongs, de nature organique ou pyriteuse, témoignent d'un milieu infesté par une microflore saprophytique à bactéries et/ou actinomycètes [68].

Ces corps — ou spores — microbiens, semblent être l'objet d'une épigénisation en pyrite du fait des conditions réductrices liées à l'éventuelle abondance du fer et du soufre dans le milieu sédimentaire [34]. Bien que très méconnue, l'importance de ce monde microbien fossile dans la transformation précoce de la matière organique vers l'état de relative stabilité chimique que représente le kérogène, est certaine. Comme précurseur du pétrole, au même titre que les vestiges algaires, il est depuis longtemps reconnu, ou soupçonné par de nombreux auteurs. En anaérobiose en effet, divers processus biochimiques tendent à produire des catabolites de plus en plus riches en C et H [13].

Depuis quelques années, l'étude de ces problèmes est entreprise par le *Comité d'Etudes Géochimiques Marines* [10], sur divers prélèvements de sédiments marins actuels et subactuels;

— **le type humique franc** semble pouvoir être distingué du précédent par des critères d'observation directe tels que : structure intime d'aspect plutôt **spongieux,** moins réfringent, sans granulations, pyriteuses ou non. Toutefois, il est souvent nécessaire d'ajouter à ces observations les données analytiques complémentaires pour assurer une diagnose valable;

— **le type gélifié** par son aspect et sa consistance est, en revanche, assez aisément interprétable comme issu du milieu terrestre. Toutefois, au niveau de très petites particules, vitrinite et inertinite sont difficiles à distinguer (voir chap. 11);

— **les types primaire, catabolique, néoformé** ne reposent pas à ce jour sur des critères définis et peuvent donc être difficilement caractérisés optiquement de façon fiable. En complément de l'observation en lumière transmise, il est souvent nécessaire de recourir à la lumière réfléchie. Quant à la fluorescence UV, elle permet d'observer la présence de substances primaires : cires, résines, caroténoïdes, et surtout des substances néoformées, en particulier les hydrocarbures.

C. Fraction remaniée (pl. 3.7, photo 1).

L'érosion des roches sédimentaires à contenu organique a pour effet de recycler celui-ci dans les sédiments, notamment marins, qui en sont issus. Ce phénomène est permanent.

Les palynologues américains ont estimé à 10 % le contingent des spores paléozoïques et mésozoïques observées dans le cortège sporopollinique des sédiments actuels dans le delta du Mississipi [88]. La même observation a été faite par les palynologues russes dans les sédiments issus de la Volga [4] et d'autres exemples sont données par divers auteurs, Wilson [88] Venkatachala et Kar [85] dans diverses formations sédimentaires. Dans le delta du Rhône, Gadel et Ragot [47] ont élevé la présence de nombreuses particules charbonneuses de PR (pouvoir réflecteur) élevé (0,5 à 7) auxquelles ils attribuent une origine allant du Tertiaire au Carbonifère.

L'autre exemple que nous retiendrons est celui des sédiments actuels de la mer de Norvège [29]. Au sein des argiles du Wurm et du post-Wurm dont le taux en carbone organique reste inférieur à 2 %, on a pu observer un important cortège de spores, pollens, Acritarches, Dinoflagellés, et débris divers, dont les âges s'étendent du Paléozoïque au Tertiaire. Cette fraction remaniée dépasse en importance celle du cortège autochtone. Son origine a pu être attribuée aux terrains anciens d'Allemagne et de Grande-Bretagne, drainés par les fleuves actuels dont les sédiments, au cours des 10 000 dernières années, se sont largement épandus dans le **Skagerrak,** et aux confins des mers du Nord et de Norvège.

Ce phénomène, ici particulièrement spectaculaire, doit être regardé comme général, dans le monde actuel et au cours des temps géologiques. Aussi l'interprétation des résultats analytiques de la géochimie organique risque d'en être parfois perturbée à défaut d'un examen des palynofaciès des mêmes prélèvements.

Nous venons d'évoquer dans ce chapitre 3, paragraphes A à C, l'ensemble souvent très hétérogène des fractions figurées ou non, toutes insolubles aux solvants organiques, qui constitue le **kérogène.**

D. Fraction soluble.

Il importe cependant de mentionner dans notre tour d'horizon, les caractères optiques de la fraction soluble. Celle d'abord qui n'est soluble qu'aux alcalis, c'est-à-dire la fraction humique et fulvique des sédiments frais ou peu évolués (voir chap. 14). Ainsi séparée puis reprécépitée, il est aisé d'en faire un examen microscopique. Ses critères sont bien ceux indiqués ci-dessus pour le type humique directement observé dans les préparations : flocons dispersés et « spongieux ». Notons que l'emploi nécessaire d'hydracides forts dans la préparation du kérogène (ou dans la plupart des préparations palynologiques) entraîne toujours, dans le cas des sédiments récents, l'élimination d'une partie de la fraction hydrolysable, d'où certaines contradictions apparentes en cours d'études, entre les teneurs en carbone organique dosées sur roche totale et la quantité de kérogène extrait.

Le plus souvent ni les méthodes de préparation du kérogène destiné à l'observation, ni celles de la palynologie ne comportent le lavage minutieux au chloroforme. Dès lors, les hydrocarbures présents dans la préparation peuvent constituer une gêne à l'observation, en particulier pour la fraction amorphe. Cependant, on peut dans ces conditions faire d'intéressantes observations sur ces hydrocarbures. La lumière transmise permet parfois d'observer la pétroléogenèse au sein de la roche-mère, par l'examen de gouttelettes (pl. 3.6, photo 14) issues de la coalescence de l'huile mouillant la frange des flocons de kérogène. Mais c'est surtout la fluorescence UV qui permet l'observation, assez spectaculaire, de ce phénomène (voir chap. 12).

IV. FRACTION MINÉRALE INCLUSE (pl. 3.8 et 3.9)

Du fait même des conditions de dépôt compatibles avec la préservation de la matière organique, la pyrite qui lui est très communément associée, est parfois intimement incluse dans les divers constituants du kérogène. Aussi est-il extrêmement difficile de se débarrasser entièrement de toute trace minérale dans le kérogène extrait de la plupart des roches. Un contrôle optique a tôt fait de nous en convaincre en révélant des cristaux de pyrite cubiques ou polymorphes au sein des flocons de matière amorphe. On observe simultanément de nombreuses pyritosphères tant dans la matière amorphe [24] que sur le tégument des microfossiles divers, voire même à l'intérieur du test de ceux-ci.

Des contrôles en microdiffraction X ont permis de vérifier la nature de ces inclusions. Celles-ci ont également fait l'objet d'études en microscopie électronique conventionnelle et à balayage, avec recherche en spectométrie X des éléments (Fe et S) composant la pyrite.

Grâce à cette technique, il a également été possible de déceler et de localiser plusieurs autres éléments dans le kérogène, notamment Ti, Ni, souvent associés au Fe dans les pyritosphères. Ces résultats ont également été vérifiés au moyen de la microsonde électronique qui a aussi permis d'observer par exemple la présence de soufre organique dans la tasmanine.

Sachant enfin que l'uranium est concentré par la matière organique, on a pu vérifier directement cette observation au moyen de la spectométrie dans les carbonates et les argiles [51].

Dans certains quartzites de l'Ordovicien cependant, le cas était ambigu. Grâce à l'autoradiographie préalable permettant la localisation des sources radioactives dans la matrice minérale, on a pu ensuite vérifier la présence d'uranium au sein de particules carbonées, disséminées au sein des grains de quartz. Cette observation est à rapprocher des associations minérales/organiques du genre **thucolites** [55] qui peuvent donc se présenter sous forme de particules remaniées dans une roche détritique [27].

V. PALYNOFACIÈS

La connaissance des principaux ensembles et sous-ensembles de constituants organiques présentés ci-dessus : nature, signification écologique, est exprimée par la notion de **palynofaciès**.

A. Définition.

Ce concept [21] intéresse l'ensemble des constituants organiques d'une roche après élimination de la phase minérale. Rapproché du **microfaciès,** il éclaire la sédimentologie de la matière organique, quelle que soit la concentration de celle-ci dans la roche considérée.

La méthode consiste à **reconnaître** les familles de constituants présents (figurés ou non) et à en apprécier la teneur relative.

A partir de ces données, il est possible :

— de connaître la richesse de la roche en constituants organiques globaux;
— d'identifier les sources du matériel organique, aérien ou aquatique;
— de caractériser le milieu de dépôt des sédiments (stagnation, transport par courants, etc.);
— d'apprécier l'importance et la diversité des influences diagénétiques subies par ce matériel.

B. Familles de palynofaciès (pl. 3.10)

L'ensemble des constituants organiques d'une roche reflète une part très significative de l'histoire génétique de celle-ci.

Ils complètent utilement à cet égard, les informations apportées par l'étude des minéraux et des (micro) fossiles minéralisés formant la roche. Le milieu de dépôt qui permet la concentration et la préservation de la matière organique est toujours aquatique. Par contre, le milieu de genèse de celle-ci est soit aérien, soit aquatique. Nous avons vu, dans le précédent paragraphe, que le matériel carboné issu de chacun de ces milieux, présente des caractères spécifiques. Une des fonctions précoces de la vie a été la protection du contenu cellulaire par des membranes, des gaines ou des téguments, en particulier chez les organismes vivant dans les environnements sujets à une grande variabilité de température, de salinité ou de pH. C'est ce que l'on observe chez les algues Chlorophycées : *Botryococcus, Tasmanites* [25], ou Dinophycées, (kystes), ou encore dans le groupe *incertae-sedis* des Acritarches (Algues probables). C'est aussi le cas des organismes saprophytes : bactéries, champignons primitifs quand ils revêtent leur forme de résistance : vie ralentie et tégument renforcé.

Il y a là une première famille de palynofaciès caractéristiques, cohérente aux points de vue biologique.

De plus, l'analyse élémentaire révèle que les substances qui la composent, présentent un rapport H/C élevé qui les situe dans la « lignée haute » du diagramme de Van Krevelen [84]. C'est la famille dite des **alginites** et substances assimilées, qui présente la même cohérence du point de vue chimique à quelques écarts près, explicables par un contact avec un milieu plus oxygéné.

Une seconde famille, assez proche de celle-ci, est celle des spores et pollens formée d'une substance — la **sporopollénine** — apparentée à la tasmanine, bien que plus riche en O_2. Cette observation ajoutée à l'argument paléontologique qui établit la filiation entre plantes terrestres et Algues, incite à regarder la structure chimique des spores végétales comme un héritage direct du milieu aquatique ancestral. Dans cette « lignée » du diagramme de Van Krevelen, l'analyse élémentaire révèle une teneur en oxygène légèrement supérieure à celle de la lignée haute.

Mais le milieu aérien est surtout illustré par la masse des tissus du phénotype des plantes terrestres essentiellement formées de cellulose et de lignine (bois, vaisseaux divers). Cet ensemble de matériaux organiques plus riche en oxygène qui forme « la lignée basse » dudit diagramme, représente la troisième famille.

La planche 3.10 montre quelques **palynofaciès typiques** illustrant ces trois familles en concordance avec les « lignées » de Van Krevelen [84].

Bien entendu, les palynofaciès mixtes sont fréquents et les matériaux issus des lignées I et III peuvent chimiquement situer leur mélange sur la lignée II, sous la forme de palynofaciès très différents de ceux représentés ici.

VI. FACIÈS D'ALTÉRATION

Les matériaux organiques décrits dans le chapitre précédent sont supposés observés à un stade de diagenèse peu avancée. Ils seraient, les uns et les autres, dans la « zone amont » des lignées de Van Krevelen, et dans le cadre central de la classification proposée figure 3.7.

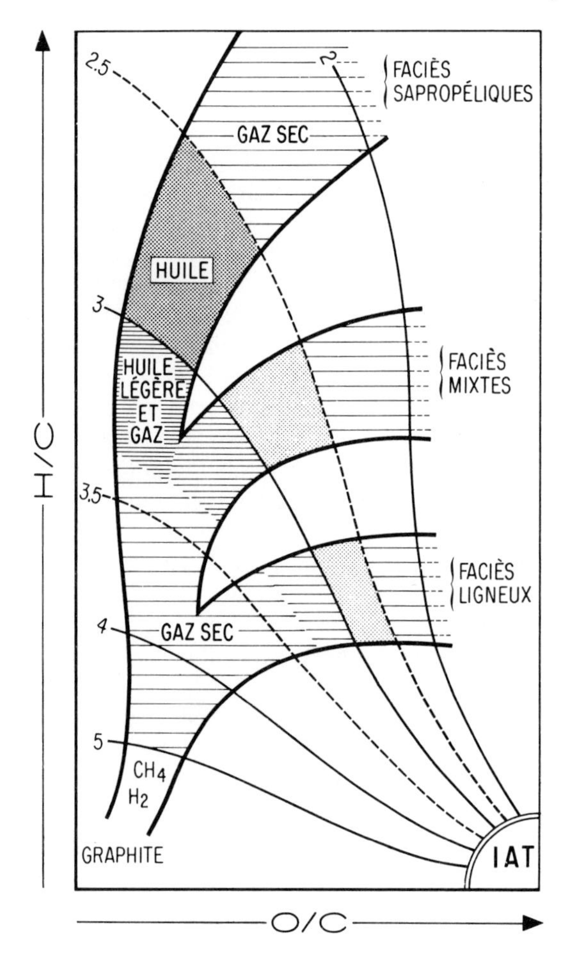

Fig. 3.7. — Formation des hydrocarbures en fonction du faciès organique et de l'indice d'altération thermique (IAT) d'après Raynaud, J.F. et Robert, P. (1976, *Bull. Centre. Rech. Pau, SNPA,* **10,** 1).

Il est donc nécessaire d'examiner différents faciès d'altération, caractéristiques de telles ou telles actions spécifiques.

A. Altérations précoces (diagénétiques).

• Gélification : c'est le processus de transformation des tissus végétaux par imprégnation d'eau et hydrolyse, réalisé en milieu réducteur et anaérobie.

• Altération enzymatique : c'est l'action dissolvante (digestion) des tissus par attaque biochimique (enzymes) des micro-organismes notamment qui, aux premiers stades de cette attaque, portent des « fleurs » de corrosion sur leur tégument [66]. Ce processus entraîne finalement la disparition des structures biologiques dont on peut reconnaître parfois le fantôme.

• Altération oxydante : c'est le cas de l'inertinite. Avec la coorongite et la kukersite la teneur en oxygène est due à des épisodes de dessication ou au milieu carbonaté.

B. Altérations catagénétiques (pl. 3.7, photos 2 à 7).

Au fur et à mesure de l'enfouissement, sous l'action principale de la chaleur, et en fonction de très longues durées, le matériel organique des roches sédimentaires subit une progressive **cuisson.** Celle-ci peut être très lente : le Cambrien de la Plate-forme russe (550 millions d'années) contient des microfossiles organiques [81] très bien conservés, ou très rapide : l'Oligocène de Basse-Provence, localement, contient des microflores carbonisées. Cette cuisson sous pression a d'importants effets sur les constituants organiques qui perdent de leur substance (effluents), de leur couleur, de leur volume (pl. 3.7, photo 7) et finalement leur structure biologique et chimique.

Au terme de cette évolution, ne demeure plus dans la roche qu'un résidu de carbone fixe qui, extrait de la roche, présente un palynofaciès du type « noir de fumée » (carbonisation et amorphisation complètes) (pl. 3.7, photo 6b) ou encore des débris noirs compacts indéterminables.

Par l'observation microscopique il est tout à fait possible d'apprécier les étapes de l'évolution catagénétique des divers constituants organiques des roches.

A côté de la mesure du PR traité au chapitre 11, l'examen en lumière transmise permet d'apprécier la couleur de l'objet observé et de la comparer à une échelle de référence calée sur celle des PR. La valeur numérique découlant de cette échelle conventionnelle est appelée « Indice d'Altération Thermique » ou « IAT » (Voir chap. 1, tableau 1). La fiabilité de la méthode dépend du choix des micro-organismes de référence dont l'extension statigraphique doit être suffisante pour assurer sa présence sur la plus grande puissance possible de terrain. Quelques exemples sont donnés planche 3.7 : Classopollis, Dinoflagellés, Tasmanites.

Les photos 7a à 7e montrent une série de spores actuelles de *Lycopodium clavatum* ayant subi un traitement thermique des plus sévères (pyrolyse de 150 °C à 450 °C avec des durées de 8 à 36 h). Ce moyen artificiel simulant la catagénèse naturelle apporte une échelle colorimétrique à paralléliser avec celle des microfossiles naturels.

On peut aussi mesurer la « transmittance » lumineuse [49] par microspectrophotométrie. Mais la rigueur qu'apporte cette méthode est contrecarrée par les conditions pratiques de l'observation souvent peu conformes aux exigences requises : homologie des microfossiles, constance de l'épaisseur du tégument mesuré.

VII. ESSAI DE CLASSIFICATION DES ROCHES ET DE LEURS CONSTITUANTS CARBONÉS

A. Présentation générale.

La classification proposée figure 3.8 tente une synthèse de l'ensemble des constituants organiques des roches, qu'ils soient rassemblés en fortes concentrations (roches carbonées) ou dispersés et dilués dans les divers types de matrices minérales. Alimentés par les débris et rejets organiques du monde vivant (biosphère), les sédiments recueillent ce stock organique qu'ils soustraient à la dégradation diagénétique (biochimique), surtout active aux stades précoces du dépôt (à l'interface eau-sédiment).

Ainsi, le rectangle central sur la figure 3.8 regroupe-t-il la matière organique fossile (kérogène) des 5 000 premiers mètres d'une série sédimentaire. L'effet de la catagénèse sur ce matériel donne naissance aux fractions organiques mobiles (effluents) et corrélativement aux résidus carbonés fixes qui perdent progressivement leurs atomes d'hydrogène pour se rapprocher du carbone pur, aux derniers termes du métamorphisme. L'ensemble des constituants organiques des roches hérités de la biosphère et, plus ou moins transformés par diagénèse, donne donc naissance à un second ensemble : les roches carbonées thermogénétiques ou catagénétiques, formé de deux sous-ensembles :

— les effluents carbonés (plus ou moins fluides) ou **bitumites;**
— les résidus carbonés (fixes) ou **carbonites.**

B. Description des termes utilisés.

1. Roches sédimentaires.

a. Sapropélites (voir paragr. II.B.2.), terme créé par Henri Potonié, 1910 [71].

Les roches de ce type, plus ou moins riches en carbone organique sont caractérisées par des constituants organiques **d'origine aquatique.**

Nous proposons le terme d'**algites** pour désigner les roches à faible dilution minérale (généralement moins de 50 %) formées essentiellement de corps algaires tels que *Botryococcus* (bogheads), *Gleocapsomorpha* (kukersite), etc. ou encore *Tasmanites* (tasmanite).

Les **laminites,** terme créé par A. Lombard, regroupent les roches à laminations organiques (varves), en alternance avec des laminations minérales.

Les « schistes bitumineux » pour la plupart (schistes-carton, « Bituminous Shales », « Green River Shales », etc) entrent dans cette catégorie où l'on observe une forte dilution minérale, de l'ordre de 60 à 90 %. L'alternance des laminations paraît observer un rythme saisonnier (saisons chaudes et froides) et leur régularité paraît d'autant plus grande que la tranche d'eau est plus faible.

Les **argiles sapropéliques,** dans une acception stricte, désignent les roches argileuses noires (souvent dénommées « black shales ») sans structure laminée distincte, et dans lesquelles la dilution du matériel organique d'origine sapropélique est très forte (de l'ordre de 95 % ou

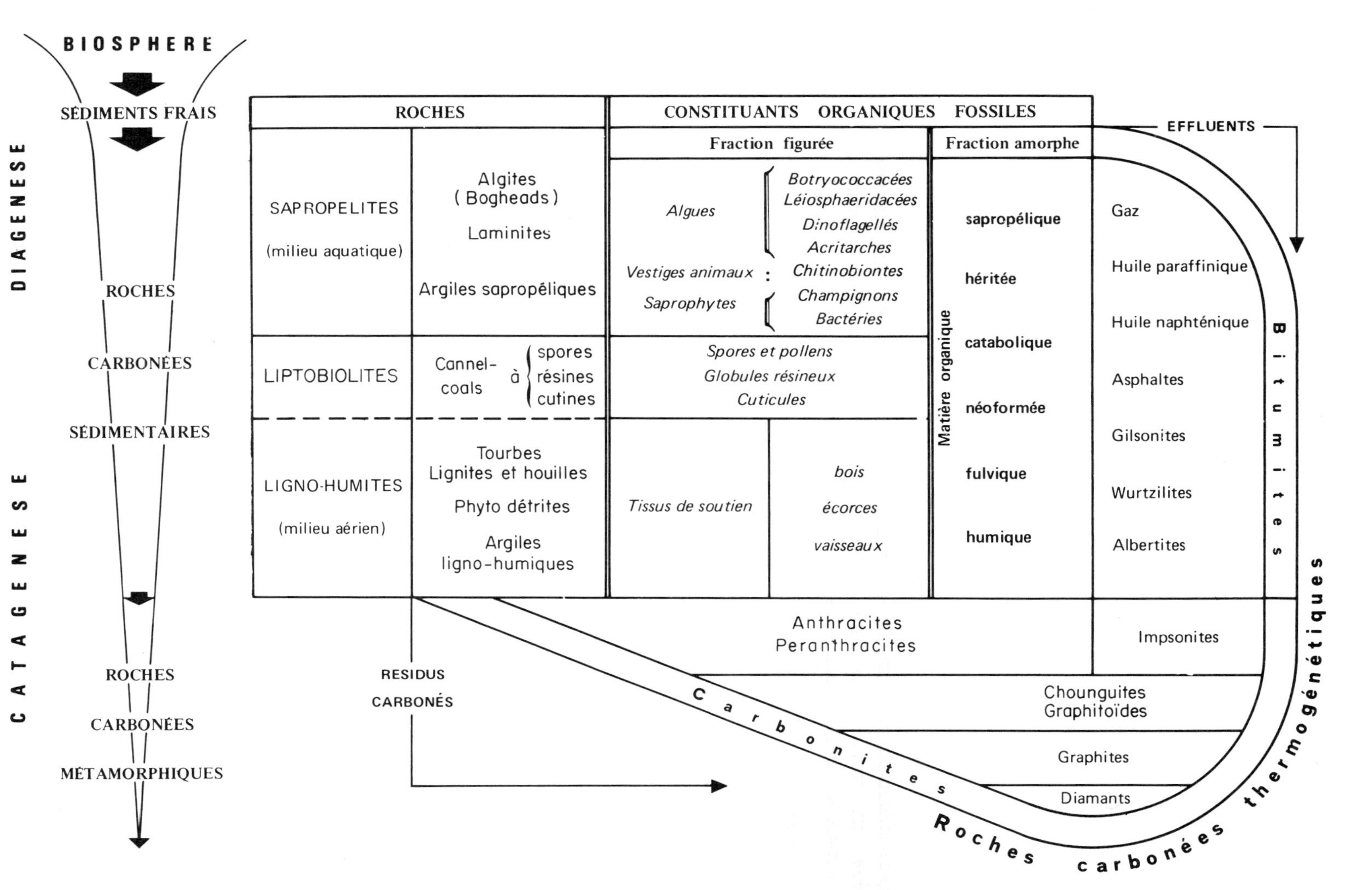

Fig. 3.8. — Essai de classification des roches et de leurs constituants carbonés.

plus). Le mécanisme de leur sédimentation sans grande régularité, s'opère dans un milieu d'énergie probablement plus grande que dans le cas des laminites organiques.

b. Liptobiolites, terme créé par Henri Potonié [71].

Les roches de ce type sont rares. Elles sont caractérisées — comme dans le cas des algites — par la prédominance de constituants (ou macéraux) tels que spores (et mégaspores) végétales (sporite), cuticules de feuilles (cutinite) ou résines végétales (résinite). La dilution par le matériel minéral est faible à très faible (moins de 50 %). Il s'agit en fait des charbons spéciaux dits « cannel coals ». Leurs constituants d'origine aérienne ont été concentrés dans le milieu de dépôt par ségrégation hydrodynamique.

c. Ligno-humites.

Nous proposons ce terme générique pour rendre compte des constituants organiques principaux, diversement concentrés dans trois principaux types de roches homologues des sapropélites, mais procédant ici d'une origine végétale aérienne, puis d'une dégradation (hydrolyse) plus ou moins avancée en milieu aquatique.

Les **lignites,** tourbes et houilles, regroupent les roches à faible dilution minérale formées principalement de vestiges ligneux noyés dans un gel humique.

Les **phytodétrites** ou schistes charbonneux, où la matrice minérale domine, incluant dans sa masse, parfois en lits millimétriques, des fragments végétaux plus ou moins fusinisés; les « black shales » et les « stériles » des gisements de charbons présentent souvent ces caractères. Leur contenu organique phytodétritique : fragments ligneux, cuticulaires, spores végétales, etc. constitue un palynofaciès typique de la lignée III des diagrammes de Van Krevelen (pl. 3.10).

Les **argiles ligno-humiques** sont des argiles noires (ou grises) dont le stock organique très faible (< 5 %) est formé essentiellement, sinon exclusivement, de matériel organique détritique amorphe ou sous forme de débris de tissus végétaux.

d. Les bitumites.

Terme générique que nous proposons pour désigner les effluents carbonés (depuis les termes plus ou moins liquides jusqu'aux termes gazeux). Ils sont issus de l'action catagénétique liée aux influences thermiques dans les zones supérieures de la lithosphère.

Aux profondeurs plus grandes les actions thermiques sont plus sévères et l'on entre dans le domaine des résidus carbonés fixes.

*
* *

Gardant l'appellation classique de « charbons » au premier type de roches carbonées (charbons humiques), et d'argiles, carbonates et autres, pour les roches sédimentaires à très faible teneur en constituants organiques, nous proposons le terme de « **kérogénites** » pour l'ensemble : sapropélites et laminites, à fort potentiel organique issu du milieu aquatique. Cela permettrait d'éviter le terme fâcheux de « schistes bitumineux » et de retrouver, dans un même ensemble cohérent, les roches capables de produire de l'huile par pyrolyse, et dont l'accumulation en certaines régions du monde constitue un objectif économique pour les décennies à venir.

2. Roches métamorphiques.

Nous proposons ici le terme de **carbonites** pour désigner l'ensemble des roches carbonées ne contenant pratiquement que du carbone : tels que les anthracites, impsonites, chounguites, graphites et diamants.

VIII. CONCLUSION

Il aura fallu beaucoup d'études et de mécomptes, avant de soupçonner l'extrême complexité de la matière première du pétrole et du gaz naturel. Sous l'appellation simplificatrice et commode de « kérogène », on désigne donc le mélange « aveugle » des constituants organiques insolubles des roches sédimentaires. C'est, sans nul doute, leur origine et leur nature si diverses qui ont introduit une ambiguïté dans l'acception de ce terme par les différents auteurs.

Par l'observation microscopique, la pétrologie a pu dominer son vaste sujet et proposer une vision cohérente de la diversité des matériaux qui constituent l'écorce terrestre. C'est aussi grâce à l'observation microscopique qu'il est possible de dominer la complexité des matériaux organiques qui forment certaines de ces roches et entrent, pour une faible part il est vrai, dans la composition des autres roches sédimentaires et métamorphiques. Cette démarche naturaliste de la connaissance géologique permet l'orientation et l'interprétation des études physico-chimiques qui d'autre part, lui apportent un indispensable complément.

L'utilisation de l'observation microscopique du kérogène désormais indispensable à la géochimie organique, représente finalement un remarquable moyen de rétrodiction géologique et de prédiction pétrolière. Ajoutée à la connaissance géologique régionale, elle ouvre de larges possibilités aux études de synthèse sédimentologique et paléogéographique des bassins sédimentaires.

BIBLIOGRAPHIE

1. Alpern, B., (1963), *in : « Pollen et Spores »,* **V, 1,** 169.
2. Alpern, B., (1976), « Les Sciences », *Alpha,* n° 106.
3. Alpern, B., Bostick, N., (1976), *Bull. CRP — SNPA,* **10,** 201.
4. Ananova, Y. N., (1960), *Bull., Ass. Naturalistes,* **65,** 3. 132, Moscou (en russe).
5. Andreev, P.F., Bogomolov, A.J., Dobryanskii, A.F. et Karsten, A.A., (1968), Transformation of Petroleum in Nature.
6. Arbey, F., (1971), *C.R. Acad, Sci. Paris,* **273,** 15, 265.
7. Bertrand, C.E., (1893), *Bull. Soc. Hist. Nat. Autun,* **VI,** 321.
8. Bertrand, C.E. et Renault, B., (1892), *C.R. Acad. Sci. Paris,* **115,** 138.
9. Bertrand, P., (1930), *Congrès Internat. Mines Métall. Géol. Appl.,* Liège, 16 p.
10. Bianchi, A.J.M., Bianchi, M.A.G., Bensoussan, M.G., Lizzarraga, M.L., Marty, D. et Roussos, S., (1977), « *Orgon-1* », éd. CNRS, 15.

11. Blakburn, K.B., (1936), *Trans. Royal Soc. Edimburgh,* **58,** 841.

12. Boneham, R.F. et Tailleur, J.L., (1972), *U.S. Geol. Survey Profess. Paper,* **800 B,** B 17.

13. Bradley, W.H., (1924), *Amer. J. Sci.,* 5th Series, **8,** 228.

14. Brooks, J. et Shaw G., (1968), *Nature,* **220,** 678.

15. Burlingame, A.L., Wszolec, P.C. et Simoneit, B.R., (1969), *in : Advances in Organic Chemistry 1968,* Pergamon Press, London, 131.

16. Busson, G., Ludlam, S.D. et Noel, D., (1972), *C.R. Acad. Sci. Paris,* **274,** 3044.

17. Busson, G. et Noel, D., (1972), *C.R. Acad. Sci. Paris,* **274,** 3172.

18. Calandra, F., (1964), *C.R. Acad. Sci. Paris,* **258,** 4112.

19. Chodat, R., (1896), *J. Botan.,* **X,** 333.

20. Combaz, A., (1962), *C.R. Acad. Sci. Paris,* **253,** 1977.

21. Combaz, A., (1964), *Rev. Micropaléontologie,* **7,** 3, 205.

22. Combaz, A., (1966), *The Paleobotanist,* **15** (1, 2) 29, Lucknow (Inde).

23. Combaz, A., (1967), *Actes Soc. Linnéenne de Bordeaux,* **104,** B29, 1.

24. Combaz, A., (1970), *C.R. Acad. Sci. Paris,* **270,** 2240.

25. Combaz, A., (1974), *in : Advances in Organic Geochemistry 1973,* **424,** Editions Technip, Paris.

26. Combaz, A., (1975), *Colloque Internat. Pétrographie de la Matière Organique des Sédiments,* éd. CNRS, Paris, 93.

27. Combaz, A., (1979), *C.R. Acad. Sci. Paris* (sous presse).

28. Combaz, A. et Poumot C., (1962), *Rev. Micropaléontologie,* **5, 3,** 147.

29. Combaz, A., Bellet, J., Poulain, D., Caratini, C. et Tissot, C., (1977), *in : « Orgon 1, Mer de Norvège »,* 139, éd. CNRS, Paris.

30. Correia, M., (1967), *Rev. Inst. Franç. du Pétrole,* XXII, 9, 1285.

31. Cramer, F.H., Diez, M. et del Carmen R., (1977), *in : Advances in Organic Geochemistry* 1975, ENADIMSA, Madrid, 891.

32. Crum-Brown A. (1912), *in : The Oil Shales of the Lothians — Part III — 2nd ed, 143 Memoirs Geol. Survey, Scotland.*

33. Cuvillier, J. et Sacal, V., (1951), Corrélation stratigraphique par microfaciès en Aquitaine Occidental, éd. Brill.

34. Debrand-Passard, S., (1975), *Geobios,* **8,** 5, 325.

35. Deflandre, G., (1961), La vie créatrice de roches, P.U.F., Paris.

36 Deflandre, G., (1967), *C.R. Acad. Sci. Paris,* **265,** 1776.

37. Deflandre, G., (1968), *C.R. Acad. Sci. Paris,* **266,** 2385.

38. Dow, W., (1977), *J. Geochem. Explor.,* **7,** 77.

39. Duparque, A., (1933), Structure microscopique des charbons du bassin houiller du Nord et du Pas-de-Calais.

40. Duparque, A., (1934), L'étude microscopique des charbons, Paris.

41. Eisenack, A., (1931), *Paleontol. Zeitsch.,* **13,** 74.

42. Erdtman, G., (1932), *Abh. Nat. Ver., Bremen,* **28,** 11.

43. Erdtman, G., (1943), An introduction to pollen analysis, *Verdoon. Sci. Books,* **12,** Waltham (Mass.).

44. Evitt, W., (1963), *Proceedings Nat. Acad. Sci.,* **49,** 158.

45. Fayol, H., (1887), *Bull. Soc. Indust. Minér., Série 2,* **XV,** 543.

46. Feys, R.C., (1958), *C.R. Acad. Sci. Paris,* **246,** 3084.

47. Gadel, F. et Ragot, J.P., (1974), *in : Advances in Organic Geochemistry* 1973, 620, Editions Technip, Paris.

48. Geoffroy Le Jeune, C.J., (1711), *Mémoires Acad. Royale Sci. Paris,* 207.

49. Gutjahr, C.C.M., (1966), *Leidse Geol. Meded.,* **38,** 1.

50. Harris, T.M., (1938), « The British Rhetic Flora », ed. British Museum.

51. Hassan, M., Selo, M. et Combaz, A., (1975), *Congrès Internat. Sédimentologie,* Nice, **7,** 69.

52. Hyde, H.A. et Williams, D.A., (1945), *Nature,* 155, 265.

53. Jansonius, J., (1976), *Geoscience and Man,* **15,** 129.

54. Jeckowsky, B. de, (1959), *Rev. Inst. Franç. du Pétrole,* **XIV,** 3, 315.

55. Jedwab, J., (1966), *Bull. Soc. Fr. Minéral. Cristall.,* **89,** 251.

56. Jemtchoujnikov, Yu. A. et Guinsbourg, A.I., (1960), Principes fondamentaux de la pétrographie des charbons, Moscou.

57. Kalmikov, G.S., (1974), *in : Problèmes de la Géologie du Pétrole,* **4,** éd. Nedra, Moscou.

58. Kiriakino, A.V., Gueneralova, V.N., Kodina, L.A. et Petrova, J.V., (1974), *Geochimya,* **6,** 904.

59. Konzalova, M., (1973), *Vestruk Ustredniho Ustavu Geologickeko,* **48,** 1, 17.

60. Kozlowski, R., (1963), *Acta Paleontol. Polsk.,* **8,** 425.

61. Kozlowski, R. et Kazmierczac, J., (1968), *C.R. Acad. Sci. Paris,* **266,** 2147.

62. Lesquereux, L., (1866), *Mem. Geol. Survey,* **2,** 425.

63. Locquin, M.V., (1977), Relations chronophénétiques entre taxons fossiles, Ecole Pratique des Hautes Etudes.

64. Louis, M., (1965), Géochimie du Pétrole, Editions Technip, Paris.

65. Maxwell, J.R., Douglas, A.G., Eglington, G. et Mac Cormick, A., (1968), *Phytochemistry,* **7,** 2157.

66. Meijer, J.F. de, (1969), *Leidse Geol. Medlegingen,* **44,** 235.

67. Moore, L.R., (1963), *Paleontology,* **6,** 349.

68. Moore, L.R., (1964), *Proceedings Yorkshire Geol. Soc.,* **34,** 3, 12.

69. Newton, E.T., (1875), *Geol. Mag.,* **8,** 338.

70. Orton, E., (1889), *U.S. Geol. Survey, Ann. Rept.* **8,** 475.

71. Potonié, H., (1910), *Die Entstehung der Kohle under Kaustobiolithe,* Borntraeger, Berlin.

72. Potonié, R., (1931), *Braunkohle,* heft **27,** 554.

73. Prévot, A.A., (1954), *La Nature,* **82,** 3229.

74. Reinsh, P.F., (1884), Erlangen, Krische, 2 vol.

75. Renault, B., (1894), *Bull. Soc. Hist. Nat. Autun,* **7,** 172.

76. Seyler, C.A., (1948), *Proceedings South Wales Inst. Engineers,* **63,** 213.

77. Shaw, G. et Yeadon, A., (1966), *J. Chem. Soc.,* **C,** 16.

78. Sidorenko, A.V. et Sidorenko, S.A., (1975), *Mémoires,* fasc. **277,** éd. Nauka, Moscou.

79. Staplin, F.L., (1969), *Bull. Canad. Petrol. Geol.,* **17,** 47.

80. Stopes, M.C., (1935), *Fuel,* **14,** 4.

81. Timofeev, B.V., (1959), *La plus ancienne flore des régions de la Baltique et sa signification stratigraphique,* Leningrad.

82. Timofeev, P.P. et Bogolubova, L.I., (1971), Recueil « Matière Organique des Sédiments Contemporains et des Sédiments Fossiles », éd. Nauka, Moscou.

83. Tissot, B., Durand, B., Espitalié, J. et Combaz, A., (1974), *AAPG Bull.,* **58,** 3, 499.

84. Van Krevelen, D.W., (1961), Coal, éd. Elsevier, Amsterdam.

85. Venkatachala, B.S. et Kar, R.K., (1969), *Micropaleontology,* **15,** 4, 491.

86. Von Post, L., (1916), *Geol. Foreningens i Stockolm Pohandengar,* 38, 384.

87. Wall, D., (1962), *Geol. Mag.,* **94,** 4, 353.

88. Wilson, L.R., (1964), *Grana Palynologica,* **V,** 3, 425.

89. Witham, H., (1883), The internal structure of fossil vegetables found in carboniferous and solithic deposits of great Britain, Edimburgh and London.

PLANCHES 3.1 A 3.10

PLANCHE 3.1
TYPES DE ROCHES CARBONÉES
(Sections perpendiculaires au plan de stratification)

Photo 1. — Houille maigre (G. × 300). Westphalien de Grande-Bretagne. (Section mince en lumière transmise). La vitrinite est dominante (rouge); jaune = exinite; noir = inertinite.

Photo 2. — Charbon de spores (cannel-coal) (G. × 300). Westphalien de Grande-Bretagne. (Section mince en lumière transmise). L'exinite, mégaspores et microspores, est dominante.

Photo 3. — Charbon d'algues (boghead) (G. × 370). Permien d'Autun (France). (Section mince en lumière transmise). Chaque îlot jaune est une colonie de *Botryococcus braunii.* Le ciment brun est constitué de matière humique.
GR 3487 — 1353/7.

Photo 4. — Kukersite (G. × 370). Ordovicien d'Estonie. Lumière transmise. Les îlots ± distincts sont également des *Botryococcacées.*
GR 401 — 1353/12.

Photo 5. — Tasmanite (G. × 240). Permien de Tasmanie. Lumière transmise. Empilement de sphères écrasées de *Tasmanites punctatus*, avec quelques cristaux dispersés de calcite.
A 16 — 1483/31.

Photo 6. — Diatomite (G × 90). Messinien de Sicile. Lumière transmise. Les microlits de Diatomées alternent ± distinctement avec ceux de matière organique et de minéraux argileux. Quelques cristaux de calcite.
GR 11339 — 1486/35.

Photo 7. — « Green River Shales ». Eocène du Colorado (U.S.A.):

a. (G. × 2,3). Surface polie, en lumière naturelle réfléchie. Les laminations à dominante organique et à dominante minérale sont parfaitement distinctes. **b.** (G. × 90). Section mince, en lumière transmise. Les taches blanches représentent les cristaux de calcite et de dolomie.
A 17 — 1486/27.

Photo 8. — « Schistes carton ». Toarcien de Lorraine (France).

a. (G. × 2,3). Surface polie, en lumière naturelle réfléchie. Les laminations moins contrastées que dans le précédent exemple sont cependant bien visibles. **b.** (G. × 40). Section mince en lumière transmise.
GR 3755 — 1361/17.

Photo 9. — « Schistes de l'Irati ». Permien du Parana (Brésil).

a. (G. × 2,3). Surface polie en lumière naturelle réfléchie. **b.** (G. × 90). Section mince, en lumière transmise.
GR 4179 — 1930/5.

(*Photos :* B. Alpern, J. Bellet, A. Combaz)

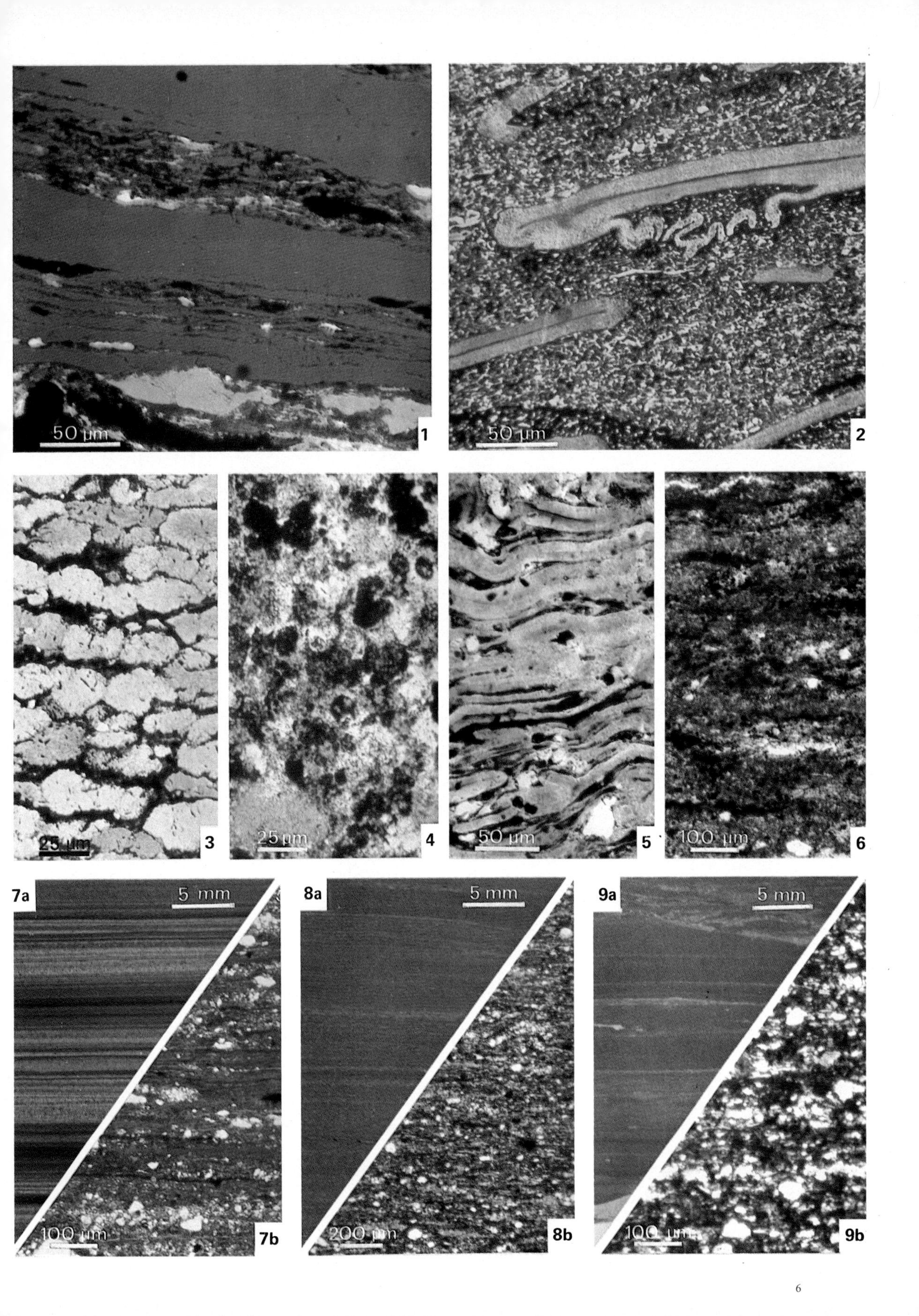
50 µm
1
50 µm
2
25 µm
3
25 µm
4
50 µm
5
100 µm
6
7a
5 mm
100 µm
7b
8a
5 mm
200 µm
8b
9a
5 mm
100 µm
9b

PLANCHE 3.2
TYPES DE ROCHES-MÈRES PLUS OU MOINS RICHES EN CARBONE ORGANIQUE
(Sections minces perpendiculaires au plan de stratification)

Photo 1. — Argiles noires radioactives (G. × 240). Silurien du Sahara Oriental.
R 7389 — 1486/20.

Photo 2. — Argiles gris-noir (G. × 90). Silurien du Sahara Occidental.
R 11346 — 1486/6.

Photo 3. — Argiles noires radioactives (G. × 90). Kimmeridgien de Mer du Nord.
LM — 1510/17.

Photo 4. — Argiles gris-noir (G. × 90). Callovo-oxfordien de Mer du Nord.
GR 5617 — 1512/2A.

Photo 5. — Marnes brun-noir (G. × 90). Cénomanien du Golfe Persique.
GR 1656 — 1018/16.

Photo 6. — Marnes laminées brun-noir (G. × 90). Givétien du Canada Occidental.
GR 4063.

Photo 7. — Marnes laminées noires à Ptéropodes (G. × 40). Frasnien du Canada Occidental.
GR 1406.

Photo 8. — Calcaire argileux (G. × 90). Dévonien du Canada.
LM — 1512/0A.

Photo 9. — Calcaire micritique gris-noir (G. × 90). Yprésien de Tunisie.
GR 2578 — 1486/22.

Photo 10. — Calcaire à Tithonelles (*Calcisphoerulidés*) (G. × 90). Cénomanien inférieur de Tunisie.
1512/5A.

Photo 11. — Marnes noires à Globigérines (G. × 40). Turonien de Casamance maritime.
R 9617 — 1019/2.

Photo 12. — Calcaire phosphaté à Foraminifères et débris de poissons (G. × 40). Turonien de Casamance maritime.

(*Photos* : J. Bellet, A. Combaz, A. Houard)

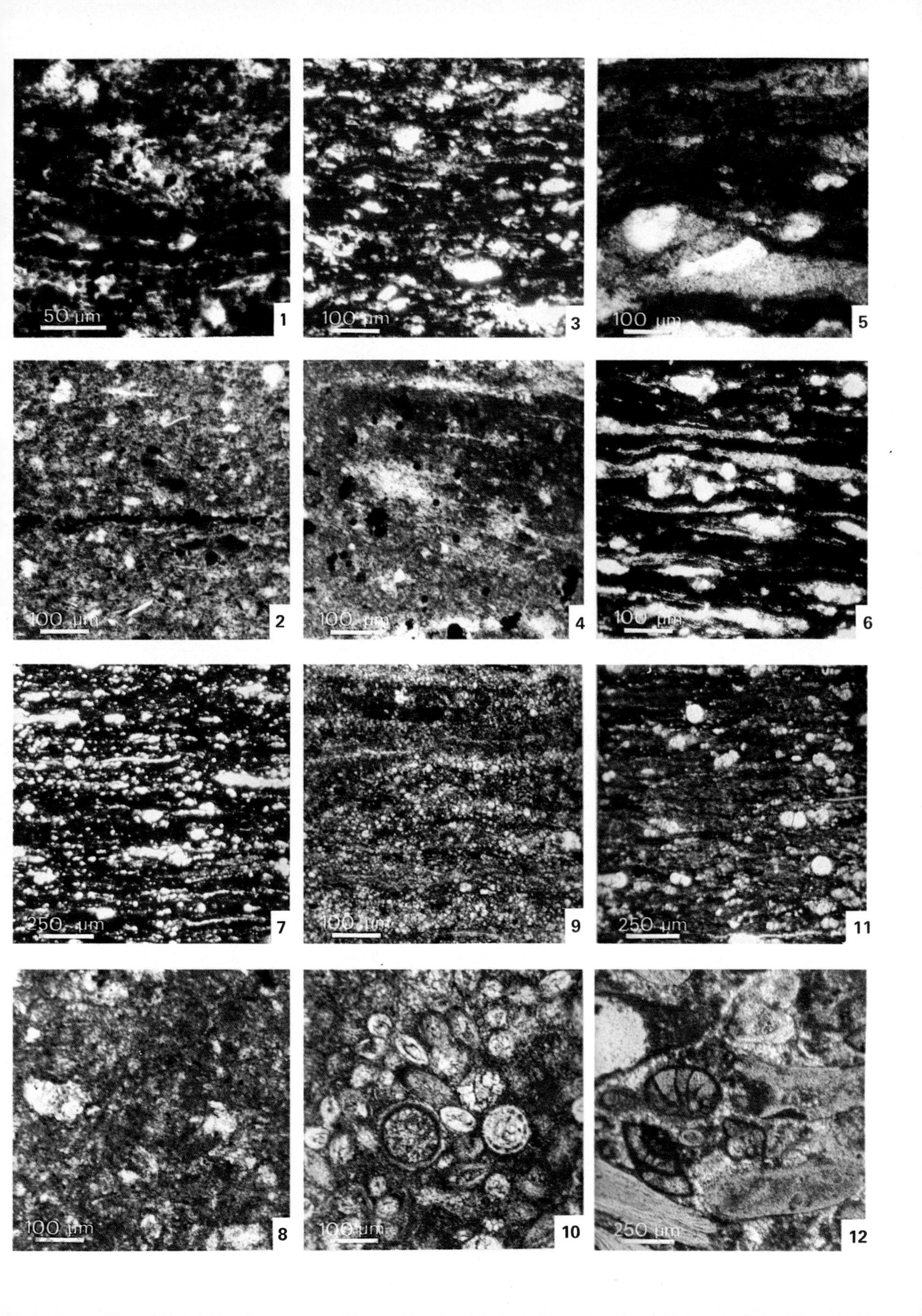
50 µm
1
100 µm
3
100 µm
5
100 µm
2
100 µm
4
100 µm
6
250 µm
7
100 µm
9
250 µm
11
100 µm
8
100 µm
10
250 µm
12

PLANCHE 3.3
MICROFOSSILES ET FRAGMENTS VÉGÉTAUX TERRESTRES
(en lumière transmise)

Photo 1. — Pollen de Gymnosperme, *Conifère* (G. × 400).
GR 3450.

Photo 2. — Pollen d'Angiosperme (G. × 900). Crétacé du Pérou.
GR 2236 — F 272-2.

Photo 3. — Pollen de Gymnosperme, *Classopolis* (G. × 500).

Photo 4. — Pollen d'Angiosperme, *Tilia* (G. × 500). Tertiaire de Mer de Norvège.
KS 02 — 4 m.

Photo 5. — Spore de Cryptogame vasculaire (G. × 370). *Hymenozonotriletes lepidophytus*. Dévonien terminal du Sahara.
R 19249 — 1488/8.

Photo 6. — Spore de Psilophytale (G. × 500). *Emphanisporites sp.* Dévonien inférieur de Libye.
R 7414 — 1489/1.

Photo 7. — Spore de Schizéacée (G. × 370). *Anemia sp.* Crétacé inférieur. Mer de Norvège.
KS 02 — 0,81 m.

Photo 8. — Spore de Cryptogame vasculaire (G. × 370). *Reinschospora*. Carbonifère de Grande-Bretagne.

Photos 9 à 12. — Spores de champignons, simples ou en cours de division (G. × 925). Holocène de la fosse de Cariaco.
1409/10 — 1405/18
1405/9 — 1406/2.

Photo 13. — Filament de champignon (G. × 370). Holocène de l'Atlantique (cône de l'Amazone).
GR 6508 — 1415/36.

Photo 14. — Fragment ligneux (G. × 240). Maestrichtien du Sahara.
R 5930 — 1489/6.

Photo 15. — Fragment ligneux (G. × 220). Tertiaire de Mer du Nord.
GR 2507 — 1253/4.

Photo 16. — Fragment de cuticule (G. × 370). Silurien de Libye.
GR 8584 — 1447/15.

Photo 17. — Fragment de cuticule (G. × 350). Dogger de Mer du Nord.

(*Photos* : J. Bellet, A. Combaz)

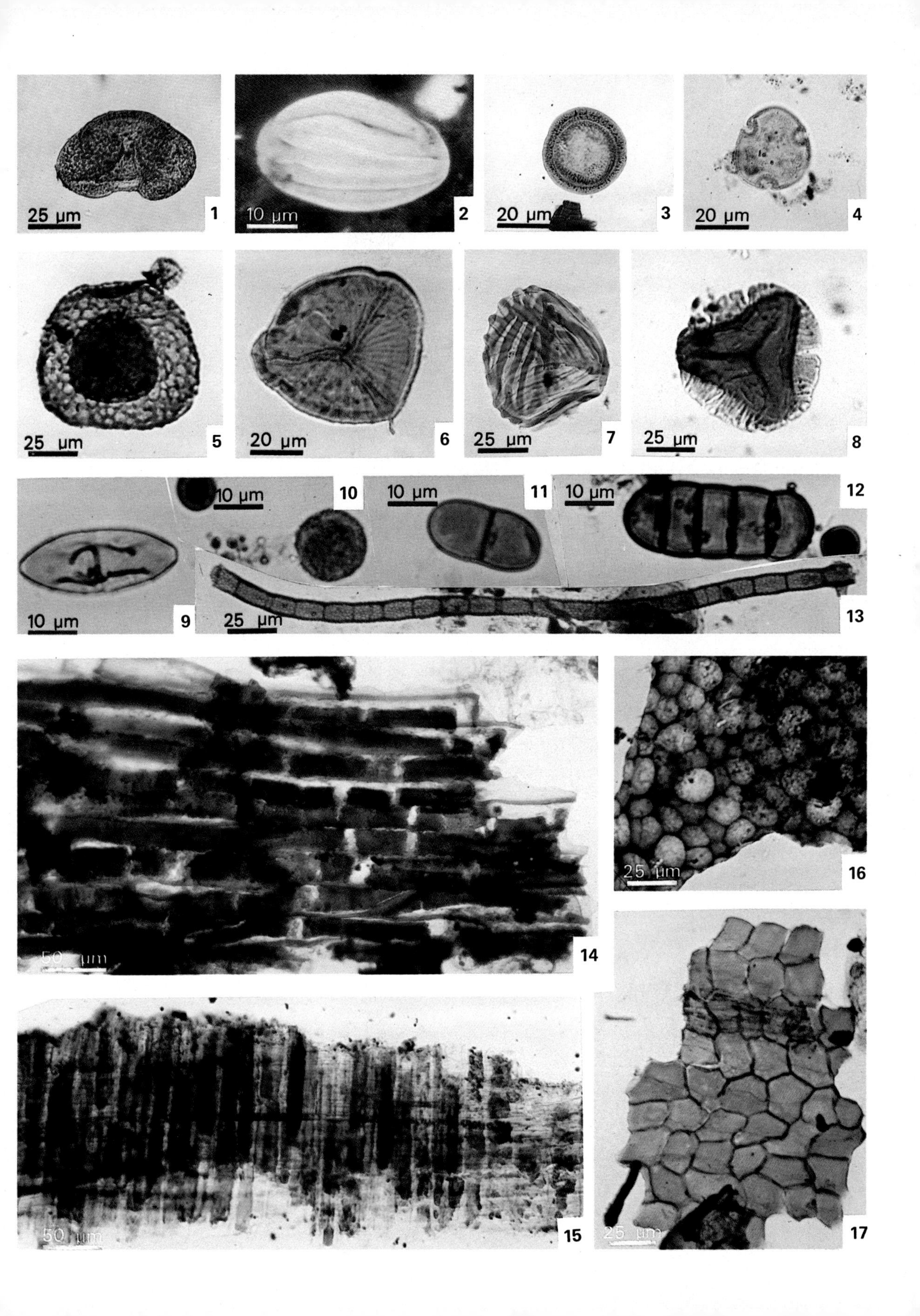
25 µm
1
10 µm
2
20 µm
3
20 µm
4
25 µm
5
20 µm
6
25 µm
7
25 µm
8
10 µm
9
10 µm
10
10 µm
11
10 µm
12
25 µm
13
50 µm
14
50 µm
15
25 µm
16
25 µm
17

PLANCHE 3.4
MICROFOSSILES ET FRAGMENTS ALGAIRES
(en lumière transmise)

Photo 1. — Acritarche (G. × 370). *Micrhystridium sp.* Dévonien supérieur du Sahara.
R 19031 — 1488/17.

Photo 2. — Acritarche (G. × 600). *Baltisphaeridium sp.* Dévonien moyen du Sahara.
R 19030 — 1488/16.

Photo 3. — Dinoflagellé (G. × 150). *Scrinodinium crystallinum.* Oxfordien de Grande-Bretagne.

Photo 4. — Dinoflagellé (G. × 250). *Deflandrea sp.* Crétacé de Libye.
GR 502 — AC 8 — 4A.

Photo 5. — Acritarche (G. × 500). *Leiofusa striata.* Silurien de Libye.
R 325A — 616/11.

Photo 6. — Acritarches (G. × 500). *Micrhystridium sp.* Silurien de Libye. Fluorescence UV sur bloc poli).
GR 3495.

Photo 7. — Algue cénobiale (G. × 400). *Deflandrastrum.* Silurien de Libye.
R 1522 A.

Photo 8. — Algue monadoïde (G. × 100). *Tytthodiscus.* Toarcien de Lorraine.
20 338.

Photo 9. — Algue monadoïde (G. × 370). *Tasmanites.* Dévonien supérieur du Sahara.
R 13190 — 1489/4.

Photo 10. — Spores d'Algues (G. × 400). *Prasinophycées.* Dénovien de Libye.
R 1674 — 619/0A.

Photo 11. — Algue coloniale (G. × 32). *Botryococcus braunii.* Permien d'Écosse.
D 10.

Photo 12. — Fragment d'algue benthique (G. × 50), section de « tige » algaire. Silurien de Libye.
LM — 627/5 A.

Photo 13. — Fragment d'algue benthique (G. × 40), section de tissu algaire liptinisé.
R 1453.

(*Photos* : J. Bellet, A. Combaz, E. Dentand, A. Houard, J.F. Raynaud, J.L. Pittion)

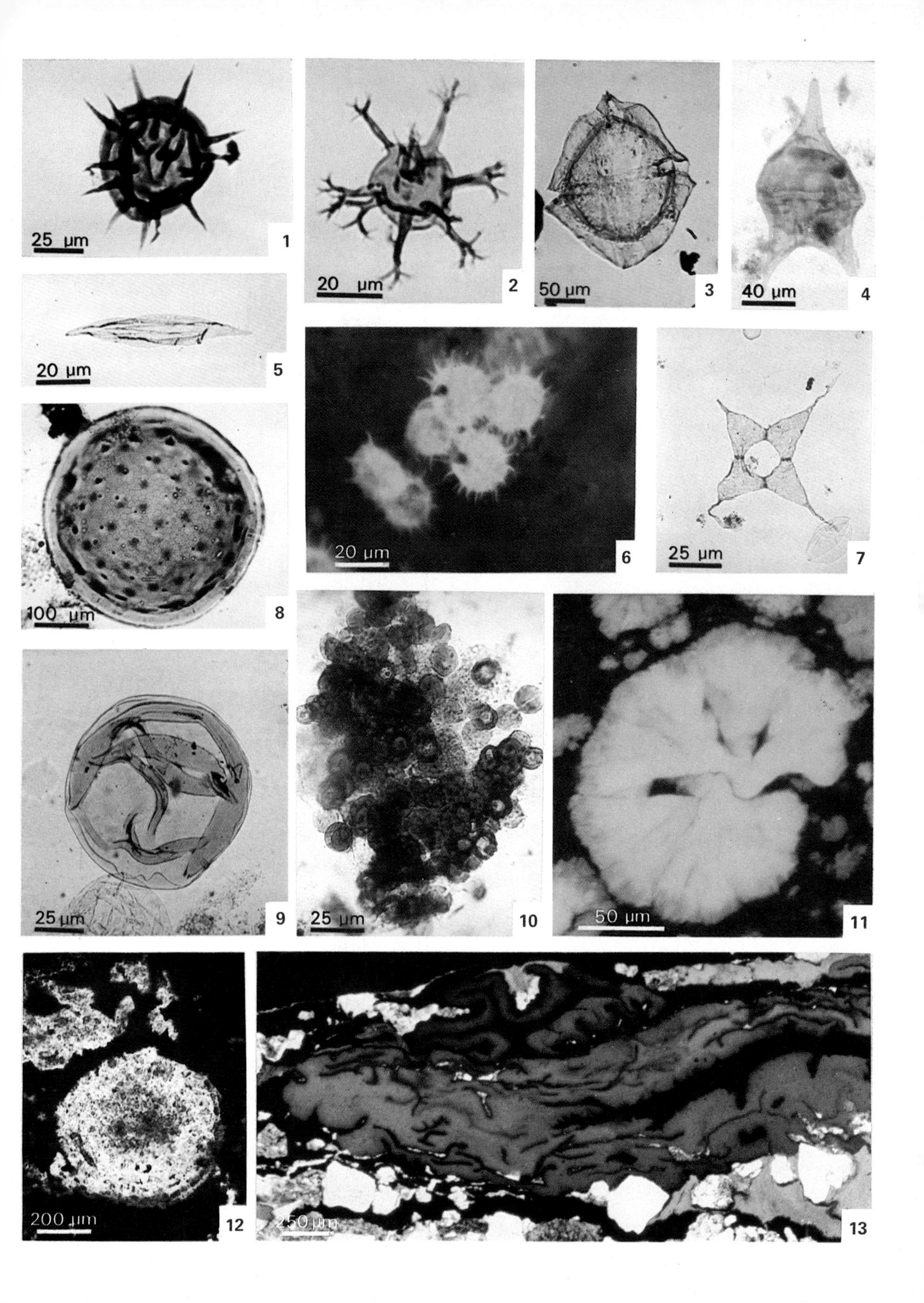
25 µm
1
20 µm
2
50 µm
3
40 µm
4
20 µm
5
20 µm
6
25 µm
7
100 µm
8
25 µm
9
25 µm
10
50 µm
11
200 µm
12
250 µm
13

PLANCHE 3.5
MICROFOSSILES ET FRAGMENTS ANIMAUX
(en lumière transmise)

Photo 1. — Sicule de Graptolithe (forme jeune) (G. × 260). Ordovicien Supérieur d'Australie.
R 9914 — 1489/9.

Photo 2. — Sicule de Graptolithe (G. × 90). Même provenance. La partie inférieure correspond à la photo précédente.
R 9914 — 1489/17.

Photo 3. — Fragment d'épiderme de *Gigantostracé* (G. × 90). Silurien du Sahara.
R 978.

Photo 4. — Fragment d'épiderme de *Gigantostracé* (G. × 370). Dévonien du Sahara.
R 17476.

Photo 5. — Scolécodonte (G. × 240). Dévonien Supérieur du Sahara.
R 3373 — 1508/13 A.

Photo 6. — Scolécodonte (G. × 90). Dévonien Supérieur du Sahara.
R 3373 — 1508/18 A.

Photo 7. — Scolécodonte (G. × 90). Dévonien Supérieur du Sahara.
R 3373 — 1508/12 A.

Photo 8. — Scolécodonte (G. × 370). Holocène de l'Atlantique Ouest.
KS4 — 2,2 — 1410/14.

Photo 9. — Microforaminifère (G. × 50). Maestrichtien du Sénégal.
R 7763 B.

Photo 10. — Microforaminifère (G. × 125). Paléocène de Libye.
GR 592 — AC 8 — 13.

Photo 11. — Microforaminifère (G. × 125). Maestrichtien du Sénégal.
R 7763 B.

Photo 12. — Chitinozoaire (G. × 200). *Desmochitina sp.*
R 2211 M — 478/5.

Photo 13. — Chitinozoaire (G. × 100). *Cyathochitina*. Ordovicien de Libye.
622/11.

Photo 14. — Chitinozoaire (G. × 200). *Urochitina sp.* Silurien Supérieur du Sahara.
R 3195 A — 472/2.

Photo 15. — Chitinozoaire (G. × 200). *Pterochitina sp.* Silurien de Libye.
R 1516 F — 478/12.

(*Photos* : J. Bellet, A. Combaz, P. Millepied)

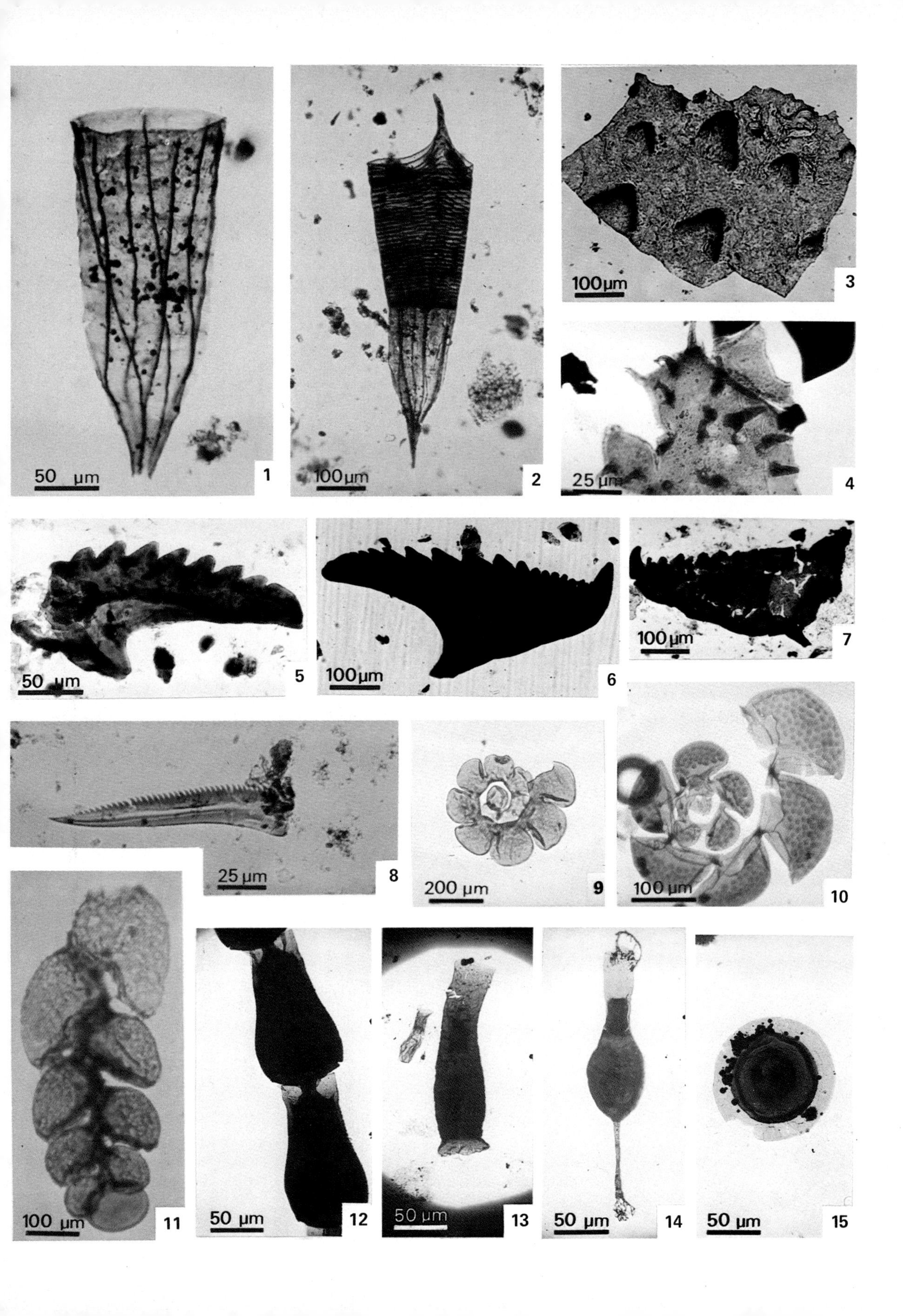
50 µm
1
100 µm
2
100 µm
3
25 µm
4
50 µm
5
100 µm
6
100 µm
7
25 µm
8
200 µm
9
100 µm
10
100 µm
11
50 µm
12
50 µm
13
50 µm
14
50 µm
15

PLANCHE 3.6

FRACTION AMORPHE

(en lumière transmise)

Photo 1. — Matière organique grumeleuse (G. × 200). Formation Mishrif (Iran).
2660/2.

Photo 2. — Matière organique grumeleuse (G. × 200). Santonien du Canada.
31405.

Photo 3. — Matière organique grumeleuse (G. × 200). Kimméridgien de Grande-Bretagne.
30 265.

Photo 4. — Matière organique granuleuse (G. × 1 400). Crétacé de Colombie.
GR 3651 — 1357/11.

Photo 5. — Matière organique granuleuse (G. × 925). Silurien du Sahara.
R 2535 — 1489/25.

Photo 6. — Matière organique spongieuse (G. × 250). Crétacé d'Australie.
R 10744.

Photo 7. — Matière organique spongieuse (G. × 270). Miocène de Kalimantan.
GR 7742 — 1509/17 A.

Photo 8. — Matière organique pelliculaire (G. × 250). Crétacé inférieur du Congo (on distingue les empreintes de cristaux de calcite).
26380.

Photo 9. — Matière organique sub-colloïdale (G. × 250). Crétacé de Syrte (Libye).
GR 548 — AC 8 17 A.

Photo 10. — Matière organique sub-colloïdale (G. × 240). Albo-cénamonien de Tunisie maritime.
GR 3737 — 1508/21 A.

Photo 11. — Matière organique gélifiée (vitrinite) (G. × 600). Dévonien moyen du Sahara.
GR 9651 — 1428/4.

Photo 12. — Matière organique gélifiée (vitrinite) (G. × 370). Callovo-oxfordien de mer du Nord.
GR 250 — 1253/9.

Photo 13. — Matière organique gélifiée (vitrinite) (G. × 180). Crétacé inférieur d'Angola maritime.
GR 2892 — 1508/15 A.

Photo 14. — Globules d'hydrocarbures précoces (G. × 500). Crétacé d'Angola.
10 966 B — AC 3/28.

(*Photos* : J. Bellet, A. Combaz, J.F Raynaud)

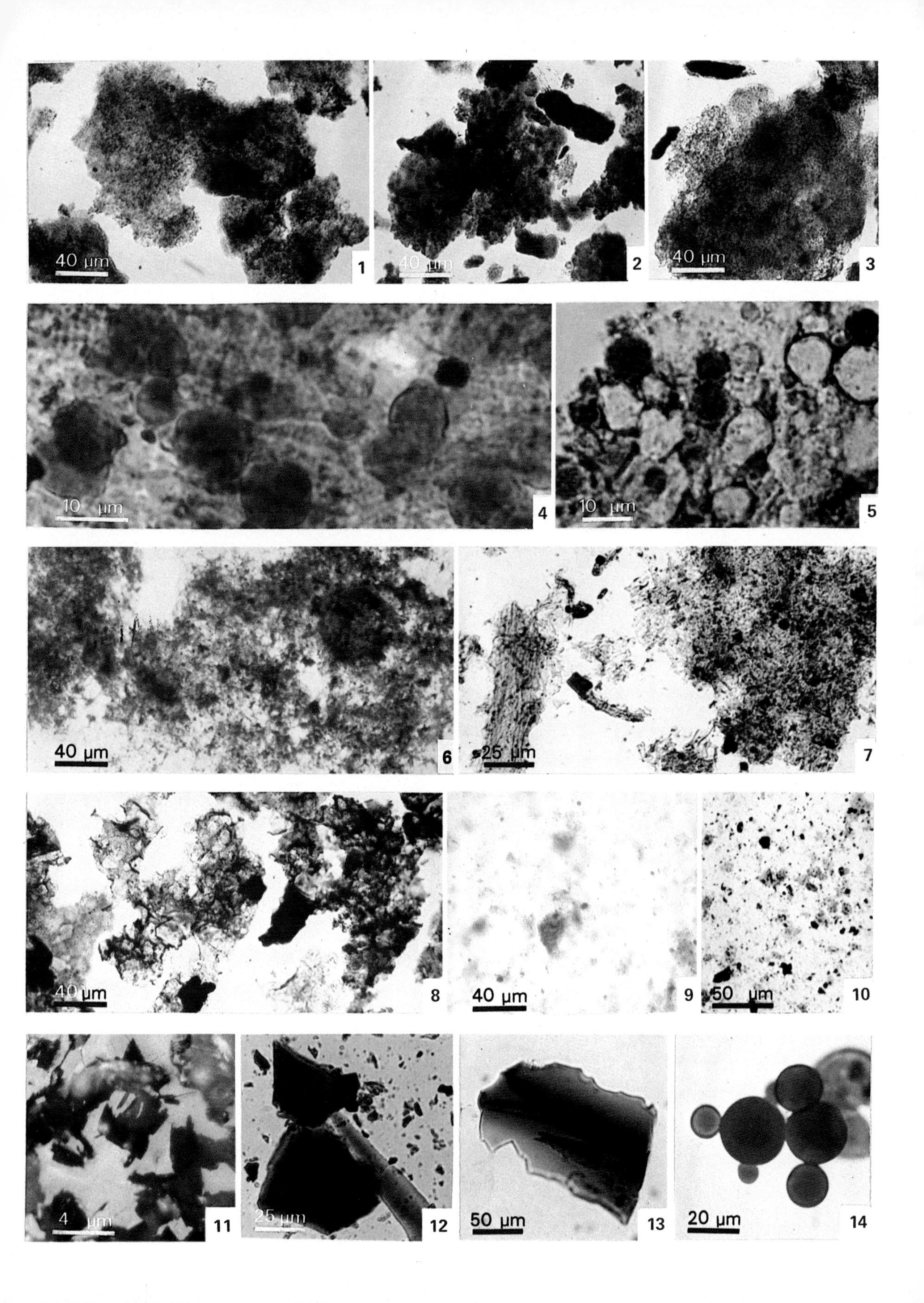

40 µm
1
40 µm
2
40 µm
3
10 µm
4
10 µm
5
40 µm
6
25 µm
7
40 µm
8
40 µm
9
50 µm
10
4 µm
11
25 µm
12
50 µm
13
20 µm
14

PLANCHE 3.7
FRACTION REMANIÉE ET ÉVOLUTION CATAGÉNÉTIQUE
(en lumière transmise)

Photo 1. — Palynofaciès d'un sédiment récent (G. × 370). Holocène de la mer de Norvège. KS 10.

a. Mélange de constituants contemporains de la sédimentation et d'éléments remaniés plus ou moins anciens. **b.** Pollens d'Angiospermes actuels et tertiaires. **c.** Pollens d'âge mésozoïque. **d et e.** Spores d'âge paléozoïque (Carbonifère supérieur). Ces microfossiles d'âges divers sont issus du même échantillon.

Photo 2. — Tétrades de *Classopollis* (pollens de Gymnosperme) à divers stades de la catagenèse (G. × 500). Jurassique de Madagascar.

a. Stade précoce antérieur à la zone de formation du pétrole. Pouvoir réflecteur (PR) $<$ 0,4 %. **b.** Stade correspondant à la zone à huile. PR $\simeq$ 0,7 %. **c.** Stade correspondant à la zone à gaz. PR = 1,3 %.

Photo 3. — Dinoflagellé *Phoberocysta neocomica* à deux stades de catagenèse (G. × 280).

a. PR V : 0,5 %; profondeur : 1 982 m; Valanginien de Hollande. **b.** PR V : 1,2 %; profondeur : 4 344 m; Valanginien de Norvège.

Photo 4. — Palynofaciès à Tasmanacées à deux stades de catagenèse (G. × 100). Silurien du Sahara.

a. Sahara oriental (zone à huile). **b.** Sahara occidental (zone à gaz).

Photo 5. — Le même faciès du Silurien saharien (G. × 60) en section mince. Ici, les *Tasmanites* sont non seulement noircies, mais craquelées (stade de production de méthane).

Photo 6. — Palynofaciès à matière organique grumeleuse à deux stades de catagenèse (G. × 100). Kimmeridgien de mer du Nord (argiles radioactives).

a. Faible catagenèse. **b.** Forte catagenèse.

Photo 7. — Stades pyrogénétiques des spores de *Lycopodium clavatum* (catagenèse artificielle). (G. × 900). Actuel.

a. État frais 1382/5. **b.** Jaune clair. PR 0,5-0,6. 1378/141. **c.** Jaune foncé. PR 0,7-0,9 (zone à huile). 1378/18 A. **d.** Brun. PR 1-1,3 (zone à gaz). 1378/24 A. **e.** Noir. PR 1,5-2 (zone à gaz). 1378/28 A.

(*Photos:* J. Bellet, A. Combaz, J.F. Raynaud, S. Jardiné)

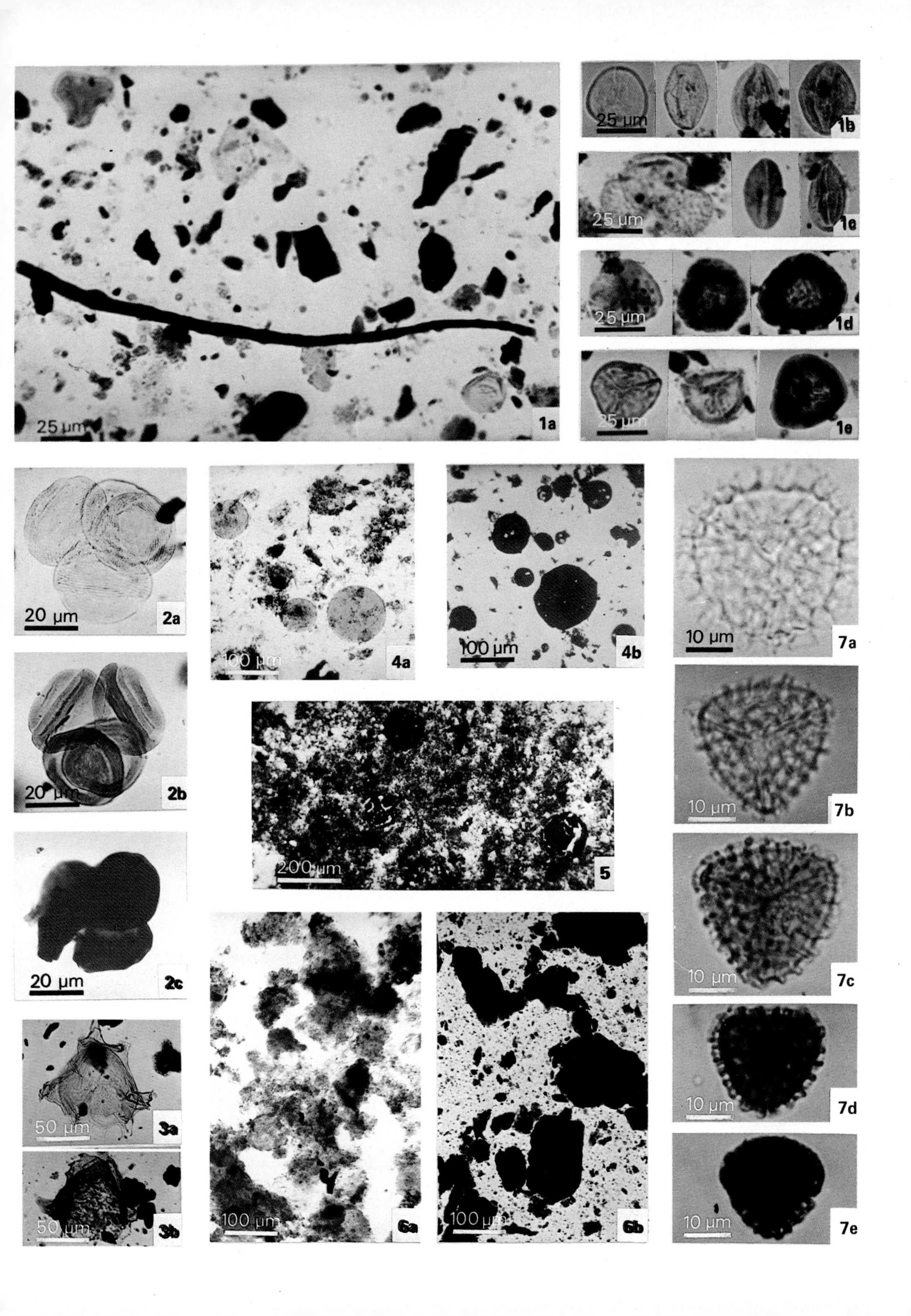
25 µm
1a
25 µm
1b
25 µm
1c
25 µm
1d
25 µm
1e
20 µm
2a
20 µm
2b
20 µm
2c
50 µm
3a
50 µm
3b
100 µm
4a
100 µm
4b
200 µm
5
100 µm
6a
100 µm
6b
10 µm
7a
10 µm
7b
10 µm
7c
10 µm
7d
10 µm
7e

PLANCHE 3.8
FRACTION MINÉRALE INCLUSE (I)
(en lumière transmise)

Photo 1. — Grappe de pyritosphères dans une *Leiosphaeridia* (G. × 500). Silurien de Libye.

Photo 2. — Inclusion pyriteuse (cubes et pyritosphères) dans une spore trilète. (G. × 500). Holocène de la plaine du Demerara (Atlantique ouest).
KS 4 — 2,8.

Photo 3. — Pyritosphère dans une colonie de *Botryococcus* (G. × 500).
965/57.

Photo 4. — *Deflandrastrum* (voir pl. 3.4, photo 7) entièrement envahi par des micro-inclusions pyriteuses (G. × 500). Silurien du Sahara.

Micrographies électroniques au MEB

Photo 5. — Pyritosphères dans un fragment de matière organique amorphe (G. × 2 000). Silurien du Sahara.
177.

Photo 6. — Pyritosphère dans un enrobage de matière organique amorphe (G. × 5 500). Silurien du Sahara.
258.

Photo 7. — « Nid » de pyritosphère déformant le bord d'une *Tasmanite* (G. × 5 500). Permien d'Australie.

Micrographie à l'électromicroscope

Photo 8. — Section et tégument de Chitinozoaire montrant à gauche l'ectoderme chitineux compact, à droite, l'endoderme : membrane en fibres entrelacées évoquant une association de protéines-collagène, et ponctuée de microcristaux de pyrite (G. × 32 000). Silurien de Libye.
R 1516.

Photo 9. — Microdiagramme Debye-Schaerer révélant la nature pyriteuse desdites inclusions dont la taille est très inférieure au micron (G. × 2 000).

(*Photos :* Cl. Caratini, A. Combaz, G. Deflandre)

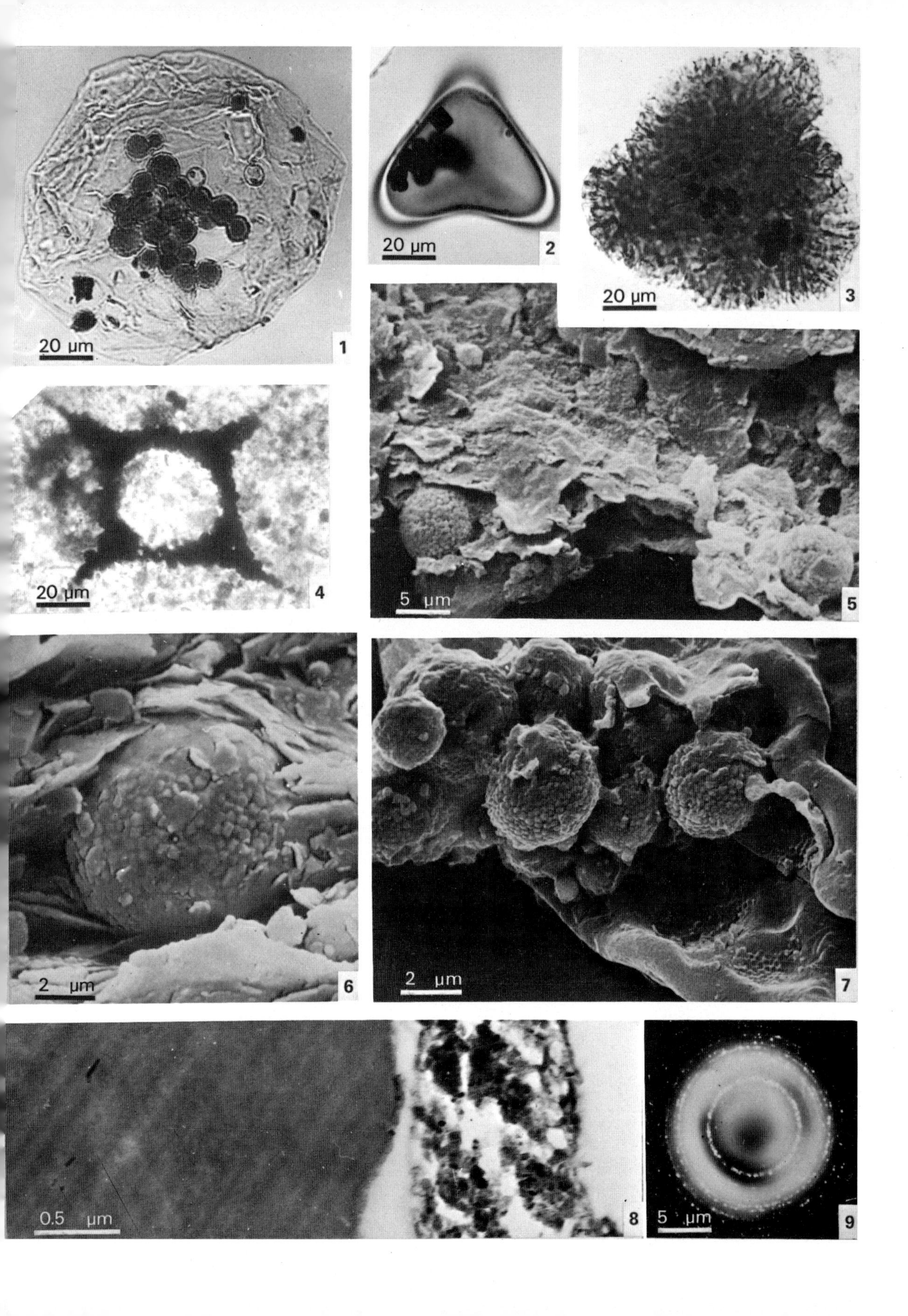
20 µm
1
20 µm
2
20 µm
3
20 µm
4
5 µm
5
2 µm
6
2 µm
7
0.5 µm
8
5 µm
9

PLANCHE 3.9
FRACTION MINÉRALE INCLUSE (II)

Photo 1. — Diagramme de diffraction X sur objets placés en tube capillaire de verre, montrant le spectre de la pyrite dans divers constituants organiques de kérogènes.

a. Dinoflagellés du Jurassique d'Australie. R 6008-106 G. **b.** Tasmanacées du Silurien du Sahara. R 2538. **c.** Chitinozoaires du Silurien de Libye. Exposition 12 heures. R 6081.

Photo 2. — Chitinozoaire (*Eremochitina*) serti de cristaux de pyrite ± distincts. R 1942.

a. Image normale au microscope électronique à balayage (G. × 330). **b.** Le même objet en balayage spectrométrique X (image Kα du carbone. Même G., 12 Kv, 0,006 μA). **c.** Le même objet en balayage spectrométrique X (image Kα du soufre).

Photo 3. — Particule carbonée incluse dans un quartzite (Ordovicien du Sahara). (G. × 330). GR 1451.

a. Section mince polie éclairage mixte. 1116/14. **b.** Image radiogénique du même objet : rayonnement α naturel.

(*Photos :* A. Combaz)

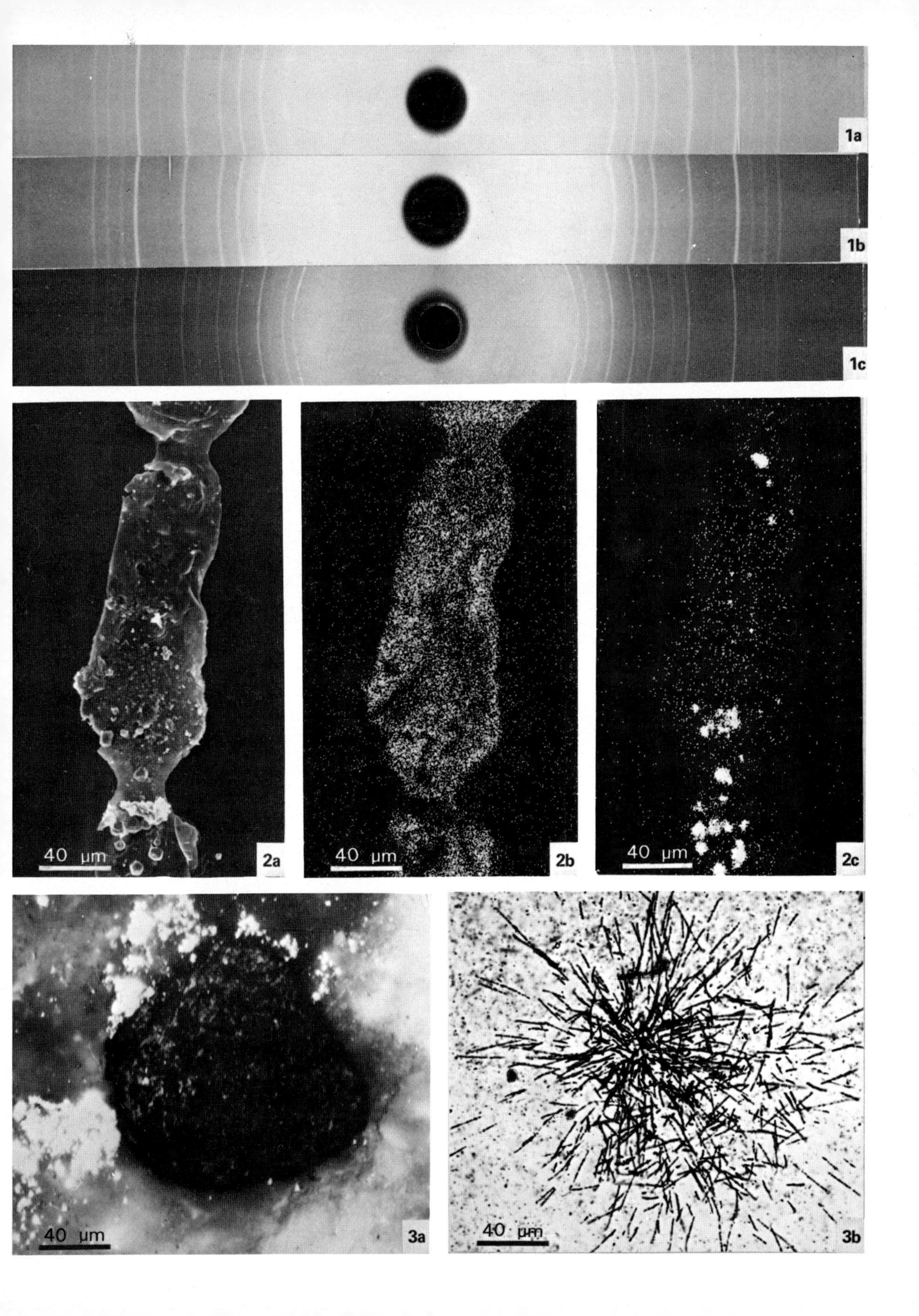
1a
1b
1c
40 µm
2a
40 µm
2b
40 µm
2c
40 µm
3a
40 µm
3b

PLANCHE 3.10
KÉROGÈNES
(en lumière transmise)

Lignée I : Kérogène algaire

Photo 1. — Crétacé de Libye (G. × 250).
GR 528 — AC 8 — 6 A.

Photo 2. — Turonien de Casamance maritime (G. × 250).

Photo 3. — Kimméridgien de mer de Norvège (G. × 160).
34 368.

Photo 4. — Kimméridgien de Grande-Bretagne (G. × 160).
30 265.

Lignée II : Kérogène liptinique

Photo 5. — A spores (Jurassique de Madagascar) (G. × 125).
33/3602 — 613 — 14.

Photo 6. — A spores (Dévonien supérieur du Sahara) (G × 90).
R 4341 — 1511/16.

Photo 7. — A pollens (Lias d'Aquitaine) (G. × 125).

Lignée III : Kérogène ligno-humique

Photo 8. — A débris végétaux (Lias d'Irlande). (G. × 160).
26/3602.

Photo 9. — A débris végétaux (Maestrichtien de Casamance maritime) (G. × 100).
R 8486 B — 1020/4.

Photo 10. — A débris végétaux (Lias de mer de Norvège) (G. × 160).
25 447.

(*Photos :* J. Bellet, A. Combaz, J.F. Raynaud)

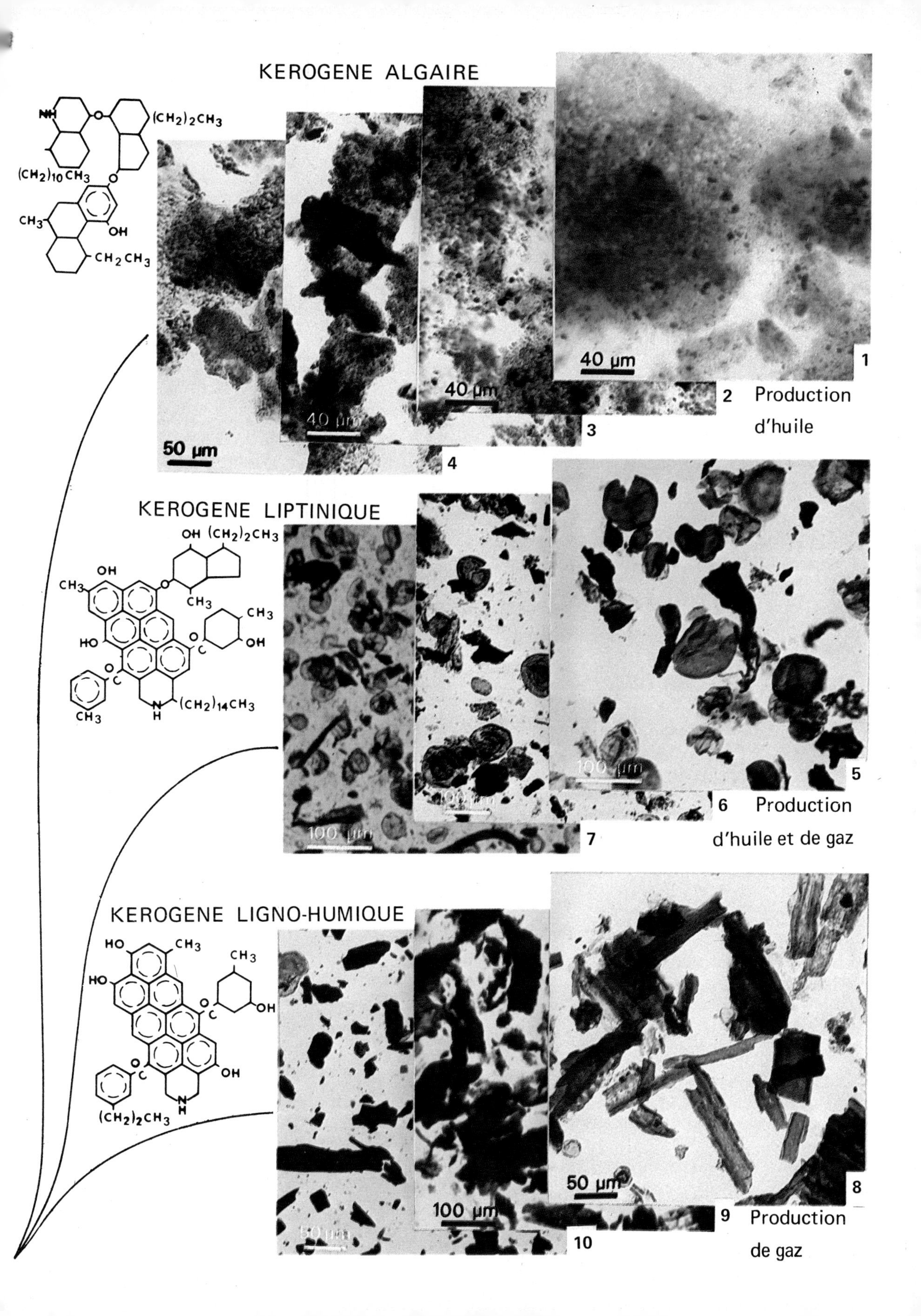
KEROGENE ALGAIRE
NH
(CH2)2CH3
(CH2)10CH3
CH3
OH
CH2CH3
40 µm
1
40 µm
2
Production
d'huile
40 µm
3
50 µm
4
KEROGENE LIPTINIQUE
OH (CH2)2CH3
OH
CH3
CH3
CH3
HO
OH
C
C
N
H
(CH2)14CH3
CH3
100 µm
5
100 µm
6
Production
100 µm
7
d'huile et de gaz
KEROGENE LIGNO-HUMIQUE
HO
CH3
CH3
HO
C
OH
OH
C
NH
H
(CH2)2CH3
50 µm
8
100 µm
9
Production
de gaz
50 µm
10

4

Elemental analysis of kerogens (C, H, O, N, S, Fe)

B. DURAND* and J.C. MONIN*

"You will know the nature of things when you know their origin and their evolution"

Heraclitus of Ephesus

I. INTRODUCTION

Researchers have long been interested in elemental analysis of sedimentary organic matter (OM). At first this kind of analysis was carried out only for coals and other forms of concentrated OM such as bogheads (algal coals) and asphalts. Few analyses were made of OM dispersed in sediments because it was hard to isolate it from the mineral phase. Forsman [12] reports that up to the 1960's about fifty analyses of the latter were available, mainly from Himus and coworkers [6, 18, 19] and from Forsman and Hunt [13]. Kerogen sensu stricto, as defined in Chapter 1, i.e. the fraction of sedimentary OM which is insoluble in organic solvents, was studied only by Forsman and Hunt [13] and only dispersed OM from ancient sediments was taken into consideration.

An increasing amount of work has since been carried out for elemental analysis of kerogens [5, 8, 9, 11, 14, 23, 24, 25, 26, 28, 29, 30, 31, 32, 36, 38, 45]. The work showed that this simple method is capable of establishing a classification of kerogens according to their ability to generate oil and gas in depth [9, 25, 36]. There was for a time no good microanalytical method for the measurement of oxygen content. Oxygen was calculated by difference after measurement of C, H, N and S, and the meaning of the "oxygen content" was often doubtful, especially for kerogens with high ash contents. Improvements in microanalytical methods for organic substances, derived from the Unterzaucher's method [41] led to the development of adequate measurements of the oxygen contents of kerogens.

*Institut Français du Pétrole. Rueil-Malmaison, France.

II. PROCEDURES

A. Current procedures of microanalysis of organic substances.

C, H, N are measured for a single aliquot part of the sample by oxidation at about 1 000-1 100° C under a stream of oxygen, or a mixture of oxygen and helium, producing CO_2, H_2O and nitrogen oxides. Nitrogen oxides are reduced into nitrogen, by passage over copper at 500° C, and excess oxygen is removed by fixation in the form of copper oxide. Detection is generally by catharometry. Separation of CO_2, H_2O and nitrogen can be carried out either by gas chromatography (GC) or by scrubbing in successive catharometers and selective filters.

If the nitrogen content is low (< 1-2%) accuracy for nitrogen measurement is poor. It may be advisable to determine C and H in a single run in accordance with the description above and to determine nitrogen in a separate run. A bigger aliquot part of the sample is then taken, CO_2 and H_2O are scrubbed before reduction on copper, and GC is used to separate nitrogen from other gases which may be released during the process.

O is measured by methods derived from Unterzaucher's method [41]. The common procedure is to pyrolyse under an inert atmosphere an aliquot part of the sample at about 1 000 °C, then to pass the freed gases through pure carbon at 1 100-1 500° C (some devices use platinized or nickeled carbon in order to lower this temperature to about 900-1 000° C), then over copper at 900° C. The whole process transforms the oxygen contained in all oxygenated molecules into CO and removes sulphur which remains fixed on copper. CO can be measured directly, either by gas chromatographic separation from other gases such as CH_4, H_2, etc. then catharometry, or by infrared (IR) detection. It can also be oxidized into CO_2 on copper oxide (300-600° C); CO_2 is measured either by coulometry or gas chromatographic separation then catharometry. For very low oxygen content (< 0.1%) transformation into CO and IR detection seems to be the most effective.

S is oxidized into SO_2 at 1 300° C in an induction furnace. The quantity of SO_2 is commonly measured by coulometry. There are two possible methods: oxidation of SO_2 in an aqueous medium into H_2SO_4, or reduction of iodine by SO_2.

B. Aspects of kerogen analysis.

The general methods must be adapted to the special case of kerogen mainly because of the four following points :

(a) Kerogens may contain water.
(b) They are heterogeneous.
(c) Their transformation into gaseous products through combustion or pyrolysis may be incomplete.
(d) They contain residual minerals, mainly pyrite.

1. Water content.

Drying kerogen before analysis is necessary since its water content modifies the H and O contents. This can easily be performed without degradation by keeping the samples at 100° C for one hour in a vacuum. At *Institut Français du Pétrole (IFP)* an appropriate device was built for sealing the sample in an argon atmosphere after drying so as to give the analyst a sample ready for analysis.

2. Heterogeneity.

Heterogeneity is partly due to the presence in kerogens of different "macerals" or microfossils with different elemental analyses. The size of these constituents may be relatively large (a spore has a 10 μ scale). It is also partly due to the mineral content which occurs in a discrete form (for instance microcrystals of pyrite; see Chapters 2 and 3). Careful crushing and homogenization is therefore advisable before analysis, but a good homogeneity cannot be assumed in all cases. The use of very small aliquot parts for analysis may thus lead to poor representativity.

3. Incomplete reactions.

Analytical procedures should be checked on relatively complex substances and not with the usual standard. Substances which are rich in fused aromatic rings are preferable. It seems likely in fact that incomplete reactions occur when the kerogen is rich in such highly condensed structures. However such standards when very pure are expensive. E. Whitehead (personal communication) suggests that hexabenzocoronene be used as a standard for controlling the quality of C and H analysis. A pragmatic, but less expensive way, is to use a series of very poor ash coals, from peat to anthracites, as standards. These are carefully crushed, homogenized and kept in small containers away from light and moisture. They are available in large quantities and their analysis is known from several measurements at the limits of the method (for instance using the highest temperatures and the longest combustion or pyrolysis time).

4. Mineral content.

Mineral content is responsible for sample heterogeneity and also interferes with the organic material. Some minerals are taken into account by the analysis either because they are unstable under the conditions required for kerogen analysis or because OM catalyzes their destruction under these conditions. Products of their destruction may also alter the silica walls of the reactors which causes a silica contribution to the analysis.

As was seen in Chapter 2, minerals which are not destroyed or only to a slight extent during isolation of kerogen, even if it is performed with care, are pyrite, marcasite (FeS_2) and the so-called heavy minerals (oxides, silicates or sulphates). Isolation of kerogen also frequently results in a neoformation of fluorides.

a. Pyrite and marcasite.

Pyrite or marcasite contributes to S measurements but also to C, H and O measurements (Table 4.1). This may be surprising, but the close association between pyrite and OM must be

TABLE 4.1
ELEMENTAL ANALYSIS OF TYPICAL RESIDUAL MINERALS OF KEROGENS
(Weight %)

Sample		C	H	O	S	N	Fe
Baryte No 1	a			1.12 4.34	2.11 1.48		
	b			16.56 17.70	0.32		
	c			27.09	1.67 0.91		
Baryte No 2	d			17.76	Not done		
Pyrite No 1 (sedimentary)		1.17	0.56	1.10	52.04 52.09	-	45.76 45.06
Pyrite No 2 (ore)		0.17	0.10	0.21	52.39 52.43	-	43.51 43.76
Pyrite No 3 (ore)		0.45	0.12	0.28	52.35 52.33	-	43.71 43.37
Ralstonite			1.18	5.13	0.65		

Duration of analysis: a = 6 min; b = 9 min; c = 11 1/2 min; d = 20 min.

TABLE 4.2
COMPARISON BETWEEN
1. ORGANIC S CALCULATED FROM Fe-MEASUREMENTS AND
2. S MEASUREMENTS AFTER AN HNO_3 TREATMENT

Bold face samples did not contain any pyrite.

Sample IFP No	1. Before HNO_3 treatment	2. After HNO_3 treatment
15760	0.8	1.1
15409	2.5	1.2
13356	0	1.0
15288	4.83	2.8
10696	4.1	3.3
15789	3.6	1.3

kept in mind. In addition, finaly divided pyrite may superficially oxidize in water or under a wet atmosphere during isolation of kerogens into hydrated iron sulphates or oxides. Iron may also catalyze the reduction of the silica walls of the reactor, thus producing an excess of oxygen. This can be avoided by putting the sample in a small platinum or silver crucible.

Iron from pyrite may also form iron oxides, with oxygen from the OM, which are not entirely reduced at the temperature of analysis, producing in this case a lack of oxygen. A special device for oxygen measurements for kerogen is therefore advisable. The general principle is as described above (paragr. II.A), but pure carbon is added to the sample. In this way iron oxides, when formed, are reduced by carbon. A more complete pyrolysis reaction is also obtained, and the whole oxygen content of OM is recovered in gases released by pyrolysis.

Presence of pyrite and/or marcasite also requires Fe measurement, because a distinction must be made between S from FeS_2 and S from OM. A functional sulphur analysis could be carried out in order to distinguish sulphides from other forms of sulphur. Such a method would be material and time consuming and in any case would be inaccurate.

Therefore, the best, although imprecise, method is to deduce from total S the quantity of mineral sulphur linked to pyrite as calculated from the FeS_2 formula, i.e. by multiplying the Fe content by 64/56. Fe analysis is generally performed by atomic absorption. Powell [14] measures the organic sulphur content of kerogen after destruction of pyrite by HNO_3. Table 4.2 shows a comparison between the values found by this method and by calculation from Fe measurements on some samples prepared at *IFP*. There is no correlation. Moreover the S content of two samples which contain nearly no pyrite is greatly modified by the HNO_3 attack. It should be recalled that modifications of kerogens due to the use of HNO_3 is very great (Chapter 2) and that there is no reason for organic sulphur not to be oxidized by HNO_3. Thus it is likely that Powell's method is not reliable.

b. Other minerals.

Heavy oxides and silicates are very stable and very probably do not contribute to organic analysis under the conditions described above.

Sulphates which are not destroyed or only slightly destroyed are mainly barites. Elemental analysis performed on pure barites under the same conditions as for kerogen shows (Table 4.1) that they contribute to oxygen and sulphur analysis. Moreover measurements are not very reproducible and depend on the duration of analysis.

Neoformed fluorides contribute to O and H measurements for they contain water of constitution. Table 4.1 shows for example an analysis of a sample of ralstonite: ($Na_x\ Mg_x\ Al_{2-x}$ $(F,OH)_6 \cdot H_2O$) containing a little thomsenolite ($NaCaAlF_6 \cdot H_2O$). They may also, because of HF or F release, attack the silica walls of the reactor.

C. Discrepancies between laboratories.

Figure 4.1a shows the correlation between C measurements and Fig. 4.1b the correlation between O measurements made on the same samples by two laboratories. C correlation is good, despite a slight systematic difference. O correlation is poor and the regression line is clearly distinct from the bisecting line. These examples demonstrate the existence of systematic errors in element analysis of kerogen, which might be great, as a result of analytical procedure. In this particular case, the difference in O values mainly comes from the methods which were used, i.e. conventional for analysis No 2, and with the addition of pure carbon to the sample in the case of analysis No 1.

Attention must also be given to the reproduciblity of measurements, and this varies a great deal depending on the laboratory.

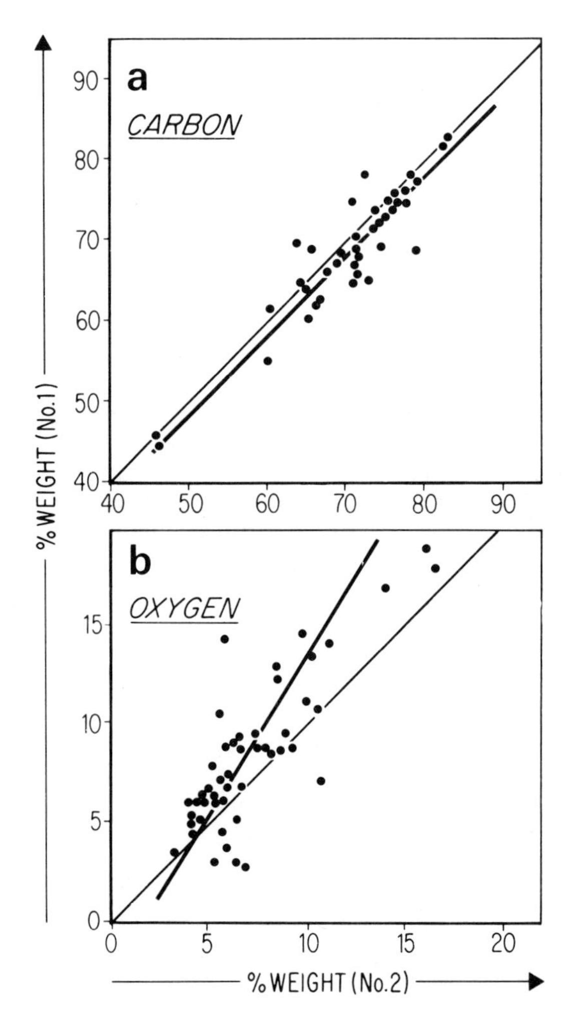

Fig. 4.1. — Correlation between the measurements made by two different analysts (No 1, No 2):

a. carbon. **b.** oxygen.

It is difficult to use the values found in the literature for comparative work done by different laboratories. It should also be recalled that authors continue to give O contents by difference despite the inadequate meaning of this calculation.

III. PROCESSING OF ELEMENTAL ANALYSIS DATA

A. Statistical results for C, H, O, N, S, Fe and mineral contents of kerogens.

We present here results ([1]) for 440 kerogen samples isolated at *IFP* by the isolation procedure described in Chapter 2 of this book.

1. Distribution of errors.

It is worthwhile to know the reliability of the measurements. The first point is the possible existence of systematic errors which can be corrected by the use of appropriate procedures and/or standards as seen above. Another point is the importance of intrinsic dispersion.

Table 4.2 shows the distribution of relative deviations for C, H, O, S, N and Fe measurements calculated as follows. Measurements were made twice on a set of randomly selected samples. The difference between the two measurements was divided by the mean value and the result expressed in percentage. It is this kind of result which is called relative deviation here.

TABLE 4.3
RELATIVE DEVIATIONS FOR ELEMENT CONTENTS AND ATOMIC RATIOS

	Mean (%)	Dispersion (%)
C	1.25	0.1 à 6.5
H	2.65	0.1 à 8.5
O	4.20	0.3 à 26
S	4.60	0.2 à 20
N	10.10	0.3 à 40
Fe	3.50	0.1 à 18
H/C	1.65	0.1 à 10
O/C	3.5	0.2 à 14

These distributions are roughly lognormal. Table 4.3 gives the mean and the maximum deviation for each element and also for the H/C and O/C ratios. Among organic elements, the best results are fortunately obtained for elements which are the most representative, i.e. C, then H, then O. The mean deviation for H/C is lower than for H. The same is true for O/C and O. This means that there is a correlation between errors. An interpretation for H/C is that H and C are measured in a single run. We have no interpretation for O/C. A lower limit exists at which analysis is no longer reliable. According to the analyst, values of element contents lower than 0.30% can be considered as having no significance.

([1]) Analyses were made by *ATX,* 24 rue Fernand-Forest, 92150 Suresnes, France.

2. *Distribution of the mineral content.*

As was said above, kerogens often contain residual minerals whose presence may alter the measurements for the organic element contents. Before commenting on this possible interference, it is worthwhile to look at the distribution of this mineral content. We shall consider separately pyrite and "other minerals". The quantity of pyrite is calculated according to the Fe content by multiplying its value by 120/56. The quantity of other minerals is calculated by taking the difference 100 − (C + H + O + N + S + Fe). This last value is of course approximate and includes other elements which may be present in kerogens, such as phosphorus, or elements added to kerogen during isolation, such as Cl of F (Chapter 2). It is also inaccurate because of the mode of calculation.

Distributions of minerals are neither normal nor lognormal:

(a) Distribution of pyrite shows (Fig. 4.2a) values up to about 70% weight; the mean value is about 18%.

(b) Distribution of "other minerals" shows (Fig. 4.2b) values up to about 30%. The mean value is about 5% weight. Negative values are found here owing to the mode of calculation. This distribution is greatly dependent on our isolation procedure. It is likely that an improvement in this procedure would partly reduce the amount of these "other minerals".

We expect to eliminate almost completely the neoformed fluorides by conducting HCl treatments at 90° C instead of 70° C (Chapter 2). This can be done by using polyethyleneglycol instead of water in the heating jackets of the device described by Fig. 2.8 in Chapter 2 of this book.

Since the presence of pyrite, barites and neoformed fluorides may lead to uncertainties in the elemental analysis of kerogens, it is worthwhile to know what amount of minerals is tolerable. We never succeeded in completely clearing up this point. A statistical treatment showed no evidence of a systematic deviation in the element contents as a function of the quantities of pyrite or of other minerals. However, in the case of some particular series for which we had a good idea of what the real elemental analyses of kerogens should have been, it appeared clearly that the presence of a large quantity of residual minerals systematically increased the H and O values.

In general, a rule of thumb was defined for kerogens isolated at *IFP*, i.e. elemental analysis is not reliable for kerogens with pyrite contents over 40% weight and/or "other mineral" contents over 15%. Moreover such kerogens are not very suitable for further analysis such as IR spectroscopy.

3. *Distribution of C, H, O, N, S, in kerogens with pyrite content under 40% and other mineral content under 15%.*

These distributions are studied for values corrected for the mineral content. The correction is made as follows. The measurements on C, H, O, N, and of organic S calculated as above (paragr. II.B.4) are summated. The value for each element is then divided by this sum, thus assuming that kerogens consist only of C, H, O, N, and S.

Distributions of these corrected values (weight %) are given for each element in Figs. 4.2c, d, e, f and g). Mean values and dispersions are also given.

The distributions of C and N are approximately normal. Those of O and S are approximately lognormal. Their examination brings out that C, H and O are by far the main components of kerogens. Presentation of weight percentages masks the importance of hydrogen which is in fact the second element by order of contribution to the molecular assemblages of kerogen.

An important fact is the existence of three modes in the distribution of hydrogen. We shall see later on that this is due to the frequency of certain types of kerogens.

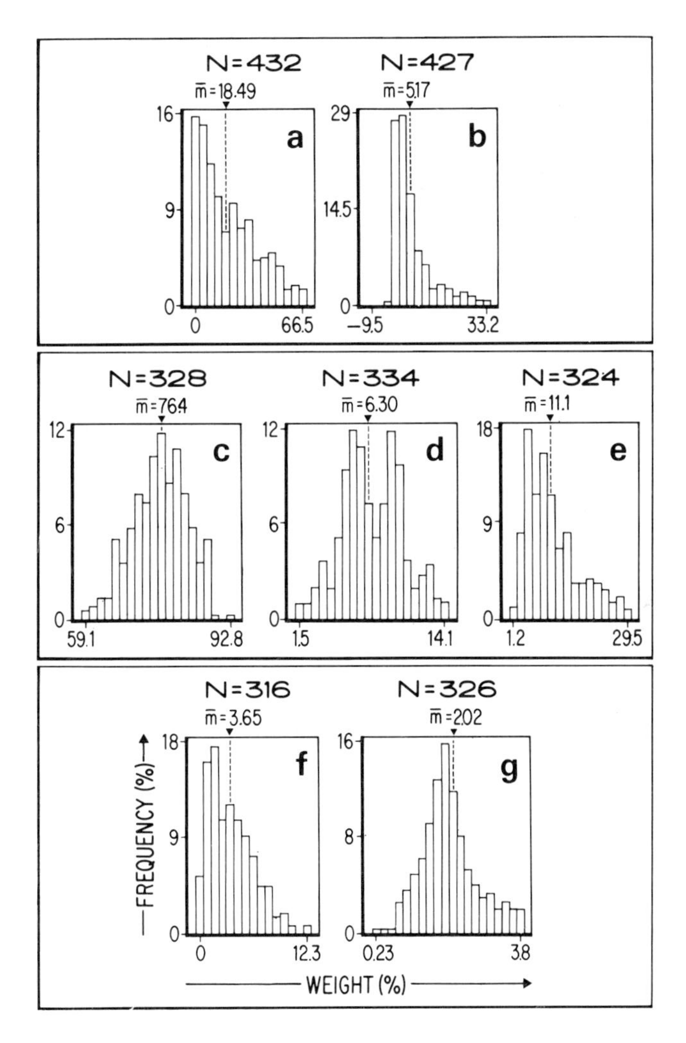

Fig. 4.2. — Distribution of mineral contents and element contents (mineral-free), arithmetic classes:

a. pyrite. **b.** other minerals. **c.** carbon. **d.** hydrogen. **e.** oxygen. **f.** organic sulphur. **g.** nitrogen.

The quantity of organic sulphur is very high, as compared to its quantity in living organisms. That means that there is a process of incorporation of inorganic sulphur in kerogen during its formation. Its distribution is bimodal. The meaning of this observation is not known at the present time.

The quantities of nitrogen are low but not negligible. The only source of nitrogen in living organisms is proteins. Since kerogens are isolated from sediments by hydrolyzing procedures, it is clear that a part of the proteinic nitrogen is incorporated in kerogens in a non-hydrolyzable form.

B. Use of C, H, O diagrams.

Since C, H and O make up most of the kerogen, the diagrams which have been proposed for the interpretation of the elemental analysis are mainly variations in combinations of these elements. The use of only two of these elements may be misleading. This in fact makes sense only for corrected values for the ash content. Because of the usually high ash contents of kerogens there may be a great deal of uncertainty as to these values. C, H and O diagrams are therefore preferable. They are of two kinds:

(a) The first kind consists in variations in a triangular diagram, as first proposed for coals by Grout [16] and developed by Ralston [27], Himus and coworkers [18, 19], Forsman *et al.* [13] and McIver [25]. C, H and O values are standardized to 100%, either weight % or atom %.

(b) The second kind was proposed by Van Krevelen [43]. It consists of an H/C vs. O/C (atomic ratios) rectangular diagram.

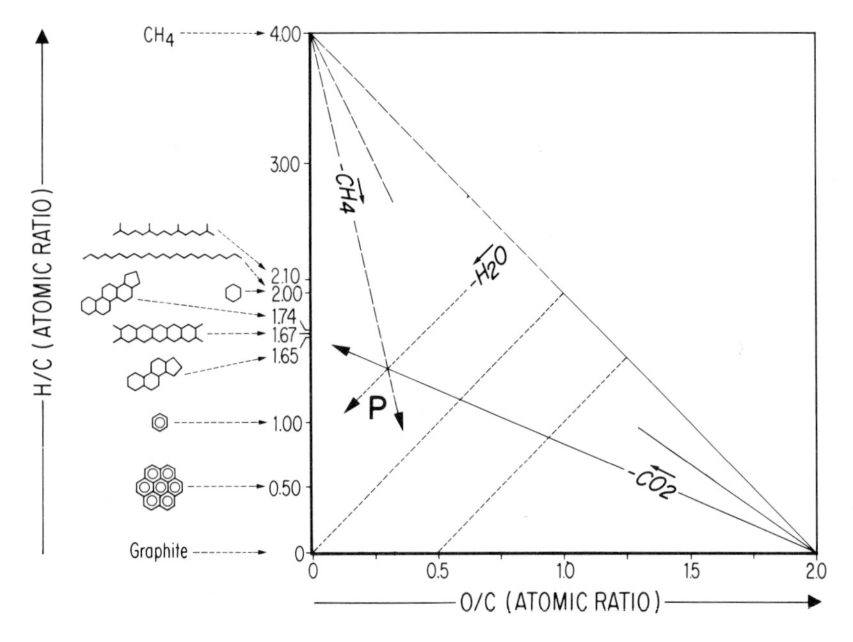

Fig. 4.3. — Van Krevelen diagram.

In these diagrams, simple reactions such as a loss of CH_4, CO_2 or H_2O are represented by straight lines joining the point representing the elemental analysis of the sample under examination to the points representing respectively the CH_4, H_2O and CO_2 compositions. In the case of the Van Krevelen diagram, the loss of H_2O is represented by parallel lines with a slope of 2 (Fig. 4.3).

In our opinion, the Van Krevelen diagram is the most suitable for processing elemental analysis:

(a) Firstly it requires no preliminary standardization of the C, H, O values.

(b) Secondly we have seen that the uncertainties as to H/C and O/C were lower than those for C, H, O values.

(c) Thirdly, as stated by Van Krevelen, it enables a rough idea to be formed of the structure of the carbon skeleton. The H/C value in fact gives a first idea of its aromaticity.

IV. INTERPRETATION OF C, H, O ANALYSES OF KEROGENS IN C, H, O DIAGRAM

When interpreting elemental analyses for kerogens, it should be recalled that they are a residue from an extraction by an organic solvent.

At the same time they are in practice obtained from the destruction of the mineral phase by hydrolysis, at least in the case of rocks which are not almost entirely made up of OM such as humic and boghead coals, asphaltite, etc.

There is little difference between elemental analyses before and after extraction since kerogen makes up most of the OM. It was shown [8,11] that extraction increases O/C and reduces H/C. This indicates that, as was to be expected, the extract was richer in H and poorer in O than the OM as a whole.

The interpretation of a shift in the diagram is therefore roughly the same whether the analyses are for unextracted OM, as is usual for coal, or for extracted material. In the first case, however, the shift is due to the formation of compounds with low molecular weights which are mobile under the given geological (or analytical) conditions, while in the second case, the formation of less mobile solvent-extracted compounds, especially oil, is taken into account.

Elemental analyses before and after hydrolysis do not differ greatly, expect in the case of recently sedimented OM (Chapter 2). The "kerogen" which is analyzed in this case corresponds hardly at all to he initial concept (insoluble fraction in organic solvents). This concept is therefore not well adapted to the recent sedimentary stage.

A. C, H, O analysis of coals as a guide for kerogens studies.

Referring to coal analyses in his diagram, Ralston [27] was the first to point out following facts which are now generally recognized :

(a) The representative points for humic coals (rich in vitrinite) are in a narrow band (Fig. 4.4).

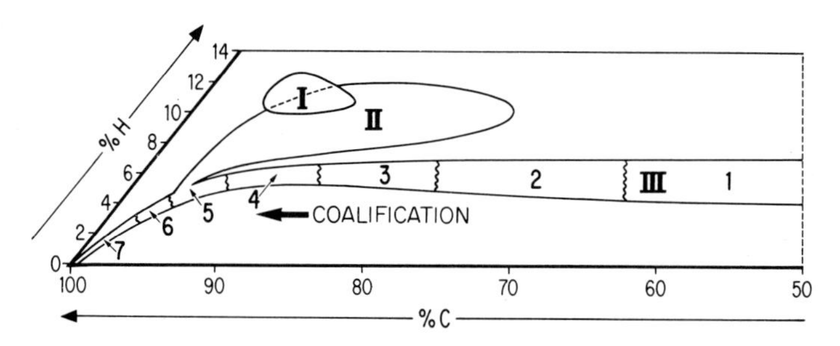

Fig. 4.4. — Location of:

I. *Botryococcus* rocks, **II.** Cannel coals, **III.** Humic coals, in the Ralston diagram.
1, 2, 3, 4, 5, 6, 7: Succession of the peat, lignite, sub-bituminous, bituminous, semibituminous, semianthracite and anthracite stages in the coal band. From Ralston [27].

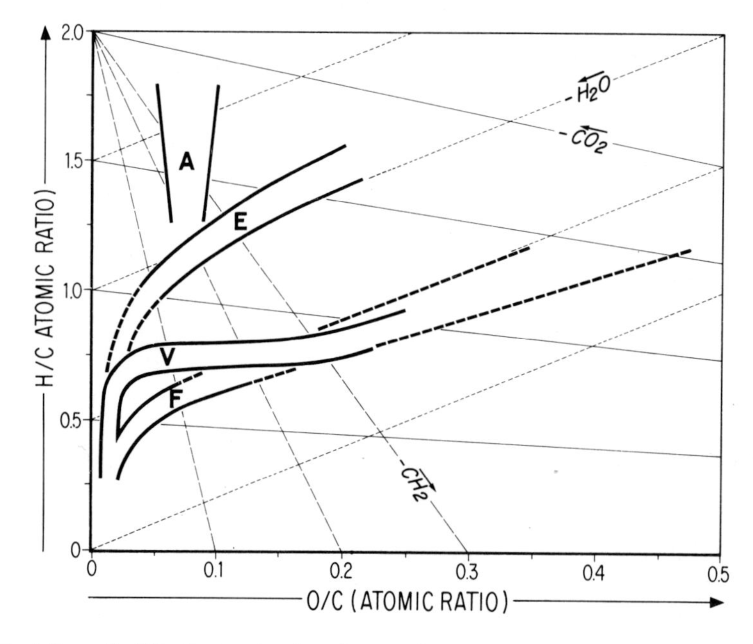

Fig. 4.5. — Coalification tracks of coal macerals:

A. alginite. **E.** exinite. **V.** vitrinite. **F.** fusinite. From Van Krevelen [43].

(b) Coals of various ranks, i.e. corresponding to increasing degrees of coalification due to burial (peat, lignite, bituminous coals, etc.) are located in successive coal-band regions, from a region with high oxygen content to a region of low oxygen content, with little overlapping. Coalification therefore results in a relative loss of oxygen and hydrogen, i.e. to carbon enrichment.

(c) Cannel coals (rich in spores) are found in a region outside of the coal band corresponding to higher hydrogen content. This region overlaps with the coal band in the semibituminous coal region.

(d) Many coal properties, such as heating power, volatile matter content and coking properties, are classified in terms of the position in the C, H, O diagram.

Similar statements were made by White [48] and Hickling [17]. The former suggested that there are three distinct maturing bands for humic coals, cannel coals and bogheads (*Botryococcus*-family algae coals).

These observations were further developed by Van Krevelen [43] who described what he called coalification tracks for coal macerals. Van Krevelen plotted elemental analyses for macerals in his diagram and showed (Fig. 4.5) that they are located in separate bands. During coalification there is also a progresssive shift along each coalification track from high H/C or O/C values towards low H/C and O/C values, i.e. towards carbon enrichment. The various coalification tracks thus merge in an undifferentiated zone near the origin of the axes.

The alginite coalification track corresponds to H/C values which are at first high and to O/C values which are low at first. On the other hand the fusinite coalification track corresponds to H/C values which are initially low and O/C values which are initially high (Fig. 4.5).

The existence of coalification bands or coalification tracks means that after formation at the humification stage the coals and macerals constituting them undergo chemical changes through preferential oxygen and hydrogen losses, i.e. carbon enrichment. The O and H probably leave either in the form of small molecules which move easily under subsurface conditions or in volatile form during operations, i.e. CO_2, H_2O and light hydrocarbons such as methane. The shape of the coalification tracks or bands suggests that the CO_2 and H_2O leave before the hydrocarbons. The O/C value falls at first more rapidly than H/C, and this means that the light oxidized molecules are formed first.

B. Evolution paths and types on the basis of C, H, O analysis.

Figure 4.6 shows C, H and O analyses of 390 kerogens isolated at *IFP* from sediments of various origins and plotted on the Van Krevelen diagram. What can be called the principal domain for kerogens can be delimited. Points are seen to be more frequent in three zones, and this corresponds to the trimodal distribution of hydrogen which was noted above (Fig. 4.2d). This Fig. cannot however be interpreted by itself.

Figure 4.7 shows a selection of analyses of Figure 4.6 made on the following bases:

(a) Series of kerogens isolated from sediments from the same deposit medium and which contained OM of the same kind at the time of deposit are considered. The latter point is proven by the various geochemical techniques used.

(b) The various samples in each series correspond to different burial values.

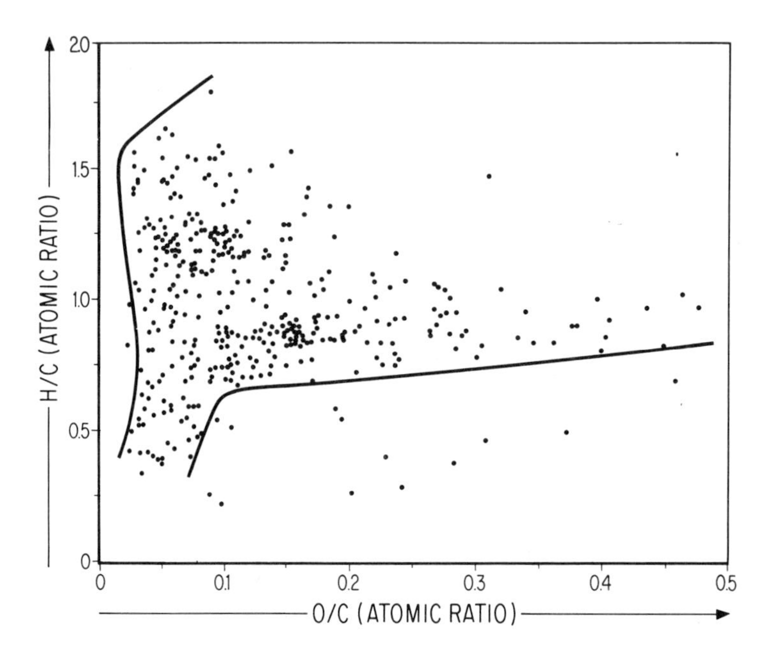

Fig. 4.6. — Principal domain for kerogens in the Van Krevelen diagram.

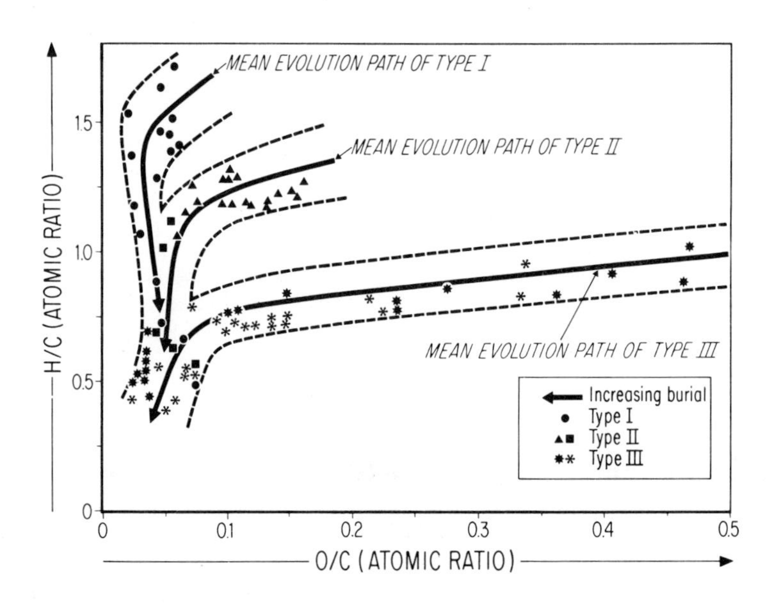

Fig. 4.7. — Position in the Van Krevelen diagram of the elemental analysis of selected series of kerogens:

Type **I:** ● Green River Shales, Uinta Basin, USA [38].

Type **II:** ▲ Lower Toarcian Shales of the Paris Basin, France [8].
■ Posidonienschiefer (Durand, unpublished).

Type **III:** ✱ Logbaba wells, Upper Cretaceous, Douala Basin, Cameroon [10].
* Humic coals [11].

Elemental analyses for these series fall into distinct bands as was the case for the various types of coals or coal macerals.

Burial is indicated by a shift from high H/C or O/C values to low values. There is thus carbon enrichment and convergence of the various series in an undifferentiated zone. The same facts have been observed with other series constituted in the same way.

Study of kerogen therefore confirms and generalizes observations for coal. The evolution resulting from burial leads to carbon enrichment for the solid phase of sedimentary OM by the formation of products which are richer in oxygen and hydrogen than the original OM.

Another type of shift can be observed when a comparison is made of the succession of kerogens isolated from initially identical OM which have been subjected to weathering. Thus Fig. 4.8 shows the shift due to outcrop weathering of OM of the Lower Toarcian in the Paris Basin. Weathering increases towards the surface. There is a considerable increase in O/C and a slight fall in H/C, i.e. oxidation.

We will call the shift of representative points for kerogens from the same kind of initial organic matter the evolution path. This corresponds to evolution under specific geological conditions i.e. burial, pedological weathering...

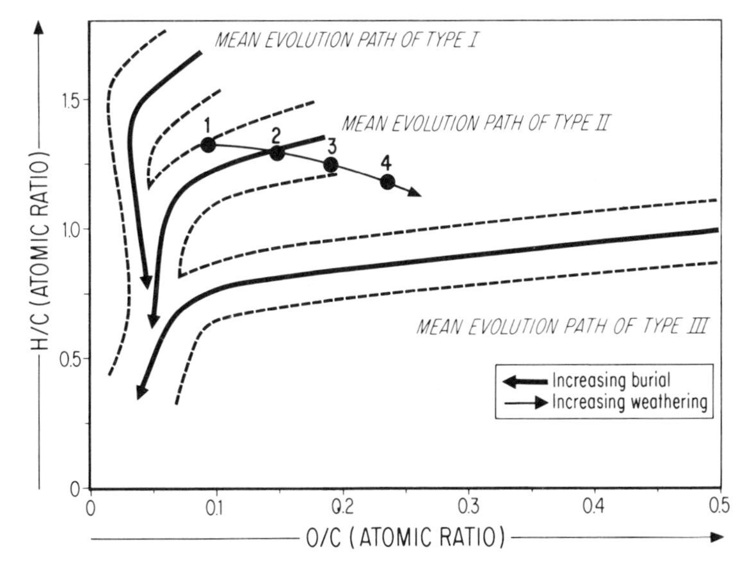

Fig. 4.8. — Displacement in the Van Krevelen diagram due to increasing weathering of a type II kerogen.

Distance from the surface: **1.** 5.3 m. **2.** 4.8 m. **3.** 1.0 m. **4.** 0.5 m.

This concept can be generalized as being the succession of physicochemical states of OM derived from the same initial OM by evolution under geological conditions. The elemental analysis of kerogens is only an aspect of these states.

It will be necessary to reconsider the concept of the nature of the OM at the time of deposit in order to define what will be called a type of OM. The nature of the OM can be evaluated only by means of existing physicochemical or petrographic techniques. Two samples have the same nature if they cannot be distinguished by the group of existing techniques. The variability of natural media and the power of resolution of analytic techniques are in fact such that from this point of view no two samples of organic sedimentary matter are identical. In addition, since the characteristics of a sample at the time of deposit has evolved since that time, it is necessary to reconstruct the initial characteristics. This becomes more difficult to the extent to which evolution is more advanced. In the course of geological evolution, OM loses its distinctive characteristics (Chapter 1). This is indicated for example by the convergence of the evolution paths in the Van Krevelen diagram.

The series described above have been constituted on the basis of many criteria. It was considered that the same sedimentation medium and the same organic materials led to a relatively constant chemical composition for the initial sedimentary OM. This was verified as far as possible by analysis of extracts, physicochemical study of kerogens, etc. Optical methods were also used to verify that the organic remains were the same. However a certain degree of uncertainty had to be admitted as to the ''nature'' of the OM or it would not have been possible to constitute any of the series. This uncertainty increased with the extent of evolution. These series are therefore based in a more pragmatic way for types of OM, i.e. OM whose nature as defined by the group of techniques was the same in the first approximation. The result of this variation within the types is that the evolution paths are bands and not lines.

The series used (Fig. 4.7) are based on three types to which reference will frequently be made in this book:

(a) **Type I is the organic content of samples from Green River Shales of the Uinta Basin (USA) studied by Tissot et al** [38]. Microscopic examination shows practically no identifiable remains of organisms. The general kerogen structure is very aliphatic (Chapter 13). The hydrocarbons formed during burial [38] or pyrolysis [42] contain *n*-alkanes of the C_{30+} range which are thought to be derived from bacterial and vegetable waxes and also anteisoalkanes probably derived from anteisoacids which are also supposed to be derived from bacterial metabolism [38]. These sediments were deposited in large lakes [15].

(b) **Type II is the organic content of shale from the Lower Toarcian in the Paris Basin (France) and the Posidonienschiefer in Germany (Liassic ϵ)** [2, 3, 8, 21, 34, 35] which constitute the same formation in France and Germany. Under the microscope it consists essentially of amorphous material probably derived from planktonic (algal) biomass. The identifiable remains of organisms (20% to 30% of the organic content) are *Tasmanaceae* and *Nostocopsis algae* [3, 34] and ligneous debris in variable proportions indicating a small terrigenous contribution. The kerogen structure includes a large share of saturated polycyclic elements. Hydrocarbons formed during burial or by pyrolysis are characterized by the abundance of cyclic structures. The sediments are marls which were deposited in a shallow epicontinental sea.

(c) **Type III is the organic content from Upper Cretaceous Shale from the Douala Basin (Cameroon)** [1, 10]. Under the microscope it appears to consist of amorphous cement containing recognizable debris of evolved plants and vitrinite splinters [10]. The kerogen contains a very large share of aromatic structures. Hydrocarbons formed in depth or obtained by pyrolysis are characterized in part by *n*-alkanes of the C_{20}-C_{35} range derived from cuticular waxes from higher plants. The sediments are silty clays from delta deposits.

These types serve only as references and other types may exist. Type I has not yet been found in other series. Because of the analogy of the chemical structure (highly aliphatic character) **we will associate with type I most of the bogheads consisting of algae of the *Botryococcus* family** although they may appear very different under the microscope.

We frequently found type II in marly carbonate sedimentation and type III in delta series. **We will assimilate humic coals with type III since all the characteristics are very close** [33].

The trimodal distribution of hydrogen which can be observed in Fig. 4.2d and also, although in a less obvious way, in Fig. 4.6, is due to the frequency of types II and III in the sediments which we studied. The inventory which we made is probably not yet sufficiently representative of the different sedimentary media so that type frequency can be determined exactly and the consequences deduced for elemental analysis. If we now examine closely the variability of the analyses within each reference type, it is possible in some cases to determine the causes of this. The work of Huc [21] shows for example that the organic content of the Lower Toarcian in the Paris Basin is a mixture of a major proportion of marine material and a minor proportion of terrestrial matter. The proportions can vary depending on the paleogeography (for example distance from coast). The position of the elemental analyses in the evolution path of type II is determined for shallow samples by the proportions of the mixture. The higher O/C values indicate a higher proportion of terrestrial material.

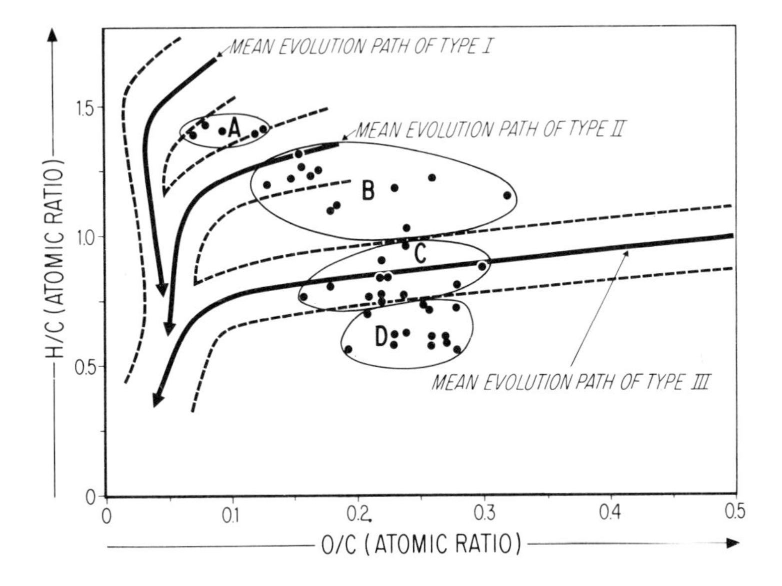

Fig. 4.9. — Kerogens from the organic matter (OM) in Cretaceous sediments of the North Atlantic, from Tissot *et al.* [40]:

A. Marine type. **B.** Mixture of marine and terrestrial types. **C.** Terrestrial type. **D.** Residual OM.

It can then be asked whether it is possible to define "purer" types than those from which our reference series are derived and whether all OM and therefore the position of points in a Van Krevelen diagram do not result from a mixture of varying proportions of these types. The question has been asked in various ways over a long period of time (Chapter 1). Thus marine OM has been opposed to terrestrial OM, sapropelic OM to humic OM, and lipids to lignin.

Recently Tissot *et al.* [40] described relatively evolved OM in Cretaceous sediments from the North Atlantic as the result of a mixture of varying proportions of a marine and a terrestrial type. They also distinguished a third contribution in the form of residual OM (Fig. 4.9).

The question of the existence of "purer" types has not yet been decided in spite of the number of samples examined. It is not of interest unless there is a small number of these types. Obviously elemental analysis cannot solve this problem all by itself.

It is clear that neither the position of the elemental analysis or even the evolution path in the Van Krevelen diagram is sufficient to characterize a type. It was seen above that the same type could correspond to different evolution paths (burial, weathering). In addition, different types could have the same evolution path. Thus the evolution paths due to burial for type II kerogens and the paths for cannel coals coincide. The former, however, are derived from a planktonic biomass and the second from spore accumulations which are lateral transitions from humic coals. They appear very differently under the microscope, and the overall chemical structure is probably very different (available data on cannel coals does not enable us to state this definitely).

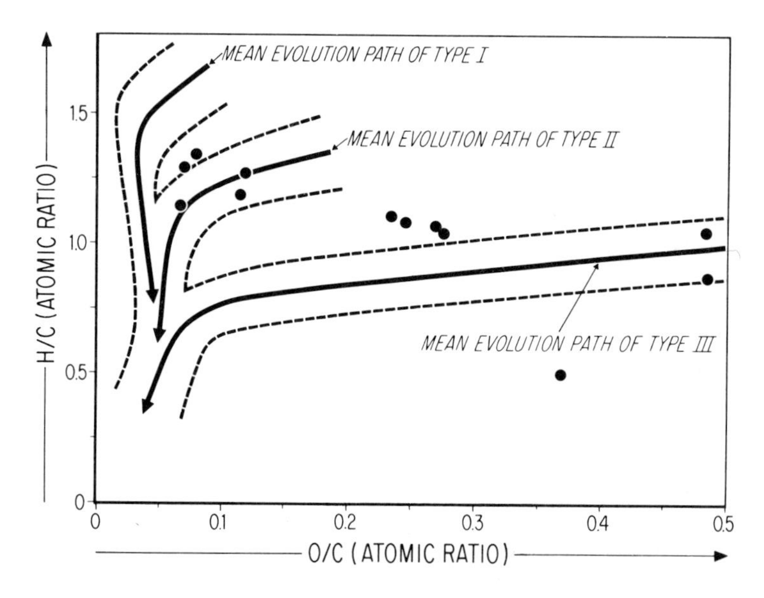

Fig. 4.10. — Location of the elemental analysis of kerogens from "amorphous" OM in a Van Krevelen diagram.

Lastly the use of C, H and O analyses in general, and of the Van Krevelen diagram in particular, makes it necessary to shift constantly from the concept of type to the concept of evolution path and back. When the type is clearly characterized this makes it possible to follow the evolution of OM under the given geological conditions and to reach a better understanding of evolution due to burial as we will see below. For the same evolutionary stage it is possible to observe type changes or mixtures of types in terms of variations in the sedimentary environment.

C. Relations between optical examinations and elemental analysis.

Exclusive use of the microscope for characterizing sedimentary OM leads to confusions which are generally due to the belief that optical examination alone can lead to a clearcut idea of the chemical structure of sedimentary OM and therefore of its elemental analysis and type.

The relations between the optical characteristics which are usually taken into account (color, transparency, reflectance, fluorescence) and elemental analysis are not linear. Moreover they are only statistical ones since the optical parameters can be greatly affected by phenomena which have little effect on the elemental analysis. Thus fluorescence is very sensitive to details of molecular structure.

Thus petrographers will classify together and will be tempted to attribute the same chemistry to materials with very different H/C. This is the case for example for macerals of the exinite family.

Van Krevelen however showed ([43] and Fig. 4.5) that elemental analysis of alginite is different from that of other macerals of this family. The danger is especially great for the so-called "amorphous" fraction of the OM which is by far the largest quantitatively in most cases and for which the safety factor of morphological criteria does not exist. The amorphous fraction is generally considered by petrographers as the OM which is richest in hydrogen. Fig. 4.10 in which 12 elemental analyses for amorphous OM are plotted shows that this is not always the case.

D. Description of evolution paths due to burial by means of Van Krevelen diagram.

1. Introduction.

The fact that kerogen is taken into consideration rather than the total OM means that (Chapter 1) a special role is given to examining relations during evolution between hydrocarbons (oil and gas) and the rest of the OM.

The following observations concern evolution in sedimentary basins before metamorphism (in the mineralogical sense of the term). They are largely founded on the study of the behavior of the three reference types I, II, and III.

It is remarkable that shifts due to burial are all indicated in the Van Krevelen diagram by carbon enrichment of the kerogen by reduction of O/C and then of H/C and thus with an H/C-O/C-diagram which converges approximately towards the carbon pole. This means that the sedimentary OM is not stable when subjected to increased temperature and pressure (in an unoxidizing medium) during burial and that most of the oxygen is bound in chemical structures which are less stable than those holding most of the hydrogen.

On the basis of changes in elemental analysis, it is possible to characterize three successive evolutionary stages (Fig. 4.11).

(a) The first stage A corresponds to a decrease in O/C which is much greater than that for H/C.

(b) The second stage B corresponds to a large decrease for H/C and stability (or slight rise) for O/C.

(c) The third stage C corresponds to slight changes in elemental analysis (which at this stage is that of very carbon-rich OM) in spite of further subsidence.

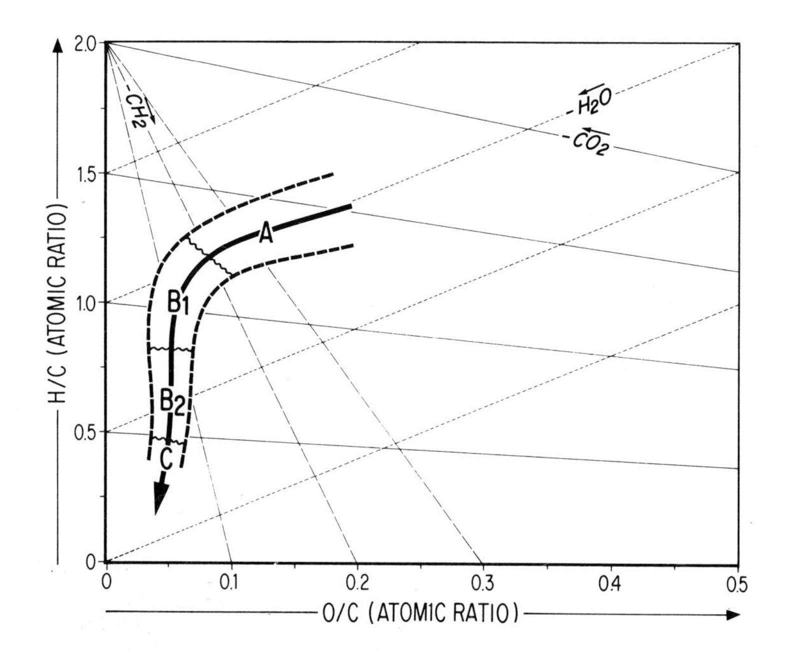

Fig. 4.11. — Evolution path in a Van Krevelen diagram.

Various methods and especially thermogravimetry (TGA, Chapter 5), IR spectroscopy (Chapter 6), electron microscopy (EM, Chapter 7), electron spin resonance (ESR, Chapter 8) and optical studies (OS, Chapters 3, 11 and 12) make it possible to determine the physicochemical changes that take place in the kerogen during the various stages which can be summarized as follows:

(a) A corresponds to the rapid disappearance of C = O functions (IR) and a large drop in fluorescence (OS) when the latter exists.

(b) B corresponds to the disappearance of saturated C – H functions (IR) and a large decrease in weight losses measured under standard conditions (TGA). It also corresponds to a rapid rise in:

. Reflecting power (OS).
. Number of aromatic protons (IR spectroscopy).
. Number of free radicals (ESR).

These characteristics indicate a rapid rise in the aromaticity of the kerogen.

(c) C corresponds to structural reorganization of the carbon skeleton of the kerogen without any appreciable increase in its aromaticity. This is indicated by a decrease in the number of free radicals (ESR) and the intensity of the aromatic (C – H) functions (IR spectroscopy) and the development which can be observed under the electron microscope of domains with dimensions of 50-1 000 Å within which aromatic layers have $\vec{c}$ axis orientations which are close.

Observation of petroleum formation phenomena in sedimentary basins ([1, 8, 10, 35, 36, 38] and Chapter 1) shows that these stages are correlated with the successive formation from OM of various categories of products which are mobile or soluble in organic solvents, (the kerogen thus represents at the end of this evolution a relatively small fraction of the initial weight of the OM):

(a) A corresponds to the formation of CO_2, H_2O and heavy heteroatomic products, i.e. resins and asphaltenes. This is often called the immature zone.

(b) B corresponds to the formation of increasingly light hydrocarbons (oil → wet gas → gas). This is often called the principal oil and gas formation phase [44]. The change in distribution is due to the increasing extent of cracking of the C – C bonds in hydrocarbons which have already been formed as well as in hydrocarbon structures still bound to kerogen. In Fig. 4.11 we have introduced a $B_1 \rightarrow B_2$ transition corresponding to maximum oil formation. This transition is not indicated in elemental analysis.

(c) No clear information is available making it possible to indicate what phase C corresponds to. It is however known that only gas can be formed at this stage. It is probable that the last heteroelements and functional groups are eliminated in the form of simple molecules such as CH_4, H_2, CO, CO_2, H_2O during this phase of structural reorganization of the carbon skeleton.

2. *Oil and gas potential.*

The position of kerogens in the H/C-O/C-diagram before displacement is related to the total quantities of hydrocarbons which will be formed during displacement per unit of weight of initial OM (oil + gas potential).

This potential is a function of the quantities of unpolyaromatic hydrocarbon structures (the polyaromatic arrangement corresponds to the more stable form of carbon which can hardly give hydrocarbons). These quantities can be estimated on the basis of the H/C ratios. It is therefore possible to predict that the oil potential raises with the H/C ratio. It is however hard to make a quantitative prediction on the basis of this simple line of reasoning. Other observations, i.e. quantities of hydrocarbons observed in sedimentary basins or behavior of initial kerogens in TGA (Chapter 5), show for the types which have been fully studied that maximum quantities of hydrocarbons (oil + gas) formed per unit weight of **initial** OM are approximately 75% for type I, 50% for type II and 30% for type III. In this case these values correlate well with the **initial** average H/C ratios for the kerogens of these three types which are respectively about 1.6, 1.3 and 0.9.

3. *Differences between types.*

The position of the evolution paths in a H/C-O/C-diagram also shows the differences in overall chemical structure of the initial kerogens and the convergence towards a polyaromatic structure with few functional groups which is about the same whatever the initial type. This convergence probably occurs both because the loss of functional groups leaves in the residue only the polyaromatic part of the initial structures and because of an overall process of creation of polyaromatic structures which do not initially exist in the OM. The relative importance of these two phenomena is not known and clearly differs depending on the initial structures.

On the basis of the H/C and O/C values it is possible to form a rough idea of the initial chemical structures.

At the beginning of the type I path or nearby, types are found with a strongly aliphatic chemical structure, while at the beginning of the type III path, or nearby, can be found types whose chemical structure consists largely of polyaromatic structures with oxygen function substitutions. These differences indicate, on the sedimentological level and in proportions which are difficult to evaluate, differences in the oxidation-reduction nature of the deposit medium in the broad sense, i.e. including transport (roughly: path I = reducing medium, path III = oxidizing medium); at the same time they indicate differences in the physicochemical nature of the progenitors (roughly: path I = lipids, path III = lignin). Obviously the structural details can be evaluated only by other methods. At the beginning of the type II path, and in general for intermediate paths between I and III, elemental analysis supplies little information on the chemical structure. The structural differences have an effect on oil formation kinetics [37] since the degradation of OM which is the origin of hydrocarbon formation takes place at a rate which is a function of the stability of the bonds making up the initial structure and the reactivity of the functions. For example bonds of the ester type will be less stable than C – C bonds. This point is discussed in detail in Chapter 13 on the basis of available data on the overall chemical structure of kerogens of types I, II and III. Observation seems to show that, all factors remaining equal (geothermal gradient, subsidence rate, etc.), hydrocarbon formation takes place earlier for types II and III than for type I.

However observation is difficult in this field since at the present time no universal indicator of the state of evolution of OM is available. Vitrinite, whose reflecting power measurement is

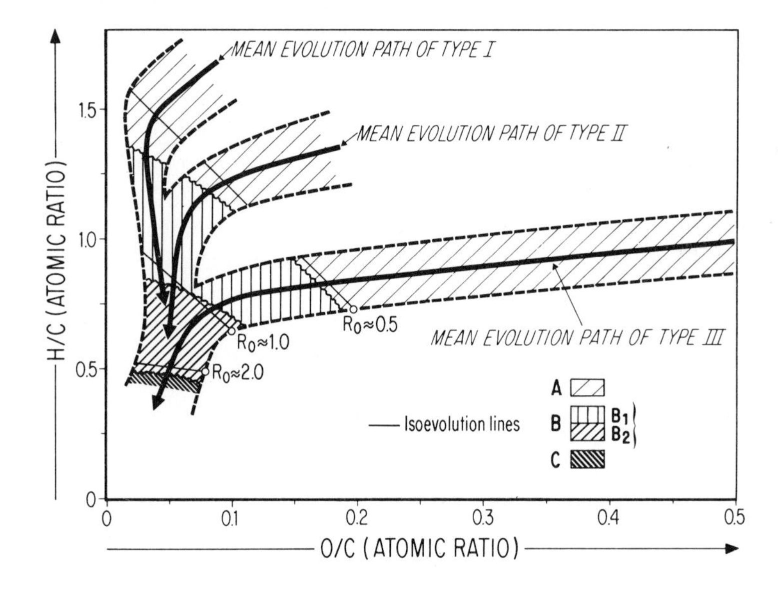

Fig. 4.12. — Differences in the kinetics of petroleum formation for types I, II and III OM:

A. Immature zone. **B.** Principal zone of oil formation. $\mathbf{B_1} \rightarrow \mathbf{B_2}$ Maximum of oil formation. **C.** Structural reorganization of the carbon skeleton.

used as an indicator for coal rank, is abundant in type III (with which we assimilate coals) but is not very representative in type II and even less so in type I which are found in marine and lacustrine series. (It should be recalled that vitrinite is theoretically derived from lignocellulosic walls of higher plant cells.)

On the basis of observation alone, comparison of the evolution of these three types by means of a single scale of evolution states cannot now be carried out without simultaneously utilizing several methods involving a good deal of uncertainty. However we show a test in Fig. 4.12. In this H/C-O/C-diagram isoevolution lines were plotted. The corresponding evolution states can be correlated for type III with the reflecting power of the vitrinite which it contains. We have superposed zones A, B_1, B_2 and C for the formation of oxygen products, oil and gas, as described above, for each of the types. This representation indicates the kinetic differences in oil formation from one type to the next. For example, the transition $B_1 \rightarrow B_2$ takes place at a less advanced stage for type III than for type I.

Under average geological conditions this difference means a difference in burial depth of about 1 km.

As we have explained, this construction is not exact. In view of the great variety of possible combinations of chemical structures making up a type, it is risky to extrapolate these conclusions to types of organic matter other than types I, II and III simply on the basis of the position of their kerogens in a Van Krevelen diagram.

4. Organic metamorphism: diagenesis, catagenesis and metagenesis.

The evolution described above takes place in a zone of temperatures and pressures which are lower than those which cause the extensive transformations of minerals known as metamorphism. During this period the minerals undergo only slight transformations [7, 22], and the OM is a much more sensitive indicator than the latter for pressure and especially temperature changes in the first kilometers of the burial depth. The roles are then reversed and the modifications of the carbon residue which is the result of organic evolution in this zone becomes only slightly sensitive in the mineral metamosphism. It is then accessible only by means of special techniques such as EM (Chapter 7).

The term "organic metamorphism" is frequently found in the American literature [20] for all the modifications of OM during burial.

We distinguished three successive zones, A, B, C, of transformation on the basis of well marked differences in the physicochemical characteristics of kerogens. These zones of "organic metamorphism" have different names depending on the authors. Tissot *et al.* [39] call zone A diagenesis, zone B catagenesis and zone C metagenesis. Here catagenesis means the main oil and gas formation phase. They also define a limit between metagenesis and mineral metamorphism. Vassoyevitch [44], the inventor of the term catagenesis, indicates by it transformations due to thermodynamic factors. This means that the beginning of zone A, where sediments which are not yet lithified are affected by factors whose effects are not yet well understood such as water chemistry, action of microorganisms, etc., is not taken into account. We will call this zone A_1 in order to distinguish it from A_2, the rest of zone A. However this subdivision is for the moment impossible to make by means of simple physicochemical criteria. In particular, it is not indicated by elemental analysis of kerogen.

Thus according to Vassoyevitch, catagenesis designates group A_2 (protocatagenesis), B (mesocatagenesis) and C (apocatagenesis). Alpern (in this book) follows the same principle and calls zone A_1 diagenesis and group A_2 + B + C catagenesis, which is then subdivided into epigenesis (A_2), mesogenesis (B) and metagenesis (C). Pelet (in this book) also distinguishes the evolution under early factors, mainly biochemical ones, which he calls early diagenesis, from the evolution under merely thermodynamic factors, which he divides into diagenesis (departure of oxygen), catagenesis (departure of hydrogen) and metagenesis.

Except for Alpern who refers explicitly to coals, most authors do not mention a precise type of OM. Since as we have seen the evolution kinetics differs a great deal depending on the types of OM, it is preferable to refer to an exact type for the zonation of "organic metamorphism" in a sedimentary basin. Type III (delta series, coals) is the best adapted since the reflecting power of vitrinite (R_o) is a relatively precise means for determining evolution. For this type the transition A $\rightarrow$ B takes place for $R_o \simeq$ 0.5-0.6; $B_1 \rightarrow B_2$ for $R_o \simeq$ 0.9-1.0; $B_2 \rightarrow$ C for $R_o \simeq$ 2-2.5. In addition the transition peat $\rightarrow$ lignite corresponding to $R_o \simeq$ 0.2-0.3 could in this case correspond to the transition $A_1 \rightarrow A_2$.

5. *Singleness of evolution paths.*

The question can be asked whether a given type of OM can follow different evolution paths during burial either because it is in different minerals or because rates of increase for temperature and pressure are very different on account of subsidence conditions. These problems have a somewhat academic nature, for it is very doubtful that it will be possible to make this kind of observations because of the difficulty of collecting coherent series. Experimental study is also not easy since it requires higher temperatures in order to reduce experimental time in the laboratory, and the result is that the reactions are changed (see below).

As concerns the first point (influence of minerals), partial observations were made which for the moment concern only shifts for kerogen elemental analyses in the Van Krevelen diagram. A comparative study of strictly identical type III OM either dispersed in 1-2% contents in clays (illite and kaolinite, a small amount of montmorillonite and mixed-layers minerals) or concentrated in coal beds between these clays was made up to maximum oil formation (transition $B_1 \rightarrow B_2$). The displacements are superposed, and this seems to indicate that the minerals of these clays did not have a great deal of influence, at least on the overall level of elemental analysis of kerogens.

It should be noted that the observation of constant characteristics in evolution paths due to burial is an argument in favor of their singleness.

6. *Simulation in laboratory.*

The analogy between the evolution of OM during burial and pyrolysis in a non-oxidizing atmosphere has long been recognized [46, 47, 48]. In both cases there is a process of carbonization principally due to the temperature. However the effect of a low temperature (20-200° C) for a long period (millions years) is simulated by a high temperature (300-600° C) for a short period. Differences in temperature (and also in pressure and confinement) are such that great differences in the details of the reactions must be expected. In particular, reactions taking place in nature at the lowest temperatures could not be reproduced in the laboratory.

At *IFP* an attempt was made to simulate evolution due to burial by pyrolyzing slightly evolved kerogens of a known type ([4], Chapter 5). A sample of approximately 100 mg was heated in a thermobalance under a nitrogen stream. The temperature was programmed at

4° C/min from room temperature up to a final temperature varying from 300 to 600° C. Examination of the pyrolysis residues obtained at increasing final temperatures by various physicochemical methods showed that they constitute a series analogous to a series of kerogens at various degrees of increasing natural evolution. In addition, in the pyrolysis of a slightly evolved kerogen three successive stages, A', B' and C' are distinguished which are very close to stages A, B, and C of the natural evolution. Oil and gas production, which takes place mainly during stage B', is roughly proportional to the H/C ratio of the initial kerogen. Optical examination [4] confirms these analogies.

As concerns the shifts in the H/C-O/C-diagram, the results are summarized in Figs. 4.13 and 4.14. Fig. 4.13 shows the comparison of the average natural evolution path (dotted line) for type III and the artificial evolution paths for 7 samples of this type (3 coals, 2 from the Douala Basin series which serves as reference for type III), 1 sample from Lower Manville Shales (Canada) and 1 sample from an Indonesian delta series (Mahakam delta). Reference points were plotted on the artificial paths at 400 and 500° C. The experiments were carried to 600° C in two cases, but shifts in the diagram are small over 500° C. They seem to us to be of little significance in view of the lack of precision of the analyses. For type III subjected to pyrolysis, the transition A' → B' took place under these experimental conditions at 375° C and the transition B' → C' at about 500° C.

Considerable differences can be noted between the shift due to burial and the shift due to pyrolysis. O/C remains high at the end of pyrolysis and this is especially true if the starting value was high (and therefore evolution due to burial was not very advanced). Artificial shifts show a retrogression corresponding approximately to the transition A' → B'. This transition is also observed for an O/C ratio which is especially high if the original value was high. The artificial shift slope in part A' is steeper than the natural shift slope in part A and this probably indicates, as was already stressed by Ralston [27], that a relatively greater amount of water is formed during A' than during A. This water formation would use hydrogen which would not then be available for hydrocarbon formation. Thus the quantity of hydrocarbons formed per weight of initial OM would be higher in nature than that formed during this type of experiment. It can thus be supposed that evolution of water under pressure would inhibit water formation and favor the elimination of O in other forms, and this might improve simulation. It seems that evolutionary processes were actually better simulated under hydrothermal conditions [43].

Figure 4.14 summarizes the observations made on samples of types I and II.

Natural and artificial shifts coincide very well for type I. As concerns type II, the artificial shift for the sample which is poorest in oxygen coincides with the natural shift while the behavior of the richest sample is quite similar to that of samples of type III. However stages A, B, C and A', B', C' coincide better.

The natural and artificial evolution paths thus seem to be further apart to the extent to which the oxygen content of the initial sample is greater and to be closer to the extent to which the hydrogen content is greater. In the case of artificial evolution, oxygen is eliminated less completely and probably, at least as concerns the beginning of evolution, by mechanisms which in relation to natural evolution favor the elimination of water at the expense of the later formation of hydrocarbons. It is however hard to verify this last point since it is impossible to establish water weight balances along the natural evolution path.

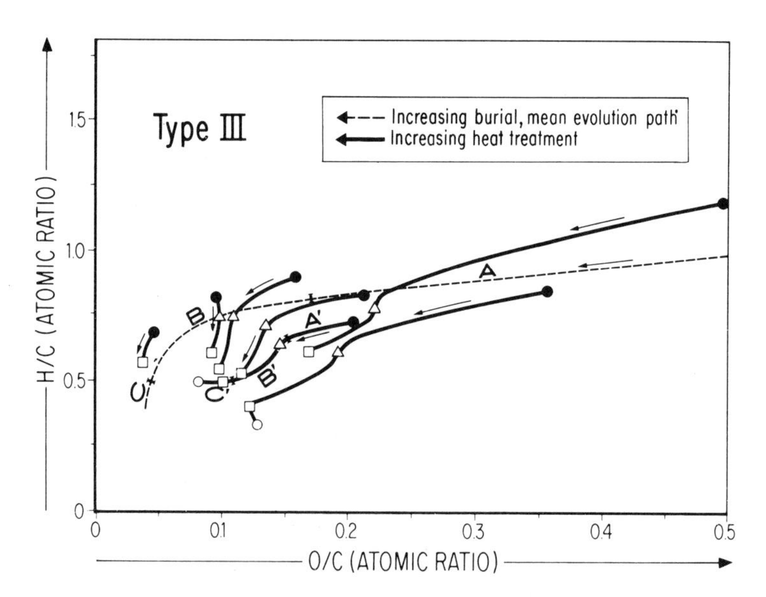

Fig. 4.13. — Artificial evolution paths of type III samples:

●: Initial sample. △, □, ○: 400° C, 500° C and 600° C chars. **A, B, C.** steps of natural evolution. **A' B' C'** steps of artificial evolution.

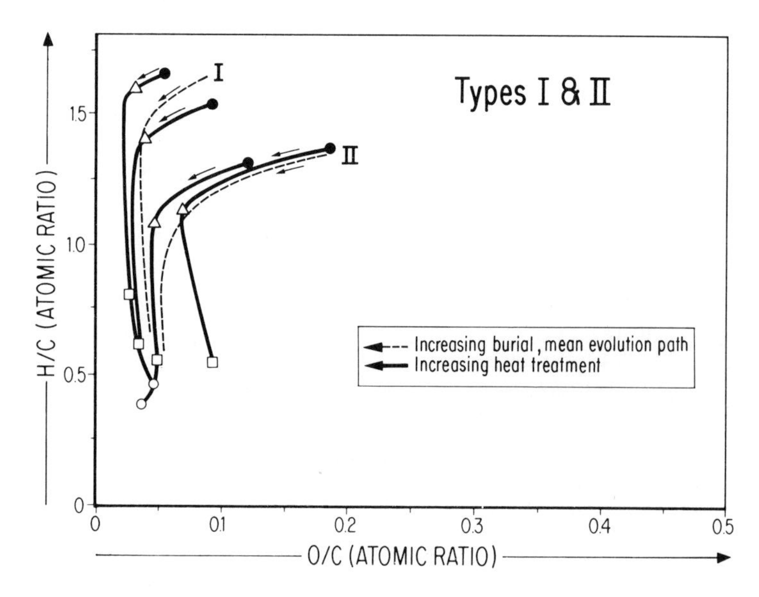

Fig. 4.14. — Artificial evolution paths of type I and II samples:

●: Initial sample. △, □, ○: 400° C, 500° C and 600° C chars.

V. SULPHUR AND NITROGEN

The accuracy for the measurements of these elements is, as was seen above (paragr. III.A.1), less than for C, H, O. Below we will present a study of types I, II and III in which we have tried to bring out the tendencies during evolution by plotting S/C or N/C for kerogen as a function of an evolution parameter. On a theoretical basis, O/C was chosen for types III and II and H/C for type I.

A. S/C.

The S/C values are very scattered for each type. At the beginning of evolution the average values are higher for type II than for types I and III. This seems to be normal in view of the fact that only type II is clearly marine and that sulphur is probably incorporated in the OM via the activity of sulphato-reducing bacteria. Sulphates are in fact more abundant in marine media.

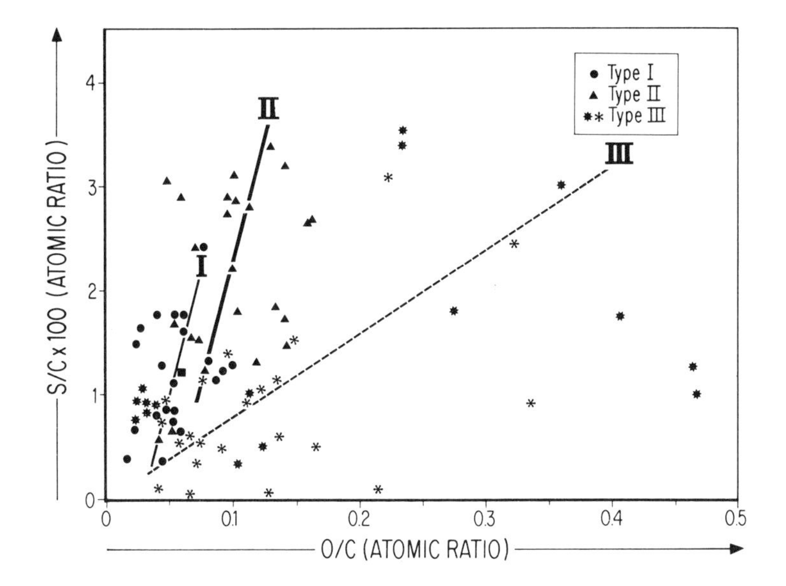

Fig. 4.15. — Elemental analysis of kerogens in a S/C vs. O/C diagram.

Type **I**: ●. Type **II**: ▲. Type **III**: * coals, * : Logbaba.

There is a correlation, although relatively poor, for types III and II between S/C and O/C (Fig. 4.15), thus showing that S/C decreases in the kerogens of these types as a function of burial. It seems to be probable that the sulphur which is thus lost is found, at least in part, in the benzothiophenes of the aromatic oil fraction. As concerns type I no correlation was observed between S/C and H/C. There is however a correlation between S/C and O/C (Fig. 4.15) which thus shows a similarity between the behavior of sulphur and oxygen.

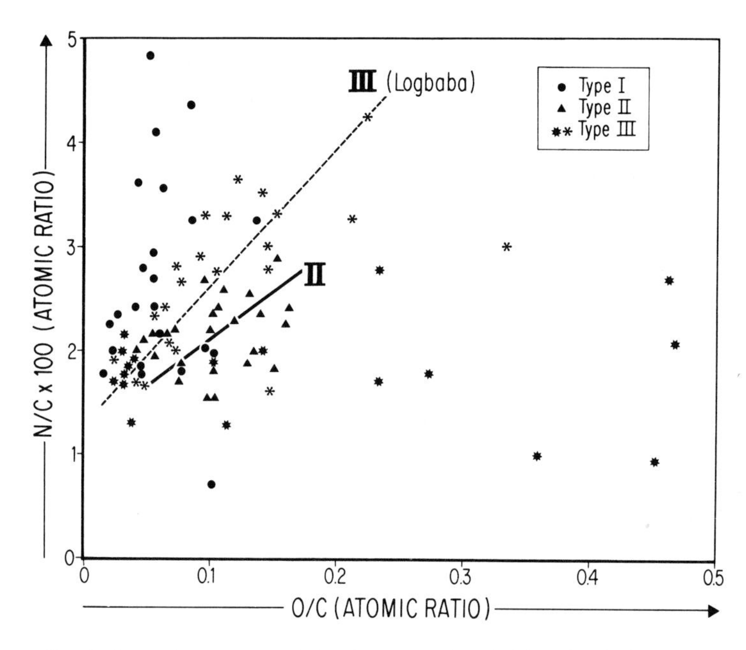

Fig. 4.16. — Elemental analysis of kerogens in a N/C vs. O/C diagram:

Type **I**: ●. Type **II**: ▲. Type **III**: ✱ coals, ∗ : Logbaba.

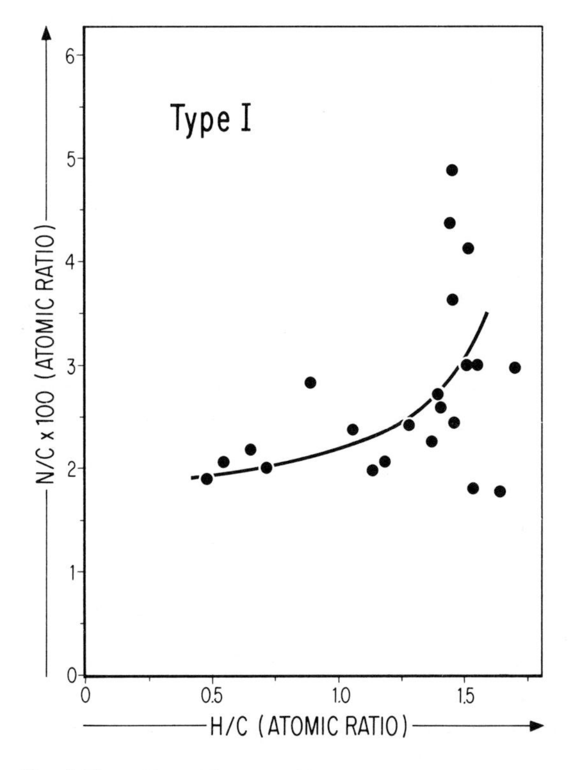

Fig. 4.17. — Type I kerogens in a N/C vs. H/C diagram.

B. N/C.

The dispersion is also very great for each type. The average values at the beginning of evolution are highest for type I. Then comes Logbaba (Type III), Type II and then coals (Type III).

There is also a relatively good correlation between N/C and O/C in the case of Logbaba and a moderately good correlation in the case of type II (Fig. 4.16). For type I there is on this occasion a correlation with H/C (Fig. 4.17).

The behavior of N/C and S/C during natural evolution is therefore not very clear in spite of a tendancy to decrease. This lack of clarity is certainly due in part to the poor accuracy of the measurements. The observed dispersions could however also be due to variability within the types as they are now defined. Nitrogen and sulphur could therefore serve to characterize subgroups within the types. The example of nitrogen seems to be the clearest, for we have seen that the Logbaba series and the coal series taken as examples of type III had very different N/C ratios.

VI. CONCLUSION.

Elemental analysis of kerogen is a method for characterizing the origin and evolution of sedimentary OM. This capacity shows that general laws exist for the arrangement of the chemical structures of the latter at the time of their formation and during their evolution under geological conditions.

Elemental analysis also establishes a framework within which other physicochemical analytic methods can be used more effectively. However too much should not be expected from it. It is not very precise and its power of resolution is therefore limited, especially in the case of organic materials which have been subjected to extensive burial and have acquired characteristics which are very close. In addition, in view of its overall characteristics it is theoretically unable to detect mixtures of different organic matters.

The Van Krevelen diagram is certainly a very practical means of interpretation, but it should be noted that this is due to other techniques for the study of kerogen. It should not then be expected to solve all problems. Confusions of types and evolution paths should especially be avoided.

Up to the present time, the analysis of the minor elements, sulphur and nitrogen, could not be situated within a coherent framework. A preliminary condition is to improve the analytic techniques.

REFERENCES

1. Albrecht, P., Vandenbroucke, M. and Mandengué, M., (1976), *Geochim. et Cosmochim. Acta* **40**, 791.
2. Alpern, B., Durand, B., Espitalié, J. and Tissot, B., (1972), *in: Advances in Organic Geochemistry,* 1971, H.R. Von Gaertner and H. Wehner ed., Pergamon Press, 1.
3. Alpern, B. and Cheymol, D., (1978), *Rev. Inst. Franç. du Pétrole,* **XXXIII**, 3, 515.
4. Alpern, B., Durand, B. and Durand-Souron, C., (1978), *Revue Inst. Franç. du Pétrole,* **XXXIII**, 6, 867.

5. Breger, I.A. and Brown, A., (1962), *Science,* 137, 221.

6. Down, A.L. and Himus, G.W., (1941), *J. Inst. Petroleum, 27,* 426.

7. Dunoyer de Segonzac, G., (1970), *Sedimentology,* **15,** 282.

8. Durand, B., Espitalié, J., Nicaise, G. and Combaz A., (1972), *Rev. Inst. Franç. Pétrole,* **XXVII,** 865.

9. Durand, B. and Espitalié, J., (1973), *C.R. Acad. Sci. Paris,* série D, t. **276,** 2253.

10. Durand, B. and Espitalié, J., (1976), *Geochim. et Cosmochim. Acta,* **40,** 801.

11. Durand, B., Nicaise, D., Roucaché, J., Vandenbroucke, M. and Hagemann, H.W., (1977), *in: Advances in Organic Geochemistry,* 1975, R. Campos and J. Goni éd., Enadimsa, Madrid, 601.

12. Forsman, J.P., (1963), "Geochemistry of Kerogen", *in: Organic Geochemistry,* I.A. Breger éd., Macmillan, New York.

13. Forsman, J.P. and Hunt, J.M., (1958), *in: Habitat of Oil,* Symposium AAPG ed. Weeks, L.G., Publ. Tulsa, 747.

14. Foscolos, A.E., Powell, T.G. and Gunther, P.R., (1976), *Geochim. et Cosmochim. Acta,* **40,** 953.

15. Fouch, T.D., (1975), *Rocky Mountain Assoc. Geologists Symposium,* 163.

16. Grout, F.F., (1907), *Econ. Geol.,* **2,** 225.

17. Hickling G., (1932), *J. Inst. Fuel* **5,** 326.

18. Himus, G.W. and Basak, G.C., (1949), *Fuel,* **28,** 57.

19. Himus, G.W., (1951), *Oil Shale and Cannel Coal.* Vol. **2,** Institute of Petroleum, London.

20. Hood, A., Gutjahr, C.C.R. and Heacock, R.L., (1975), *AAPG Bull.* **59,** 986.

21. Huc, A.Y., (1978), *Géochimie organique des schistes bitumineux du Toarcien du Bassin de Paris,* Thesis, Université de Strasbourg.

22. Kubler, B., (1966), *in: Colloque sur les étages tectoniques à la Baconnière,* Neuchâtel, 105, Neuchâtel Univ.

23. Laplante, R.E., (1974), *AAPG Bull.,* **58,** 1281.

24. Long, G., Neglia, S. and Favretto, L., (1968), *Geochim. et Cosmochim. Acta,* **32,** 647.

25. McIver, R.D., (1967), Proceedings of the 7th World Petroleum Congress, **2,** 25, Mexico.

26. Marchand, A., Libert, P. and Combaz, A., (1969), *Rev. Inst. Franç. du Pétrole,* **XXIV,** 3,20.

27. Ralston, O.C., (1915), *US Bureau of Mines,* Techn. Paper N° 93.

28. Robinson, W.E. and Stanfield, K.E., (1960), *US Bureau of Mines Inf. Circ.* No 7968.

29. Robinson, W.E., (1969), *in: Organic Geochemistry,* G. Eglinton and M. Murphy ed., Springer Verlag Press, 181.

30. Sato, S., (1976), *The Science Reports of the Tohoku University* **XIII,** 2, 85.

31. Saxby, J.D., (1970), *Chem. Geol.* **6,** 173.

32. Smith, J.W., (1961), *US Bureau Mines, Rep. Invest.* N° 5725.

33. Stach, E., Mackowsky, M.Th., Teichmüller, M., Taylor, G.H., Chandra, D. and Teichmüller, R., (1975), *Stach's Textbook of Coal Petrology,* Gebrüder Borntraeger, Berlin.

34. Teichmüller, M. and Ottenjahn, K., (1977), *Erdöl und Kohle,* **30,** 387.

35. Tissot, B., Califet-Debyser, Y., Deroo, G. and Oudin, J.L., (1971), *AAPG Bull.,* **55,** 2177.

36. Tissot, B., Durand, B., Espitalié, J. and Combaz, A., (1974), *AAPG Bull.,* **58,** 3, 499.

37. Tissot, B. and Espitalié, J., (1975), *Rev. Inst. Franç. Pétrole,* **XXX,** 5, 743.

38. Tissot, B., Deroo, G. and Hood, A., (1978), *Geochim. et Cosmochim. Acta,* **42,** 1469.

39. Tissot, B. and Welte, D.H., (1978), *Petroleum Formation and Occurrence,* Springer Verlag Press.

40. Tissot, B., Deroo, G. and Herbin, J.P., (1978), *in: Implications of Deep Drilling. Results in the North Atlantic,* Maurice Ewing Series, **3.**

41. Unterzaucher, J., (1940), *Ber. Deut. Chem. Ges.* **73 B,** 391.

42. Vandenbroucke, M., Durand, B. and Hood, A., (1977), paper presented at the *8th Internat. Meeting Organic Geochemistry,* Moscou.

43. Van Krevelen, D.W., (1961), *Coal,* Elsevier ed., Amsterdam.

44. Vassoyevitch, N.B., Korchagina Yu I. Lopatin N.V. and Chernyshev V.V., (1970), *Int. Geol. Rev.,* **12,** 11, 1276.

45. Welte, D., (1969), *in: Handbook of Geochemistry,* Vol. **II/1,** 6-L-22, K.H. Wedepohl ed. Springer-Verlag Press.

46. White, D., (1915), *J. Washington Acad. Sci.* **6,** 189.

47. White, D., (1916), *Bull. Geol. Soc. Amer.,* **28,** 1917.

48. White, D., (1925), *Trans. Amer. Inst. Miner. and Metall. Eng.*

5

Thermogravimetric analysis and associated techniques applied to kerogens

C. DURAND-SOURON *

I. INTRODUCTION

Coals have been among the first organic compounds which have been studied by thermal analysis, mainly thermogravimetry and differential thermal analysis. An overview of the investigations carried out till 1960 is given by Van Krevelen [11]. Lawson made a more recent review [4]; he cites many references and insists on the need to carefully control the experiments, an indispensable requisite if one wishes to compare the results of experiments of different origin with any success. Amongst the difficulties encountered he also mentions the heterogeneity of coal, the variations in thermal properties due to its fusibility and the creation of volatile matter. Notwithstanding these limitations, the two techniques mentioned are no doubt a precious tool, and they are frequently applied for the comparative study of coals and the different phases of their thermal evolution.

To the best of our knowledge, kerogen has been very little studied by these techniques. Posidonia shales studied under oxidizing atmosphere by Von Gaertner and Schmitt [13] are the closest example. At the *Institut Français du Pétrole* (*IFP*), the first systematic experiments were performed in 1969. A standard method of operation has been established. Several hundreds of samples of various origin have thus been investigated with these rather fast techniques which have yielded criteria on the nature and the evolution of kerogen. In addition, a number of samples have been analyzed by complementary methods: elementary analysis (EA), infrared (IR), mass spectrometry (MS), electron microscope (EM), electron spin resonance (ESR), after having first undergone a preparatory treatment derived from the standard method. This was done to better analyze the transformation of kerogen when heated and to determine the simulation conditions of natural transformation (catagenesis). It is only mentioned here for the record; other Chapters will give further details on this subject. We shall put the emphasis first on the results and conclusions which *IFP* has obtained from the thermal analysis of kerogen, in particular from the thermogravimetric analysis TGA within the temperature range 25-600° C, a rather narrow range but which proved in pratice to be the

*Institut Français du Pétrole. Rueil-Malmaison, France.

most rewarding. The possibilities offered by the combination of thermal analysis and MS will be discussed. Almost instant data are thereby obtained on the nature of the products generated, and this enables a better interpretation of thermograms to be made.

II. INFLUENCE OF EXPERIMENTAL CONDITIONS

A. General.

TGA in its narrow acception [10] continuously measures the variations in the weight of a sample which undergoes a linear increase in temperature. If the temperature remains constant during measuring, the method is called "isothermal thermogravimetric analysis". The time derivative of the thermogravimetry curve is abbreviated to DTG. Differential thermal analysis (DTA) is a detection and a measure of the quantity of heat produced or absorbed either by a chemical reaction or a physical process in a sample. This is done under a programmed temperature increase. Combining TGA and DTA in the same apparatus is of interest because the results simultaneously obtained on the same sample under the same conditions can be perfectly compared. For this reason we used such apparatus for most of our experiments. To be able to adjust the parameters it is useful to know which technique is emphasized [6], here it is in general TGA.

The analysis of the effluents released by a pyrolysis may be done by various techniques. We have chosen MS [7]. Because at the time we did not dispose of a high resolution spectrometer coupled with a thermobalance, we simulated the pyrolysis within the source of the mass spectrometer. The experimental conditions differ slightly and will be discussed separately.

In the simple case where a sample, when heated, decomposes through one or more known stoechiometric chemical reactions, various factors influence the kinetics of the reactions but not the end results. These factors are mainly type of apparatus, sampling, (size, form, compaction), ambient atmosphere during the experiment and way of heating. For kerogen, however, these factors cause more complex effects.

B. Apparatus.

The geometry of the crucibles, the size of the reaction chamber, the confinement or the circulation of gas, the method for measuring variations in weight (asymmetrical or symmetrical balance), the location of the thermocouples in the sample... all these "geometric" conditions vary from one apparatus to another.

Precise information on the apparatus we have used is therefore worth being recorded. It concerns a Mettler TA 1 thermoanalyzer for measuring the TGA and the DTA, except where mentionned otherwise, with:

(a) Platinum microcrucibles for 5 mg samples.

(b) Platinum crucibles of 0.5 to 1 cm^3 for 50 mg samples and larger.

(c) The crucibles are generally not covered and thus in communication with the reaction chamber, the volume of which is about 1 liter.

(d) Generally we used a gas flow of roughly 10 liters/hr; for gases other than air, we created a vacuum (10^{-1} torr) before the experiment; the experiments "under vacuum" refer to an initial pressure of about 10^{-4} torr.
(e) Pt-PtRh thermocouples, located on the rod which bears the crucibles.
(f) Single-plate balance system.

When using an asymmetrical balance one should not forget to account for the apparent increase in weight due to the variation in Archimedean pressure when working under increasing temperatures with small-sized samples. This variation can be estimated empirically. It will suffice to correct the experimental curve, by taking the origin of weights at a given temperature; we have chosen 150° C so that this value should also serve to correct the eventual absorption of atmospheric humidity.
In our device, thermocouples are not within the sample but in contact with the crucible. Results obtained this way cannot be as good as those provided by a specific apparatus.

C. Sampling.

Kerogen is *a priori* not a single compound but a more or less homogeneous mixture of fine particules of various origin (*cf.* Chapters 1 and 3). Besides organic matter (OM) it may in particular contain some minerals which are extremely difficult to eliminate. This heterogeneity poses the problem of whether sampling is reproductible, as illustrated in Fig. 5.1 for samples of 5 and 60 mg of the same kerogen. It appears that the heterogeneity of the sample has only rather weak influence on the experiments under the conditions selected. One notices however, that the spread of measurements is narrower for samples of 5 mg than for those of 60 mg, though statistically a series of samples of 60 mg is more homogeneous than a series of samples of 5 mg. Another factor should therefore be considered, i.e. the size of the sample.

D. Size of sample.

Whereas heat transmission is quasi instantaneous in a sample of small size, it needs a certain lapse of time for a larger sample. Similarly, the diffusion of the products generated during heating takes some time which is not negligible for larger samples. In the meantime, secondary reactions could occur which increase or decrease the total loss of weight. This is shown in Fig. 5.2 from which it is deduced that the size-effect of a sample differs from sample to sample. This cannot be neglected when TGA is used as a "preparatory technique" for other analytical methods.

E. Ambient atmosphere.

The ambient atmosphere plays a multiple role. It plays a part in heat transfer. It assists in diluting and eliminating the reaction products, if one works in a gas stream or under continuous vacuum, and this will influence the development of the reaction. Finally, it could also form part of a reaction, either with the original sample or with the products generated by an immediately prior reaction. Such a case will be studied when discussing the oxidation of kerogen. The examples shown in Fig. 5.3 concern only gases which do not take part chemically in

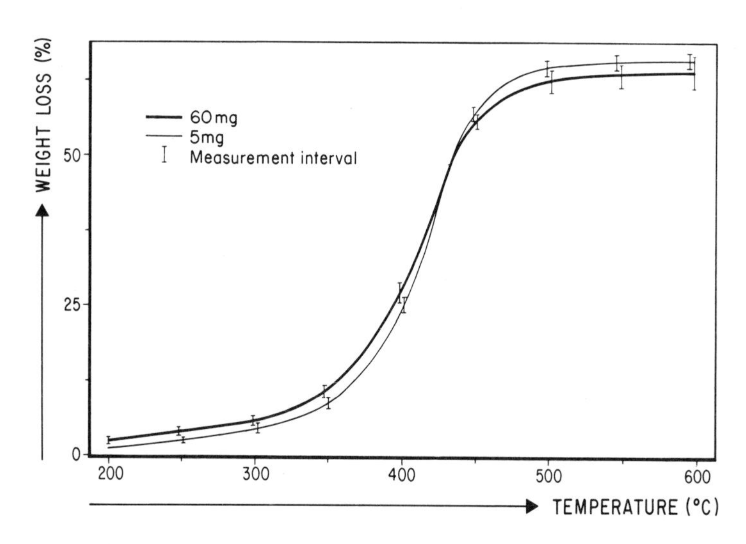

Fig. 5.1. — Reproducibility of sampling for 5 and 60 mg samples of the same kerogen analyzed under similar conditions (nitrogen circulation of 10 liters/hr, rate of heating: 4° C/min.). The spread of measures shown corresponds to extreme values for a group of 15 measurements.

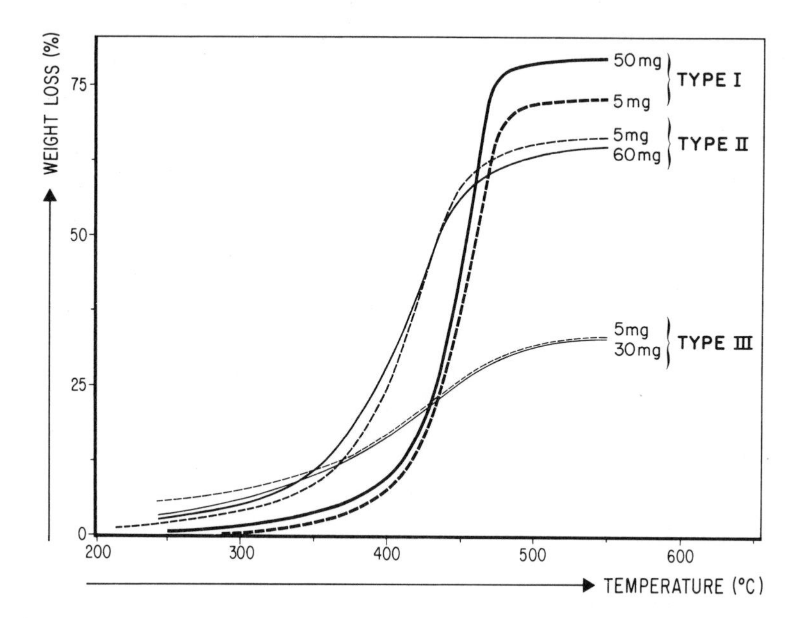

Fig. 5.2. — Effect of the size of a sample for three different types of kerogen analyzed under similar conditions (nitrogen circulation of 10 liters/hr, rate of heating: 4° C/min.).

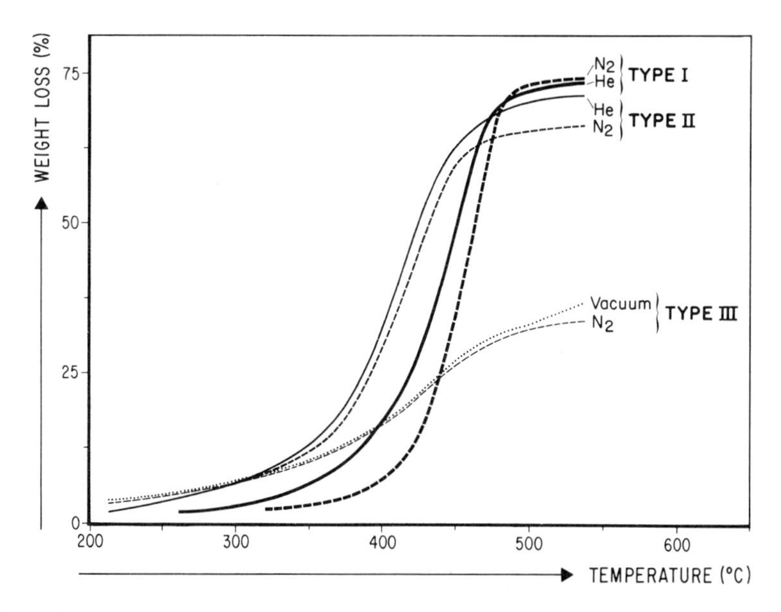

Fig. 5.3. — Effect of the nature of the ambient, non-reactive gaseous atmosphere on the pyrolysis of three different types of kerogen analyzed under similar conditions (sample of 5 mg, rate of heating : 4° C/min., circulation of gas at the rate of 10 liters/hr or under secondary vacuum).

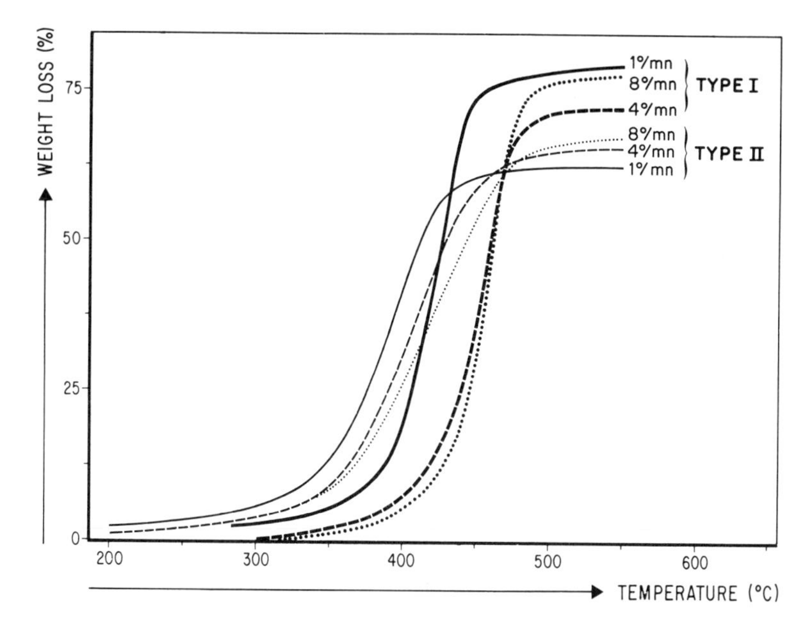

Fig. 5.4. — Influence of the rate of heating (1, 4 and 8° C/min.) on the TGA curve of two different types of kerogen analyzed under similar conditions (sample of 5 mg, nitrogen circulation of 10 liters/hr).

the reactions (nitrogen, helium and working under vacuum). They reveal that the total loss of weight and the shape of the curves are influenced, in various proportions, according to the type of kerogen, by the nature of the gas stream, but in general the curves remain quite comparable.

F. Heating rate.

Figure 5.4 illustrates the differences in the way and to what extent two kerogen samples decompose under three different heating rates. Not only is there a shift in temperature curves, which is a usual phenomenon for simple samples, but the amount of final decomposition varies as a function of the heating rate, in the opposite way for the two samples.

G. Conclusions.

We have demonstrated that the experimental factors such as size of the sample, ambient atmosphere, and rate of heating not only influence the kinetics of the decomposition of kerogen but also the final decomposition rate. The consequences are twofold:

(a) The experimental conditions under which the analyses are carried out should be clearly defined.
(b) Indications on decomposition schemes are provided.

These two points are explained as follows:

1. Standard conditions.

(a) To counter the size-effect of the sample, we operate on samples of 5 mg, except for "preparative" pyrolysis which needs larger samples and for which the curves recorded would allow an eventual re-calibration for temperature.
(b) The ambient atmosphere is usually nitrogen, a gas that can easily be obtained and which does not interfere with the decomposition reactions. The curves obtained closely resemble those recorded with other non-reactive ambient atmospheres.
(c) The rate of heating has been set at 4° C/min. which enables two analyses to be made in one day, reaching 600° C, together with detailed records; this would be more difficult to obtain with a higher rate of heating.

2. Schemes of decomposition.

A standard decomposition scheme shows that the phenomenon is limited only by the quantity of the product available and that it reaches a limit whatever the experimental conditions may be. One of the factors that could retard its kinetics is thermal diffusion.

This is not sufficient to explain what really happens in the kerogen, in particular why the same external factors, namely the nature of the gas, the size of the sample and the rate of heating, could have different consequences depending on the nature of the original kerogen and more precisely, why they could increase or decrease the extent of final decomposition. Recent studies [12] on similar products have demonstrated that reactions could indeed follow different courses depending on the experimental conditions.

Until the interpretation of these phenomena have progressed further, one should be extremely precise as to the conditions under which the experiments are carried out, always trying to closely check their effects on the decomposition phenomena.

H. Mass Spectrometry.

We have already explained [7, 8] the technique used to analyze the effluents, i.e. by pyrolysis of the kerogen within the source of a mass spectrometer. A few points should briefly be recalled. The pyrolysis in question takes place in a high vacuum, and it could therefore show a certain shift in temperature as compared to phenomena recorded otherwise. The rate of temperature increase is not so well defined as for the thermoanalyzer, and this could also provoke a shift. Moreover, the maximum temperature for the apparatus we used is limited to 500° C. Finally, the actual analysis of the effluents depends on a rather delicate calibration concerning only molecular fragments with an m/e ratio below 120. Eventual heavier fragments are not taken into consideration. The results obtained are thus necessarily incomplete; this should not be neglected in the interpretation. More complete results only can be obtained by means of "preparations" and an analysis of fractions separated beforehand. This cumbersome and complex process is not systematically applied. It is also certain that coupling the thermobalance with a high-resolution spectrometer would have made it possible to obtain the results exposed here under better conditions.

III. EXPERIMENTS PERFORMED UNDER A NON-OXIDIZING ATMOSPHERE

A. Description of the thermograms.

We shall describe here the results of the experiments with TGA, used under the standard conditions defined earlier (nitrogen flow and heating rate of 4° C/min., between 25 and 600° C).

Figure 5.5 shows the decomposition curves recorded for immature kerogen (vitrinite reflectance of about 0.5) of three different types. The end decompositions are quite different, but studying the shape of the curves leads to distinguishing three differents stages on all thermograms:

(a) A first stage, till 350-400° C, involves a rather small loss of weight.

(b) A second stage, between about 350 and 500° C, reflects an important loss of weight, with an inflexion, generally; sometimes a weak DTA phenomenon is obtained.

(c) A third stage, above about 500° C, again shows a very slight loss of weight.

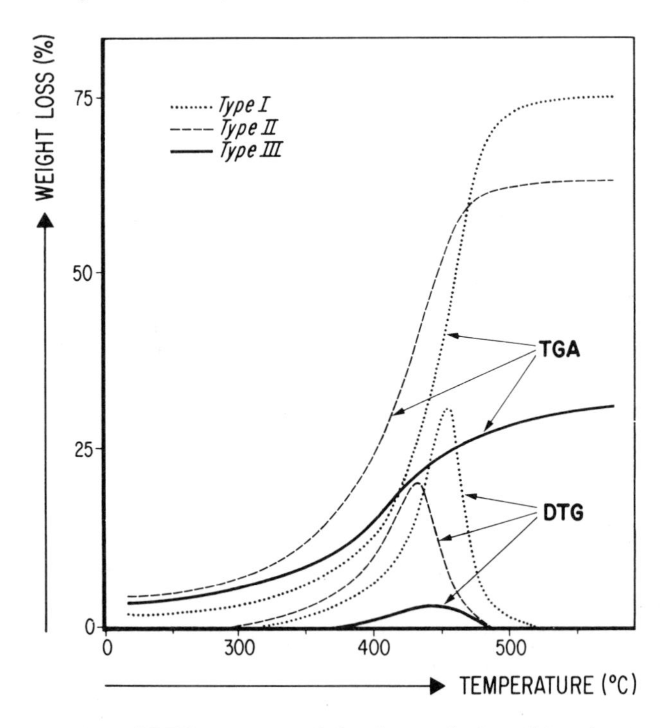

Fig. 5.5. — TGA and DTG curves, recorded under standard conditions for three types of immature kerogen (vitrinite reflectance: 0.5).

B. Analysis of residues of pyrolysis.

With a view to determining the limits between the stages mentioned and, more precisely, to knowing what they mean physically, two types of analyses have been carried out, one of the residue and the other of the products generated. It was necessary to make "preparative" pyrolyses for the study of the residue. The techniques used were EA, IR spectrometry, ESR, EM. The results are discussed in other Chapters of this study. Briefly, it amounts to:

1. First stage.

Decrease of the oxygen content, progressive disappearance of IR bands corresponding to oxygenated functions (C = O most obvious).

2. Second stage.

Decrease in hydrogen content, progressive disappearance of IR bands corresponding to saturated C – H, increase of the reflectance and of the number of free radicals as detected by ESR.

3. *Third stage.*

Little change in EA, structural reorganization under EM. Decrease in number of free radicals, as detected by ESR and of the bands which characterize aromatic protons as detected by IR spectrometry.

C. Analysis of volatile products.

The MS analysis of the products generated also distinguishes the first two stages, bearing the reservations in mind we mentioned earlier (Fig. 5.6):

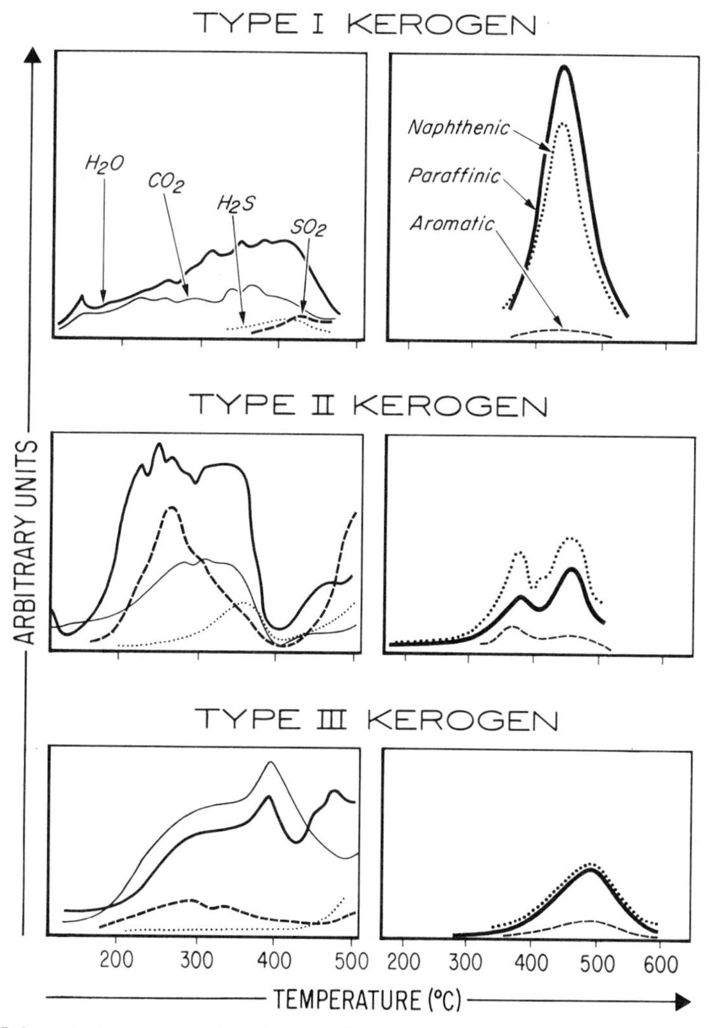

Fig. 5.6. — Instantaneous release in MS of volatile compounds for three different types of kerogen (sample of 3 mg, heating under vacuum within the source of the mass spectrometer at 5° C/min.). There is no coherent relation between the arbitrary units.

1. *First stage.*

Release of water, carbon dioxide, eventually SO_2 and H_2S in smaller quantities; very few hydrocarbons. In some cases hetero-atomic fragments appear near the end of the first stage, as shown by mass doublets. The most frequent heteroatoms are mainly oxygen, then come sulphur and nitrogen.

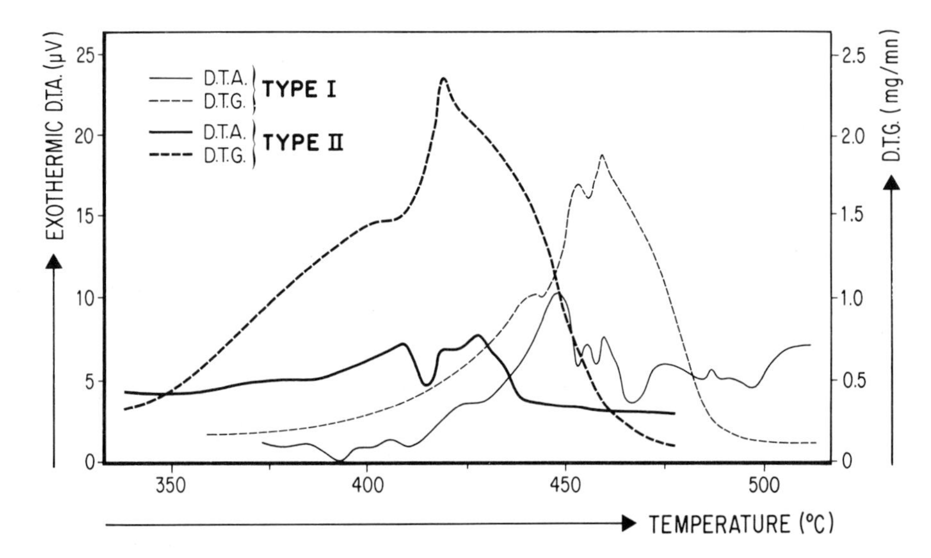

Fig. 5.7. — Examples of bimodal DTG curves accompanied by a DTA effect (Samples: type I kerogen = 50 mg, type II kerogen = 70 mg. Rate of heating: 4° C/min. Nitrogen circulation of 10 liters/hr).

2. *Second stage.*

Release of H_2O and CO_2 in smaller quantities and hydrocarbons (C_xH_y fragments) in great quantities; sometimes, emanation of sulphur compounds (H_2S, SO_2, S_x) during the decomposition of pyrite; sometimes release of methane at the end of the stage. It is possible [7], according to the distribution of the fragments, to classify the hydrocarbons generated into different "families": paraffinic, naphthenic, aromatic. Usually, the different families are generated nearly simultaneously. This release is sometimes bimodal (which could also correspond to a bimodal DTG curve and to a DTA effect) (Fig. 5.7). The phenomena related to these effects have not been interpreted with certainty, but two hypotheses can be put forward:

(a) Fusion of the kerogen, preceded and followed by the release of hydrocarbons.

(b) Generation of relatively heavy products which then surround the kerogen particles, momentarily preventing hydrocarbons from escaping (hindrance to diffusion process).

D. Interpretation of curves.

Based on these various observations it is possible to deduce some information from the aspect of a TGA curve:

(a) A strong loss of weight in the first stage corresponds to a release of oxygenated products and to a kerogen rich in oxygen.

(b) A strong loss of weight in the second stage means a strong release of hydrocarbons and hence a kerogen rich in hydrogen.

The relative importance of these two stages thus makes possible a classification of kerogen according to its nature and degree of evolution; this closely ressembles the classification that would have been obtained for instance from elemental analyses. It furthermore helps to classify kerogen according to its potential to artificially generate hydrocarbons. These observations led to the development of "quick methods" for evaluating the petroleum potential of the organic matter in rocks [2, 3], methods which are used notably in petroleum exploration and for the evaluation of oil-shale resources. But a study of the TGA curves also serves to follow the natural maturation of kerogen during burial. Besides, it can be demonstrated that pyrolysis, carried out under the conditions used here for TGA, achieves in most cases quite a good simulation of natural maturation.

IV. EFFECT OF NATURAL MATURATION. SIMULATION

During its geological evolution, OM generates hydrocarbons in variable quantities depending on its original type (Chapter 13). This generation occurs within a rather well determined maturation domain that can be detected by the variation of certain properties, in particular vitrinite reflectance. We shall examine the effect of this maturation on the thermal decomposition curves of kerogen of the same type. But, since temperature and time are the most important factors in this maturation a simulation by TGA of the natural maturation of kerogen may be considered valid.

A. Effect of natural maturation for the same initial type.

We have chosen as an example a type II kerogen series, represented by samples from the lower Toarcian shales of the Paris Basin and of Germany (Fig. 5.8). This evolution is characterized first by a small increase in the rate of decomposition, then by a decrease which occurs first at the level of the first stage (deoxygenation) and then of the second stage (dehydrogenation). Moreover, this second stage undergoes a shift towards the high temperatures, reflecting an OM which becomes more and more difficult to decompose. In MS (Figs. 5.9 and 5.10) one also notices a decrease in quantity of O – and S – compounds released (first stage) then (second stage) a weak increase followed by a decrease in hydrocarbons. The release of heteroatomic compounds is discernable only for the two least evolutive terms of the type. Methane is detected only towards the end of the second stage for all samples.

B. Comparison of different types.

The relative importance of the stages depends on the kerogen type being analyzed. The first stage remains of importance till a degree of evolution that corresponds with a vitrinite reflectance of approximatively 0.7 for type III, while for kerogen of type I, the oxygen content of which is lesser, it remains weak right from the beginning. On the other hand, the second stage, which is always important for type I, is relatively weak for type III. It results in a rather easy distinction between kerogen types, based on TGA curves for slightly maturated kerogen, but this distinction is not reliable for maturated kerogen. It should nevertheless be remarked that, altough the same (chemical) hydrocarbon families are present in each type, their distribution is somewhat different, depending on the original OM. This distribution could assist in distinguishing the type of kerogen but it is not a very sensitive criterion and concerns, it must be remembered, only the rather light hydrocarbons.

C. Simulation.

It has been verified that the residues from the preparative pyrolysis of immature kerogen follow an "evolution path" comparable to that of a natural maturation; further details may be found in other Chapters of this book. Natural and artificial evolutions paths are close to each other for type I and II, for type III larger differences occur. In other words, at laboratory time scale, the natural maturation of kerogen can be reproduced in a rather satisfactory way. As a consequence products generated have the same overall composition in both natural and artificial evolution all along the evolution path. It may thus be expected that, given a type of kerogen, the least mature kerogen generates the greatest quantity of hydrocarbons. We explained previously that the overall balance of hydrocarbon decomposition products cannot easily be apprehended by mass spectrometry. However, it can be assessed, knowing, on the

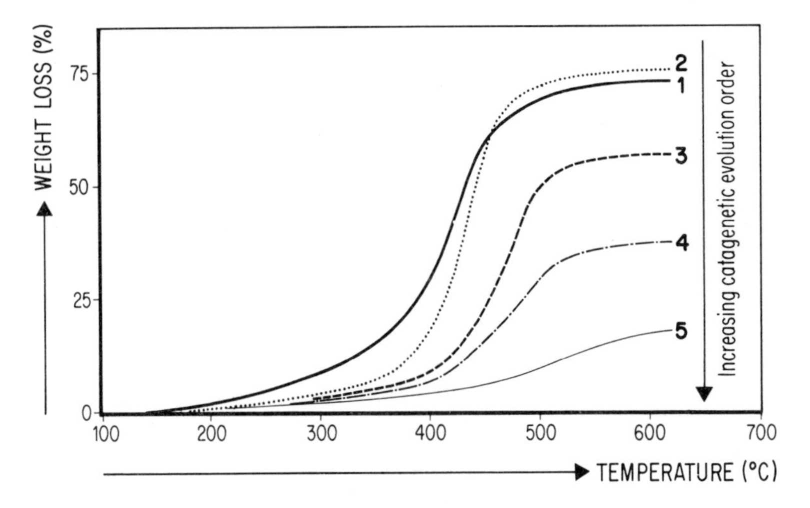

Fig. 5.8. — TGA curves plotted under standard conditions for samples of type II kerogen showing increasing maturation.

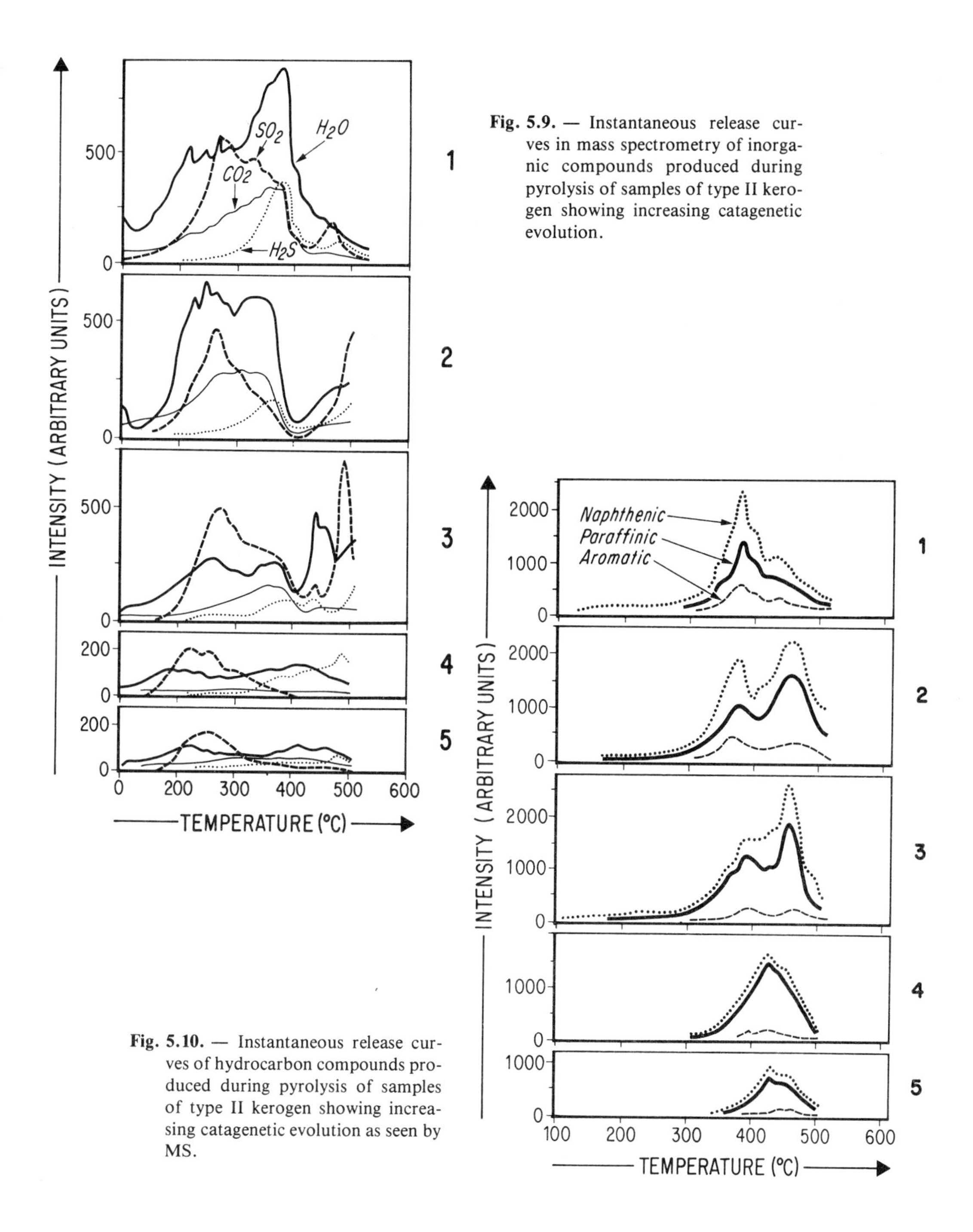

Fig. 5.9. — Instantaneous release curves in mass spectrometry of inorganic compounds produced during pyrolysis of samples of type II kerogen showing increasing catagenetic evolution.

Fig. 5.10. — Instantaneous release curves of hydrocarbon compounds produced during pyrolysis of samples of type II kerogen showing increasing catagenetic evolution as seen by MS.

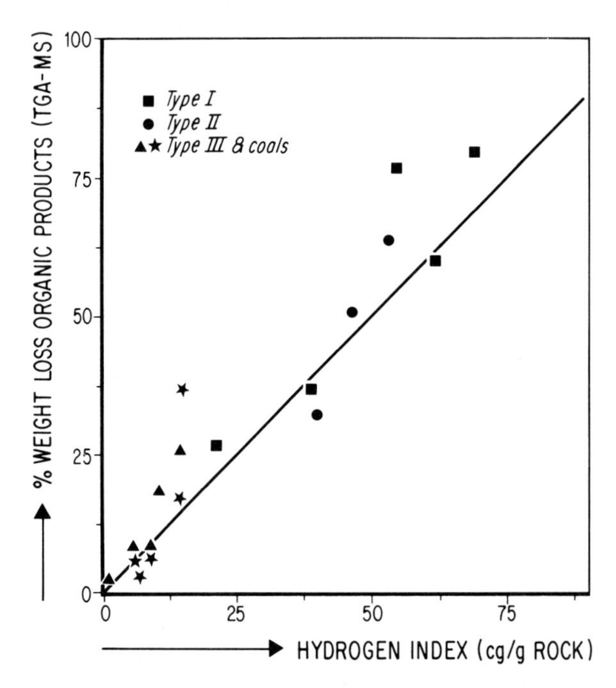

Fig. 5.11. — Comparison between the "organic products" released in MS (difference between the total weight loss measured in TGA and the quantities of "inorganic products" H_2O, CO_2, SO_2, H_2S measured in MS) and the "hydrocarbons" measured by FID (flame ionization detector) through "quick pyrolysis method".

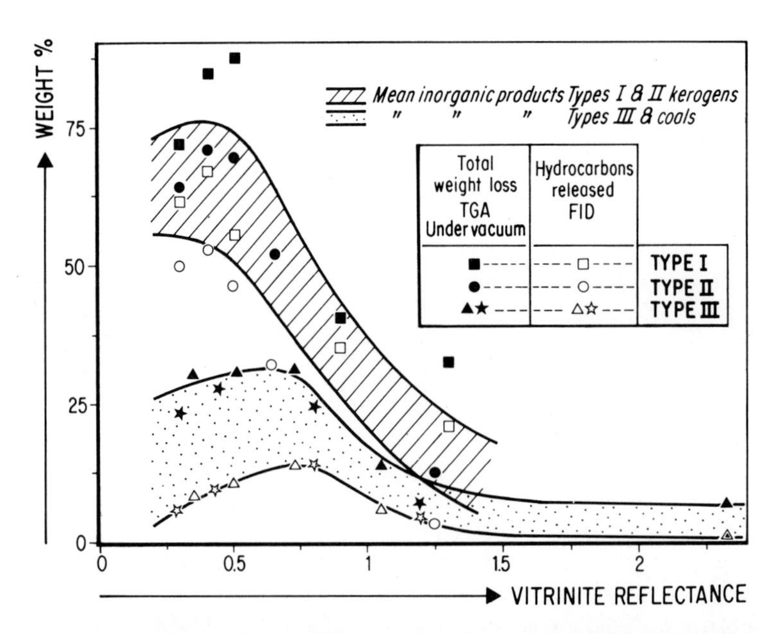

Fig. 5.12. — Variations in the quantities of volatile matter generated during pyrolysis as a function of vitrinite reflectance for kerogen of different types (the areas between the lower limit of each zone (▨ ░) and X-axis represent the mean quantities of hydrocarbons generated when heating kerogens).

one hand, the total rate of decomposition (measured by the thermobalance) and, on the other hand, the total quantity of inorganic products, H_2O, CO_2, SO_2, H_2S, (measured by mass spectrometer). The difference amounts to the total organic products, including hydrocarbons represented in part by the instantaneous decomposition curves in mass spectrometry. A better approximation can be obtained from the data "quick pyrolysis methods" [2, 3]: the quantity of hydrocarbon compounds (per weight unit of the initial kerogen) released during a pyrolysis experiment is measured by passing through a flame ionization detector. Corrections must be made owing to experimental conditions which are not exactly the same as the standard we used here, i.e. stream of helium, heating rate 25° C/min. The agreement between data calculated by difference and data from quick pyrolysis methods is quite good (Fig. 5.11). This shows that the difference between the total weight loss measured by TGA and the total quantity of inorganic products measured by MS consists mainly of hydrocarbon products. It appears (Fig. 5.12) that this total quantity of hydrocarbons generated during the TGA run passes through a maximum for a degree of maturation that varyies according to types of kerogens between vitrinite reflectance of 0.5 and 0.7. In other words, under pyrolysis conditions, the least mature kerogen does not produce the most hydrocarbons. This is particularly evident for type III. To obtain a maximum production of hydrocarbons under these pyrolysis conditions, it would thus seem necessary that the kerogen should have undergone a certain evolution, corresponding to a deoxygenation. It should be noticed that in type III, where immature kerogen is the most oxygenated, this effect is greater. This is an important remark for defining the simulation conditions for the natural maturation of kerogen; it has also a practical value for determining how OM should be treated with a view to obtaining hydrocarbons.

V. EXPERIMENTS PERFORMED UNDER A HYDROGEN ATMOSPHERE

The first experiments carried out with type II [1] kerogens showed that the curves obtained in either a hydrogen or a nitrogen medium were similar, with a loss of weight clearly greater in a hydrogen medium. This phenomenon has been confirmed with other types of kerogen. There is always this greater loss under a hydrogen atmosphere (Fig. 5.13). This difference sometimes shows up in the first stage of decomposition, but it occurs mainly during the second stage. Its magnitude is particularly great for kerogen of type I for which the loss of weight during the second stage is already very great under neutral atmosphere. It also appears to be greater for mature than for immature kerogen. However, we have not been able to establish a precise relation between these observations and other characteristics of kerogen.

One of the reasons explaining this difficulty is the well known fact that hydrogen reactions become efficient only under pressure and in the presence of a catalyst.

For the conditions under which we operate, the physical properties of hydrogen probably play a role (connected with diffusion phenomena) and possibly a certain chemical action as well.

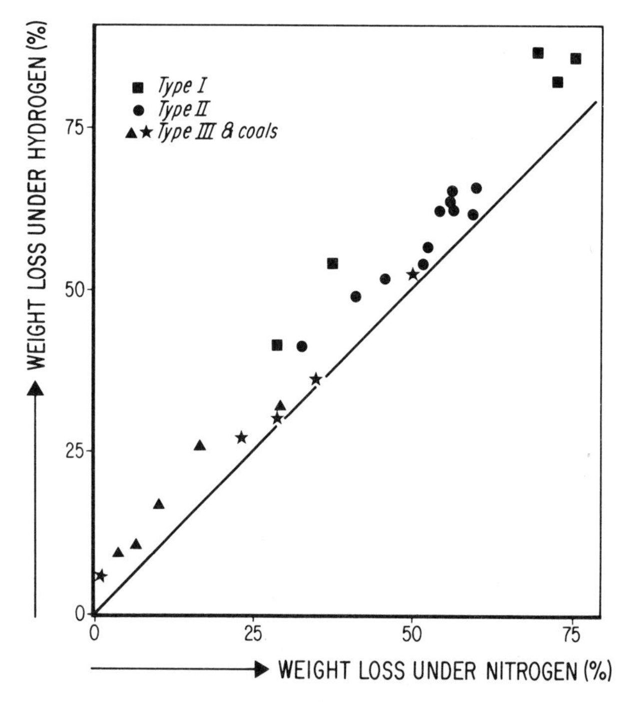

Fig. 5.13. — Comparison of the loss of weight under a hydrogen and a nitrogen atmosphere for kerogen of different types (gas circulation: 10 liters/hr. Rate of heating: 4° C/min. Sample of 5 mg).

VI. EXPERIMENTS PERFORMED UNDER AN OXIDIZING ATMOSPHERE

Whereas the oxidation of coal has been the subject of numerous studies, this is not general for kerogen [4]. We have described the phenomena [1] registered by heating kerogen samples of type II in air and oxygen, but these studies have not been pursued systematically.

The results are briefly reviewed (Fig. 5.14) :

(a) In air with a sample of 5 mg and a heating rate of 4° C/min., a nearly total decomposition occured in two stages, each accompanied by an exothermic DTA peak, with the first one, between 270 and 300° C, generally being flat and not so strong and the second, between 335 and 390° C, sharper and more intense.

(b) In oxygen with a sample of 5 mg and a heating rate of 2° C/min., a more abrupt loss of weight occurs, accompanied by only one exothermic DTA peak between 330 and 370° C.

It might also happen, in some cases, with air, that one could observe either a slower loss of weight, without a well-marked DTA peak, or on the contrary an abrupt loss of weight, similar to what happens in an oxygen atmosphere. The analysis of the factors which influence these phenomena, as expressed in the shape of the curves, has not been sufficiently investigated to be able to determine whether it bears on one or various preponderant factors. In any

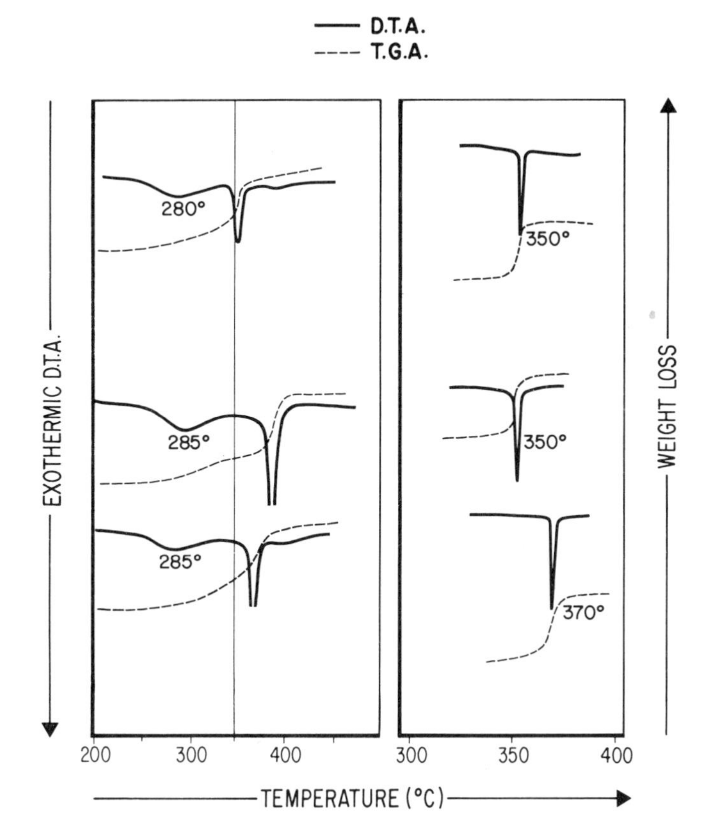

Fig. 5.14. — TGA and DTA curves with circulation of air (Rate of heating: 4° C/min.) and of oxygen (Rate of heating: 2° C/min.) for three samples of type II kerogen (Sample: 5 mg).

case it appears of even greater importance, when working with oxygen, to more clearly define the conditions under which the experiments are carried out than for other media. It should be mentioned besides that, for rock samples in which diffusion imposes a slowdown of the phenomena (Fig. 5.15), the reproductibility tends to improve.

Referring to similar investigations on coal, the first peak should correspond to the oxidation of aliphatic chains and the second to those of aromatic cycles [5]. It is not sure however that the existence of the double DTA peak implies the existence of two type of compounds [14].

If a more complete study proved this to be true, it would permit a good characterization of the chemical structures which could be helpful in determining rock criteria, useful in petroleum exploration.

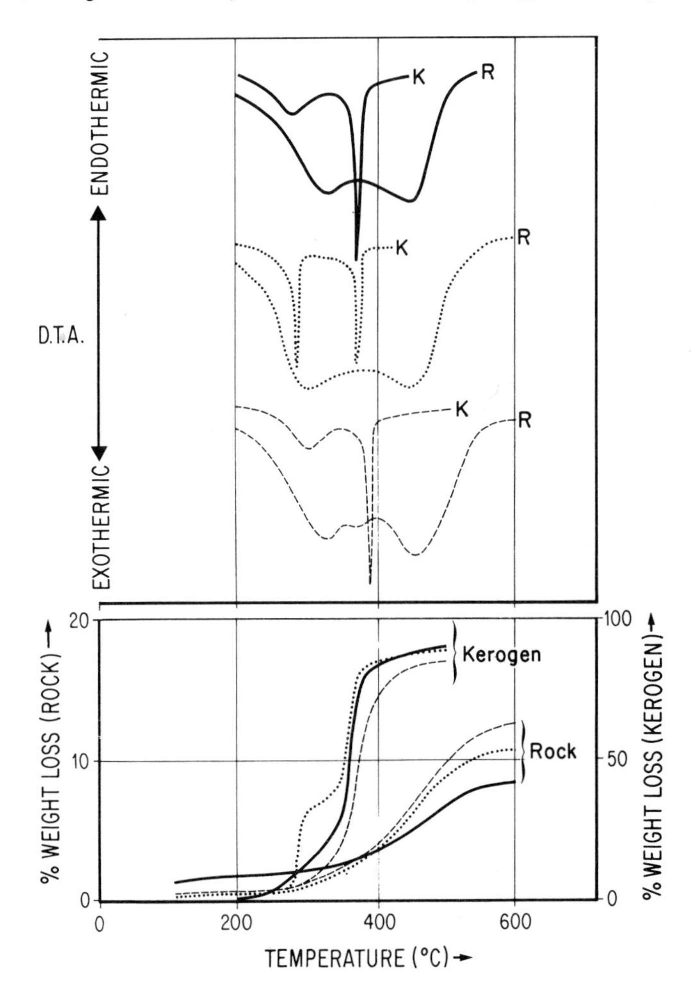

Fig. 5.15. — Comparison of the TGA and DTA curves with circulation of air for three rocks containing kerogen of type II and for the corresponding isolated kerogen (Rate of heating: 4° C/min. Air circulation of 10 liters/hr. Rock samples of 100 mg. Kerogen samples of 5 mg).

VII. CONCLUSIONS

We have discussed the results that may be obtained when studying kerogen with some thermal methods. We have deliberately remained in the empirical domain, the raw material being too complex to be completely and correctly formulated at present. However a statistical exploitation of this type of results and of other observations did enable a mathematical model to be built simulating the generation of hydrocarbons from kerogen during catagenesis [9].

With TGA one disposes of a method characterizing kerogen by the shape of the curves, the total loss of weight at 500° C and the relative importance of the first two stages of decomposition. This analysis could be carried out more quickly by increasing, for instance, the rate of heating, though bearing in mind that standard conditions will imperatively have to be set. It implies, however, a preparation of the kerogen which can be avoided only by working on rock material. This is done by "quick pyrolysis methods" [2, 3] which could be usefully completed by studies under oxygen, the fundamental aspects of which remain to be investigated, and for which the DTA could supply elements of interest.

From a structural point of view, the order of release of the products of decomposition, hence of the energies that link the different chemical "compounds" that constitute the "molecule", leads to information on the raw material being studied and its diagenetic evolution, in combination with the different techniques discussed in the other Chapters. Moreover, one notices the analogy between the artificial evolution, caused by heating, and the natural maturation, that generates hydrocarbons. This rather satisfactory simulation of the generation of hydrocarbons merits to be studied in greater detail for the deoxygenation phase, in particular for type III kerogen and coals. This forces us to be prudent, in respect to the generation of methane from kerogen for example. Indeed, we detected it only towards the end of the generation of hydrocarbons, for all types of kerogen studied, and this would seem to indicate that also in nature the generation of methane from kerogen could only be a late phenomenon. Admittedly, if methane was generated earlier in nature, it would be possible that our methods could not correctly simulate this formation.

REFERENCES

1. Espitalié, J., Durand, B., Roussel, J.C. and Souron, C., (1973), *Rev. Inst. Franc. du Pétrole,* **XXVIII,** 1, 37.
2. Espitalié, J. and Durand, B., (1976), US Patent n° 3, 953, 171 Ap 27, 1976.
3. Espitalié, J., Laporte, J.L., Madec, M., Leplat, B., Poulet, J. and Boutefeu, A., (1977), *Rev. Inst. Franç. du Pétrole,* **XXXII,** 23.
4. Lawson, G.J., (1970), *Differential Thermal Analysis,* **1,** 705, McKenzie Ed., Academic Press, London.
5. Marinov, V., (1975), *J. of Thermal Analysis,* **7,** 333.
6. Maurer, R. and Wiedemann, H.G., (1969), *Thermal analysis,* **1,** 177, R.F. Schwenker and P.D. Garn Ed., Academic Press, London.
7. Souron, C., Boulet, R. and Espitalié, J., (1974), *Rev. Inst. Franç. du Pétrole,* **XXIX,** 5, 661.
8. Souron, C., Boulet, R., and Espitalié, J., (1975), *Proceedings of the 7th International Meeting on Organic Geochemistry,* 797, Enadimsa, Madrid.
9. Tissot, B. and Espitalié, J., (1975), *Rev. Inst. Franç. du Pétrole,* **XXX,** 5.
10. Vallet, P., (1972), *Thermogravimétrie,* Gauthier-Villars, Paris.
11. Van Krevelen, D.W., (1961), *Coal,* Elsevier, Amsterdam.
12. Villey, M., Oberlin, A. and Combaz, A., (1976), *C.R. Ac. Sci. Paris,* **282,** Série D, 1657.
13. Von Gaertner, H.R. and Schmitz, H.H., (1963), *Proceedings of the 6th World Petroleum Congress,* **1, 21,** 355.
14. Yang, R.T., Steinberg, M. and Smol, R., (1976), *Anal. Chem.,* **48,** 12, 1696.

6

Characterization of kerogens and of their evolution by infrared spectroscopy

P.G. ROUXHET*, P.L. ROBIN* and G. NICAISE**

I. INTRODUCTION

Infrared (IR) spectra of complex organic solids such as coals, chars, kerogens, humic substances and some natural polymers have been presented in various publications [7, 44, 23, 19, 20, 40, 36, 15, 37, 28]. They show a limited number of rather broad bands which are due to well given chemical groups and can be assigned on the basis of numerous spectra of simple substances [4, 3, 2]. However they do not offer the possibility of having a fingerprint of the sample, which once made IR spectroscopy a major tool of organic analysis.

Moreover all these carbonaceous solids essentially give the same bands; the spectra of coals indeed offer much similarity with those of dispersed organic matter (OM) such as kerogen [23, 15, 28] and humic substances [36]. Therefore the data available for coals [7, 44, 19, 20, 40] can be used to identify the absorption bands of kerogen but, at first sight, the information offered by the spectra is considerably limited.

In fact, the spectra differ from each other by the intensity of the various bands. For instance the spectra of coals change regularly according to the rank and reflect the main chemical modifications occurring as the burial depth increases; they have also been used to study the carbonization processes and various types of industrial carbons.

A valuable use of IR spectroscopy for the characterization of these solids thus requires considering the band intensity which can be quantified by the measurements of absorption coefficients. If this is done, the infrared spectra provide a sort of semi-quantitative functional analysis of the complex solid, examined as a whole.

In this Chapter the attribution of the various absorption bands of kerogen will be presented but emphasis will be put on the quantitative use of the spectra. The experimental aspects will be discussed and the possibilities of the method will be illustrated by a comparison of

*Université Catholique de Louvain. Groupe de Physico-Chimie Minérale et de Catalyse. Louvain-la-Neuve, Belgique.

**Institut Français du Pétrole. Rueil-Malmaison, France.

samples of different origins and different degrees of evolution. The kerogens will also be compared systematically to typical coals.

II. DESCRIPTION OF THE SPECTRA

Representative spectra of kerogens are given in Figs. 6.1 and 6.2. They refer to samples belonging to the three typical series I, II and III described in Chapter 4 and characterized by different degrees of evolution; spectra of coal samples (IV) are also presented. The joined Van Krevelen diagrams allow to locate the selected samples on the evolution paths of the organic matter.

A. Baseline.

The baseline of the spectra ascends frequently from low wavenumber to high wavenumber, particularly from 2 000 to 4 000 cm^{-1}; this is due to increase of scattering, as the radiation wavelength approaches the particle size. The effect is stronger as the material is darker; therefore, as already observed for coals [7, 44, 19], the slope of the baseline rises as the degree of evolution becomes higher.

B. Bands of foreign substances.

Kerogen concentrates often show sharp bands at 425 and 350 cm^{-1} which are due to pyrite; however, they are not troublesome as they fall in a wavenumber range which is not very useful for the study of kerogen. Reference has been made in Chapter 2 to the use of IR spectra to check the purification of kerogen. General data on the infrared absorption of inorganic constituents of rocks can be found in the books of Farmer [16], Van der Marel and Beutelspacher [43].

Chloroform, which is used in the preparation of kerogen concentrate, gives a very intense absorption at 750 cm^{-1}. A peak due to chloroform is present in spectra of samples Ia and Id in Fig. 6.1.; usually it does not appear after treatment at 100° C under vacuum.

C. Bands of molecular water.

Molecular water may contribute to the broad OH stretching band around 3 430 cm^{-1}, the bending mode contributing to the band at about 1 630 cm^{-1}. Experiments performed with coal [19] have shown that the moisture content increases upon prolonged grinding; excessive grinding of pure KBr may also give rise to incorporation of moisture.

In practice, the coarsely ground KBr powder should be kept in the oven and further ground when mixed with the kerogen; blank pellets made of pure KBr are used to check that water contamination of KBr is negligible.

Storing KBr pellets for at least 7 days in a dessicator is sufficient for removing moisture held by kerogen [30] and can be used as a standard procedure if the intensity of the bands at 3 430 and 1 630 cm^{-1} has to be considered. Heat treatment of the kerogen + KBr mixture would give faster drying but the temperature should not exceed 140° C; heating kerogens around 180° C has been found to produce drastic spectral changes due to oxidation.

The moisture content of kerogen and coal is related to their oxygen content and not to their surface area [30, 48]; water is mainly retained by hydroxyl groups [5, 35] and can probably diffuse through the bulk organic material, which should be considered more like a sponge than like a rigid solid.

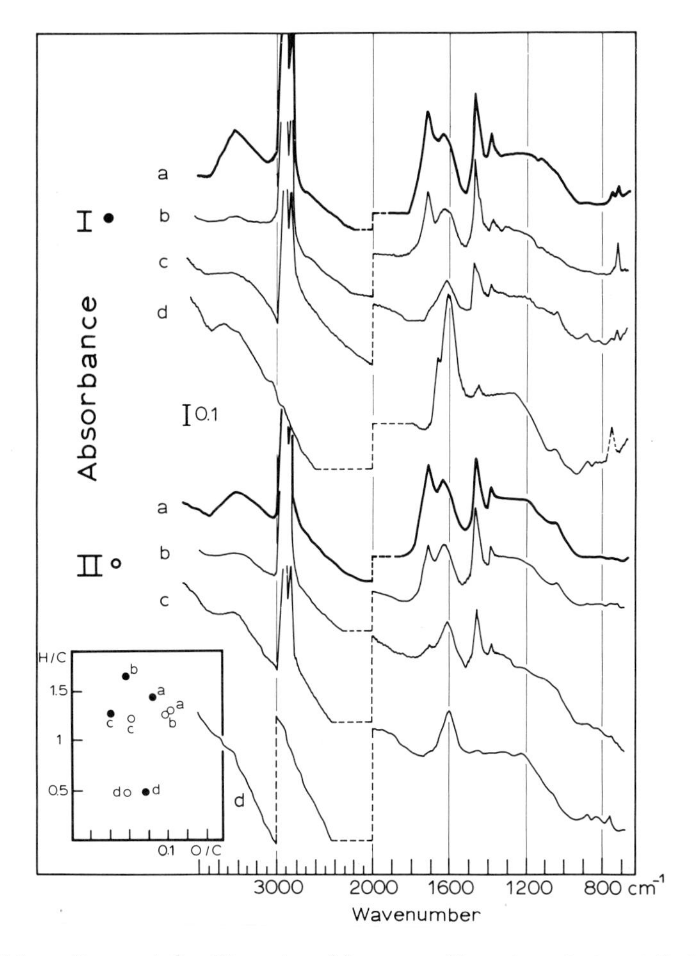

Fig. 6.1. — Representative IR spectra of kerogens with an important contribution of algae in the parent material; the elemental composition is plotted on the joined Van Krevelen diagram:

Type **I:** ●. Type **II:** ○.
IFP No: **Ia:** 15774. **Ib:** 15760. **Ic:** 15785. **Id:** 15794.
IIa: 10968. **IIb:** 11300. **IIc:** 11304. **IId:** 14090.

D. OH stretching band.

The OH stretching vibration is responsible for the broad band below 3 600 cm^{-1} giving a maximum around 3 430-3 400 cm^{-1}. Because the baseline is not horizontal and because of superposition of CH stretching bands between 3 000 and 2 800 cm^{-1}, the low wavenumber end of the band does not always appear very clearly; however it frequently extends below 3 000 cm^{-1}. Extension of the OH band toward low wavenumbers reflects the presence of stronger hydrogen bonding, possibly due to carboxylic acids. It is generally more pronounced in shallow materials and, in some cases (samples IIIa, IVb in Fig. 6.2.), a distinct maximum is observed around 3 200 cm^{-1}, in addition to that at 3 400-3 430 cm^{-1}.

E. Bands due to alkyl groups.

The peaks around 2 930 and 2 860 cm^{-1} are due mainly to the asymmetric and symmetric stretching of alkyl CH_2 groups, expected at 2 926 and 2 853 $\pm$ 10 cm^{-1} [4,2]. The CH_3 groups are expected to give asymmetric and symmetric stretching bands at 2 962 and 2 872 cm^{-1}. They are responsible for broadening of the band at high wavenumber, which sometimes appears as a shoulder at 2 960 cm^{-1} in spectra of coals and kerogens of low hydrogen content and high CH_3 content (kerogens of type III). The absorption due to symmetric stretching of CH_3 groups is responsible for a decrease of the valley separating the two peaks.

The band at 1 455 cm^{-1} is due to asymmetric bending of CH_2 and CH_3 groups. The band at 1 375 cm^{-1} is mainly due to symmetric bending of CH_3 groups; the CH_2 groups only give a negligible contribution at that wavenumber, except when they are close to an oxygenated function [18]. The apparent height of these bands may be misleading [19]; in fact they superpose to a broad absorption band which will be discussed below.

The relative abundance of CH_3 vs. CH_2 groups is reflected by the half band width of the peak at 2 930 cm^{-1}, by the depth of the valley separating the peaks at 2 930 and 2 860 cm^{-1}, and by the relative intensity of the bands at 1 375 and 1 455, or 1 375 and 2 930 cm^{-1}. Such parameters are not very accurate and can only be used to detect trends among series of samples. They show that the CH_3/CH_2 ratio tends to be higher in kerogens with lower hydrogen contents. If typical shallow materials are considered, it appears that coals and kerogens of type III have a higher CH_3/CH_2 ratio than kerogens of type I and II; on the other band catagenetic evolution leads to an increase of this ratio [28].

The depth of the valley between the peaks at 1 455 and 1 375 cm^{-1} may vary from one sample to another; in some cases, particularly in coals and in kerogens of series III, a small peak is resolved around 1 395-1 400 cm^{-1}. This absorption may be attributed to CH deformation of $C-(CH_3)_2$ or $C-(CH_3)_3$ groups [4, 25].

The spectra of some kerogens of high hydrogen content (samples Ia, b, c in Fig. 6.1) present a sharp band at 720 cm^{-1} which is due to skeletal vibration of straight chains with more than 4 CH_2 groups [4].

F. CH vibrations of unsaturated groups.

All coals and kerogens having reached a certain degree of evolution give absorption bands at 870, 820 and 750 cm^{-1} which are due to out of plane deformation vibrations of aromatic

CH groups, and correspond respectively to isolated, two adjacent and more than two adjacent hydrogen atoms [2]. The three bands have about the same intensity, whatever are the degree of evolution and the nature of parent OM. The stretching band of the aromatic CH groups is expected between 3 000 and 3 100 cm^{-1} but it is much weaker than the out of plane deformation band; it is only observed as a distinct band in the spectra of coals such as samples IVd and e [38] in Fig. 6.2.

Bands below 1 000 cm^{-1} in the spectra of shallow kerogens may be due to CH deformation of alkene groups; however they are extremely weak and have not been studied systematically.

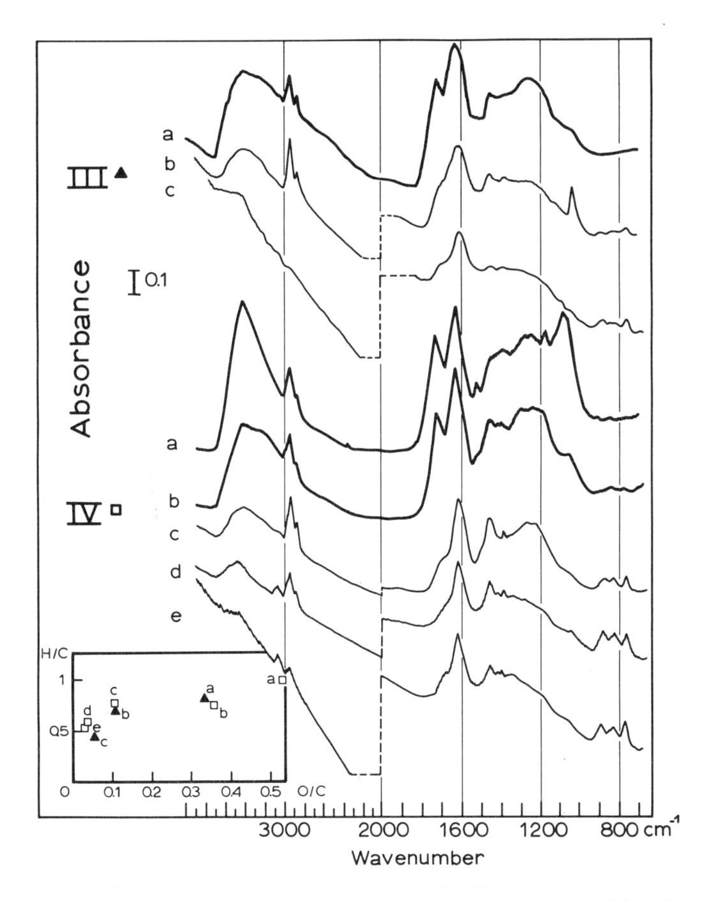

Fig. 6.2. — Representative IR spectra of coals (**IV**, □) and kerogens with an important contribution from higher plants in the parent material (**III**, ▲); the elemental composition is plotted on the joined Van Krevelen diagram.

IFP No: **IIIa:** 19288. **IIIb:** 12938. **IIIc:** 13381. **IVa:** peat 15557. **IVb:** lignite 15288. **IVc:** bituminous coal 15291. **IVd:** bituminous coal 20526. **IVe:** bituminous coal 19874.

G. Band near 1 710 cm^{-1}.

This band is due to C = O stretching of carbonyl (aldehyde, ketone) and carboxyl (acid, ester) [7, 4]. It is fairly broad and does not show details allowing to differentiate various components. It is quite intense for shallow materials but, at high degree of evolution, it may be absent or reduced to a weak shoulder. The band increases strongly when kerogens are heated in air at 200° C (unpublished observations); for coal, it also tends to develop because of oxidation resulting from grinding in air [19].

H. Band at 1 630-1 600 cm^{-1}.

This band has been extensively discussed in the literature of coal (see review by Robin and Rouxhet [29]) and, in fact, may contain contributions of different types of vibrations. Besides bending vibrations of molecular water, which can be removed or at least strongly reduced by aging the KBr pellets in a dessicator, absorption in that region may be assigned to C = O stretching of quinones bridged to acidic hydroxyl and to C = C stretching of olefins, aromatic rings and polyaromatic layers, eventually arranged in stacks of 2-3 units (Chapter 7).

In shallow material the band is fairly broad and centered around 1 630 cm^{-1}; for more evolved samples, it is shifted to 1 600 cm^{-1}. The shift seems to result mainly from a displacement of the high wavenumber side of the band, the low wavenumber side remaining in the range of 1 550-1 600 cm^{-1}. These modifications go along with the decrease of the band at 1 710 cm^{-1} and may be attributed to the removal of oxygenated functions.

I. Unresolved absorption between 1 800 and 930 cm^{-1}.

The spectra of kerogens and coals show unresolved absorption extending from the 1 710-1 630 cm^{-1} bands to 930 cm^{-1}, with a maximum around 1 200 cm^{-1}. The high wavenumber region (above about 1 400 cm^{-1}) is presumably due to extension and overlap of the bands already described, attributed to C = O and C = C stretching, and to C – H deformation. The low wavenumber region may be attributed to C – C skeletal vibrations of aliphatic and aromatic fragments, to C – O stretching of ether and alcohol groups and to C – OH deformation of alcohol.

The intense absorption observed between 1 000 and 1 100 cm^{-1} in the spectra of shallow material such as peat goes along with an intense OH stretching band and may be attributed to alcohol functions. The spectra of evolved products often show a small peak or a shoulder around 1 030 cm^{-1}, which may possibly be due to C – O vibration of aryl-alkyl ethers [6]; the C – O absorption of aryl-aryl ethers is expected to occur at higher wavenumber.

J. Comparison with other natural compounds.

The spectra of **humic acids** show the same features as those of kerogens and coals [36]. This is illustrated by spectrum e of Fig. 6.3, which presents much similarity with the spectra of lignite (sample IVb in Fig. 6.2) or shallow kerogens originating from higher plants (sample IIIa in Fig. 6.2); note however that the band at 1 710 cm^{-1} is appreciably more intense.

For many humic acids, having a high nitrogen content (3-6%) the bands at 1 710 and 1 630 cm^{-1} are masked by an intense band at 1 660 cm^{-1}, which is accompanied by a band at 1 540 cm^{-1}, as illustrated by spectrum d of Fig. 6.3. They are attributed to amide I band (CO stretching) and amide II band respectively [2]. When the humic acids are submitted to acid treatment, the residue does no longer show these bands but the modification of the OH stretching band indicates the presence of more carboxylic acids [9, 22].

Asphaltenes originating from kerogen rocks also give spectra similar to kerogens [8]. The saturated CH stretching bands are usually more intense, in agreement with a higher hydrogen content; the broad absorption band shows more details and is relatively less intense. As illustrated by spectrum a of Fig. 6.3, a peak is frequently observed at 1 660 cm^{-1}, in addition to the bands already described at 1 710 and 1 630 cm^{-1}. For kerogens the same absorption is presumably present but is not resolved with respect to the two other overlapping bands; it appears as a distinct shoulder in the spectrum Id of Fig. 6.1. It is likely not due to amide I band, because of the low nitrogen content; possible attributions are quinonic groups or aro-

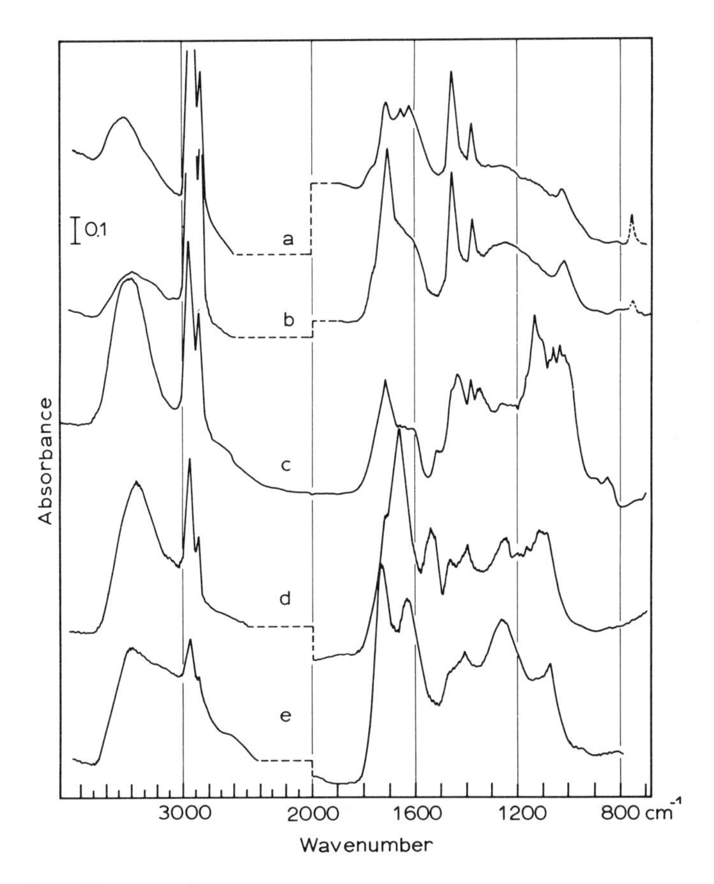

Fig. 6.3. — IR spectra of:

a, b. asphaltenes associated to kerogen IIa (**a.** new preparation. **b.** aged preparation). **c.** sporopollenin isolated from spores of lycopode (IFP No GF 5242). **d, e.** humic acids ([9], **d.** Leman. **e.** Guinée).

matic acids with hydroxyl substitution in adjacent position, allowing intramolecular hydrogen bonding. Spectrum b of Fig. 6.3 has been obtained for an aged preparation of the same asphaltene; it is clear that oxidation has produced an increase of the absorption in the range of 1 000-1 300 cm^{-1} and of the band of carbonyl and carboxyl groups at 1 710 cm^{-1}.

A frequent approach for investigating the properties of carbonaceous solids is to study model compounds; pyrolysis of the latter is often used to simulate industrial processes based on high carbon solids or catagenetic evolution [44, 19]. **Sporopollenin** isolated from pollen or spores is considered as an interesting model compound for the parent organic matter which leads to formation of kerogens with a high content fo aliphatic groups [24, 45]. The spectrum of a typical sporopollenin, isolated from spores of lycopode, is presented in Fig. 6.3 (c). It shows many features commented before; like peat and lignite, it has a very intense OH stretching band and a high absorption between 1 000 and 1 100 cm^{-1} attributed to alcohol functions; the strong band around 1 140 cm^{-1} would rather be due to ether functions [2].

K. Band near 1 510 cm^{-1}.

Peat and lignite (Fig. 6.2, IVa and b) show a band or a shoulder near 1 510 cm^{-1}, which must have an assignment different from amide II band of humic acids. A similar absorption feature is also observed in the spectrum of sporopollenin. It is attributed to the C=C vibration of aromatic rings, like the band at 1 630 cm^{-1} [4, 2, 21]. In more evolved kerogens and coals the band is no longer resolved but is probably spread out over a broader spectral range and included in the broad absorption band discussed above.

III. EXPERIMENTAL ASPECTS

A. Sample preparation.

The KBr pellet method is the most common one for obtaining IR spectra of powdered solids and has been applied with success to kerogens and coals. The role of the medium is not only to hold the powder but also to reduce light scattering, which decreases as the refractive index of the medium is closer to the refractive index of the imbedded particles. The refractive index of kerogens in the visible region ranges from 1.5 to 1.8, according to the carbon content [17]; this suggests that potassium bromide ($n \simeq 1.54$) is a suitable medium. Alkali halides of higher refractive index, such as cesium iodide ($n = 1.74$ around 3 500 cm^{-1}), might possibly be more adequate for samples of high carbon content but have been rarely used.

For some spectra, the connection of the bands at 2 930-2 860 cm^{-1} with their baseline on the low wavenumber side is quite different from the high wavenumber side, giving an impression of strong asymmetry as illustrated by spectrum Ic of Fig. 6.1. The same phenomenon is frequently observed for the band of pyrite around 425 cm^{-1}. It is due to a Christiansen effect [14, 27]: in the vicinity of an absorption band the refractive index may change abruptly and the effect of light scattering may be stronger on one side of the band than on the other side. However the Christiansen effect has never been a big problem in our spectra.

The procedure followed in our laboratory consists in grinding first the kerogen in an agate mortar; an equivalent amount of coarsely ground KBr is then added and the two components are ground together; further addition of KBr is made by fractions equivalent to the amount of material already in the mortar and at each step the mixture is ground before another fraction is added. This procedure is required to insure a homogeneous distribution of kerogen in KBr. Better results are obtained with the agate hand mortar than with mechanical ball grinders especially made for KBr pellet preparation.

When band intensities are measured, the mixture must be pressed into a pellet of regular shape by using an appropriate die. When the spectra are used only for qualitative purpose, an alternative procedure consists in replacing the die by a piece of cardboard with a suitable hole in which the powdered mixture is placed and pressed; this procedure is less time consuming and allows to handle and store the KBr pellet fixed in the piece of cardboard, on which references can be written. According to our experience, it is not required to evacuate the die and powder during pressing.

As discussed above, storing the pellets in a dessicator for one week may be used as a standard drying procedure.

The effective thickness of the layer of kerogen varies between 0.7 and 3 mg/cm^2 depending on the type of sample; kerogen concentration in KBr is typically of the order of 1%.

B. Spectrograph.

Light scattering and continuous absorption reduce the intensity of the light reaching the detector in spectral regions without absorption band, in other words the overall transparency of the pellet. As the spectrograph has to work with a low intensity level, it is necessary to use an instrument of good performance and to set it carefully in order to obtain undeformed spectra, especially if the intensity of absorption bands is measured.

The slit must be small enough to avoid artificial broadening of the absorption bands, which leads to a loss of resolution and also gives rise to an error on the band area; for that purpose the spectral width corresponding to the chosen slit width should be 5-10 times smaller than the half band width of the sharpest absorption band to be recorded. Provided this is insured, the slit must be large enough, so that the intensity of the light reaching the detector is not too weak. Amplification or gain must be set to have a sufficient output signal. Using a smaller slit may thus require higher amplification; note that the output signal is roughly proportional to the square of the slit width and proportional to the gain. The time constant is chosen to keep an acceptable noise, the latter depending on the amplification. The scanning speed is selected, taking into account the width of the absorption bands and the time constant.

C. Background absorption.

Let us first consider a spectrum recorded in transmittance. If the reference beam is not attenuated ($I_r = I_0$), the background or baseline of the spectrum $T_b = I_b/I_0$, reflects the non specific loss of intensity resulting from scattering and continuous absorption. The effect of absorbing species is to further decrease the transmittance by a factor T_s. The total transmittance is then:

$$T_t = I/I_0 = (I/I_b)\,(I_b/I_0) = T_b\,T_s$$

where T_s is expected to follow the Beer-Lambert law. The position of the baseline on the chart paper may be changed by introducing a comb in the reference beam; the resulting spectrum is such that the ratio T_t/T_b remains constant for the absorption bands, but consequently their height and shape are modified.

On a spectrum recorded in absorbance, the contribution of specific absorption is measured by the difference between the total absorbance measured in an absorption band and the absorbance of the baseline: $A_s = A_t - A_b$. Displacement of the latter may not affect the height or the shape of the band and the height is expected to follow the Beer-Lambert law.

When spectra are compared, comparison of band heights and band shapes are always useful, in addition to band position. Therefore, even if the band intensities are not measured as such, the spectra should be recorded in absorbance.

It is interesting to note that a good correlation has been found between the background absorbance measured at 2 000 cm^{-1} and the carbon content, for various types of sedimented organic matter [23].

It has been checked that the background absorbance does not influence appreciably the band shape and height. This was done by recording the spectrum of KBr pellets of a kerogen ($\approx$ 1.6 mg/cm^2) with various amounts of graphite added. Of course, as the amount of graphite increased, the reference beam had to be more and more attenuated in order to keep the baseline at the same level of the chart paper. Simultaneously spectrograph setting had to be adjusted: as compared to the pellet without graphite, the pellet with 0.30 mg/cm^2 of graphite required a slit about twice larger, which was the highest value acceptable for recording the CH stretching bands, and a gain 5 times larger, which required a larger response time to keep the noise reasonably low. With more graphite, proper recording of the spectrum was practically impossible due to the low overall intensity level.

D. Measurement of band intensity.

1. General procedure.

When an absorption band, recorded in absorbance, does not overlap with another, its intensity may be measured by its height with respect to the baseline. However a recommended procedure is to measure its area or integrated absorbance, tracing the baseline by interpolation from both sides. The band area is less sensitive than the band height to instrument deformations and errors related to baseline tracing. Moreover, the study of minerals [33, 32] has shown that the use of band areas allows to compare the absorption coefficients of a band in various samples, even if the band shape varies.

The Beer-Lambert law may thus be used under the following form:

$$\int_{\text{band}} A \, d\nu = K.d$$

where: d is the thickness of the absorbing sample, which for a powder may be conveniently expressed as an effective thickness in mg cm^{-2} and K is an integrated absorption coefficient, characteristic of the solid, which is given in wavenumber $\times$ absorbance unit $\times$ mg^{-1} $\times$ cm^2, or mg^{-1} cm. When similar solids are compared, K is expected to be proportional to the content of the chemical groups responsible for the band considered. If the thickness d taken into consideration is corrected for the ash content, the absorption coefficient K is characteristic of the carbonaceus fraction of a kerogen concentrate.

According to the above discussion on background absorption, the values of K are not influenced by difference of overall transparency.

2. *Pellet homogeneity.*

The application of Beer-Lambert law requires that the absorbing sample presents a constant thickness with respect to the light beam. The meaning and the consequences of this requirement can be easily understood by considering the effect of making a hole of 0.1 cm^2 in a sample of 1 cm^2 area exposed to the beam. Let this sample have a background transmittance T_b and a total transmittance T_t and give an absorption band characterized by $T_s = T_t/T_b$ or an absorbance $A_s = A_t - A_b$. The consequence of the hole is that the apparent reduction of the light intensity due to specific absorption is equal to:

$$T_{sa} = \frac{0.9\, I + 0.1\, I_0}{0.9\, I_b + 0.1\, I_0}$$

which is larger than $T_s = I/I_b$. It follows that the apparent absorbance of the band is smaller than the true absorbance.

The deviation increases as I decreases, therefore the effect may be responsible for negative deviation of a Beer-Lambert plot.

If an absorption band is scanned, the perturbation is stronger as I becomes smaller with respect to I_b ; therefore such a hole tends to produce a flattening of the absorption bands and an increase of their half height width.

The effect is more marked if I_b is smaller with respect to I_0; therefore it is more critical for samples which give high scattering than for samples having a good overall transparency.

Detailed discussion of powdered materials can be found in the literature [14]. For practical purpose, the absorbing powder dispersed in a KBr pellet may be considered to have a reasonably homogeneous effective thickness if the particles are homogeneously distributed and if the particle size is much smaller than the average effective thickness. Note that for a specific weight of 1, an effective thickness of 1-2 mg/cm^2 which is frequently used for kerogen, corresponds to 10-20 μ.

3. *Band surfacing.*

In the spectra of kerogens and coals, many bands overlap and the measurement of absorption coefficients requires separation into individual bands and separation from the baseline before band surfacing. The procedure used in our laboratory is illustrated by Fig. 6.4.

The measurement of the total absorption coefficient of the CH stretching bands (2 930-2 860 cm^{-1}) K_{al}, on one hand, and of aromatic CH deformation bands (870, 820, 750 cm^{-1}) K_{aro}, on the other hand, does not give any problem (Table 6.1).

For the OH stretching band (K_{OH}) difficulties arise in the case of strongly scattering samples [30]. The standard procedure we have used for all the kerogens and coals consists in drawing the baseline over a broad spectral range and measuring the area of the band sketched between 3 700 and 2 700 cm^{-1}.

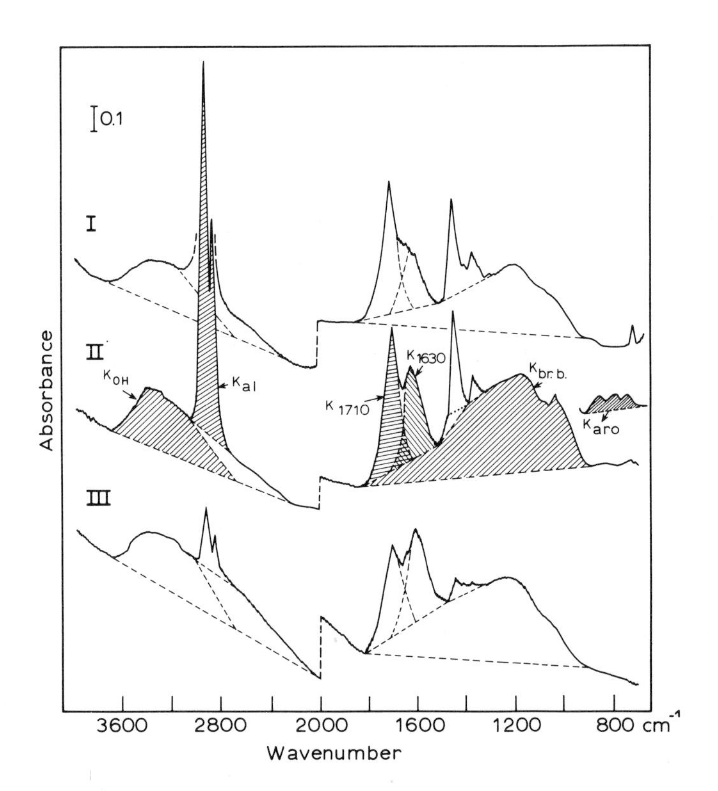

Fig. 6.4. — Typical spectra illustrating the band separation for the measurement of absorption coefficients.

TABLE 6.1
ABBREVIATIONS FOR THE PARAMETERS DETERMINED FROM THE SPECTRA

K	: Absorption coefficient in wavenumber × absorbance unit × mg^{-1} × cm^2.
K_{OH}	: OH stretching band, 3 700-2 700 cm^{-1}.
K_{al}	: CH stretching bands of alkyl groups, 2 930 – 2 860 cm^{-1}
K_{1710}	: C=O stretching band, centered around 1 710 cm^{-1}.
K_{1630}	: band at 1 630 – 1 600 cm^{-1}, due to C=O stretching of bridged quinones and C=C stretching of olefins, aromatic rings and polyaromatic rings.
K_{1455}	: asymmetric bending of CH_2 and CH_3 groups at 1 455 cm^{-1}.
$K_{br.b}$	: broad band extending between 1 800 and 930 cm^{-1}.
K_{aro}	: aromatic CH deformation bands at 870, 820 and 750 cm^{-1}.
I_{C-OR}	: difference between the oxygen content (%) and 1/2 K_{1710}, index of oxygen present in functional groups other than carbonyl and carboxyl.

For the bands at 1 710 and about 1 630 cm^{-1}, the baseline was traced as a straight line from about 1 800 cm^{-1} to the adjacent walley near 1 500 cm^{-1}; the two bands are separated from each other as indicated in the Figure.

The absorption between 1 800 and 930 cm^{-1} is quantified as a single broad band, the baseline of which is taken as a straight line between these two wavenumbers; the resulting absorption coefficient was called $K_{br.b}$. For its measurement, the separation from the bands at 1 455 and 1 375 cm^{-1} is not critical. However determining the absorption coefficient of the latter is more delicate; account must be taken of the shape of the bands and the results should be used with caution. When the two bands are well separated, their baseline can be traced between the points at which absorbance changes steeply as indicated by the dotted line in Fig. 6.4, spectrum b; in other cases it seems more adequate to consider the presence of three bands, the central component around 1 400 cm^{-1}, being assigned to deformation of $C-(CH_3)_2$ and $C-(CH_3)_3$ groups; in some samples the intensity of these bands is so weak that its measurement is not reliable.

4. *Practical evaluation.*

It has been pointed out that the spectra of coals and kerogens are nearly a limit case for the use of band intensities. It appears also that the procedure used for surfacing the bands presents some arbitrary aspects. However, if the procedures are standardized and applied with care, these limitations are not critical, particularly when the purpose is to compare various samples.

In practice good straight lines are obtained when integrated areas are plotted as a function of the effective thickness d. A given experimentalist, working carefully, may obtain standard deviation on K of $\pm$ 5 to 10%. However deviations from one experimentalist to another may be more important; in extreme cases K values may differ by a factor of 2. Comparative determinations have shown that the main source of deviation is due to the KBr pellet preparation, particularly grinding, and not to the steps of recording or surfacing the absorption bands.

According to our experience, hydrogen rich kerogens, for instance shallow kerogens of types I and II, are more difficult to grind and to disperse in a homogeneous way than coals and kerogens of type III. However, as the latter are darker, they give more light scattering and the requirements on homogeneity and particle size are higher.

Grinding in liquid nitrogen may be a useful procedure; however it complicates sample handling and is not suitable when very small amounts of material are available. Using an interferometer instead of a dispersive spectrograph may improve the accuracy of K value, by allowing to record spectra of samples with a larger effective thickness and making thus less critical the requirement of homogeneous effective thickness; however these instruments are about 5 times more expensive than good performance dispersive spectrographs.

E. Complementary use of chemical treatment.

Determining the contribution of various chemical functions in the infrared bands of kerogen has been attempted [29] by measuring the variation of absorption coefficient, ΔK, which results from the application of chemical treatment transforming selectively certain functional groups into groups absorbing at different wavenumber. The methods are inspired by the procedures of functional analysis of coal, the classical methods of analysis such as titration, gas volumetry and elemental analysis being replaced here by recording the infrared spectra.

This approach has been applied to the band at 1 710 cm^{-1}, using reactions of oximation of carbonyl, amidation of acids, saponification and reduction to estimate the contribution of carboxyl groups, carboxylic acids and esters respectively [31]. These contributions can be tentatively converted into percent of oxygen present in the corresponding functions, on the rough basis that 1% oxygen present as carbonyl, acid or ester is responsible for an increment of absorption coefficient ΔK_{1710} of about 2 wavenumber × absorbance unit × $mg^{-1} \times cm^2$. The difference between the oxygen content (%) corrected for the ash content, and the half of K_{1710}, called I_{C-OR}, is considered as an index giving a rough estimate of the oxygen present in other functional groups such as ethers and non carboxylic hydroxyls.

Reduction by $LiAlH_4$ is thought to remove the contribution of bridged quinones from the band at 1 630 cm^{-1} [29,31]. Bromination was applied in order to remove the contribution of olefins from the same band; however various reactions may take place besides addition of bromine to olefins [46,47]. When present, the band at 1 520 cm^{-1} is removed by bromination (unpublished data) and in all cases the absorbance near 1 500 cm^{-1} is decreased. This suggests that the absorption of aromatic rings may be decreased or even wiped out by bromination; consequently the part of the band at 1 630 cm^{-1} which is affected neither by reduction nor by bromination is tentatively attributed to polyaromatic structures.

Attempts to use chemical treatment for distinguishing the contributions of various types of OH groups to the OH stretching band have not been successful.

IV. COMPARISON OF KEROGENS OF DIFFERENT ORIGINS AND DEGREES OF EVOLUTION

A. Overall view.

The spectra of shallow kerogens presented in Figs. 6.1 (Ia,IIa) and 6.2 (IIIa) reflect the differences of elemental composition, due to the origin of the material. Kerogens of types I and II, the parent material of which contained an important contribution of algae and which are characterized by a high H/C ratio, give intense bands due to CH_2 and CH_3 groups. Kerogens of type III, mainly originating from higher plants, give spectra very close to those of coals of similar O/C ratio.

The same figures illustrate the close parallelism of evolution between the spectra and the elemental composition, observed for coals [7, 10] and kerogens [28, 11, 41]. The first step of evolution, which is marked by a decrease of the O/C atomic ratio, provokes a strong reduction of the band of carbonyl and carboxyl groups at 1 710 cm^{-1} and of the OH stretching band. The second step, which gives a strong decrease of the H/C ratio, goes along with the removal of the bands of CH_2 and CH_3 groups and corresponds to liberation of oil.

The same trends are followed when the evolution is simulated by pyrolysis of shallow kerogens under inert atmosphere, by temperature programming of 4° C/min [28]. However differences between catagenesis and pyrolysis will appear below, after more detailed examination of the spectra.

The measurements of absorption coefficients, the meaning of which is recalled in Table 6.1, and the eventual combination with chemical treatment increase strongly the amount of information provided by the IR spectra. This will be illustrated by data covering a broad range of samples, with respect to the parent organic matter and the degree of evolution. They concern kerogens of types I, II and III mentioned before and originate from our recent work [28, 31, 34]; quantitative data obtained for coals studied by Durand *et al.* [10] are also included; data on free radicals originate from an electron spin resonance (ESR) study by Durand *et al.* [12].

In order to facilitate a comparison of kerogens according to their origin and to their degree of evolution, the figures will give plots of absorption coefficients *K*, calculated with respect to the OM and thus corrected for the ash content, as a function of the reflectance *R* of vitrinite in oil, which roughly indicates an overall degree of evolution of the OM. (Chapter 11).

B. Correlations concerning hydrocarbonous functions.

Figure 6.5 presents the correlation between the absorption coefficient K_{al}, reflecting the concentration of saturated hydrocarbonous moieties, and the hydrogen content; the low value of the slope as compared to alkanes, is imputed to the vicinity of C=C and C=O bonds. This correlation does not show any significant deviation according to the nature of

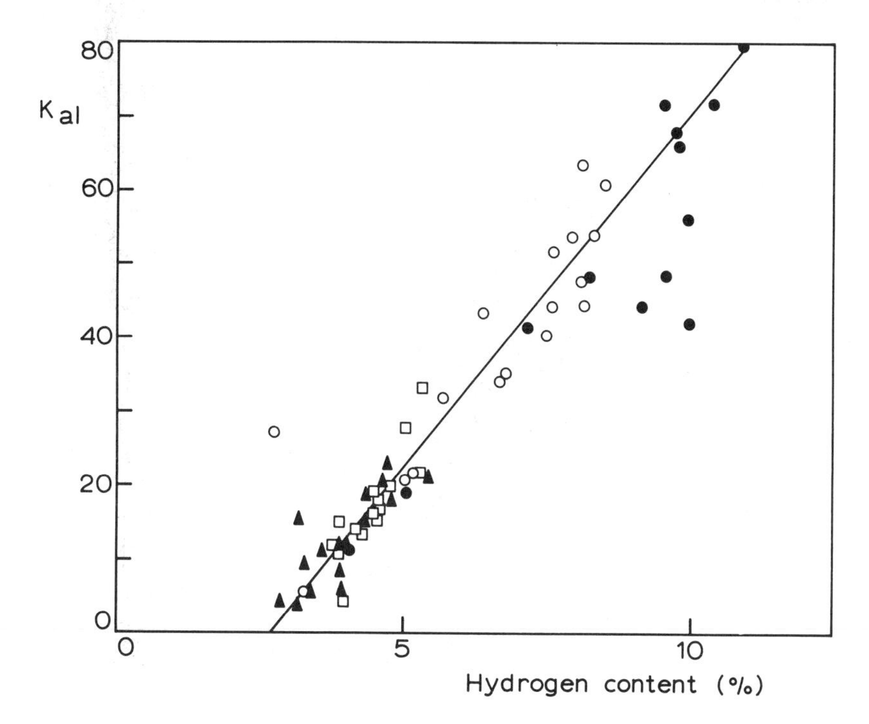

Fig. 6.5. — Correlation between K_{al} and the hydrogen content (weight %):

●: kerogens of series I. ○: kerogens of series II. ▲: kerogens of series III. □: coals.

the parent material or the degree of evolution; however the ratio K_{al}/K_{1455}, which decreases under the influence of surrounding C = C and C = O bonds, indicates that the latter are more important in products originating from higher plants [28]. Moreover a detailed analysis of the shape of the stretching and deformation bands indicates variations of the CH_3/CH_2 ratio which have been mentioned in paragr. II.E.

A single linear relationship has also been observed between the intensity of the aromatic CH band of various types of sedimented organic matter and the percent of aromatic carbon deduced from elemental composition and density [23].

The non zero intercept in Fig. 6.5 shows that the residual hydrogen (H/C $\simeq$ 0.4) present in strongly carbonized material may not be accounted for by CH_2 and CH_3 groups similar to those present in shallow materials. There is little information on the chemical nature of the residual hydrogen; it may not be entirely attributed to OH or aromatic CH groups similar to those observed in less evolved kerogens [28].

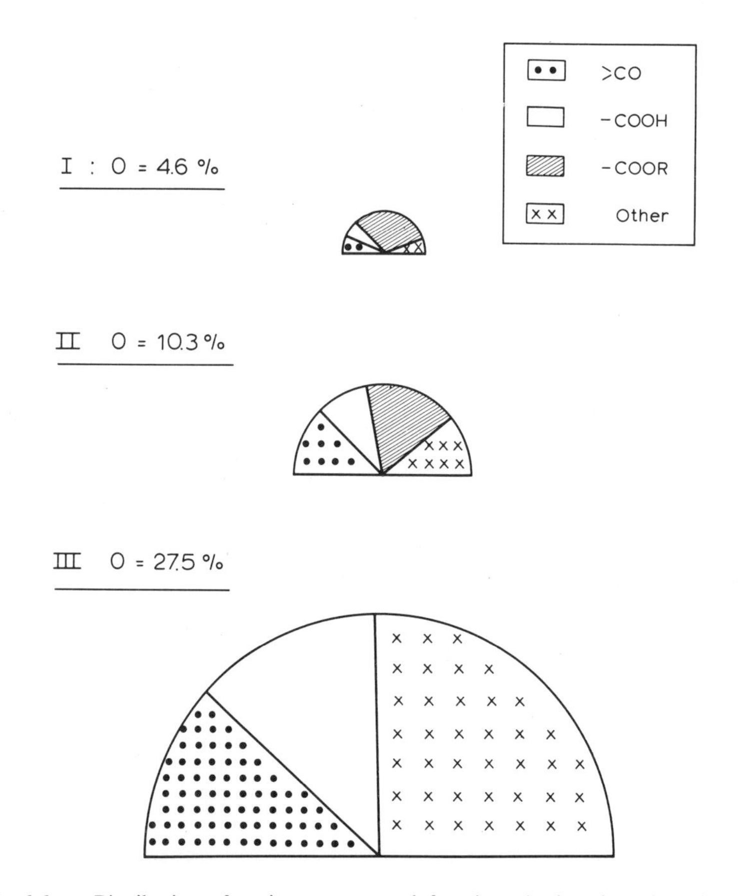

Fig. 6.6. — Distribution of various oxygenated functions (carbonyl, carboxylic acid, ester, other) in shallow kerogens belonging to the three series studied:

I: IFP No 15769. **II:** IFP No 10968. **III:** IFP No 19288 [31].

C. Comparison of shallow kerogens and coals.

The differences in the intensity of CH_2 and CH_3 bands have been pointed out above. Figs. 6.7 and 6.10 show that the values of K_{1710}, K_{OH} and I_{C-OR} are much higher in lignites and kerogens of type III than in kerogens of types I and II.

Combination of chemical treatment and infrared analysis has allowed to estimate the contributions of carbonyl groups, of acids and of esters respectively, to the band at 1 710 cm^{-1} [31]. The results obtained for shallow kerogens are illustrated by Fig. 6.6. The total oxygen content, the content of carbonyl and carboxylic acid groups, and the content of oxygenated functions other than carbonyl or carboxyl increase according to the sequence I < II < III. The relative abundance of ester groups is appreciable in kerogens of types I and II but negligible in kerogens of type III. The presence of many carboxylic acids in the latter is in agreement with the shape of the OH stretching band which indicates the presence of strongly hydrogen bonded hydroxyl (Fig. 6.2).

The absorption coefficient K_{1630} (Fig. 6.9) is much higher for lignites and shallow kerogens of type III than for shallow kerogens of type I and II. This difference concerns all the contributions to the band; that of quinones, of olefins and aromatic entities, of polyaromatic structures [31].

The absorption coefficient $K_{br.b}$ (Fig. 6.9) shows the same differences as K_{1630} according to the type of parent material. The higher values shown by lignites and kerogens of type III are due to a higher content of unsaturated hydrocarbon structures (particularly aromatic and polyaromatic) and oxygenated functions.

D. Effect of catagenesis.

The sharp decrease of the oxygen content in the region below $R = 1\%$ is accompanied by a drop of the content of carbonyl and carboxyl groups (Fig. 6.7). A detailed analysis has shown that carboxylic acids are the most sensitive, while the decrease of carbonyl groups extends over the whole range of R-values investigated, a fraction of them being preserved up to advanced degrees of catagenesis. These changes are accompanied by a sharp decrease of the OH stretching band (Fig. 6.10). For material originating from higher plants the content of saturated hydrocarbonous structures (K_{al} in Fig. 6.8) increases due to relative enrichment resulting from the loss of strongly oxygenated compounds and subsequently decreases. The same trend is observed for series I and II when K_{al} is plotted as a function of the burial depth.

Upon further evolution (R from 1 to 2%, corresponding approximately to bituminous coals), liberation of residual groups absorbing at 1 710 cm^{-1}, mainly carbonyl, and hydrocarbons proceeds further (Figs. 6.7. and 6.8).

These chemical modifications are accompanied by the formation of aromatic CH groups, which is closely parallel to the formation of free radicals (Chapter 8), as illustrated by Fig. 6.8. The increase of aromatic CH groups and free radicals reflects the disappearance of substituents of aromatic rings.

The evolution of K_{OH}, I_{C-OR}, K_{1630} and $K_{br.b}$ differs appreciably according to the parent organic matter. For coals and kerogens of type III, the parameter I_{C-OR}, which reflects the relative abundance of hydroxyl but also of ether groups, decreases strongly (Fig. 6.10).

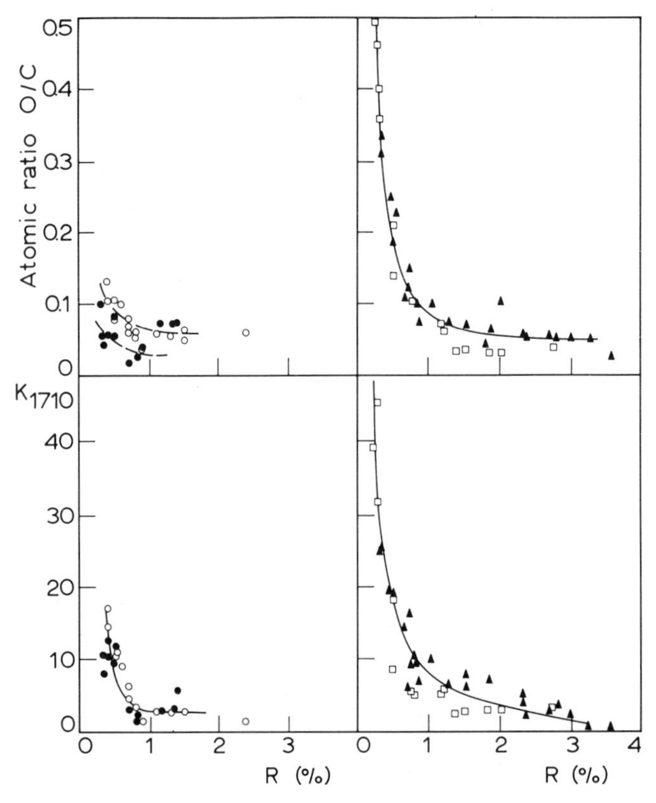

Fig. 6.7. — Evolution of the O/C atomic ratio and K_{1710} as a function of reflectance R:

●: kerogens of series I.
○: kerogens of series II.
▲: kerogens of series III.
□: coals.

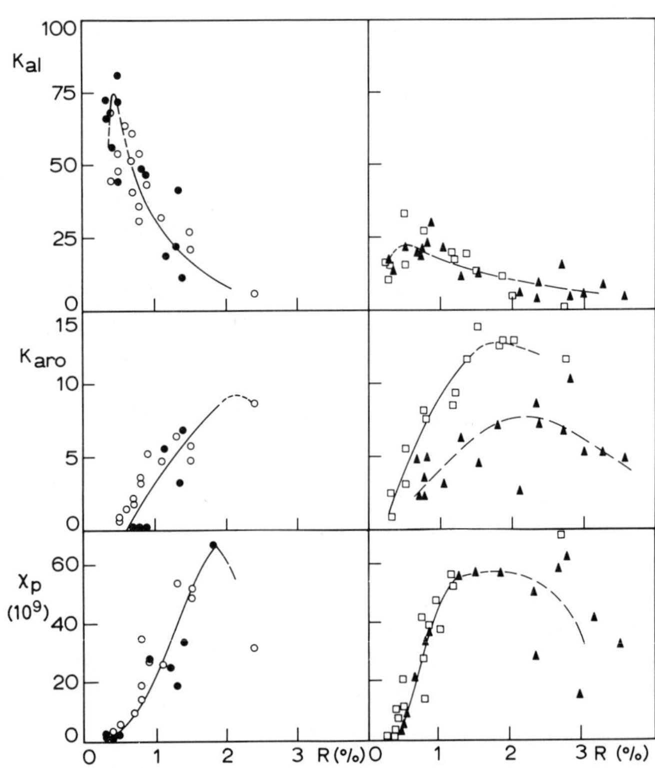

Fig. 6.8. — Evolution of K_{al}, K_{aro} and the paramagnetic susceptibility ($\chi_P \times 10^9$ in emu CGS g^{-1}) as a function of reflectance R:

●: kerogens of series I.
○: kerogens of series II.
▲: kerogens of series III.
□: coals.

Simultaneously there is a marked decrease of $K_{br.b}$ and K_{1630} (Fig. 6.9) and of the effect of reduction and bromination on the 1 630 cm^{-1} band [31] (data for coals to be published); this is due to removal of contributing oxygenated functions and olefinic bonds and to elimination of aromatic rings, either through liberation of aromatic compounds or through condensation into polyaromatic entities characterized by lower specific absorption coefficients. For kerogens of type I and II, these modifications occur possibly to a lesser extent but I_{C-OR}, K_{OH}, K_{1630} and $K_{br.b}$ rise appreciably below $R = 1.5\%$; this is attributed to a relative enrichment resulting from the release of a great amount of hydrocarbons.

In the evolution range corresponding to R-values of 1.5-2, the kerogens of different origins and coals reach about the same elemental composition (H/C $\simeq$ 0.5; O/C $\simeq$ 0.05). They are also very similar with respect to more detailed chemical features such as the content of aro-

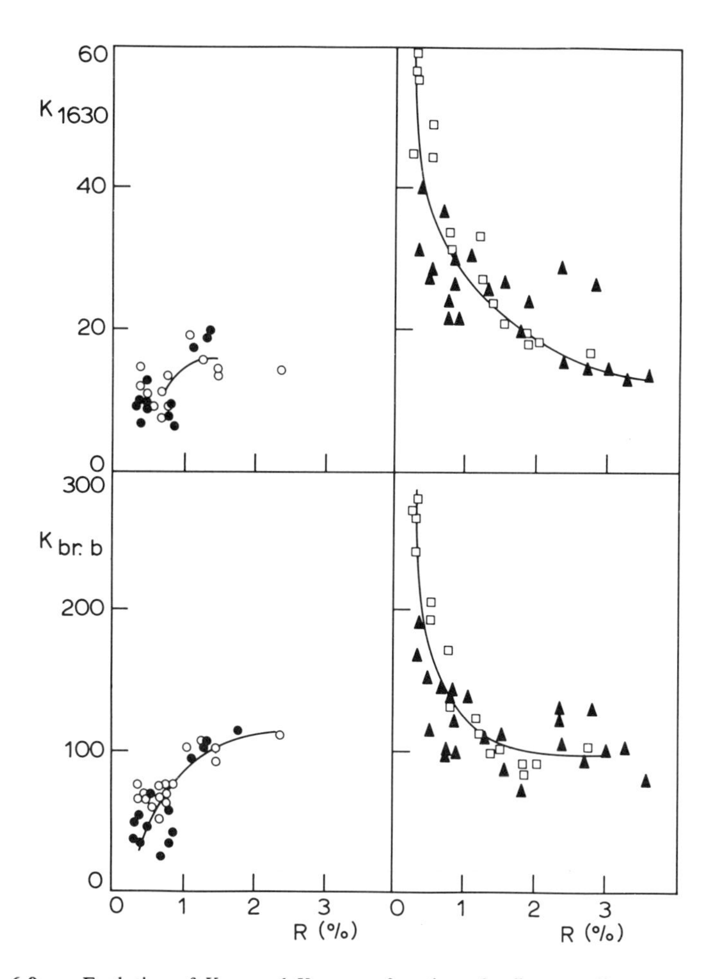

Fig. 6.9. — Evolution of K_{1630} and $K_{br.b}$ as a function of reflectance R:

●: kerogens of series I. ○: kerogens of series II. ▲: kerogens of series III. □: coals.

matic CH groups and free radicals (Fig. 6.8), of hydroxyl and other C–OR groups (Fig. 6.10) and the characteristics of the carbonaceous skeleton (Fig. 6.9); the residual band at 1 630 cm^{-1} is practically insensitive to bromination. At that stage the material contains mainly stacks of 2 or 3 polyaromatic layers bearing H atoms, short aliphatic chains, OH and other C–OR groups. The most significant difference found between kerogens of different origins is the size of the clusters in which these polyaromatic stacks are oriented more or less parallel to each other (Chapter 7).

Above *R*-values of about 2, which correspond approximatively to the rank of anthracite, the baseline of the IR spectra is much deformed by scattering and surfacing the OH stretching band is difficult. Besides the latter, the band at 1 600 cm^{-1}, the broad band extending from 1 800 to 930 cm^{-1} and the bands of aromatic CH are the only ones still present in the spectrum. They do no longer change appreciably, however the few data available indicate that their intensity tends to decrease slightly as reflectance increases.

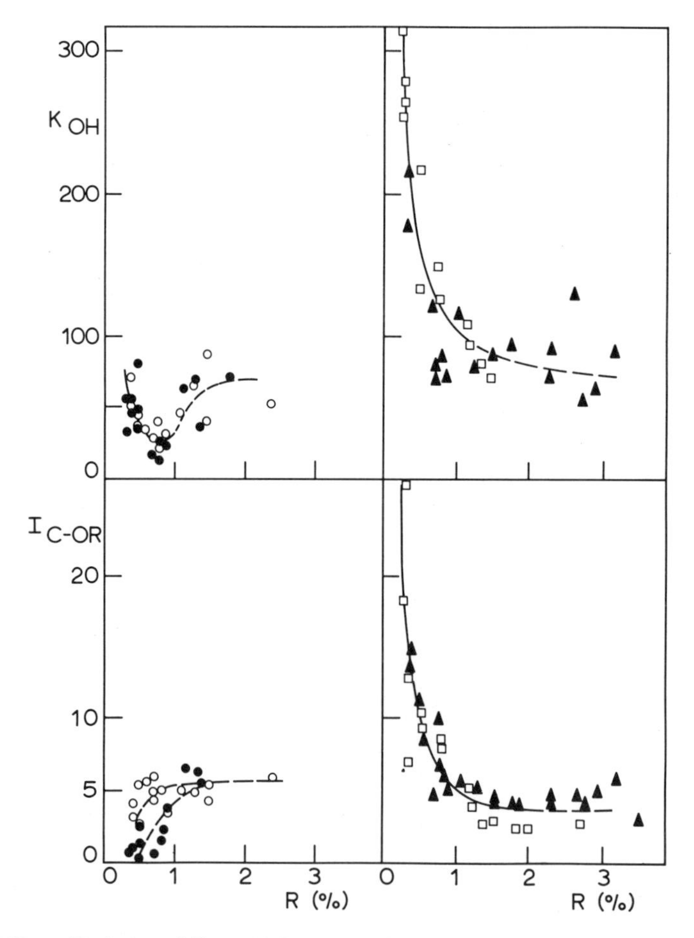

Fig. 6.10. — Evolution of K_{OH} and I_{C-OR} as a function of reflectance *R:*

●: kerogens of series I. ○: kerogens of series II. ▲: kerogens of series III. □: coals.

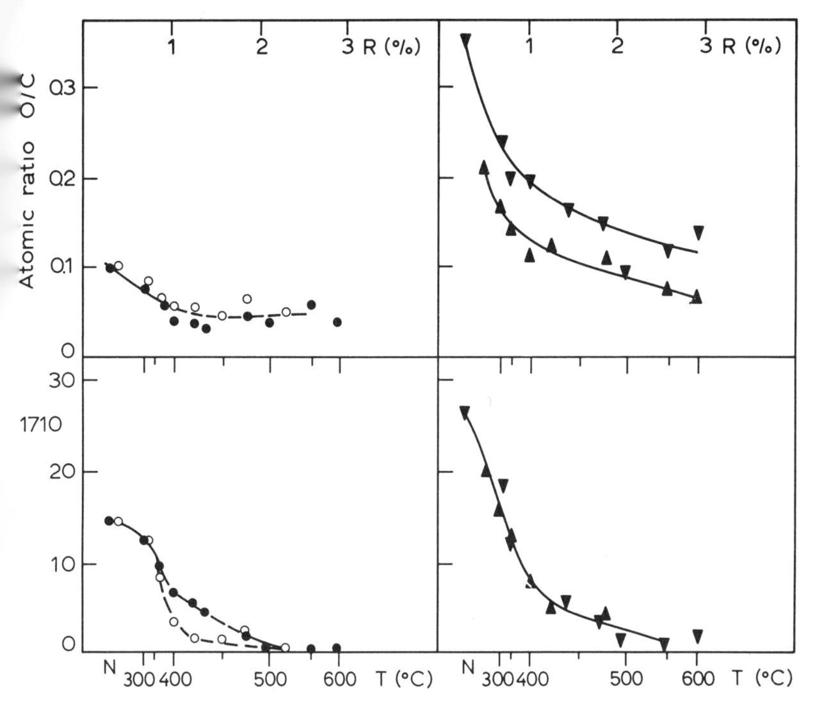

Fig. 6.11. — Effect of pyrolysis of shallow kerogens on the O/C atomic ratio and K_{1710} :

● : kerogen of type I.

○ : kerogen of type II.

▲ ▼: kerogens of type III.

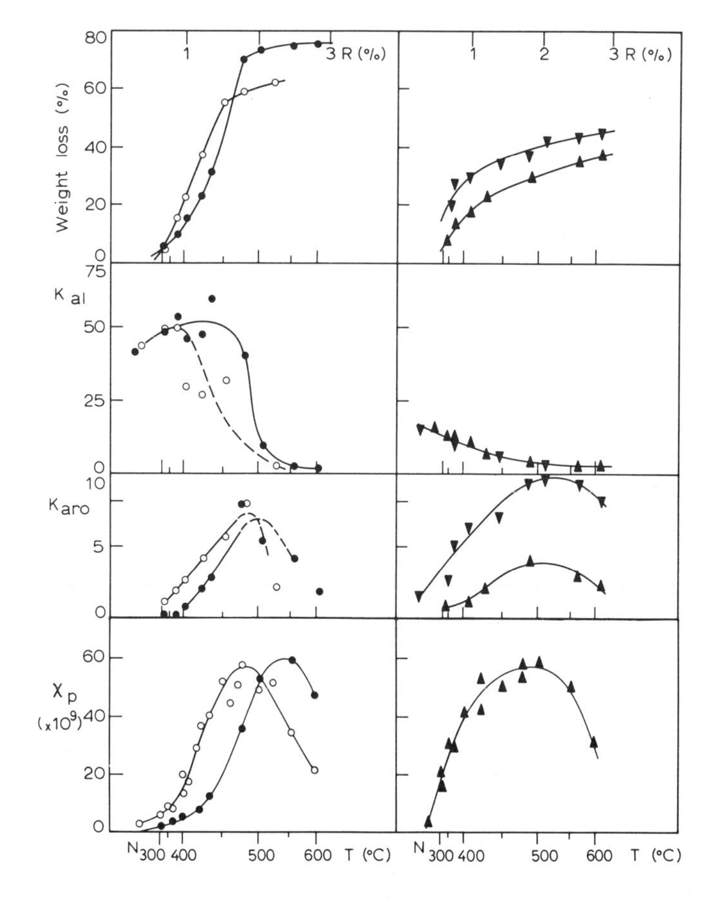

Fig. 6.12. — Weight loss curves of shallow kerogens and effect of pyrolysis on K_{al}, K_{aro} and the paramagnetic susceptibility ($\chi_p \times 10^{-9}$ in emu CGS g^{-1}):

● : kerogen of type I.

○ : kerogen of type II.

▲ ▼: kerogens of type III.

E. Effect of pyrolysis.

The effect of pyrolysis on the IR spectra of various types of kerogens are illustrated by Figs. 6.11, 6.12, 6.13, and 6.14, in which the various parameters are plotted as a function of the final temperature reached during pyrolysis of shallow material with programmed temperature (4° C/min). In order to allow a rough comparison between these figures and those illustrating the effect of catagenesis, the temperature scale is not linear but fits a reflectance scale, through a calibration obtained by pyrolysis of a shallow kerogen of type III.

1. Similarity with catagenesis.

Many aspects of the modifications of the spectra are similar for natural evolution and for pyrolysis of shallow kerogens:

(a) The elimination of saturated hydrocarbons is preceded by a marked decrease of the content of carbonyl and carboxyl groups (Figs. 6.11 and 6.12).
(b) The carboxylic acids are very sensitive while the removal of carbonyl groups is extended over a broad range of temperatures [31].
(c) The loss of hydrocarbons is accompanied by the formation of aromatic CH groups which are lost upon further evolution (Fig. 6.12).
(d) The evolution of the relative abundance of aromatic CH groups is similar to that of free radicals (Fig. 6.12.)

2. Particular aspects.

When **kerogens of type I and II** are heated, K_{al} shows a slight increase and decreases only when a high proportion of carbonyl and carboxyl groups has been removed, as indicated by a comparison of Figs. 6.11 and 6.12. However this may be misleading because the heated material has suffered a weight loss shown in Fig. 6.12; normalizing the band area not to the amount of analyzed organic matter but with respect to the amount of material which had to be heated to provide the analyzed OM, shows that hydrocarbonous functions are lost before the sharp decrease of K_{al}. This loss does not provoke an appreciable change of the H/C atomic ratio and may be attributed to liberation of tars with about the same composition as kerogen, arising from bond breakage due to decomposition of carbonyl and carboxyl functions; this interpretation is in agreement with a mass spectroscopic study of the products evolved [39].

The increase of K_{1630}, $K_{br.b}$ and K_{OH} (Figs. 6.13 and 6.14) is clearly due to a relative increase of concentration of the concerned functional groups, due to removal of hydrocarbons. It is concomitant with the formation of aromatic CH groups and free radicals (Fig. 6.12). The absorption coefficients K_{1630}, $K_{br.b}$, K_{OH}, K_{aro} and the concentration of the free radicals reach a peak value which is very close to the value obtained at similar reflectance under natural conditions. The temperature at which the maximum occurs corresponds approximately to the formation of clusters in which the stacks of polyaromatic molecules are more or less parallel to each other (Chapter 7).

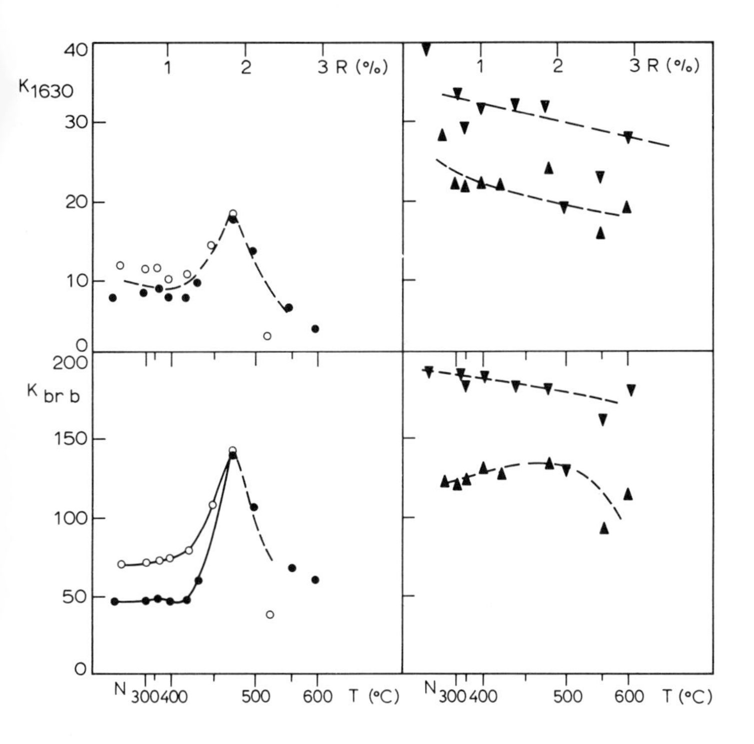

Fig. 6.13. — Effect of pyrolysis of shallow kerogens on K_{1630} and $K_{br.b}$:

● : kerogen of type I.
○ : kerogen of type II.
▲ ▼: kerogens of type III.

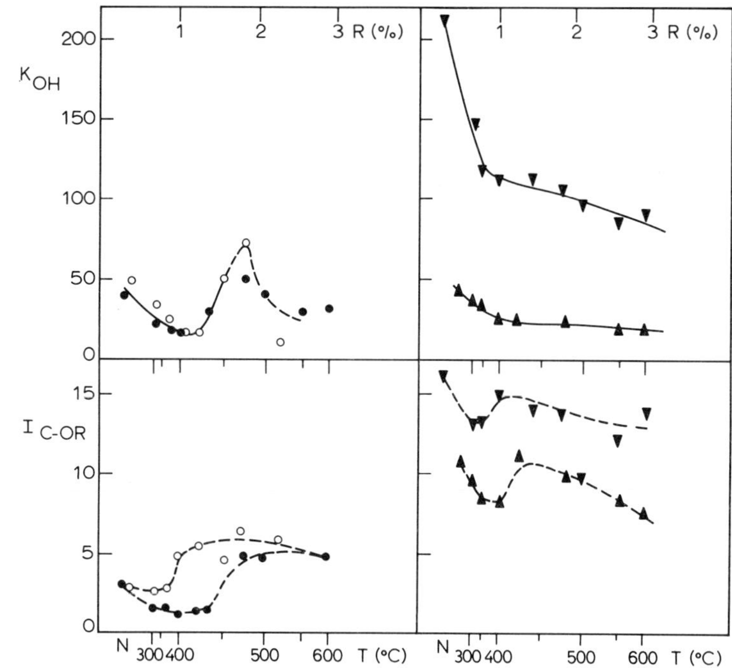

Fig. 6.14. — Effect of pyrolysis of shallow kerogens on K_{OH} and I_{C-OR}:

● : kerogen of type I.
○ : kerogen of type II.
▲ ▼: kerogens of type III.

A treatment at higher temperature provokes the removal of free radicals, aromatic CH groups and OH groups. The simultaneous decrease of $K_{br.b}$ and K_{1630} is attributed to a diminution of the specific absorption coefficients resulting from improved molecular orientation within the clusters or formation of C – C bonds between adjacent stacks, both effects increasing the delocalization of π electrons. Elimination of oxygenated functions may also be partly responsible for the decrease of $K_{br.b}$.

When **kerogens of type III** are heated, the oxygen content decreases but keeps a level appreciably higher than in naturally evolved kerogens or kerogens of type I or II submitted to a similar heat treatment. Comparison between Figs. 6.11 and 6.14 shows that, among the various oxygenated functions, this is due to hydroxyl or ether groups and not to carbonyl or carboxyl groups. Comparison of the series obtained by pyrolysis of two different kerogens of type III indicates that elimination of OH and other C – OR functions at high temperature is less effective for kerogens having initially a high content of these groups.

Keeping large K_{OH} and I_{C-OR} at elevated temperature goes along with maintaining large values of K_{1630} and $K_{br.b}$ and with formation of the clusters at higher temperature (Chapter 7).

This influence of oxygenated functions on the structural reorganization of pyrolyzed carbonaceous material has also been observed by comparing the pyrolysis of **model substances** such as sporopollenin and lignite, which behave like kerogens of type I or II and kerogens of type III, respectively. Moreover, heating sporopollenin in air at 200° C provokes the incorporation of oxygen and, upon subsequent pyrolysis, the product follows an evolution similar to that of lignite [45, 26].

The effects of pyrolysis described above may depend on the experimental factors such as the rate of heating and the sample thickness and packing.

3. Difference with catagenesis.

A marked difference between catagenesis and pyrolysis is thus found for the evolution of oxygenated functions other than carbonyl and carboxyl:

(a) Catagenesis leads to a material, the chemical composition of which is practically independent of the starting material.

(b) On the contrary, removal of C – OR functions by pyrolysis is less complete for samples which had originally a high content of these functions and this tends to block the reorganization of the carbonaceous skeleton.

V. CONCLUSION

The IR spectra of kerogens and coals show practically always the same spectral features; however, when used in a quantitative manner, they provide an interesting approach for investigating the chemical composition of these carbonaceous materials. Important advantages are that the sample can be analyzed as it is, without complicated and time consuming treatments, and that a very small amount of material is sufficient for obtaining spectra. The spectra can also be used to analyze the effects of chemical treatment affecting selectively some functional groups.

Practical limitations concern the precision of the measured absorption coefficients; therefore characterization of a given sample should always be done using various KBr pellets of different concentrations. Another limitation of the method is that various types of functional groups contribute to a given absorption band and that the unitary absorption coefficient of these molecular groups may vary according to their environment.

The data presented above show that, despite limitations, the method offers valuable information. In fact many of these limitations are not critical when the purpose is to compare samples or series of samples. Instead of separating the complex solid into various components for a detailed analysis, IR spectroscopy allows to characterize it as a whole, by a limited number of parameters, the absorption coefficients, which, while empirical to some extent, have a fairly clear chemical meaning.

A. Natural samples and geological aspects.

The IR spectra of shallow kerogens differ strongly according to the nature of parent material. The differences concern the content of saturated hydrocarbonous structures, the hydroxyl content, the relative abundance of carbonyl groups, esters and carboxylic acids and the intensity of the bands containing an important contribution of aromatic rings and polyaromatic structures.

Catagenesis leads to a porous residual solid of high carbon content; the chemical composition of the latter is independent of the nature of the parent organic matter, which however influences the size of the clusters formed by the stacks of polyaromatic layers. The sequence of spectra obtained for samples of increasing burial depth offers a picture of the main chemical modifications taking place upon catagenesis:

(a) Removal of carboxyl and carbonyl groups.
(b) Removal of saturated hydrocarbonous groups.
(c) Formation and removal of aromatic CH groups, closely parallel to that of free radicals.
(d) Evolution of hydroxyls and ether groups.

These modifications and their relationship with the liberation of hydrocarbons [11, 41, 13, 42] are schematically illustrated by Fig. 6.15, which is based on data for series III; the content of extractable OM, related to the organic carbon, has been determined by Albrecht *et al.* [1]:

(a) The sharp decrease of the band at 1 710 cm^{-1} (K_{1710}), due to decomposition of carbonyl and carboxyl groups with formation of H_2O and CO_2, takes place in the non nature zone.

(b) The begining of the oil formation zone corresponds to the decrease of the CH stretching band (K_{al}), and, in that way, it can be localized without perturbation arising from migration phenomena.

(c) The end of the oil formation zone corresponds about to the maximum abundance of aromatic CH groups (K_{aro}).

(d) The dry gas formation zone follows the oil formation zone and corresponds to elimination of aromatic CH and residual aliphatic groups.

The IR spectra of a series of kerogens from a given sedimentary basin allow thus:

to identify the type of parent material,
to evaluate the degree of maturation of the OM,
and consequently to estimate the oil potential of the sediment.

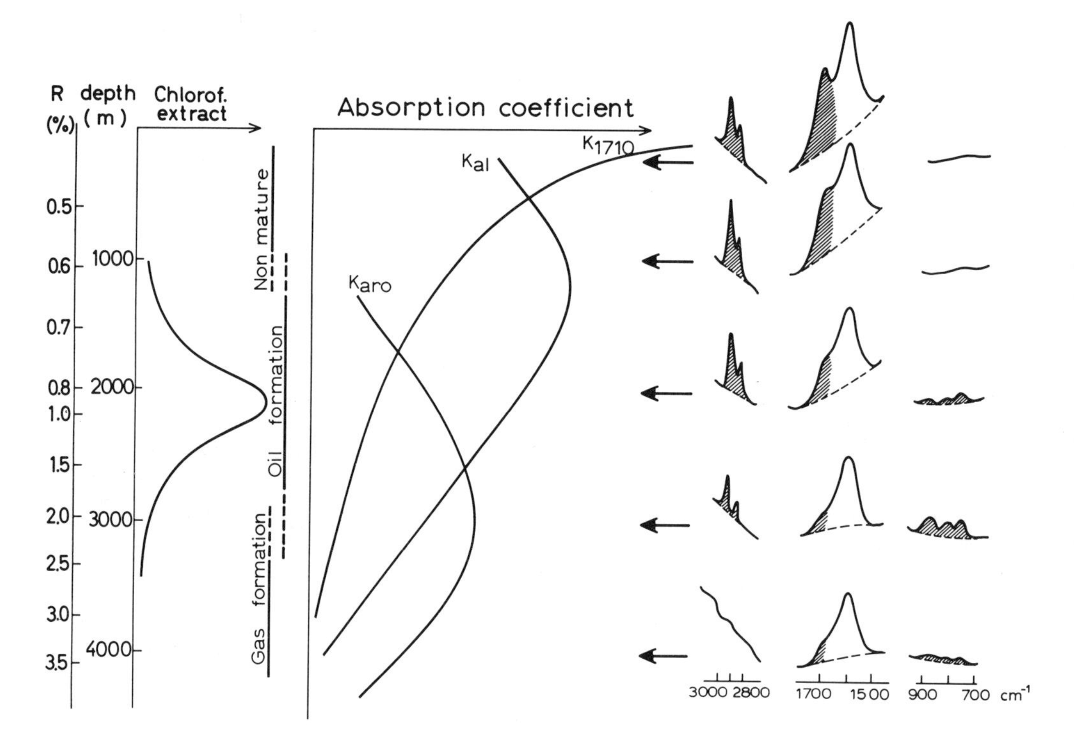

Fig. 6.15. — Schematic illustration, based on data for series III, of the relationship between the evolution of infrared spectra during catagenesis and the liberation of hydrocarbons:

right: infrared bands. **center:** evolution of absorption coefficients. **left:** content of the rock in OM extractable by chloroform related to the organic carbon.

They may also help the geologist in identifying anomalous samples and understanding complex situations originating for instance from alteration processes or detritism. However IR spectroscopy alone is not suitable for determining both the type of parent material and the degree of maturation on a single sample; for that purpose, other techniques as exemplified by other Chapters of this book, are more fruitful.

B. Pyrolyzed samples and industrial aspects.

IR spectra allow to estimate the oil potential of bituminous shales; they can also be used to characterize other types of carbonaceous solids such as coke, chars, carbon black, carbon deposited on the surface of solid catalysts.

The spectra offer a valuable tool for investigating the influence of oxygenated functions on the behaviour of carbonaceous matter at high temperature. This is of great concern in coke production, as the mechanical properties of the material at high temperature are strongly influenced by the relationship between elimination of oxygenated functions and elimination of hydrocarbonous structures [44]. The oxygenated functions may also be important in other applications such as coal gasification or liquefaction, carbonization processes and pyrolysis of bituminous shales or organic solid waste: they are expected to play a role in the influence, on structural properties and reactivity of carbon, of factors such as the heating rate, the exposure to oxygen at some stage, the thickness or the agitation of the bed of solid particles.

REFERENCES

1. Albrecht, P., Vandenbroucke, M., and Mandengue M., (1976), *Geochim. et Cosmochim. Acta,* **40,** 791
2. Avram, M., and Mateescu, G., (1972), "Infrared Spectroscopy " Wiley ed.
3. Bellamy, L.J., (1968), "Advances in Infrared Group Frequencies", Chapman and Hall ed.
4. Bellamy, L.J., (1958, 1975), "The Infrared Spectra of Complex Molecules", 2d ed, Methuen, 3rd ed, Chapman and Hall.
5. Blom, L., Edelhausen, L., and Van Krevelen, D.W., (1957), *Fuel,* **36,** 135.
6. Briggs, L.H., Colebrook, L.D., Fales, H.M., and Widman, W.C., (1957), *Anal. Chem.,* **29,** 904.
7. Brown, J.K., (1955), *J. Chem. Soc.,* **744** and **752.**
8. Castex, H., (1977), *Institut Français du Pétrole, réf.* **25 169.**
9. Debyser, Y., Leblond, C., Dastillung, M., and Gadel, F., (1977), *Proceedings 8th Internat. Meeting Organic Geochemistry, Moscow,* to be published.
10. Durand, B., Nicaise, G., Roucaché, J., Vandenbroucke, M., and Hagemann, H.W., (1977), *in: Advances in Organic Geochemistry, 1975,* ed. Campos, R., and Goni, J., Enadimsa, Madrid, 601.
11. Durand, B., and Espitalie, J., (1973), *C.R. Acad. Sci. Paris,* **276,** 2253.
12. Durand, B., Marchand, A., Amiell, J., and Combaz, A., (1977), *in: Advances in Organic Geochemistry, 1975,* ed. Campos R., and Goni, J., Enadimsa, Madrid, 753.
13. Durand, B., and Espitalié, J., (1976), *Geochim. et Cosmochim. Acta,* **40,** 801.
14. Duyckaerts, G., (1959), *The Analyst,* **84,** 201.
15. Espitalié, J., Durand, B., Roussel. J.C., and Souron, C., (1973), *Rev. Inst. Franç. du Pétrole,* **XXVIII,** 37.
16. Farmer, V.C., (1974), "The Infrared Spectra of Minerals", Mineral. Soc., London.
17. Forsman, J.P., and Hunt, J.M., (1958), *in: Habitat of Oil,* AAPG Symposium, Tulsa, 747.
18. Francis, S.A., (1951), *J. Chem. Phys.,* **19,** 942.
19. Friedel, R.A., (1966), *in: Applied Infrared Spectroscopy,* ed. Kendall D.N., Reinhold, 312.
20. Friedel, R.A., and Carlson, G.L., (1972), *Fuel,* **51,** 194.
21. Hergert, H.L., (1960), *J. Organ. Chem.,* **25,** 405.
22. Huc, A.Y., (1973), "Contribution à l'étude de l'Humus Marin et de ses Relations avec les Kérogènes", Thesis, Université de Nancy.
23. King, L.H., Goodspeed, F.E., and Montgomery, D.S., (1963), *Mines Branch Research Report* **R114,** Dept. of Mines and Technical Survey, Ottawa, Canada.
24. Marchand, A., Libert, P., Achard, M.F., and Combaz, A., (1974), *in: Advances in Organic Geochemistry,* 1973, Tissot B., and Bienner F., ed. Editions Technip, Paris, 117.
25. McMurry, H.L., and Thornton, V., (1952), *Anal. Chem.* **24,** 318.
26. Oberlin, A., Villey, M., and Combaz, A., (1978), *Carbon,* **16,** 73.
27. Prost. R., (1969), *Ann. Agron.,* **20,** 547.
28. Robin, P.L., Rouxhet, P.G., and Durand, B., (1977), *in: Advances in Organic Geochemistry, 1975,* ed. Campos, R., and Goni, J., Enadimsa, Madrid, 693.
29. Robin, P.L., and Rouxhet, P.G., (1976), *Rev. Inst. Franç. du Pétrole,* XXXI, 955.
30. Robin, P.L., and Rouxhet, P.G., (1976), *Fuel,* **55,** 177.
31. Robin, P.L., and Rouxhet, P.G., (1978), *Geochim. et Cosmochim. Acta,* **42,** 1341.
32. Rousseaux, J.M., Gomez Laverde, C., Nathan, Y., and Rouxhet, P.G., (1973), *in: Proceedings 1972 Internat. Clay Conference,* ed. Serratosa J.M., Div. Ciencias C.S.I.C., Madrid, 89.

33. Rouxhet, P.G., (1970), *Clay Minerals,* **8,** 375.
34. Rouxhet, P.G., and Robin, P.L., (1978), *Fuel,* **57,** 553.
35. Schafer, H.N.S., (1972), *Fuel,* **51,** 4.
36. Schnitzer, M. and Khan, S.V., (1972) "Humic Substances in the Environment", Marcel Dekker.
37. Shaks, I.A., and Faizulina, E.M., (1974) "Infrared Spectra of Fossil Organic Matter", Leningrad.
38. Shih, J.W., Osawa, Y., and Fujii, S., (1972), *Fuel,* **51,** 153.
39. Souron, C., Boulet, R., and Espitalie, J., (1974), *Rev. Inst. Franç. du Pétrole,* **XXIX,** 661.
40. Speight, J.G., (1972), *Applied Spectroscopy Rev.,* **5,** 211.
41. Tissot, B., Durand, B., Espitalie J., and Combaz, A., (1974), *AAPG Bull.,* **58,** 499.
42. Vandenbroucke, M., Albrecht, P., and Durand, B., (1976), *Geochim. et Cosmochim. Acta,* **40,** 1241.
43. Van der Marel, H.W., and Beutelspacher, H., (1976), "Atlas of Infrared Spectroscopy of Clay Minerals and their Admixtures", Elsevier, Amsterdam.
44. Van Krevelen, D.W., (1961), "Coal", Elsevier, Amsterdam.
45. Villey, M., Oberlin, A., and Combaz, A., (1976), *C.R. Acad. Sci. Paris,* **282,** 1657.
46. Vitorovic, D.K., and Pfendt, P.A., (1974), *Ann. Acad. Brasil Cienc.,* **46,** 49.
47. Vitorovic, D.K., Krsmanovic, V.D., and Pfendt, P.A., (1977), *in: Advances in Organic Geochemistry, 1975,* ed. Campos, R., and Goni J., Enadimsa, Madrid, 717.
48. Youssef, A.M., (1974), *Carbon,* **12,** 433.

7

Electron microscopic study of kerogen microtexture. Selected criteria for determining the evolution path and evolution stage of kerogen

A. OBERLIN*, J.L. BOULMIER* and M. VILLEY*

PART ONE
KEROGENS AMONG CHARS. HOW TO DETERMINE MICROTEXTURE (MICROSTRUCTURE)

I. INTRODUCTION

Organic matter (OM) either pyrolyzed or progressively buried in sediments, is degraded. Tars escape first, then gases (H_2, CH_4) as heat-treatment temperature (HTT) increases or as burial progresses (catagenesis then metagenesis, Chapter 4). Simultaneously the carbon content of the residue increases until it is almost pure carbon. This latter can eventually be transformed into graphite (progressively or abruptly). In any case changes occur both in atomic structure and in texture (microstructure).

Diffraction techniques have been commonly applied to the study of carbonification ([1]), carbonization and graphitization. In the past, X-ray diffraction was used first, then electron diffraction and finally all the derived image formation techniques such as conventional trans mission electron microscopy (CTEM). A critical review of these techniques will first be made.

The validity and relevance of high resolution microscopy techniques applied to kerogen analysis can be judged only, if the position of kerogens among natural carbonaceous mate-

---

([1]) Though unusual, the word carbonification has been used to distinguish laboratory transformation (carbonization) from natural one [41].

*Laboratoire Marcel Mathieu. ER 131 du CNRS. UER Sciences. Orléans, France.

rials (coals, anthracites, graphites) is considered and if these natural materials are studied within the scope of chars and industrial carbons.

II. DIFFRACTION TECHNIQUES

A family of lattice planes in a crystal can be compared to a line grating. The scattering of X-rays or electrons by atoms forming matter is similar to the scattering of a monochromatic light wave falling at normal incidence on the grating. Hence the resulting diffraction patterns can also be compared. The diffraction pattern of a line grating is formed by sets of parallel fringes, the periodicity of which depends on the distance between two lines of the grating (grating spacing). The larger the spacing, the nearer the fringes. In addition, if the grating becomes smaller, i.e if the total number of lines decreases, the fringes broaden. Using a reciprocal space simplifies the pattern interpretation. The space coordinates are reciprocal to the object space ones, so that every increase in any parameter in the object space causes a decrease in the corresponding parameter in the reciprocal space.

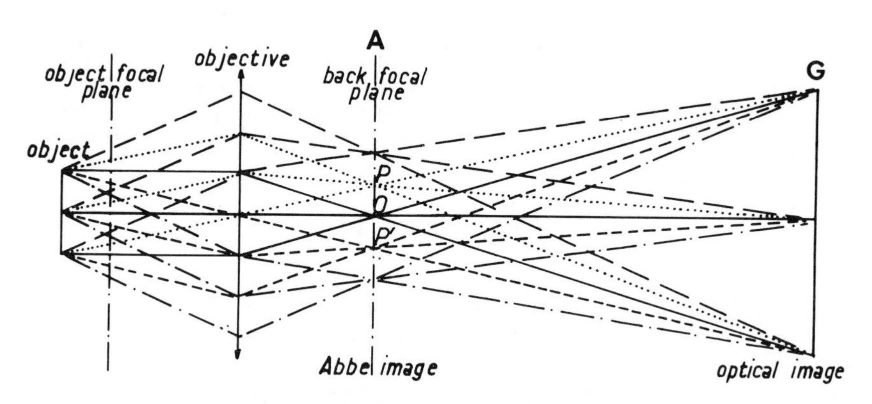

Fig. 7.1. — Ray path through a convergent lens.

Since the diffraction pattern is produced by interference (first Fourier transform), the object is restored if all the beams it issues reinterfere (second Fourier transform). In other words, the object's atomic structure can be obtained if the diffraction pattern of the diffraction pattern is performed (Fourier synthesis). In light optics, the first Fourier transform, i.e the diffraction pattern, can be observed on a screen far away from the object (fringes at infinity), but the Fourier synthesis is very easily carried out through a convergent lens. The object is set slightly beyond the focal plane of the lens (Fig. 7.1). The diffraction pattern is formed in the back-focal plane A of the lens, and the diffraction pattern of the diffraction pattern is the optical image formed in the Gaussian plane G. Such a double transformation is automatically performed by the lens and cannot be obtained without it. This is the most important difference between X-rays and electrons, and different techniques derive from this property. As a matter of fact X-rays cannot be focused by a lens while electrons can.

A. X-ray diffraction.

X-rays are photons and are thus sensitive only to the electron shell of atoms. As a result, the atomic scattering factor, i.e. scattered wave intensity, is low. Hence, the sample volume must be at least a few tenths of a cubic millimetre to obtain intense patterns. On the other hand, when a crystallographer wants to restore the atomic structure, he has to use mathematical treatment to transform the diffraction pattern since no lens exists for X-rays. Structures can be calculated in two ways: either by successively transforming every *hkl* reflection or by continuously transforming the entire reciprocal space.

Whatever choice is made, the same difficulties arise. The data can only be statistical since first a relatively large amount of sample is needed; then it is impossible to directly restore the object space. Therefrom every individual unit-cell cannot be described and is replaced by an average one which may be quite different.

B. Electron diffraction.

Electrons, being electrically charged particles, are not only sensitive to the electron cloud of atoms but to their nucleus as well. This results in a high atomic scattered amplitude. The intensity of a scattered beam is about 10^8 times higher than the corresponding X-ray one. An intense pattern can then result from a volume of sample as small as 0.1 μm^3. In addition, convergent lenses such as electromagnetic ones are available for electrons. When using such a lens, the ray path is the one described in Fig. 7.1, and the lens automatically performs the double Fourier transform, i.e. the Fourier synthesis of the object. For this very reason, an electron microscope is a suitable tool for determining both structure and microtexture.

To the objective lens represented in Fig. 7.1 are added projection lenses used for magnifying the images formed either in focal plane A of the lens or in Gaussian plane G. The enlarged images are recorded either on a fluorescent screen or on a photographic plate (the maximum magnification is about 10^6). The object is included in the pole-piece of the objective lens. The smaller the pole-piece gap, the better the resolving power ($\approx$ 5 Å). Hence the object has to be small. It also has to be thin, due to the low penetrating power of electrons inside matter. Only powdered samples or thin-sections can thus be examined. They are deposited on a supporting carbon film (less than 100 Å thick) covering a small copper grid ($\phi \approx$ 3 mm, thickness $\approx$ 2/100 mm, grid opening $\approx$ 100 μm). In the case of low rank chars, coals or kerogens, samples are ground in liquid nitrogen to avoid artefacts due to plasticity. A 10^{-5} - 10^{-6} torr vacuum is maintained in the apparatus to provide the necessary electron free-path. Due to the high vacuum and electron impact, only materials which tolerate both vacuum and electrons can be examined since electron action is at the same time chemical (reducing effect), thermal (heating effect) and electrical (charging of an insulating object). In spite of the smallness of the pole-piece gap, sophisticated devices have been worked out to perform tilting and rotation of the object (goniometer stage).

Such a conventional transmission electron microscope (CTEM) may be used in many ways. As in X-ray diffraction, one can consider either the interference of many beams (phase contrast) or only one beam at a time (amplitude contrast). Beside these ways, the electron microscope can also be used in two additional ways. In the first and simplest one it acts as a light microscope with high resolving power, and its applications are restricted to morphology. The second way is the selected area electron diffraction (SAD). The microscope is used as a diffractometer, i.e. with its projection lenses, it records the diffraction pattern formed in back focal plane A of the objective lens (Fig. 7.1). This does not contain any basic difference from X-ray diffraction except the very small size of the sample. Owing to an intermediate image aperture (set in the Gaussian plane) (1) the SAD pattern obtained on the fluorescent screen is that of an object less than one micrometer in diameter and a few angstroms thick.

1. Dark-field imaging.

a. Principle.

The principle of this technique appears in Fig. 7.2. An aperture (objective aperture) is set in the back-focal plane A of the objective lens. It is small enough to allow only a given *hkl* scattered beam to pass through. In Gaussian plane G, the regions issuing the *hkl* beam appear bright on a dark-field (DF). The optical image thus obtained is the *hkl* DF image. If the aperture does not intercept any beam, the optical image is entirely dark. If the regions issuing the *hkl* beam are reduced in size, the scattered beam becomes broader. The bright regions in the

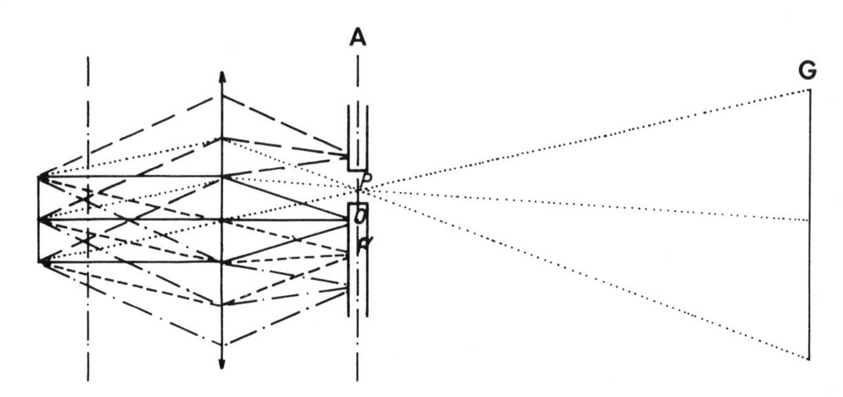

Fig. 7.2. — Ray path through the objective lens in dark-field (DF).

DF image become small but they still appear bright. They remain visible and can be measured, provided that their size is larger than the resolving power of the microscope. The objective aperture therefore has to be large enough to reduce its own diffraction effect. The diffraction pattern of the objective aperture appears in the Gaussian plane as a central bright spot surrounded by dark and bright rings (Fig. 7.3a). The intensity profile of this Fig. is represented in Fig. 7.3b. Since this pattern is superimposed on all image points, it limits the resolving power. To obtain DF images corresponding only to a given *hkl* beam, the objective aperture has to be small. Hence it is necessary to take into account the size of the central diffraction peak of the aperture. Usually the width at half maximum of the central diffraction peak (Airy disc A) is considered. It is given by the formula:

$$A = \frac{0.51\,\lambda f_o}{a}$$

(1) This is equivalent to setting a smaller field-limiting aperture in the object.

where: f_o is the objective lens's focal length and a the radius of the aperture. As an example A is 10 Å if $a = 3$ μm; it is only 2 Å if $a = 15$ μm.

In practice, the aperture is not moved in the back focal plane since the resolving power would be lowered [15]. It is set paraxial, the object is tilted to bring the *hkl* planes at the Bragg angle (Fig. 7.4a), and then the incident beam is tilted to bring the $\bar{h}\bar{k}\bar{l}$ reflection paraxial (Fig. 7.4b) so as to pass through the aperture.

b. *Molecules imaging.*

A modified DF technique has been elaborated to study disordered materials [23, 35, 36, 37]. In such samples, the deviation from the Bragg angle can be very great since the scattered beam is very broad, i.e the reciprocal *hkl* nodes are so elongated that they may even form infinite *hk* lines. If so, during the tilting of the incident beam there is no need to tilt the object. From this, it is possible to radially explore the diffraction pattern by progressively tilting the incident beam. This is equivalent to a progressive move of the aperture relative to the pattern and can be illustrated by Fig. 7.5. In this Figure the contour of a 5 μm aperture (in diameter) has been superimposed in various positions on the SAD pattern of a medium rank kerogen. Since this pattern is an image of the reciprocal space, its scale corresponds to $1/d$ and the unit to be used is thus 1 $Å^{-1}$. The aperture is 0.1 $Å^{-1}$ in apparent diameter and its Airy disc diameter is 13 Å. The successive positions of the aperture correspond first to bright-field (aperture centered on the incident beam) then to DF with the aperture respectively centered on 0.29; 0.37; 0.50; 0.67 and 0.84 $Å^{-1}$ (i.e. respectively corresponding to 3.4; 2.7; 2.0; 1.49 and 1.20 Å in the object space). In addition, a given *hkl* Debye-Scherrer ring can be explored by moving the incident beam along a circle [32, 33]. Finally the entire reciprocal space can also be explored [24] by using a goniometer stage which tilts the object relative to the incident beam. In this case, the Bragg condition is successively fulfilled for families of lattice planes oriented diversely in the sample. It can be said, in short, that any periodic unit (diffracting domain) larger than the resolving power of the microscope will be projected as a bright domain on a DF, providing the aperture lets one of its scattered beams pass through. This technique is thus particularly well adapted for studying molecules, such as aromatic ones for instance.

Since the bright domains are assumed to represent elementary units inside the sample, it has to be verified that they are real objects. Hence they have to occupy a fixed position in the three-dimensional fragment observed. They thus have to remain at the same place in the DF image whatever the aperture size may be and to become fuzzy when out of focus [34]. The electron beam must also be verified to see that is does not induce any visible transformation during the observation time. If all these conditions are fulfilled, size histograms of the bright domains can be drawn.

c. *Lattice plane spacing measurement.*

Reciprocally, if the scattered beams are not visible in the SAD pattern, they can be noted by the position of the aperture corresponding to the occurence of bright domains in DF images. In practice, the tilting of the incident beam is stopped when bright domains appear in DF. Then a picture is taken which superimposes the aperture contour on the

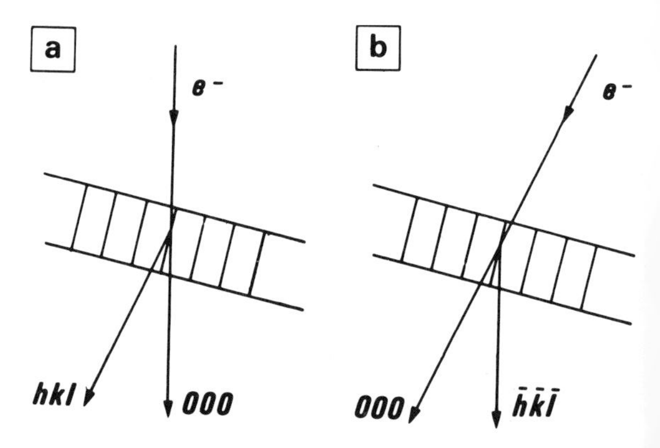

Fig. 7.4. — Tilted DF

a. *hkl* planes brought to the Bragg angle by tilting the crystal. **b.** $\bar{h}\bar{k}\bar{l}$ planes brought to the Bragg angle by tilting the incident beam.

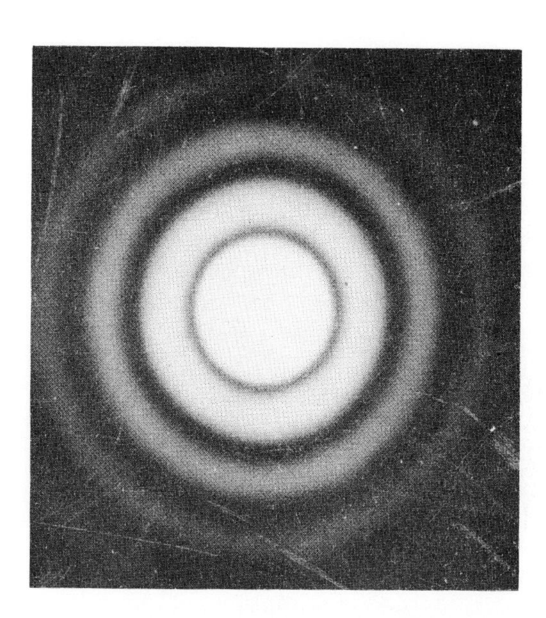

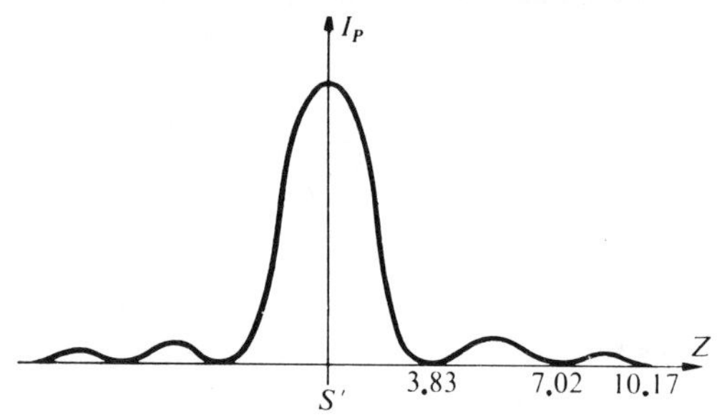

Fig. 7.3. — Diffraction pattern of the objective aperture

a. Picture. **b.** Intensity profile.

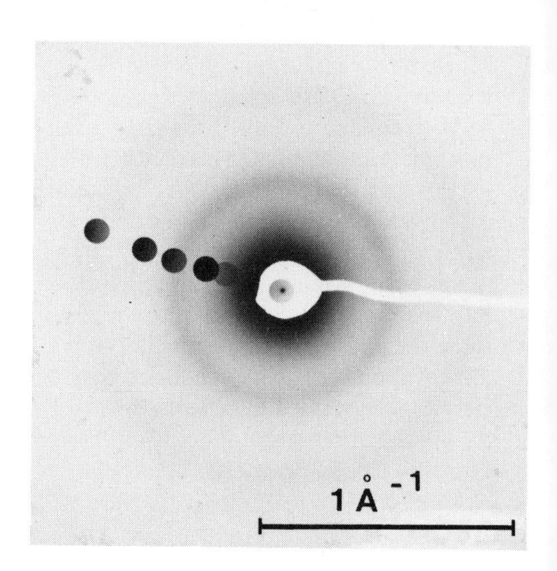

Fig. 7.5. — Superimposition of the aperture contours and of the SAD pattern of a medium rank char.

SAD pattern. Another micrograph of a suitable standard is then made. The value of $|\vec{s}| = 1/d_{hkl} \approx 2\theta/\lambda$ which corresponds to the aperture center can thus be measured ([1]). If a very small aperture is chosen, the resolving power is low and the bright domains are fuzzy, but the d_{hkl} measurement is fairly precise. If the 0.1 Å^{-1} aperture (Fig. 7.5) is used for measuring $d_{00.2}$ of graphite, the error has been experimentally determined as :

$$3.3 < d < 3.5 \text{ Å, i.e. the error is } \pm 0.009 \text{ Å}^{-1} \qquad [36].$$

2. *Lattice fringes.*

Providing that the crystal observed is very thin and the lens perfect, the image contrast approximates the projection of the object's atomic structure down the optical axis. If the transfer function of the objective lens (which introduces phase shifts) is taken into account

Fig. 7.6. — 00.2 lattice-fringes in graphite, representing aromatic layers seen edge-on.

(i.e. if proper defocusing of the objective lens in chosen), information about the local structure of a crystal can be obtained on the scale of a unit cell. For most current materials such as

---

([1]) 2θ is the angle between the incident and the scattered beam and, λ is the wave-length. Bragg's law $2\, d_{hkl} \sin\theta = \lambda$ may be approximated as such since for electrons λ is very small (0.037 Å for 100 kV).

metals, carbonaceous materials, etc., this technique has to be restricted to the interference between only two beams to form an image of a family of lattice planes (lattice fringes). To eliminate the undesirable beams introducing a back-ground noise into the image, an aperture is set in the SAD pattern plane. It lets through both the *hkl* and incident beams which interfere if the focus of the objective lens is suitable. In the optical image a set of *hkl* fringes appears representing the projection of the *hkl* planes on the observation plane. Fig. 7.6 (to be compared with Fig. 7.8a) shows as an example, the 00.2 fringes of a graphite crystal. Since the Bragg angle is very small ($\theta_{00.2} \approx 0.3°$ for 100 kV), the 00.2 reflection occurs when the aromatic layers are practically parallel to the incident beam, i.e. they are seen edge-on.

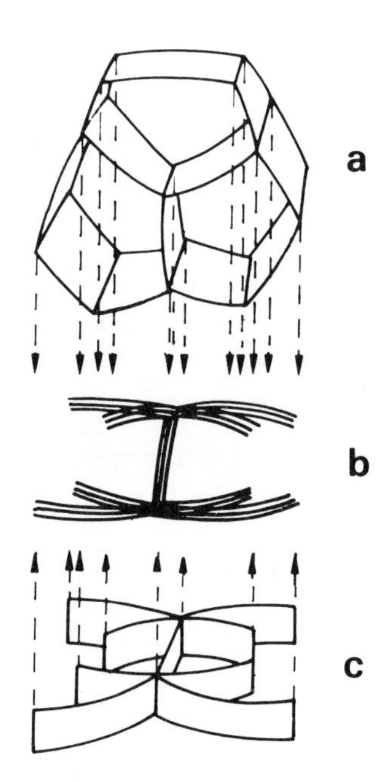

Fig. 7.7. — **a.** Schematic representation of a foam-like microtexture. **b.** 00.2 lattice fringes. **c.** Wrong restoration of the object from image (**b**).

Nevertheless it has to be kept in mind that lattice fringes are projections of the lattice planes on the observation plane along the lens's optical axis. As a result, many details are superimposed due to the thickness of the sample. At the same time many details remain unseen since the planes at an oblique angle to the supporting film are not at the Bragg angle. Let us consider, as an example, polyhedral bubbles with walls formed of aromatic layers parallel to the bubble interfaces. These bubbles are represented in Fig. 7.7a, but the aromatic layers which form the walls have been omitted. The 00.2 lattice fringes are represented in Fig. 7.7b. If no other projection of the microtexture is available, the restoration of the object will be wrong and ribbons will be assumed to form the sample (Fig. 7.7c). This example clearly demonstrates the absolute necessity of using more than one way in CTEM and, moreover, the necessity of total exploration of the available reciprocal space.

III. CARBONACEOUS MATERIALS

A. Graphite.

Graphite is a natural crystal defined by its three dimensional atomic structure, the basic elements of which are flat aromatic layers 3.354 Å thick (Fig. 7.8a). It is characterized by its hexagonal symmetry and its parameters a = 2 461 Å, c = 6 708 Å (Fig. 7.8b). The X-ray pattern of graphite is represented in Fig. 7.9.

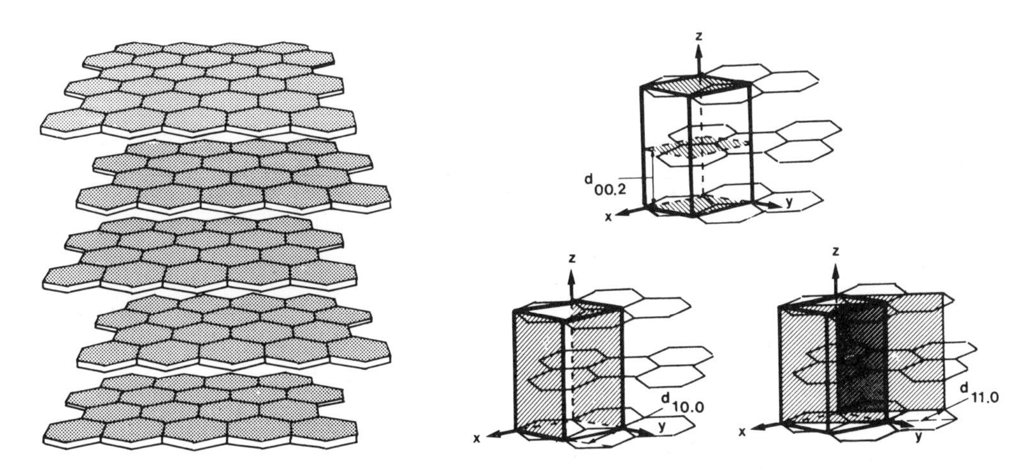

Fig. 7.8. — **a.** Graphite lattice. **b.** Graphite unit-cell associated with examples of diffracting planes.

B. Carbonization and graphitization.

Pyrolysis of various organic matters carried out in condensed phase (solid or liquid) results in almost pure carbon residue at about 1 000° C. Oxygen-poor and hydrogen-rich organic precursors, such as pitch for instance, produce carbon which can be progressively transformed into graphite by heat-treatment up to 3 000° C (graphitizing or soft carbons). On the contrary, oxygen-rich organic matter such as cellulose produces a carbon which cannot be transformed into graphite (non-graphitizing or hard carbons). Carbons heat-treated between 1 000 and 3 000° C have been studied both by X-rays and CTEM.

1. X-ray analysis.

Figure 7.9 represents the X-ray pattern of a pitch-coke sample heat treated at about 1 000° C. In this pattern *00.l* reflections appear. No *hkl* reflections are observed. They are replaced by diffuse and asymmetric *hk* bands appearing close to the graphite *hk.0* reflections

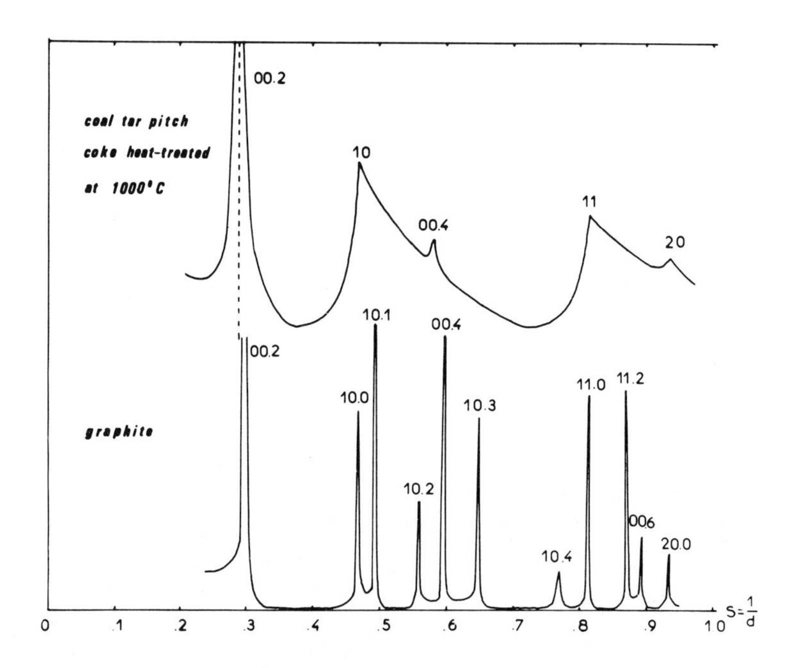

Fig. 7.9. — X-ray diffractometer recording for graphite and for turbostratic carbon heat-treated at about 1000° C.

(their maximum is shifted towards the wide angles). Warren first [53] then Franklin [11] deduced from such patterns the existence of stacks of parallel carbon layers. In these stacks, the superimposed layers are not arranged in ordered sequence as in graphite but at random. They are slightly rotated around an axis perpendicular to them. Such an arrangement is a turbostratic structure. It is equivalent to the structure of a single layer, i.e. it has only a two-dimensional crystalline order responsible for the occurrence of *hk* bands in the X-ray pattern. The *00.l* Bragg reflections indicate that the carbon layers are parallel, with an almost constant average spacing. Up to 1 600° C all carbons are turbostratic.

Over this temperature in graphitizing carbons, an increasing number of pairs of layers tends to follow the graphite order. This results in a progressive ordering along the perpendicular to the layers. The *hk.l* reflections correspondingly appear as broad maxima along *hk* bands. They then become sharper and sharper. During progressive graphitization [20] *00.l* diffraction lines can be used to determine $\overline{N}$ (mean number of layers per stack) and $\bar{d}_{00.2}$ (mean interlayer spacing). The degree of crystalline order may be deduced solely from the shape and intensity of *hk.l* maxima along *hk* bands (mainly when $h - k = 3n$). Warren and Franklin have attributed very small sizes to the stacks in carbons heat treated at 1 000° C. They have found about 2 to 4 layers per elementary stack less than 15 Å in diameter. When graphitization progresses, first, the number $\overline{N}$ increases very quickly, followed by the diameter L_a of the layers. Simultaneously $\bar{d}_{00.2}$ decreases from 3.44 Å to 3.35 Å.

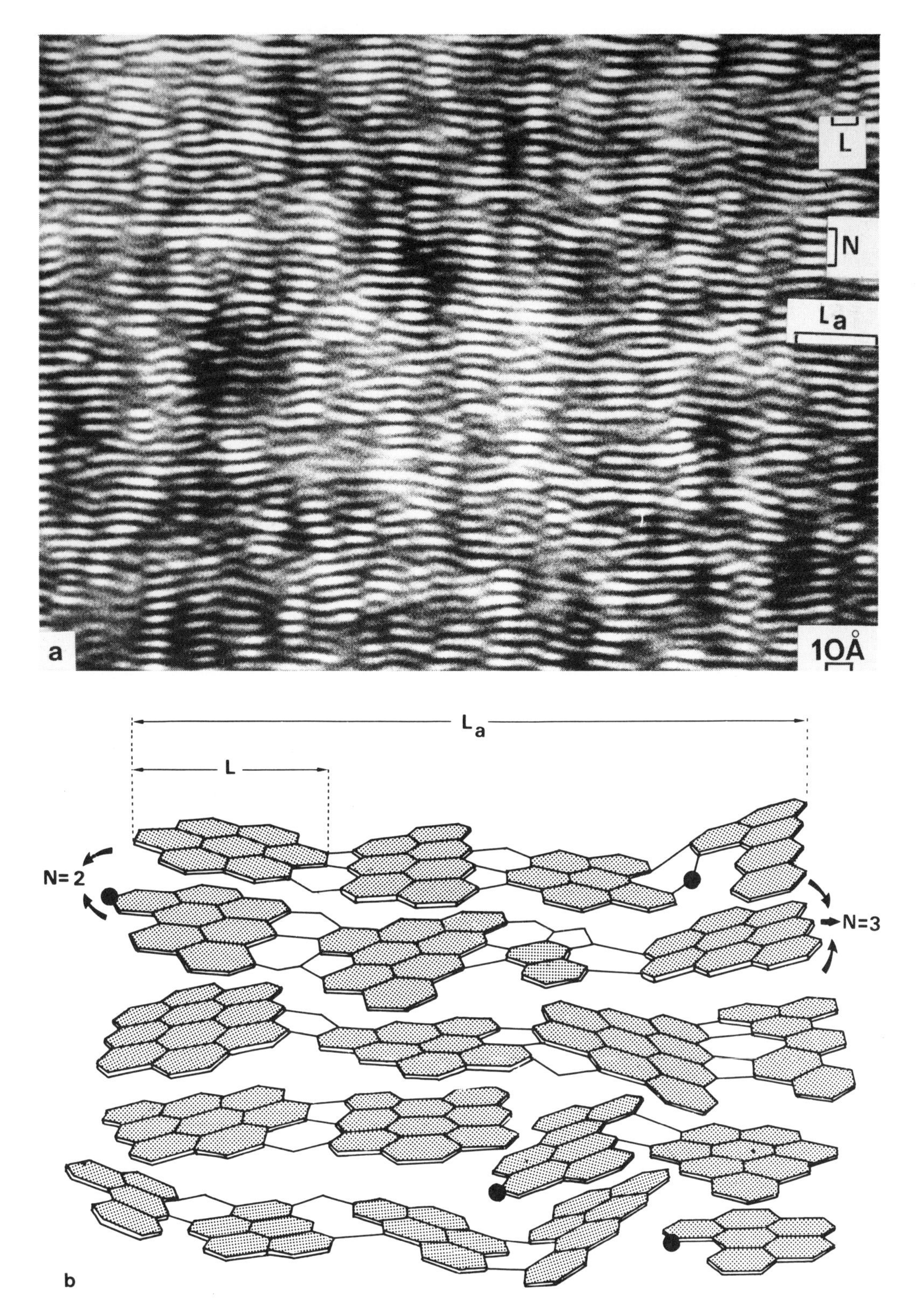

Fig. 7.10. — **a.** 00.2 lattice fringes of a soft carbon heat-treated at 1000° C. **b.** Structural model of turbostratic structure.

2. *Electron microscopic analysis.*

Graphitization studies have also been made by CTEM [38, 39, 40, 46], and the data obtained fit well with those of X-rays. Micrographs such as Fig. 7.10a can be used to establish a structural model for turbostratic graphitizing carbons (Fig. 7.10b). Stacks of two to three aromatic layers having a diameter L less than 10 Å form the elementary bricks of the samples. These small mosaic elements are associated approximately in parallel, with tilt and twist boundaries where defects accumulate (such as dangling bonds, tetrahedral bonds or heteroatoms). They thus form larger wrinkled layers having a diameter L_a. The zigzag structure of distorted carbon layers, joined to a very long-range parallel preferred-orientation has been acquired below 500° C during formation of the mesophase from the liquid state. The mesophase is made of spherical liquid crystals a few micrometers in size. They contain more or less parallel aromatic molecules [6]. During the pregraphitization stage ($\leq$ 2 000° C) N is the only parameter which varies (it increases from 2 to more than 40) while L_a remains small (about 200 Å). The defects frozen-in at the tilt and twist boundaries between mosaic elements are responsible for the stability of the zigzag structure, hence for the lack of growth [17]. Above 2 000° C a sudden dewrinkling of the layers occurs, due to the disappearance of defects, so that the tilt and twist boundaries can disappear. A rapid growth of L_a from 200 to more than 5 000 Å is thus favored in a way similar to annealing [17]. It can be assumed that the long-range molecular orientation produced in the mesophase stage has prepared for the coalescence of aromatic layers within a long-range.

In non-graphitizing carbons, no crystalline order appears whatever HTT may be. This lack of graphitization is not revealed by X-ray data alone. CTEM shows [40] that these carbons, though made of the same mosaic elements, have no long-range molecular orientation. For such materials, the carbonization process does not occur in the liquid state as in soft carbons, but in the solid state. Hence the release of volatile matter produces a foam-like microtexture, pushing the elementary units back to the pore walls. The molecular orientation is thus reduced to the volume of the wall which is very small (pores of about 100 Å size have walls much smaller). The diameter of the wrinkled carbon layers produced under such conditions cannot exceed 50-70 Å, and numerous grain-boundaries appear where defects are frozen in. These boundaries also correspond to sudden changes in layer direction. Above 2 000° C the carbon layers contained in the pore walls become dewrinkled as in soft carbons (Fig. 7.11) to be compared to Figs. 7.6 and 7.10), but any graphitization is inhibited by the smallness of the dewrinkled layers.

Until recently the very early stages of carbonization could not be studied easily since samples are less and less carbon rich, less aromatic and richer in non-aromatic groups. In X-rays patterns the *00.l* reflections become increasingly broader and fainter. High orders and even *hk* bands themselves disappear. Qualitative studies, though popular, give spurious results, and more sophisticated mathematical processing has to be used for quantitative studies. It will be shown in the following paragraphs that DF aromatic molecules imaging can be used to study low rank chars [24, 46]. These are made of the same bricks already observed for higher HTT, but they are connected and separated by non-aromatic groups.

C. Other kinds of industrial carbon.

When heat-treated, gas phases lead to carbon as do liquid or solid precursors. Hydrocarbons, for instance, when decomposed on a hot surface, produce carbon layers which lie parallel to their support (**pyrocarbons**). The higher the temperature, the better the orientation. Carbon is also produced at high temperature (> 800° C) either in flames or during incomplete hydrocarbon pyrolysis. The particles thus formed (**carbon-blacks**) are made of concentric carbon layers [8, 14] (onion-like texture).

Likewise, **catalytic carbons** should also be mentioned. Numerous active impurities (Fe, Ni, Co, Si, SiO_2, etc.) are known which produce more or less crystalline carbons at a relatively low HTT (≤ 2 000° C). These carbons, improperly called graphite, practically always show abnormal interlayer spacing (> 3 354 Å) and they may be crystalline or turbostratic as well [30].

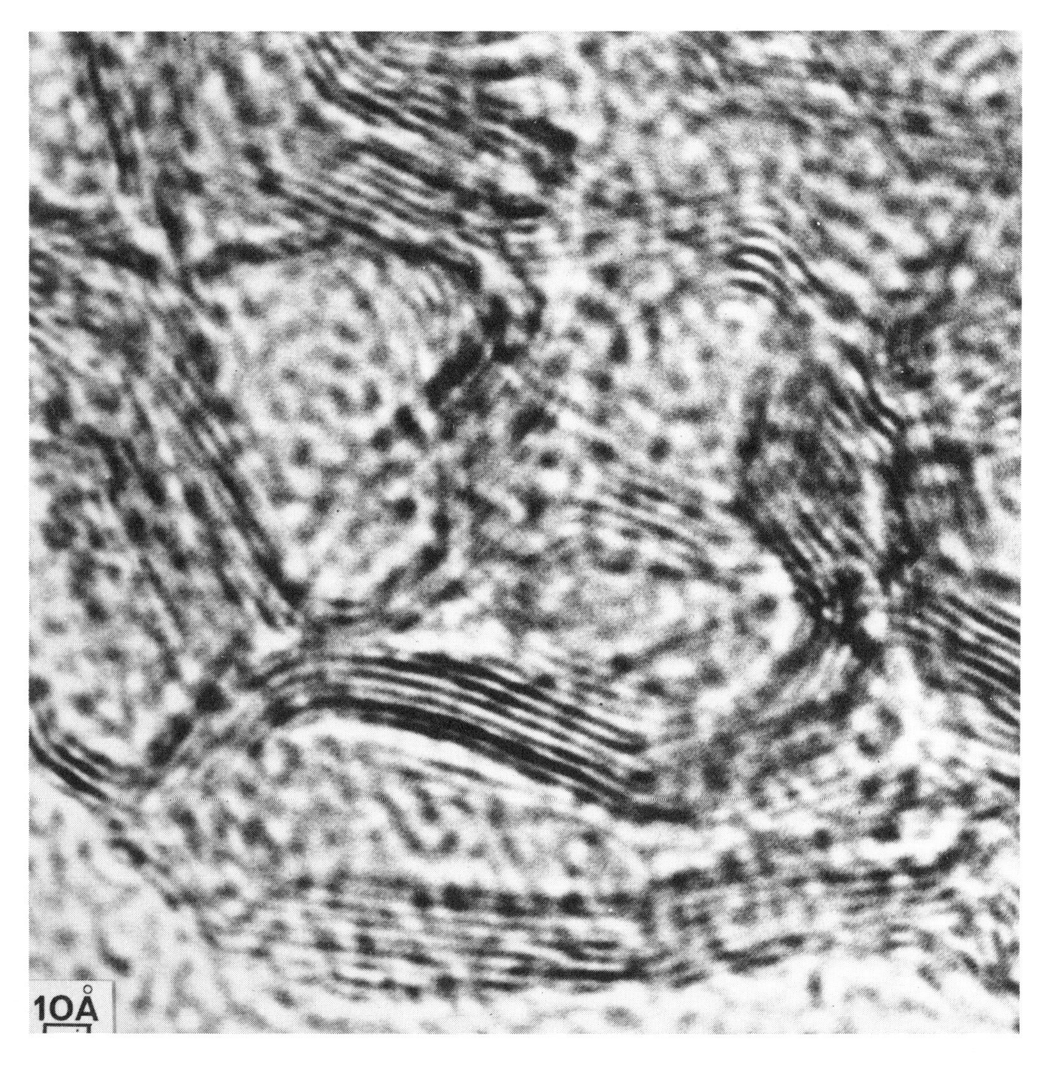

Fig. 7.11. — Non-graphitizing carbon microstructure (00.2 lattice fringes).

D. Natural carbons.

All natural carbon-rich materials in sedimentary series hitherto studied are either turbostratic carbons or graphites [12, 21, 41]. Except for very special cases such as the proximity of igneous intrusions [21], they never show signs of progressive graphitization, i.e. of intermediate stages. Heretofore no *hk.l* reflections have been observed, even for high rank coals such as anthracites. It is only in metamorphic series that some patterns seem to be similar to partially graphitized carbons [3]. Unfortunately X-ray data found in the literature are somewhat confusing. Many authors [13, 18, 19] do not take into account all the diffraction data including all the diffraction lines, forgetting that the study of only one or a few reflections can often be misleading.

In fact, CTEM [28, 39, 40] has shown that natural carbon-rich samples are usually similar to industrial carbons. No crystalline order appears in these samples, except if heat-treated above 2 000°C. Most of them are not too different from non-graphitizing carbons, except for some bitumen samples (in the petrographic sense of the term) which probably belong to graphitizing carbons [48].

In conclusion, both synthetic and natural products seem to follow a similar evolution whether they are carbonized or carbonified. This suggests that kerogens and coals are natural chars which could be studied by CTEM to determine their microtexture. In addition, thermal simulation of the natural evolution of kerogen should be profitably done in order to emphasize or explain their properties.

PART TWO
EXPERIMENTAL RESULTS AND DISCUSSION

I. NATURAL SERIES OF KEROGEN

A. Nomenclature.

Kerogens here studied are classified according to their elemental analysis (type I, II, III) as developed in Chapter 4. The list, nomenclature, chemical analysis and equivalent vitrinite reflectance of some of the samples are given in Tables 7.1.a,b and c (see also Figs. 7.30 to 7.32). In the first column, natural series are numbered I, II, III. A number characterizing the sample is given as the index. The heat-treated series are numbered I', II' and III', while the number in the index is replaced by the HTT. For instance I'_{420} is the least evolved kerogen of series I heat-treated at 420° C. In the second and third columns the atomic ratios H/C and O/C are respectively given. The equivalent vitrinite reflectance (R_o) is given in the fourth column [1].

---

[1] Reflectance is an electronic property strictly specific of a given material. It depends first on the chemical composition, then on the crystalline organization. Hence it should be necessary to measure it on the main and characteristic part of the kerogen.

TABLE 7.1
EXPERIMENTAL DATA FOR KEROGENS
(a) SERIES I AND I', (b) SERIES II and II', (c) SERIES III AND III'

Series I	H/C	O/C × 100	R_o	Series I'	H/C	O/C	R_o
I_1	1.57	10.08	0.30				
I_2	1.64	4.63	0.50				
to				I'_{420}	1.42	3.76	1.24
I_{10}	1.18	2.44	0.85				
I_{11}	1.17	7.84	1.30				
				I'_{470}	0.87	4.59	1.84
I_{12}	0.80	8.62	1.4				
I_{13}*	0.64	12.22	1.2	I'_{500}*	0.64	3.23	2.09
I_{14}*	0.48	7.83	1.8	I'_{600}*	0.49	3.59	2.88

(a)

Series II	H/C	O/C	R_o	Series II'	H/C	O/C	R_o
II_1	1.32	10.3	0.4				
				II'_{300}	1.21	8.93	0.68
to				II'_{435}	0.88	5.22	1.38
II_{11}	0.84	5.81	1.1				
II_{16}*	0.66	5.27	1.40				
				II'_{470}*	0.62	6.41	1.84
				II'_{500}*	0.55	5.91	2.09
II_{19}*	0.47	5.85	2.40				
II_{21}*	0.44	5.76	2.70	II'_{600}	0.44	5.19	2.88

(b)

Series III	H/C	O/C	R_o	Series III'	H/C	O/C	R_o
III_1	0.83	21.39	0.51				
to				III'_{300}	0.76	16.90	0.68
III_5	0.73	10.41	0.82				
				III'_{500}	0.51	11.28	2.09
				III'_{600}*	0.52	6.71	2.86
III_{10}*	0.44	5.97	2.33				
III_{13}*	0.40	4.97	2.81	III'_{800}	—	—	3.00
III_{14}*	0.43	2.34	3.6				

(c)

(asterisks correspond to molecular orientation occurrence)

B. Minor phases.

Strictly speaking, kerogens are always inhomogeneous and always contain "impurities" such as graphite particles, "carbon-blacks", mineral impurities (pyrites, rutile, anatase or zircon) or even particles at various degrees of catagenesis. If their proportion does not exceed 10-15%, the samples can be considered as homogeneous. An electron microscope is a particularly suitable tool for studying the minor phases present in kerogens, since particles less than

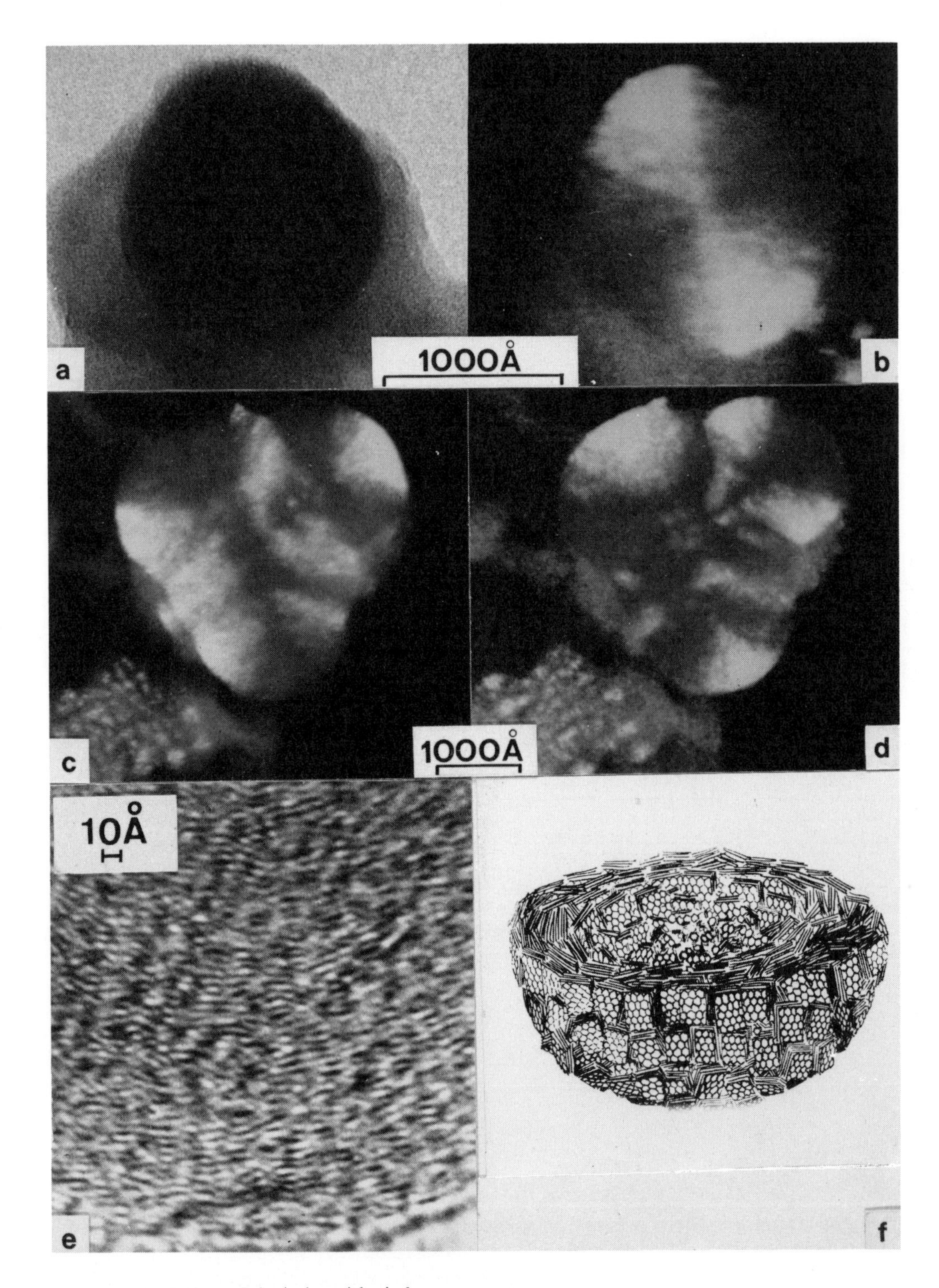

Fig. 7.12. — Spherical particles in kerogen.

a. BF (Bright field) picture. **b.** 00.2 DF picture. **c.** and **d.** particles with the objective aperture moved 90° along 00.2. **e.** 00.2 lattice fringes. **f.** Scheme of the microtexture.

1 μm can be described. Many particles are always studied for any sample, and both major and minor phases are thus described.

1. Graphite.

Microcrystals of graphite have been observed mainly in series III (a few percent), and they are probably of detrital origin.

2. "Carbon-blacks".

Some particles, roughly circular in shape (Fig. 7.12a), light up very strongly when a small aperture is radially moved in the SAD pattern. In this case, the aperture positions correspond to 00.2, 10 and 11 scattered beams of carbon. In 00.2 DF bright regions are shaped as two symmetrical sectors (Fig.7.12b). If the aperture is moved along the 00.2 Debye-Scherrer ring, the sectors turn at the same angle in the same direction (Figs. 7.12c and 7.12d). The aromatic layers are thus concentric in the particle, as shown by 00.2 lattice imaging (Fig. 7.12e). In fact, the particles are spheres containing concentric sheets of carbon layers. This onion-like texture (Fig. 7.12f) is a characteristic feature of carbon-blacks [8, 14]. Such particles were first observed in coals and anthracites. A few percent are found in kerogen series III (Fig. 7.13), they are less numerous in series II [29] and up to now have never been observed in

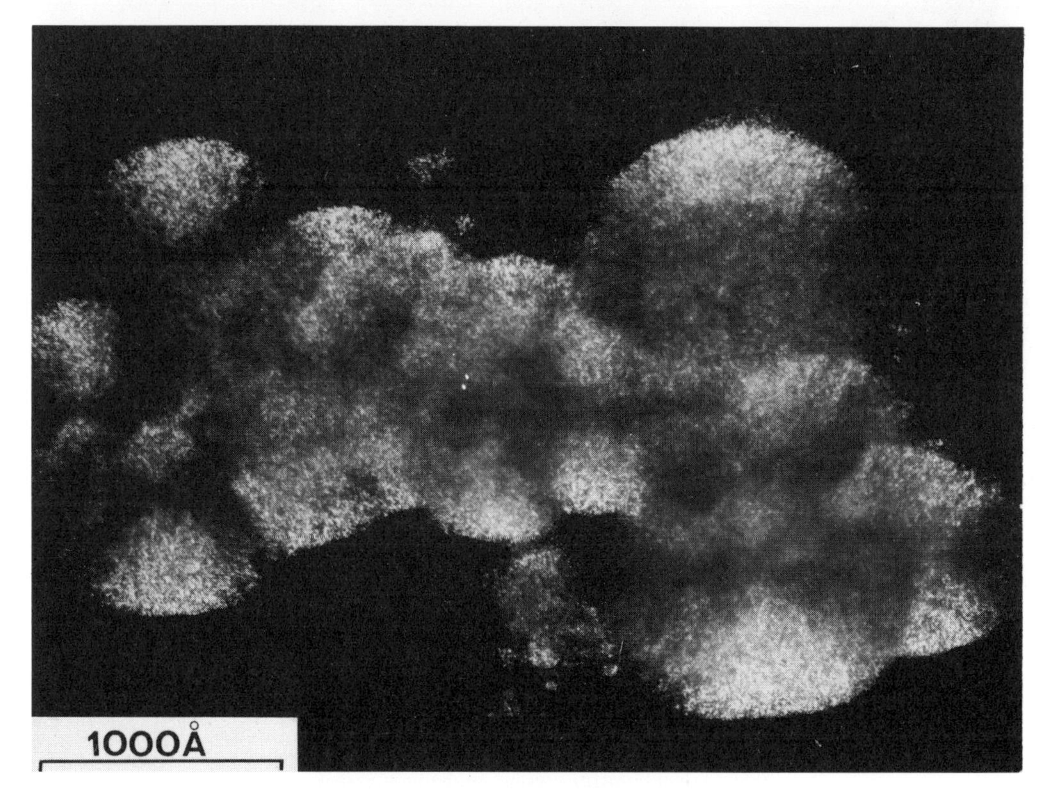

Fig. 7.13. — Black-like particles in a kerogen of series III.

series I. Statistical studies must be undertaken to correlate the occurrence of these particles to paleogeographic conditions. Experimental studies also have to be made to determine whether or not only high temperature conditions are able to produce these phases.

3. *Microcrystals of pyrites.*

They are frequently found in kerogens (Fig. 7.14a). When the aperture is set on the 00.2 scattered beam of carbon, a thin bright rim occurs around each crystal, which indicates that it is enclosed within a carbon shell (Figs. 7.14b and 7.14c). In this case, the carbon has a low cristallinity which is deduced from the fact that the elementary bright domains are very small and dim. It should be noticed that pyrites crystals (especially if larger than 1 μm) are very sen-

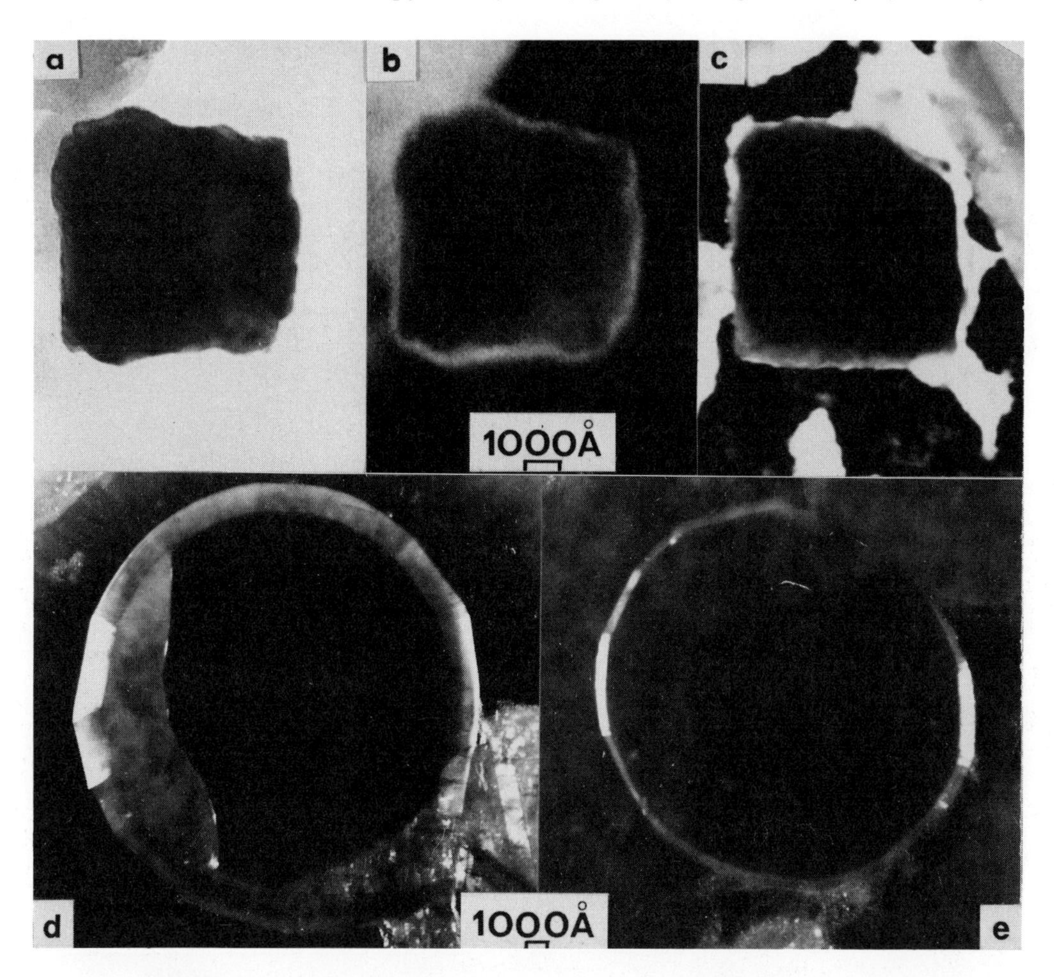

Fig. 7.14. — Pyrite crystals

a. BF picture. **b.** and **c.** 00.2 DF picture showing the shell of carbonaceous product around a pyrite crystal. **d.** and **e.** Catalytic shell of carbon after partial decomposition of pyrites (00.2 DF).

sitive to the electron beam. If this latter is too strongly focused on the sample, pyrites decompose, very suddenly loose their crystalline shape (cubes or cubooctahedrons) and become sherical. At this moment they move very fast on the supporting film in a Brownian motion. Catalytic transformation of carbon occurs immediately when pyrites reach either a kerogen particle or the carbon supporting film. This results in a shell of highly organized catalytic carbon [30] which either remains around the moving sphere (Fig. 7.14d) or may be empty (Fig. 7.14e). The sphere sometimes happens to explode and form smaller ones moving on the supporting film.

4. Titanium oxides.

Elongated crystals of rutile and anatase have been found in some recent sediments [25] and surprisingly are surrounded by carbon shells as pyrites do (Figs. 7.15a and 7.15b).

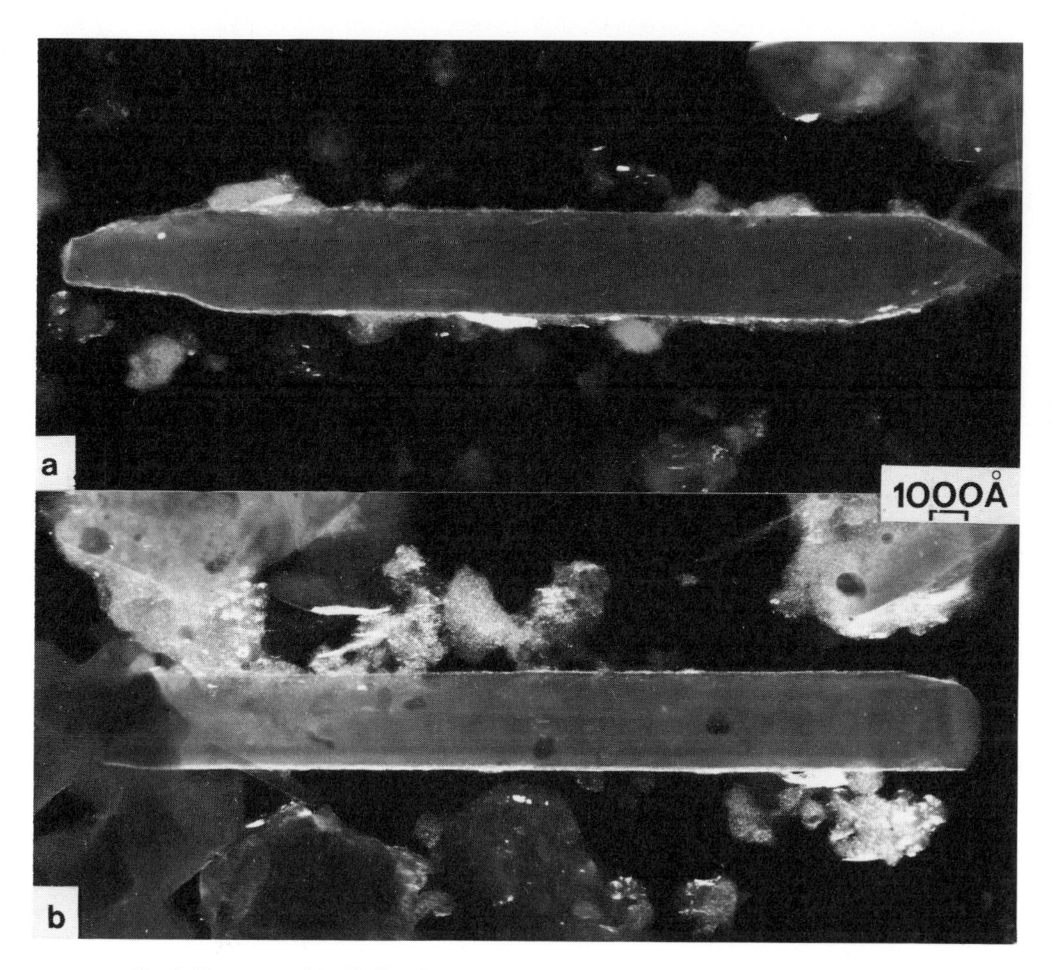

Fig. 7.15. — **a.** and **b.** Shells of carbonaceous products around titanium oxide crystals (00.2 DF).

5. *Other components.*

In recent sediments and also in some fusinite samples, a new kind of carbon morphology has recently been found. As shown by SAD patterns, the material is of high rank. It is porous with very large pores filled with small crystals (compare Fig. 7.16a in brigt field (BF) to 00.2 DF image Fig. 7.16b). Their almost regular network gives them a sponge-like appearance

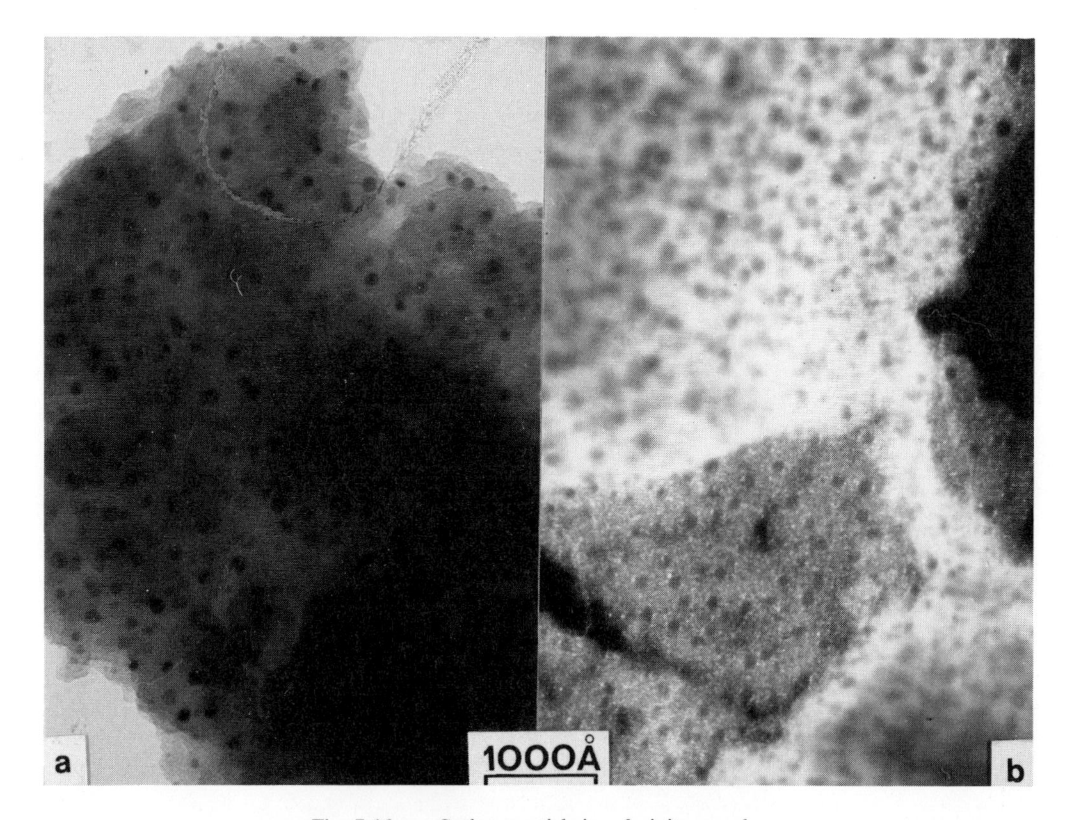

Fig. 7.16. — Carbon particle in a fusinite sample.

a. BF picture. **b.** 00.2 DF picture.

which is easy to recognize. The crystals have not been identified yet because their amount, with respect to the amount of carbon, is too small to give clear SAD patterns. Calcium and sulfur have been detected in the particles by using an electron microscope equipped with a non-dispersive X-ray spectrometer ([1]). X-ray analysis carried out on the ashes of such samples indicates the occurrence of anhydrite (SO_4Ca) crystals. These porous carbons have been noticed so recently that it can be assumed they are more common than presently expected.

([1]) Thanks are due to the Laboratoire de Minéralogie Cristallographie, *Université de Paris VI,* for access to the apparatus.

C. Major phase (low degree of catagenesis).

A single aromatic molecule can be approximately compared to a small graphitic layer. A set of such molecules randomly scattered in a small particle produces a diffraction pattern made from 10 and 11 bands. The smaller the molecules, the fuzzier the bands. If aromatic molecules happen to be piled up in parallel, an additional 00.2 ring appears. If the sample is formed by very small aromatic stacks (molecules with 5-10 aromatic rings, piled up by twos) only halos are observed in the SAD pattern (Fig. 7.5). Nevertheless in DF images, every stack appears as a bright domain about 5-10 Å in size, if any part of its scattered beam passes through the objective aperture. Reciprocally, when bright domains light up in the DF image, the aperture is certainly set right on the corresponding scattered beam in the SAD pattern. It thus becomes very easy either to image aromatic molecules or to demonstrate the occurrence of aromatic molecules by taking advantage of this reciprocity.

1. Aromatic molecule stacks.

Producing such an image has been applied to every kerogen sample, starting with kerogens at a low degree of catagenesis [4, 27, 28]. In DF images bright dots appear for three positions of the aperture, first in the range $0.13\,Å^{-1} - 0.32\,Å^{-1}$ ($3.1\,Å < d < 8\,Å$), then at about $0.50\,Å^{-1}$ ($d = 2\,Å$), and finally at $0.84\,Å^{-1}$ ($d = 1.2\,Å$). Outside of these positions, the optical image is completely dark. This indicates the occurrence of scattered beams close to the 00.2,10 and 11 beams of carbon, i.e. the occurrence of stacks of aromatic molecules. These latter are seen edge-on in 00.2 DF and more or less obliquely in 10 or 11 DF [28]. Since very small apertures ($0.1\,Å^{-1}$ in diameter) were used in these experiments, the micrographs obtained are not suitable for measuring molecule sizes. For obtaining more precise data [4, 34], apertures of various sizes were used, respectively centered on $0.23\,Å^{-1}$, the average $d_{00.2}$ value (Fig. 7.17a) and $0.50\,Å^{-1}$ the 10 value (Fig. 7.17b). An apparatus designed especially for this was used to evaluate the bright dot diameters. Precision of the measurements is about 10%. As an example, Fig. 7.18 shows a size histogram of molecule stacks in kerogen III_{14}. From these data the aromatic layer diameter can be estimated to be less than 10 aromatic rings, while there are no more than two layers per stack. The same results are obtained by lattice imaging. 00.2 lattice-fringes are approximately 7-8 Å long and are associated in groups of two to three. This shows that kerogens contain elementary units very similar to those found in carbons (Fig. 7.10b). However such units are not closely connected to each other as in turbostratic carbons but are connected and also largely separated by non-aromatic compounds (Fig. 7.22).

2. Interlayer spacing spreading.

a. Electron microscopy data.

The $0.13\text{-}0.32\,Å^{-1}$ range where bright dots are found indicates a very great spreading of the $d_{00.2}$ interlayer spacing (3.1-8 Å). This result has to be explained, since this set of values cannot be compared to 3.44 Å, the minimum $d_{00.2}$ value in turbostratic carbons.

Three kinds of bright dots can be roughly recognized depending on the value of the interlayer spacing. The first minimum value of $|\vec{s}| = 1/d$ corresponds to the largest spacing d_3

(first domains to light-up). Then d_2 corresponds to the maximum number of bright dots lit at the same time. The last maximum value of $|\vec{s}|$ represents the minimum d_1 spacing (last visi-

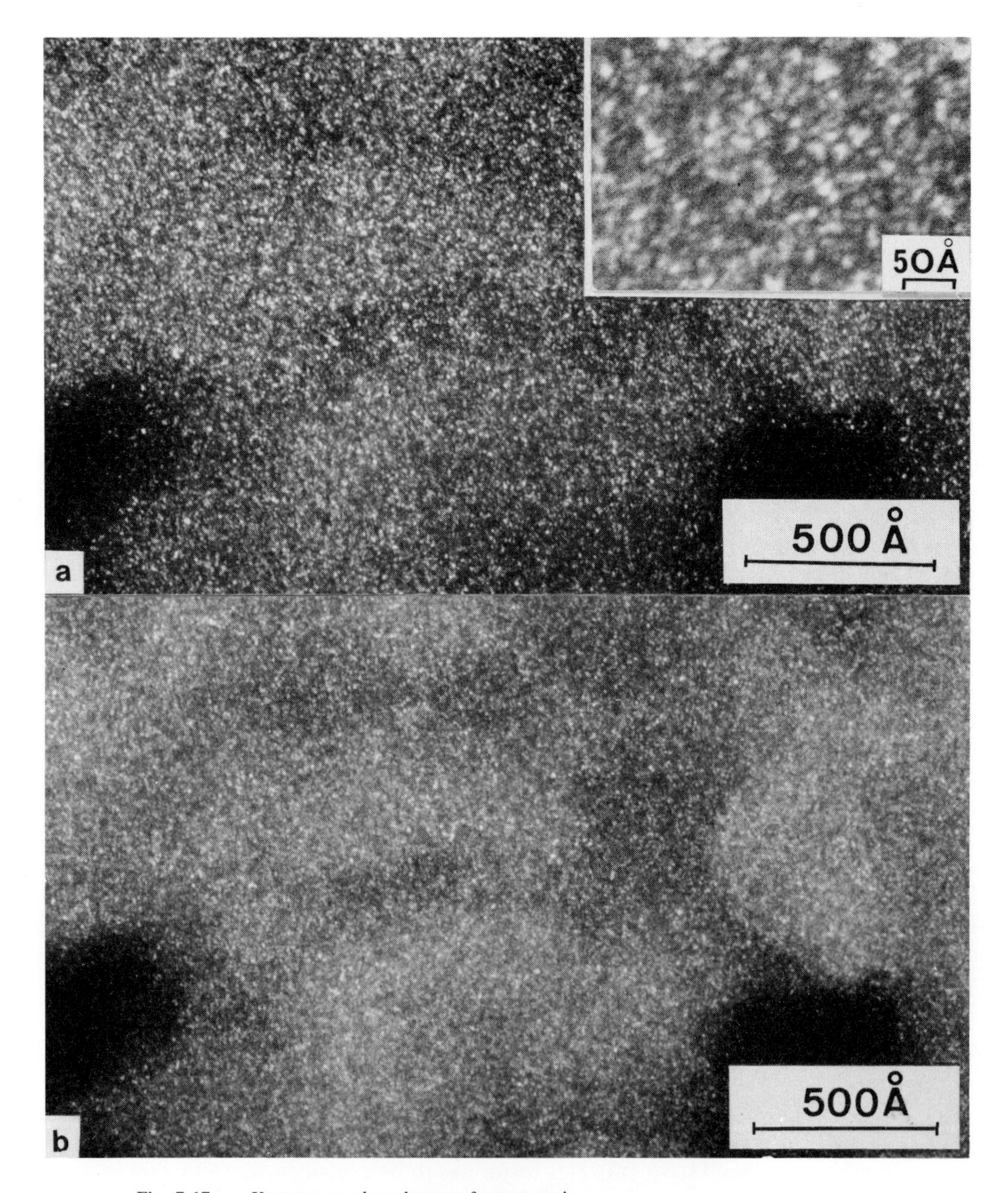

Fig. 7.17. — Kerogen at a low degree of catagenesis.

a. 00.2 DF picture; insert: enlarged detail of (**a**). Each bright dot is a stack of aromatic molecules seen edge-on. **b.** 10 DF picture.

ble bright dots). The results are presented in Table 7.2 for the three series of kerogens (natural and heat-treated).

TABLE 7.2
INTERLAYER SPACING SPREADING
(asterisk correspond to molecular orientation occurrence)

(a)

Series I	d_1	d_2	d_3
I_2	3.1-3.5	4.5-5.9	5.0-8
I'_{420}	3.1-3.4	4.5-5.1	6.6-8
I_{10}	3.1-3.4	4.8-5.7	6.5-8
I_{11}	3.1-3.4	3.6-4.6	4.6-6
I'_{470}	3.1-3.4	4.2-5.0	5.5-6
I_{12}	3.0-3.5	3.8-4.1	4.6-5.6
I'_{500}*	3.1-3.4	3.7-4.1	4.6-5.6
I_{13}*	3.1-3.4	3.9-4.5	4.8-5.4
I'_{600}*	3.1-3.4	3.9-4.0	4.1-5.1

(b)

Series II	d_1	d_2	d_3
II_1	3.1-3.8	4.0-5.0	6-8
II'_{300}	3.3-3.6	4.2-5.0	6-8
II_{11}	3.1-3.4	4.0-4.5	6-8
II'_{435}	3.2-3.6	3.6-4.7	5.5-6.5
II_{16}*	3.2-3.6	4.1-4.5	6-6.5
II'_{470}*	—	—	—
II'_{500}*	3.1-3.4	3.6-4.4	5.0-6.0
II_{19}*	3.1-3.4	3.6-4.3	4.4-6.1
II_{21}*	3.2-3.5	3.4-3.9	4.2-5.1
II'_{600}*	3.2-3.4	3.7-4.2	4.5-5.5

(c)

Series III	d_1	d_2	d_3
III_1	3.1-3.4	3.9-5.2	5.0-7.4
III'_{300}	3.3-3.6	3.8-5.0	6.2-7.6
III_5	3.0-3.4	3.8-4.8	5.0-6.6
III'_{500}	3.1-3.3	3.9-4.9	4.6-7.2
III_{10}*	3.1-3.4	3.8-4.1	4.1-5.7
III'_{600}*	3.1-3.4	3.5-4.0	4.4-6.9
III_{13}*	3.1-3.4	3.6-4.0	4.1-4.9
III'_{800}*	3.1-3.4	3.6-4.0	4.2-5.1

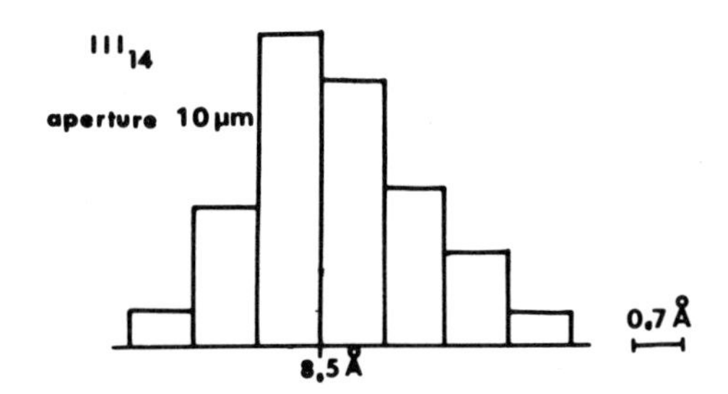

Fig. 7.18. — Size histogram of aromatic molecules in kerogen III_{14} (0.2 Å^{-1} aperture).

The lowest d_1 spacing varies only stlightly in the various kerogens, even in high rank ones (3.8-3.1 Å), but the occurrence of interlayer spacings smaller than 3.44 Å has to be explained. The bright dots observed in DF pictures have been interpreted as stacks of aromatic molecules less than 10 Å in size. Such small units scatter very broad beams and can thus be greatly deviated from the Bragg position. Consequently the interlayer spacing measured is too small. For instance if an elementary stack with a 3.44 Å interlayer spacing is tilted at its maximum out-of-Bragg position, the resulting experimental spacing is 3.16 Å. The d_2 range (about 4-5 Å) is always relatively wide in kerogens of low rank which means that this most frequent spacing cannot be precisely determined. Finally, the d_3 variation is due to the fact that the spacings of the last visible bright dots vary at random from one region of the particle to another. The histogram of the interlayer spacing frequency deduced from these results is asymmetric, and its maximum is not sharp (Fig. 7.19a).

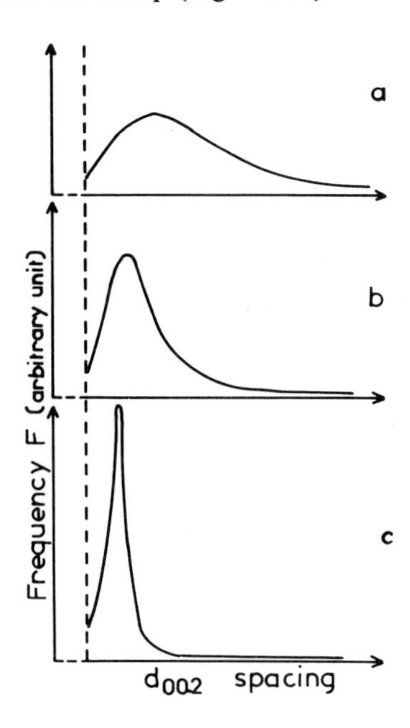

Fig. 7.19. — Schematic representation of the interlayer spacing spreading.

a. In the initial kerogen. **b.** After heat-treatment at 600° C.
c. In an anthracene char heat-treated at 1 000° C.

If all the beams scattered in the 0.32-0.13 $Å^{-1}$ region are used to form lattice-fringes, small groups of 2 to 3 fringes approximately 7-8 Å long are obtained. Fraunhofer diffraction may be used for measuring the fringe spacing ([1]). The lattice-fringe picture is used as an object and is placed beyond the focal plane of a convergent lens. A laser beam falls on this object and the ray path is equivalent to what is presented in Fig. 7.1. The resulting Fraunhofer diffraction pattern is recorded in the back focal plane of the lens [1] and used for the fringe spacing measurements. It shows widespread spacing values varying from 3.3 to 6 Å which fit well with the DF data.

The occurrence of bright dots in a 0.25 to 0.13 $Å^{-1}$ range of scattering has been attributed to unusual interlayer spacing. Such anomalies can be due to (and also stabilized by) the fixation of non aromatic groups at the boundaries of the graphitic layers [7]. However, the scattering corresponding to $|\vec{s}| < 0.25$ $Å^{-1}$ can also be interpreted in an other way, i.e. it can represent the scattering of non-aromatic molecules such as alicyclic ones [2, 10, 54, 55]. Neither DF techniques nor lattice-imaging can help in making a choice between these two hypothesis ([2]). Indirect chemical reasoning alone can be used to make the decision.

b. *Mixture of aromatic and aliphatic molecules.*

Let us consider a kerogen formed of a mixture of polyaromatic and aliphatic molecules. It can easily be demonstrated that aliphatic molecules correspond to $(H/C)_{aliph} = 2$ while aromatic ones correspond to $(H/C)_{arom} \leq 1$. If p is the aliphatic carbon fraction, $(1-p)$ is the aromatic carbon fraction also defined as the aromaticity factor f_a. With $(H/C)_{total}$ being the average number of bonds per carbon atom, we can write:

$$p(H/C)_{aliph} + (1 - p)(H/C)_{arom} = 2p + f_a(H/C)_{arom} = (H/C)_{total}$$

The corresponding data are given in Table 7.3 (columns 1 to 4). The two last columns first indicate some possible aromatic molecules (represented in Fig. 7.20) then the possible percentage of m molecules (Fig. 7.20). This molecule was chosen because it is similar in size to the bright dots observed in DF pictures. Table 7.3 shows that, when $(H/C)_{total}$ is equal to 1.3, for instance, p_{max} is only 0.65. If $p = 0$, $(H/C)_{arom} = 1.3$ which is impossible since the smallest possible aromatic molecule (a in Fig. 7.20) corresponds to $(H/C) = 1$. Molecules a appear only if p is at least 0.3. When $p = 0.65$, 35% of the aromatic molecules would have to be entirely pure carbon, which corresponds to an infinite graphitic layer (here designated as graphite). Obviously graphite is never found in such materials. Hence molecules smaller in size obviously have to be considered, such as m in Fig. 7.20 (last column of Table 7.3). When $(H/C)_{total} = 1.3$, the percentage is already 46%; it reaches 100% when $(H/C) = 0.5$.

Considering this discussion, it seems reasonable to assume that DF and lattice-fringe images are entirely due to polyaromatic molecules, whenever the (H/C) ratio is equal to one or less. It results that bright dots corresponding to non aromatic molecules can be suspected only in series I and II at the lowest degrees of catagenesis (Table 7.1).

([1]) Thanks are due to Mr. Fontanel of *Institut Français du Pétrole* who took the pictures.

([2]) It is not possible either to separate the results due to very large spacings from those due to interlayer porosity.

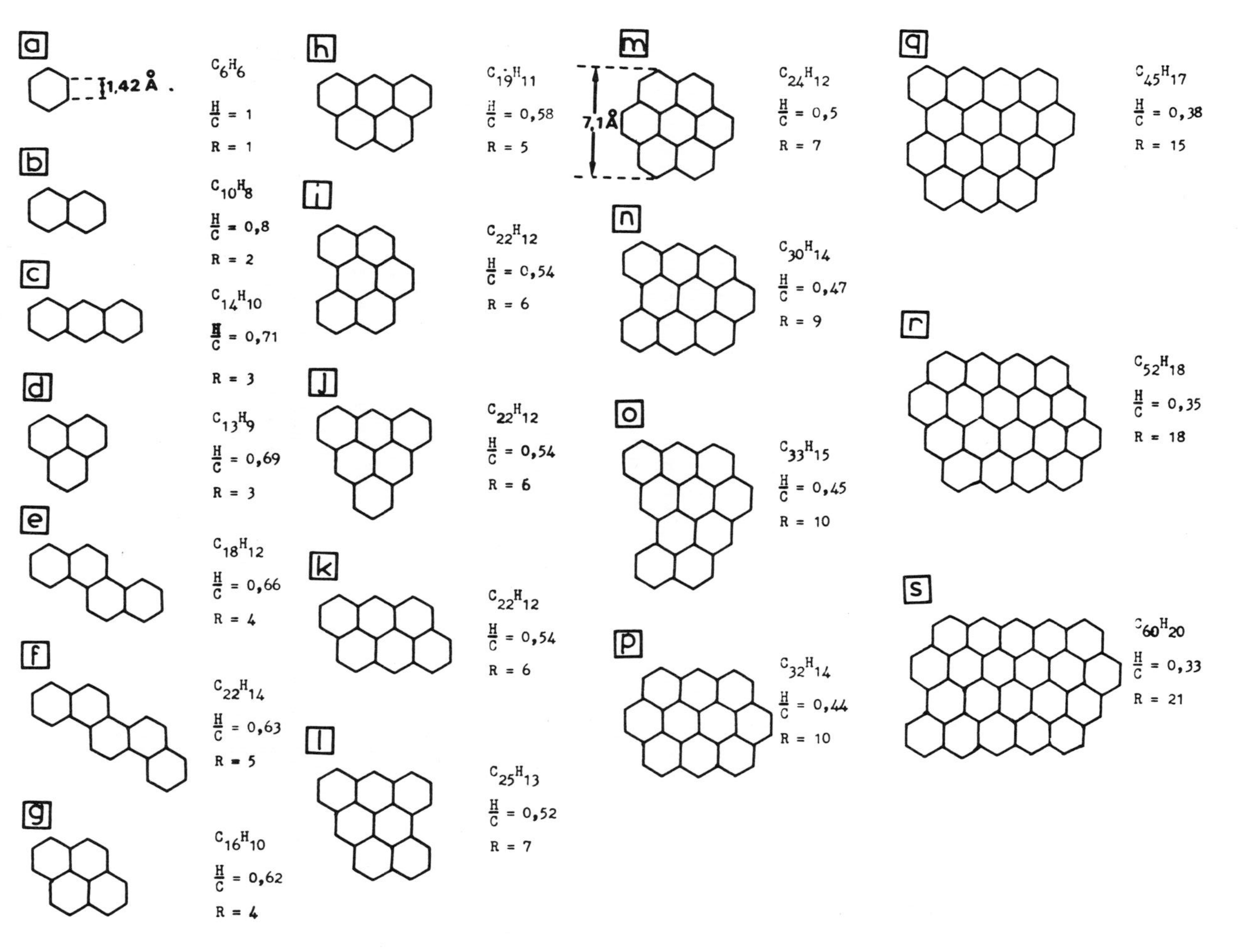

Fig. 7.20. — Some possible structures of polycondensed aromatic molecules.

TABLE 7.3
POSSIBLE POLYAROMATIC MOLECULES

$(H/C)_{total}$	p	f_a	$(H/C)_{arom}$	Possible aromatic structures*	m Molecules (%)
1.30	0.00	1.00	1.30	imposs.	46
	0.10	0.90	1.22	imposs.	
	0.20	0.80	1.13	imposs.	
	0.30	0.70	1.00	a	
	0.40	0.60	0.83	b	
	0.50	0.50	0.60	f,g,h	
	0.60	0.40	0.25	s	
	0.65	0.35	0.00	graphite	
1.00	0.00	1.00	1.00	a	66
	0.10	0.90	0.89	b	
	0.20	0.80	0.75	c	
	0.30	0.70	0.57	h	
	0.40	0.60	0.33	r,s	
	0.50	0.50	0.00	graphite	
0.80	0.00	1.00	0.80	a	80
	0.10	0.90	0.78	b	
	0.20	0.80	0.50	l,m,n	
	0.30	0.70	0.29	s	
	0.40	0.60	0.00	graphite	
0.50	0.00	1.00	0.50	l,m,n	100
	0.10	0.90	0.33	r,s	
	0.20	0.80	0.13		
	0.25	0.75	0.00	Graphite	

* See structures in Fig. 7.20.

3. Orientation of the molecules.

It has been shown that in 00.2 DF pictures the bright dots are homogeneously distributed within any kerogen sample when the degree of catagenesis is low (Fig. 7.17a). Distribution

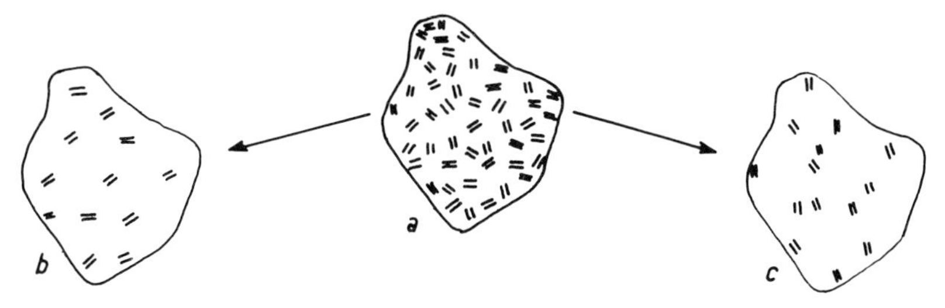

Fig. 7.21. — **a.** Scheme of a kerogen particle containing aromatic layer stacks at random. To be clear, only stacks at the Bragg angle are represented by a set of two lines. **b.** and **c.** 00.2 DF images with the objective aperture moved 90° along 00.2.

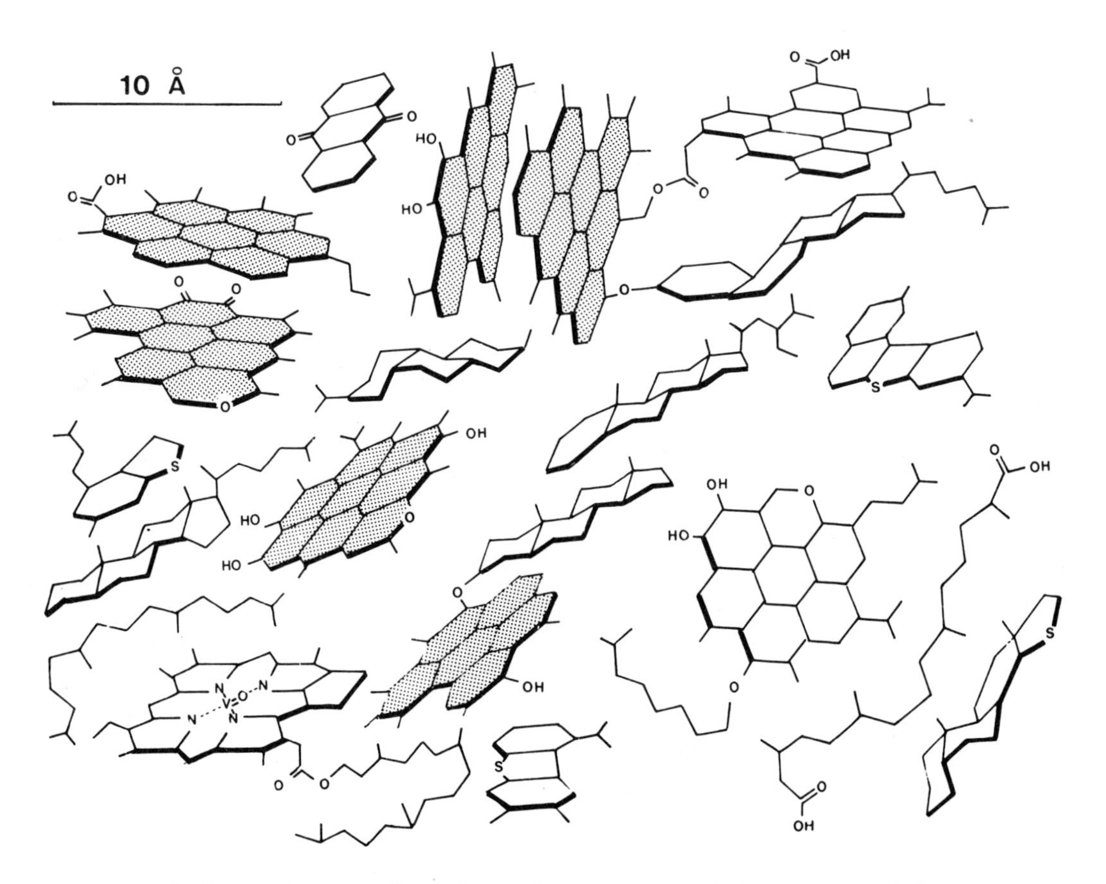

Fig. 7.22. — Structural scheme of a type II kerogen at an evolution stage corresponding to the beginning of oil formation (equivalent reflectance of vitrinite: 0.6) Aromatic molecules are plane (by delocalization of π electrons). Stacks of aromatic layers capable of producing diffraction patterns are shaded: in the volume of kerogen represented here, there are two of them in one direction and one in the perpendicular sense. Non aromatic structural blocks have been introduced by B. Durand and M. Vandenbroucke on the basis of results from elemental analyses of type II kerogens and from NMR data on corresponding NSO compounds. Principal structural characteristics of the kerogen represented here are:

Weight% of elements: C = 83.13 H = 8.10 O = 5.88 S = 1.57 N = 0.69 V = 0.62. Atomic H/C = 1.17. Atomic O/C = 5.31%. Aliphatic C = 28%. Naphthenic C = 21%. Aromatic C = 51%. Density = 1.2.

remains homogeneous for every position given to the aperture on the 00.2 ring. This indicates a random distribution of stack orientations. Fig. 7.21 shows, in the upper part, a particle *a* of kerogen containing stacks of two aromatic molecules distributed at random. Only the stacks seen edge-on (i.e. at the Bragg angle) have been represented by a set of two lines. In b and c are represented the stacks appearing as bright dots in 00.2 DF. From b to c the 00.2 ring has been moved 90 degrees in relation to the objective aperture, i.e. two portions of the 00.2 Debye-Scherrer ring 90° apart from each other have been used for molecule imaging. In any case a homogeneous distribution of bright dots is obtained.

4. *Conclusion.*

Size histograms of bright domains are sharp, as demonstrated in Fig. 7.18, for instance. From their average size they can be assimilated with *h* to *p* molecules in Fig. 7.20. Some molecules pile up by twos (or threes) and form turbostratic stacks. These elementary units show a wide spreading of inter layer spacing. The minimum d_l is close to the turbostratic carbon value and is the same in all kerogens. In samples at a low degree of catagenesis the spacing spreading ranges from 3.4 to more than 8 Å. These stacks are distributed at random in kerogen and connected by non-aromatic groups. A structural model deriving from the above data is represented in Fig. 7.22. It has been completed by using the model of kerogens established by Tissot *et al.* [47]. By comparing this model with Fig. 7.10b, it can be deduced that similar aromatic elementary units are present in both kerogens and chars.

D. Characteristic features of catagenesis and metagenesis.

1. *Bright dot size.*

As catagenesis progresses there is no growth of the individual aromatic stacks, and the same elementary domains are observed in all three series.

Smallness and absence of growth of the graphitic layers piled up in stacks are features common to all chars and carbons. Even in graphitizing carbons the mosaic units contained in the wrinkled carbon layers remain unchanged up to 2 000° C (Fig. 7.10b). These results fit well with the data deduced from other techniques such as chemical analysis, combustion-heat or density measurements.

Chemical analysis leads to a determination of the aromaticity factor f_a and of the aromatic molecule size as well. The number R of aromatic rings in a structural unit which contains n carbon atoms can then be derived from combustion-heat or density measurement data [49]. Such data have been obtained for vitrinites and exinites (Table 7.4). Kerogens with an elemental composition close to vitrinites (series III) and to exinites (series II) are also represented in this Table. It first appears that the elementary units are small (less than 10 rings), then that their size remains almost constant as catagenesis progresses. Incidently if Table 7.3 is considered again, it can be noticed that the aliphatic molecule proportion has to decrease as $(H/C)_{total}$ decreases, since the sizes of the aromatic molecules remain constant.

TABLE 7.4
STRUCTURAL PARAMETERS FOR VITRINITES AND EXINITES [49]
AS COMPARED TO KEROGENS

Vitrinites				Equivalent Kerogen Sample
H/C	f_a	n	R	
0.86	0.70	17	4	III_1
0.79	0.78	17	4	
0.77	0.80	19	4.5	
0.75	0.83	19	4.5	III_6
0.73	0.86	20	5.0	III_7
0.68	0.88	20	5.3	
0.65	0.90	20	5.6	
0.59	0.92	20	6.0	III_{13}
0.50	0.96	20	6.3	III_{14}
0.44	0.99	20	6.6	
0.37	1.00	30	10	

Exinites			Equivalent Kerogen Sample
H/C	f_a	R	
1.07	0.61	4	II_5
0.9	0.73	4	II_8
0.8	0.83	4.5	II_{11}
0.69	0.87	5	II_{13}
0.64	0.9	6	II_{18}
0.58	0.92	6	II_{19}

2. *Interlayer spacing spreading.*

It decreases as the degree of catagenesis increases. Hence the histogram becomes sharper (Table 7.2 and Fig. 7.19b). However it is not as sharp as the soft carbon one represented as a comparison in Fig. 7.19c. The progressive decrease in spacing spreading is due to the progressive disappearance of the non-aromatic groups fixed at the boundaries of the aromatic elementary units. This favors the occurrence of layer couples which are both parallel and have a constant spacing.

3. *Molecular orientation [5,26].*

When evolution due to burial reaches a given stage (which is about the transition between catagenesis and metagenesis), a sudden transformation occurs in 00.2 DF images. The bright dots aggregate into clusters (Fig. 7.23a).

This means that, inside the corresponding region of the kerogen particle, some diffracting domains have approximately the same orientation in space, since their 00.2 scattered beams pass altogether through the aperture opening. The occurrence of bright dot clusters therefore corresponds to a molecular orientation developed locally [4, 5, 28] as is schematically shown in Fig. 7.24 (this figure has to be compared to Fig. 7.21; the same assumptions have been made for both drawings). Such a molecular orientation has been found in all series of kerogens (natural or heat-treated). In Tables 7.1 and 7.2 it is noted with an asterisk in the first column. This molecular orientation can be fully described only if its extension is determined in the three space dimensions and if the misorientation of its elementary units is also evaluated [24, 50]. More detailed explanation will be given in the following, preceding the results obtained on kerogens.

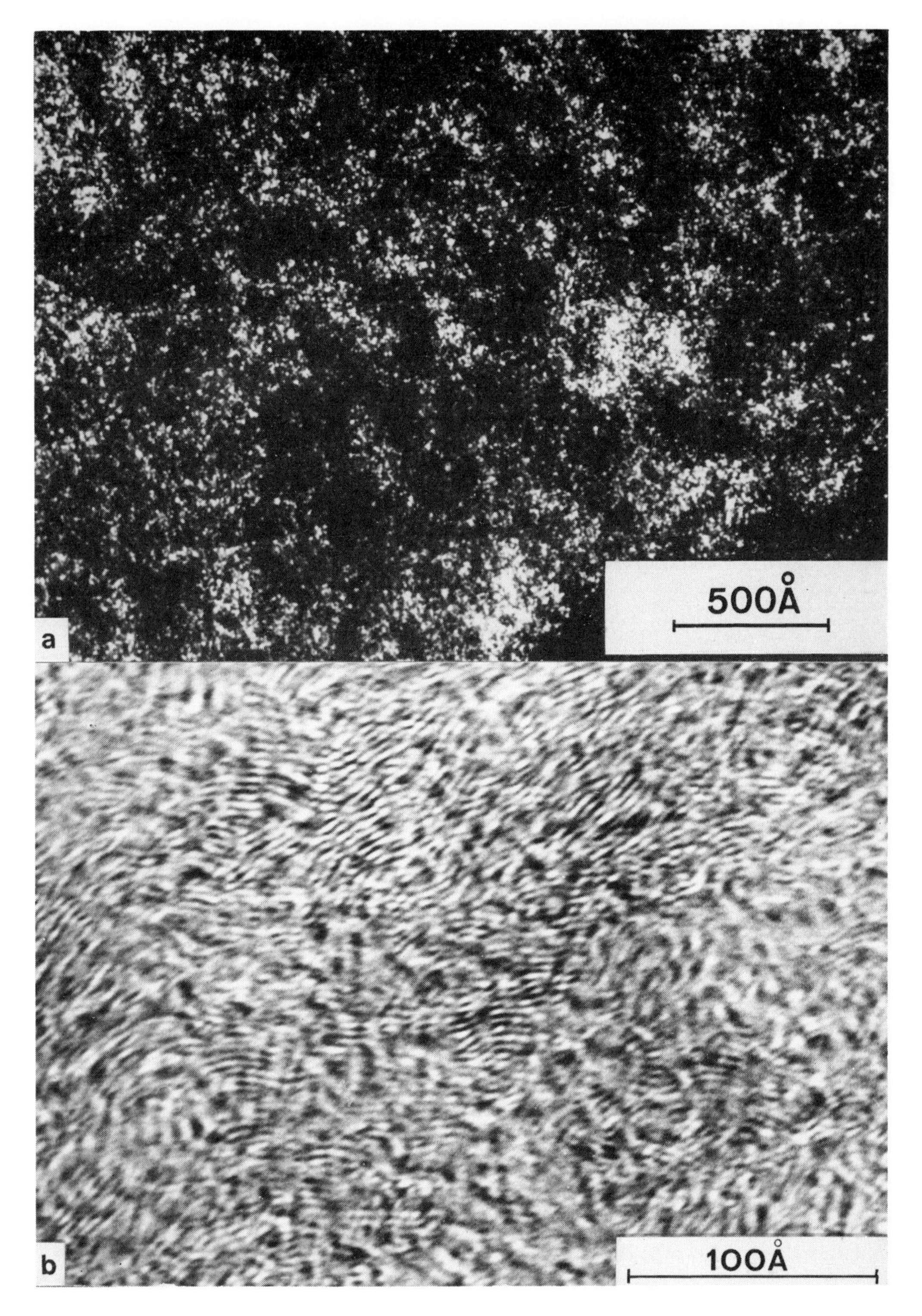

Fig. 7.23. — Series II kerogen (II_{21}, highly evolved)

a. 00.2 DF. **b.** 00.2 lattice-imaging.

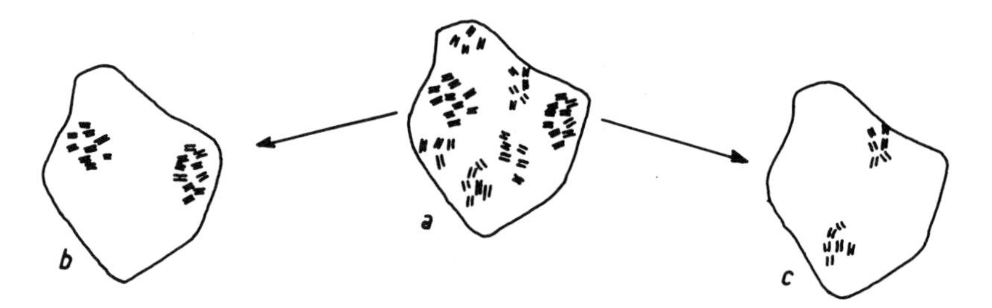

Fig. 7.24. — **a.** Scheme of molecular orientation inside a kerogen particle (same conventions as in Fig. 7.21). **b.** and **c.** 00.2 DF images with the objective aperture moved 90° along 00.2.

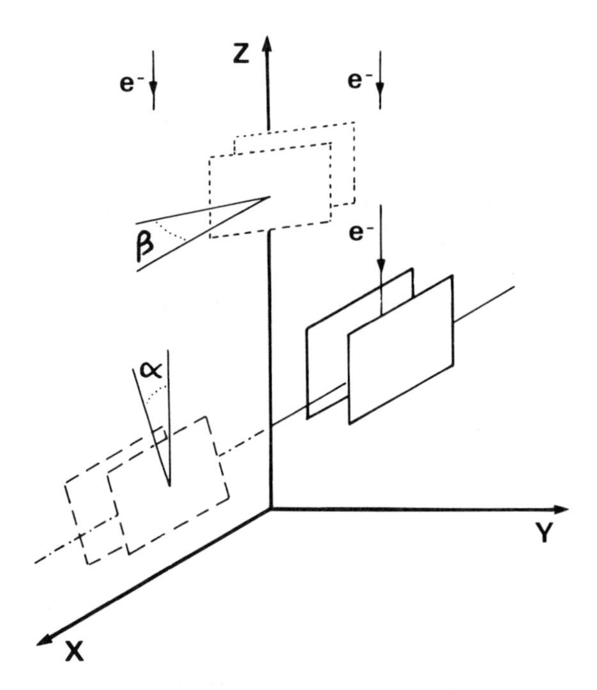

Fig. 7.25. — Orientations of aromatic layer stacks relatively to the incident beam.

Figure 7.25 schematically represents stacks of two aromatic layers. The first one (solid line) is in the Bragg position for the 00.2 reflection. Therefore the electron beam direction is along OZ (as seen part.1, paragr. II.B.1,$\theta_{00.2} \approx 0.3°$ for 100 kV). The second stack (dashed line) is tilted around OX of an angle α. The third (dotted line) is twisted around OZ of an angle β. Let us consider the direction $\vec{c}$ perpendicular to the layers. It will be along OY if all the layer stacks are parallel to ZOX. The preferred orientation should thus be a perfect parallel orientation. If some misorientation develops, the $\vec{c}$ axes would be distributed around OY inside a cone having half angles α and β. For determining the degree of orientation, the frequency distribution curves of the $\vec{c}$ axes have to be determined along OX (tilt) and OZ (twist). The degrees of orientation $\Delta\theta$ are the half width at half maximum of these curves. In the case of a fibrous orientation, the layers are only parallel to an axis, the fiber axis (OX in Fig. 7.25). The $\vec{c}$ axes are thus distributed at random around OX, i.e. their frequency is the same in any direction around OX.

However their twist around OZ can be described by means of a frequency curve as above. All possible intermediates can be found. If the aperture used in 00.2 DF imaging is infinitely small, the stacks imaged as bright dots would be only those in the kerogen particle which occupy the position of the solid line stack relative to the beam (Fig. 7.25). By moving the aperture along the 00.2 ring, first the relative orientation of all the clusters would be determined, then the twist of every stack would be evaluated. By tilting the kerogen particle by using a goniometer stage, the tilting of the stacks would be evaluated. Unfortunately, in practice the aperture is not infinitely small. It allows a rotation around OZ approximated by the 00.2 ring portion intersected by the opening projection in the SAD pattern (for instance the 0.1 $Å^{-1}$ aperture in Fig. 7.5 tolerates ± 10°). In the same way it allows a deviation from the Bragg angle (tilt) calculated as ± 32° in Fig. 7.5. It thus appears necessary to use apertures of various sizes for exploring the 00.2 ring.

a. Three dimensional extension of molecular orientation.

Figure 7.26a represents the DF obtained when the 00.2 ring is in the starting position relative to the 0.2 $Å^{-1}$ aperture (± 20° in twist, ± 42° in tilt). Clusters of bright domains can be observed in some regions while others appear completely dark. These dark zones can be explained either by their lack of preferred orientation or by a preferred orientation which differs in direction from that of the bright regions. The second alternative happens to be the right one, since the bright regions disappear as others appear when the 00.2 ring is moved relative to the objective aperture (Fig. 7.26a should be compared with Fig. 7.26b). The appearance and disappearance of the bright zones appear very progressive and can be compared to the undulous extinction observed by polarized light microscopy in some quartz crystals. (Incidently, when the 0.2 $Å^{-1}$ aperture is replaced by the 0.1 $Å^{-1}$ one, the intensity of the clusters noticeably decreases but their shape does not change; hence the twist of the stacks is large inside each cluster). The dimensions of the clusters have been estimated by image analyzing techniques [1] (image analyzing computer) but till now cannot be measured precisely.

The extension of the molecular orientation has been determined in the XOY plane which leads to an average diameter. It is also necessary [4] to determine the average thickness by tilting the particle around OY (Fig. 7.25). In this case, the stacks at the Bragg angle always remain in this position and the cluster never disappears. Let us now consider two clusters l_0 apart and h in thickness (Fig. 7.27). If a tilting angle i is chosen, the distance becomes:

$$l_i = l_0 \cos i - h \sin i.$$

With i being known and l_i being measured, the thickness h can be deduced. In the three series of kerogens the clusters have been found to be isometric in shape (their thickness is about the same as their diameter). The lattice-fringe images (Fig. 7.23b) also show the occurrence of molecular orientation.

b. Misorientation of the elementary units.

Until now, the frequency curves of the distribution of the $\vec{c}$ axes in a kerogen cannot be precisely established since the intensity changes of the bright dot clusters are not easily measured. However it is possible to find the maximum intensity of a cluster as the particle is tilted

---

[1] Thanks are due to Mr. Gateau and Mr. Prevosteau at *Bureau de Recherches Géologiques et Minières,* (BRGM) Orléans, who made the measurements.

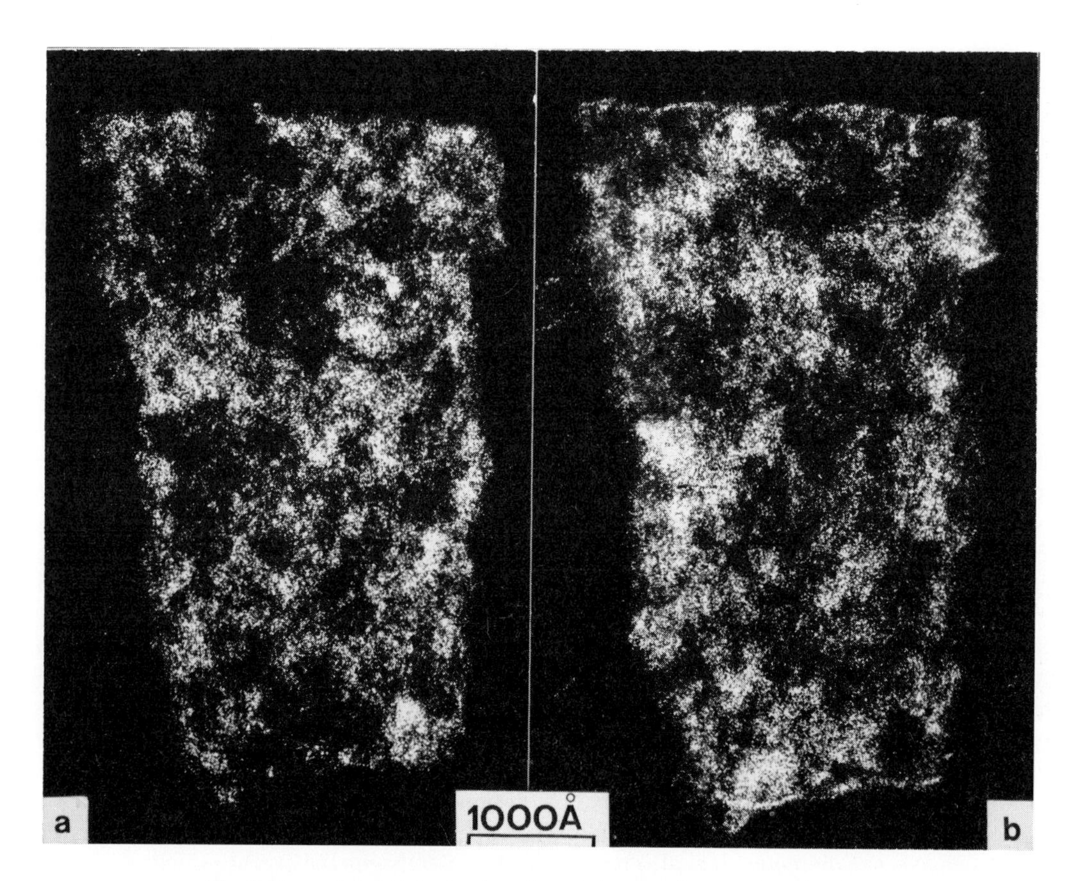

Fig. 7.26. — Azimuthal misorientation of clusters of aromatic molecules.
a. and **b.** objective aperture moved 90° along 00.2.

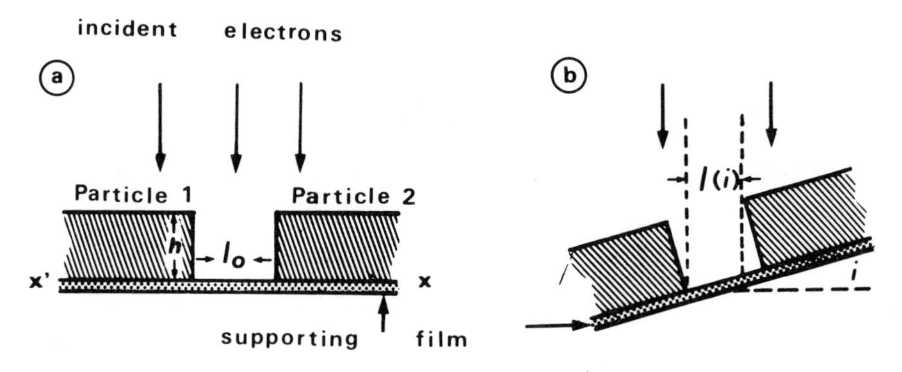

Fig. 7.27. — Determination of cluster thickness.

around OX. Then it is possible to determine whether or not this intensity decreases as the α tilting angle increases. Hence the range where the intensity remains high can be evaluated. The position where the cluster practically disappears can be determined as well. In a second stage, the position which corresponds to the maximum cluster intensity remains fixed, and the twist β can be evaluated by exploring the 00.2 ring with the 0.1 $Å^{-1}$ aperture.

In the example in Fig. 7.26, the clusters practically disappear when the tilting angle is about 50° or the aperture rotation 50° as well. The intensity decreases slowly and remains high through about 30°. Since α and β are approximately the same and are below 50°, the preferred orientation is thus parallel. However since the decrease in intensity is slow, the degree of orientation is very low ($\Delta\theta \approx 30°$).

As metagenesis progresses, cluster sizes do not change markedly, although their brightness progressively increases ([1]). Such an increase in brightness is due to an improvement in the preferred orientation (i.e an increase in the degree of orientation) ($\Delta\theta \approx 15°$).

At the highest stage of carbonification corresponding to the highest degree of metagenesis, the possible structural model for kerogens is not very different from what is shown in Fig. 7.10b. In this Figure, the elementary mosaic units are represented as being included in a wrinkled carbon layer and are all approximately parallel to one other. They obviously represent a molecular orientation. Its extension is about 100 Å in size in this example. This figure can easily be imagined either to be repeated for a few micrometres, thus simulating graphitizing carbons, or to be limited to about 50 Å as in non graphitizing carbons. In kerogens the size is found to vary from 100 Å to 1 000 Å. An additional argument is thus given which shows the close similarities between natural and synthetic carbons. A comparison between Figs. 7.10b and 7.22 shows that, in the first case (high rank kerogens), the aromatic molecules are closely and tightly attached to each other by defective boundaries, while in low rank kerogens they are widely separated by other compounds.

4. *Size of the clusters.*

In series I, II and III the size of the clusters ranges from less than 100 Å to 250 Å. Up to now, larger sizes were found only in the case of a sample located near a dolerite inclusion (1 000-2 000 Å). The smallest sizes are always found in samples attributed to series III. Other series will be discussed further.

II. THERMAL SIMULATIONS (CRITERIA FOR A DISTINCTION BETWEEN SERIES)

As previously shown, kerogens behave similarly to pyrolyzed organic matter. Hence thermal simulation can be applied for answering two questions. First, does the progressive heat-treatment of the less evolved kerogen leads to a reproduction of the whole series? Then, does the heat-treatment of various kerogens supposed to belong to the same series confirms the hypothesis? If this can be done successfully, the series an unknown kerogen belongs to can be determined.

([1]) It should be noted at this point that the contrast of the bright dots and of the clusters themselves is so low when orientation appears that the cluster size tends to be found too small. For the same reason of contrast, when the clusters are small, a too high temperature or a too high degree of catagenesis has to be reached to make them visible.

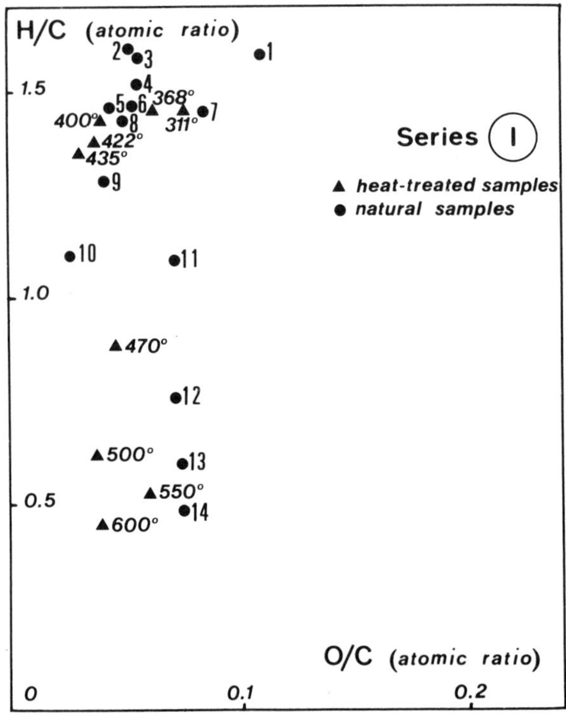

Fig. 7.28. — Van Krevelen diagram for series I. Natural and heat-treated kerogens.

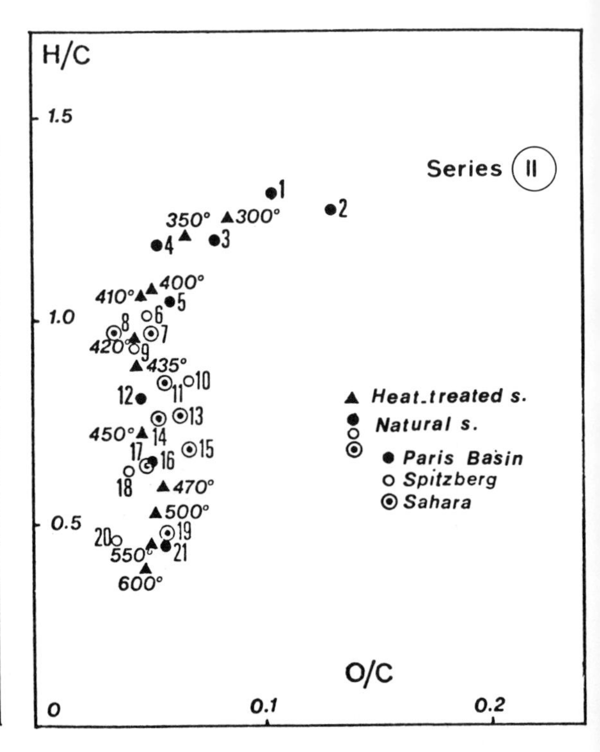

Fig. 7.29. — Van Krevelen diagram for series II. Natural and heat-treated kerogens.

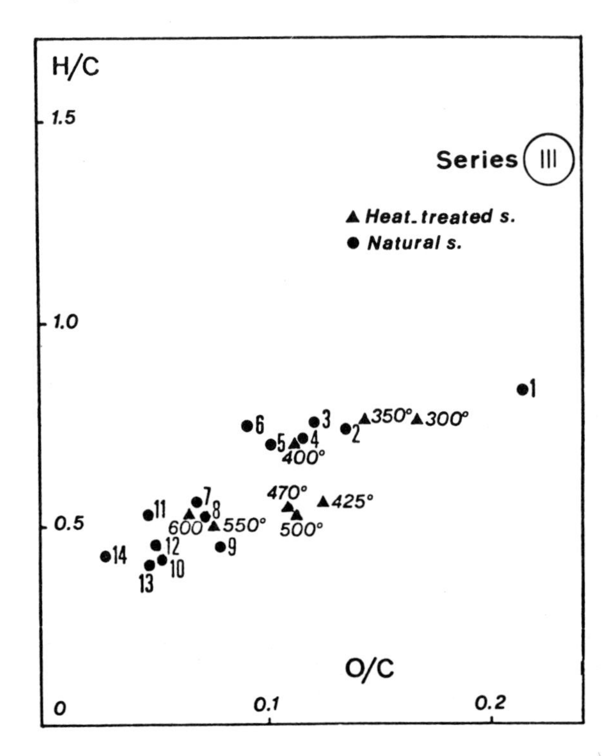

Fig. 7.30. — Van Krevelen diagram for series III. Natural and heat-treated kerogens.

A. Heat-treatment of the less evolved kerogen.

The same textural and structural features are usually observed in natural and heat-treated kerogens which produce neighboring representative points in the Van Krevelen diagram (Figs. 7.28 to 7.30). These Figures and Tables 7.1 and 7.2 show that interlayer spacing spreading decreases after heat-treatment as it does during catagenesis. Molecular orientation also occurs in both natural and synthetic series as indicated by asterisks in Tables 7.1 and 7.2. It should be noted that molecular orientation occurs in I'_{500}, II'_{470}, III'_{600}. (the lack of contrast of the clusters as they just appear is common to both natural and heat-treated series as it has been mentioned previously) $\Delta\theta$ is about 30° when molecular orientation appears, it is only 15 at 1 000° C.

From series I' to series III' or III the molecular orientation decreases from about 1 000-2 000 Å to 100 Å (Figs. 7.31a and 7.31b should be compared with Figs. 7.23 and 7.32). Hence, when considering these figures, **the extent of molecular orientation can be used as a specific criterion for making a distinction between series.** Series I (or I') should lead to a highly extended molecular orientation ($\geq$ 1 000 Å). On the contrary oxygen-rich substances (series III or III') should not show any orientation larger than 100 Å, while series II (or II') should be intermediate (> 200 Å).

Unfortunately an important discrepancy occurs when the extent of molecular orientation is compared in natural and heat-treated series. Among the samples attributed to series I (Table 7.1 and Fig. 7.28) those in which molecular orientation occurs (I_{13} and I_{14}) reveal reduced orientations. Their extent is intermediate between II and III. A smaller discrepancy is found between series II' issued from the heat-treatment of II_1 and series II. The size of the orientation here is about 150 Å, i.e. smaller than the value found for II_{21} (>200 Å).

For establishing a valid distinction between series is should be necessary to carry out treatment of many intermediate kerogens, either ones belonging to one of the three series or additional samples belonging to intermediate series. As a matter of fact, it is particularly important to have a worth while criterion to apply, since for highly evolved samples all other techniques are non longer accurate.

B. Heat-treatment of intermediate kerogens.

1. Series I.

I_{11} and I_{12} after heat-treatment lead to a small extent of orientation which tends to place them between series II and III. On the contrary I_{10} leads to an orientation of 1 000 Å, which is very similar to the one represented in Fig. 7.31. Additionaly, heat-treatment of other low rank kerogens classified in series I by chemical and geological data leads to a highly extended molecular orientation (1 000 Å). For instance Kerosen shales are typical series I.

2. Series II.

Kerogens such as II_1, chosen as the initial kerogen of series II to be heat-treated, comes from the NE of the Paris Basin (Toarcian shales). It has been found recently [16] that only

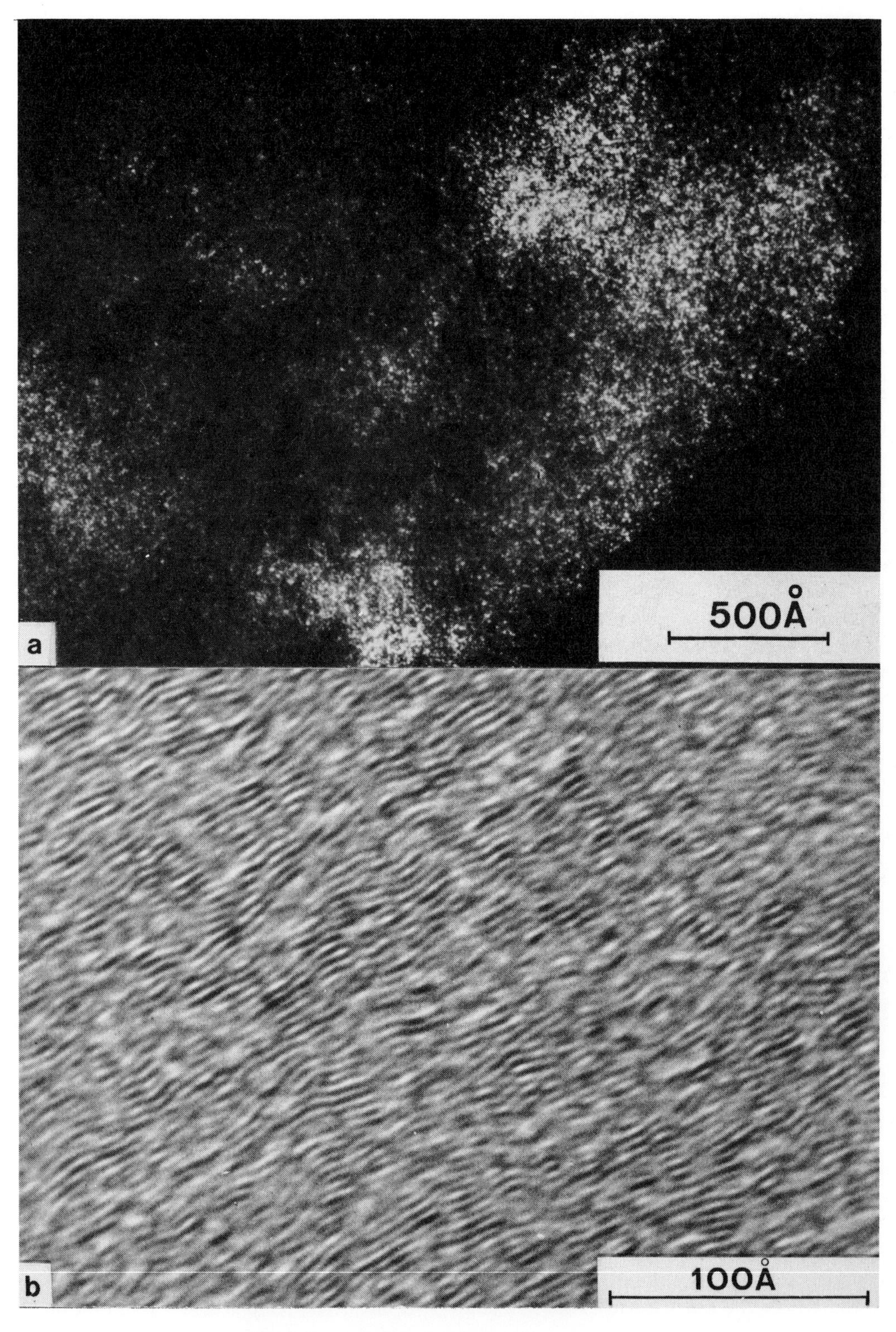

Fig. 7.31. — Series I kerogen (I'_{600})

a. 00.2 DF. **b.** 00.2 lattice-imaging.

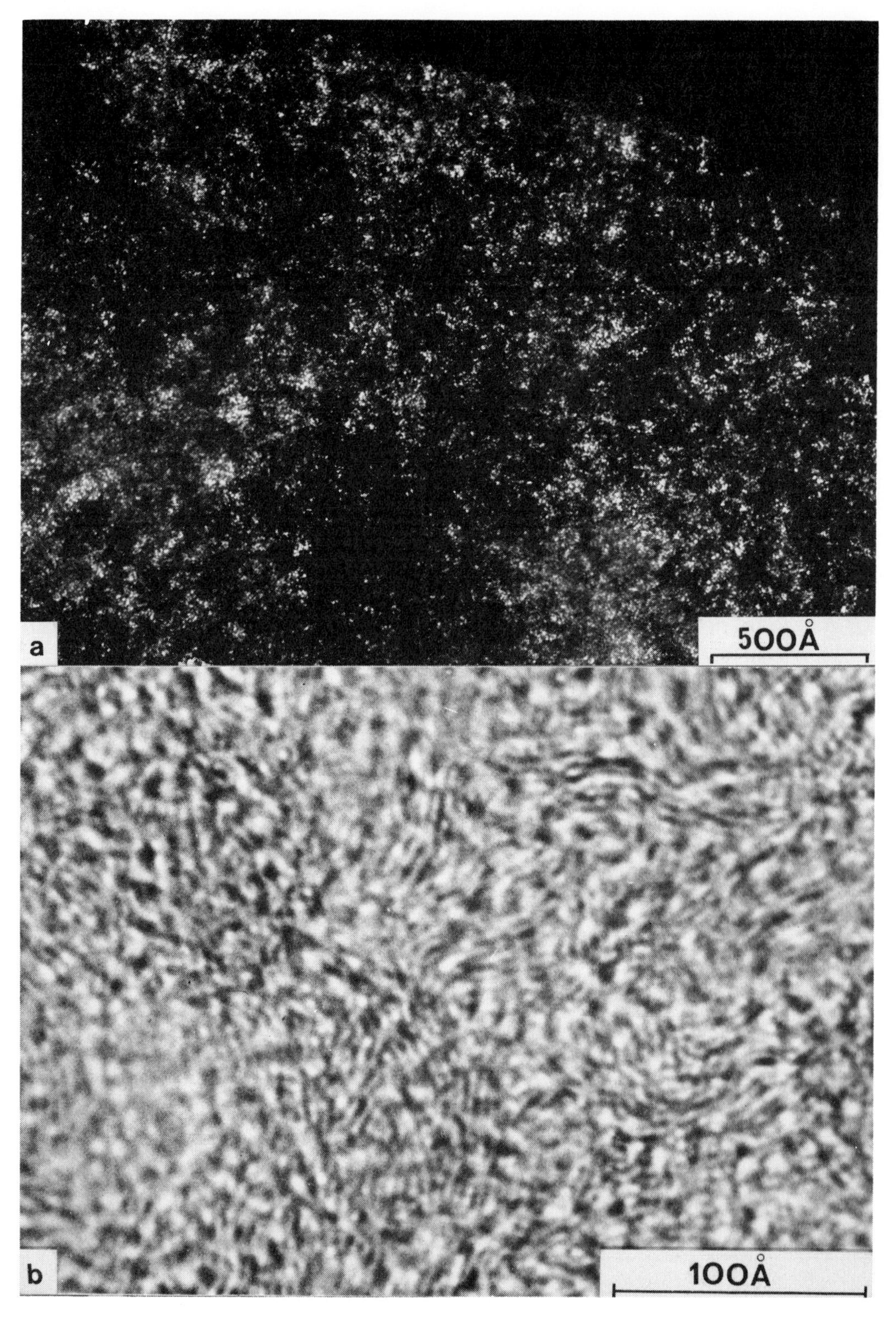

Fig. 7.32. — Series III kerogen (highly evolved)

a. 00.2 DF. **b.** 00.2 lattice-imaging.

southern samples belong to a typical series II. If such samples are heat-treated, the extension of the molecular orientation is again > 200 Å as for natural series. Other low rank kerogens typically classified in series II, such as Ypresian sediment from Tunisia, behave like II_{21} when heat-treated i.e the orientation is > 200 Å.

3. Series III.

All series III were correctly classified, including additional samples such as Cretaceous sediment from the West Alberta Basin, Canada.

C. Comparison with coals.

The samples studied here are kerogens from vitrinite coals (Fig. 7.33). They are formed by the same elementary aromatic stacks (less than 10 Å in size) connected by non aromatic groups. A wide interlayer spacing spreading occurs for low rank coals, and it decreases as rank increases. However it remains more noticeable than in kerogens at a similar stage (even in anthracites). First randomly dispersed in the fragment studied, the aromatic stacks agreggate and molecular orientation appears at the anthracite stage. It is practically in the same range as the III series kerogens ($R_o > 2$, H/C ≈ 0.45, O/C ≈ 0.05). The extent of the molecular orientation is also the same, i.e about 100 Å. Unfortunately these samples are heterogeneous. Most of them contain a relatively large amount of graphite and particles at various stages of catagenesis, probably all of detrital origin. A complete geochemical study [9] of this coal series has emphasized the close similarities between type III kerogens and kerogens from coals as confirmed by detailed electron microscopy studies of anthracites [39].

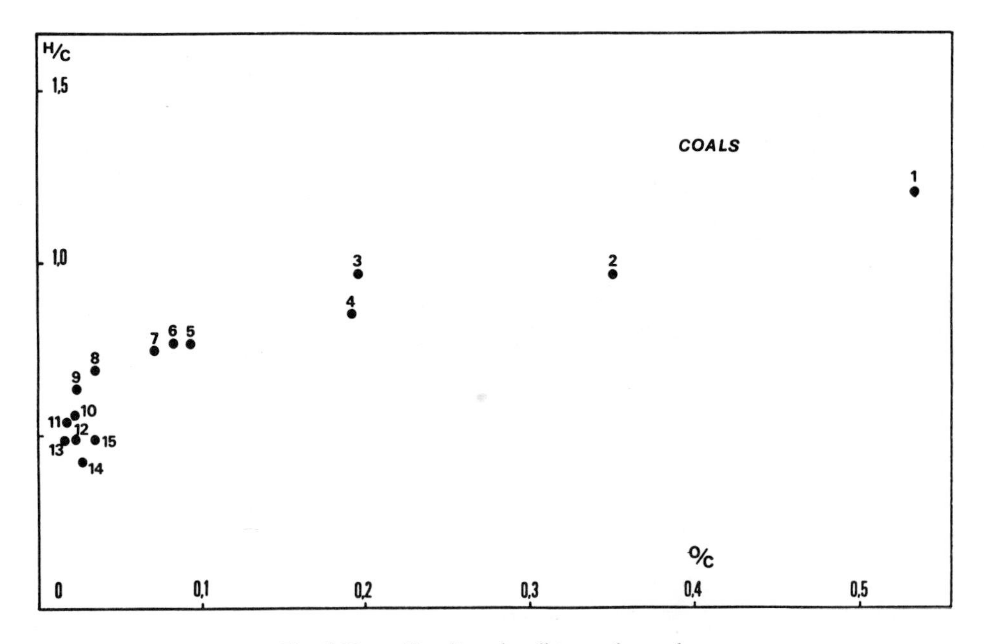

Fig. 7.33. — Van Krevelen diagram for coals.

D. Discussion and conclusion.

The heat-treatment of kerogens suggest that the extent of molecular orientation can be used as an absolute criterion for distinguishing between kerogen types, since the largest clusters (≥ 1 000 Å) correspond to series I', the medium (250 Å) to series II' and the smallest (100 Å) to series III'.

It should be noted that all oxygen-rich kerogens have been quite correctly classified as belonging to series III. For all others, when consideration of extent of the molecular orientation disagrees with geological data, it always tends to displace the samples from series I or II to series III. If it is so, an excess of oxygen is always found in chemical analysis. It is the case for all medium and highly evolved natural samples attributed to the series I and it is also the case of natural samples of NE Paris Basin.

Hence it can be misleading to attribute the extent of molecular orientation exclusively to the effect of the chemical composition of the parent material. It is necessary to take into account the physico-chemical conditions of catagenesis such as the effect of an oxidizing environment. It seems highly probable that such an effect should be responsible for the lowering of the extent of molecular orientation described above. It can also be said that **molecular orientation is probably closely related to the oil potential of a kerogen** whatever its type of history may be.

The great importance of oxidation has recently been emphazized by analyzing the surface alteration of the sample II_1 which increases the oxygen content of the sample [22]. Another recent study also shows the importance of oxidation. In the Precambrian kerogens (from bitumens and shale) from Oklo (Gabon), the widely extended molecular orientation in the zone prevented from oxidation is entirely inhibited in the oxidized samples [48]. However the stage of evolution at which oxidation has the strongest effect is not yet clear. Further experiments have to be performed on heat-treated kerogens and pyrolyzed materials as well.

From the above data it may be deduced that CTEM is a worthwhile technique, and the evaluation of the extent of molecular orientation is a worthwhile criterion as well. However these investigations devoted purely to the study of kerogen (whether heat-treated or not) cannot fully explain the mechanism of catagenesis. As a matter of fact, many factors cannot be considered separately when only natural samples are studied, such as the influence of bacteria the inhomogeneity of precursors or detrital additions. Likewise the effect of the physico-chemical conditions during each stage of catagenesis, though recognized, cannot be studied experimentally. At least the kerogen microstructure can be partially attributed to such effects. For stipulating the role played by each of these factors it is necessary to pyrolyze simpler and purer substances than kerogens although they have to be chosen with a similar elemental composition [31, 50, 51, 52].

PART THREE
PYROLYSIS AS A MODEL OF CATAGENESIS

I. MECHANISM OF PYROLYSIS

A. Natural substances.

Sporopollenin (H/C = 1.56; O/C = 0.34) has been chosen since it produces a relatively high amount of tars. It also contains a fair amount of oxygen, and its elemental composition is similar to poorly evolved series II kerogens. Lignite (H/C = 1.12; O/C = 0.44) has been chosen since it contains a high amount of oxygen, and it can be considered as similar to the initial kerogen in series III.

In both cases 140 mg were packed in a covered graphite crucible and heat-treated in a N_2 flow at 4° C/min. Differential thermal analysis (DTA) was performed during heat-treatment. Samples corresponding to various representative points of the DTA curve were subjected to rapid quenching under inert gas outside of the furnace. They were used for weight loss measurements (WL), chemical analysis, infrared spectrometry (IR), electron spin resonance (ESR) and CTEM studies [31, 50, 51, 52].

The carbonization path of sporopollenin is represented by dashed line in the Van Krevelen diagram in Fig. 7.34. DTA, WL and DWL (differential weight loss curves) are represented in Fig. 7.35. The DTA curve for this material has two parts. A first endotherm (point 2), which corresponds to the first DWL maximum, is mainly related to the departure of oxygen groups as deduced from Fig. 7.34. The IR absorption coefficients (Fig. 7.36) for OH, carbonyl and carboxyl groups show that, in this domain, their number gradually decreases and that they finally disappear. The second endotherm which coincides with the second DWL maximum corresponds to the departure of tar. Secondary endothermic peaks numbered as 4 and 5 and a diffuse accident 6 follow on the endotherm. Visual observation as well as optical and scanning microscopy studies show that sporopollenin softens at 436° C (point 4) and then melts. Plasticizing occurs slightly before the second DWL maximum which corresponds to the maximum tar departure rate. Maximum tar departure can be noticed on the DTA curve as the accident at point 6. A strong outgassing is observed in the material (similar to boiling point) which extends from melting until nearly the end of tar departure. Numbered samples on the DTA curve were studied in DF high resolution EM. At 460° C (asterisk on the DTA curve in Fig. 7.35) the bright dots gather in clusters about 1 000 Å in size, i.e. the molecular orientation is produced at this temperature. For HTT higher than 460° C, the size of the elementary stacks and the extent of orientation remain constant. Therefore the bright dot cluster contrast increases, indicating an improvement in the preferred orientation from $\Delta\theta \lessapprox 30°$ to 15° at 1 000° C.

Sporopollenin is extremely sensitive to experimental conditions. For instance, if heat-treated under pressure (5 bars) its molecular orientation extent increases to more than 2 000 Å. The factors which lead to tar retention affect this material in the same way and also

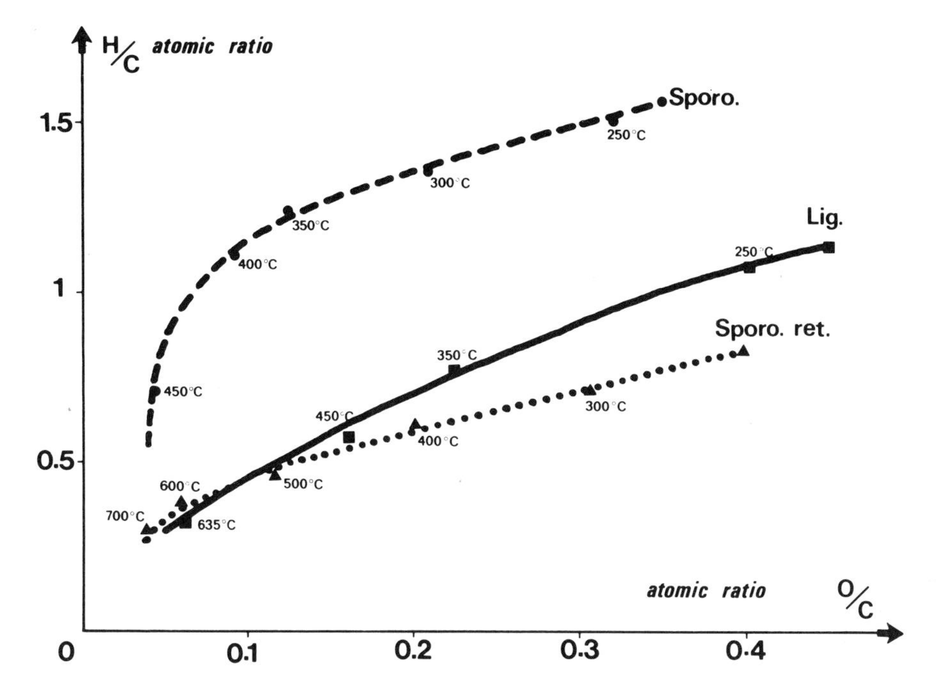

Fig. 7.34. — Van Krevelen diagrams for natural substances. Lignite is represented by a solid line, sporopollenin by a dashed line, oxidized (reticulated) sporopollenin by a dotted line.

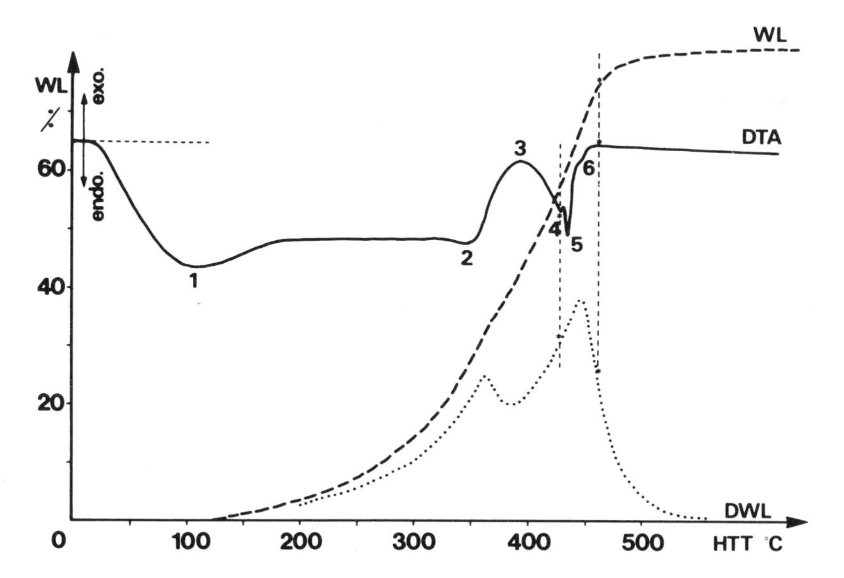

Fig. 7.35. — Sporopollenin heat-treated at atmospheric pressure. Solid line: differential thermal analysis. Dashed line: weight loss. Dotted line: weight loss rate. The asterisks note the occurrence of softening and of molecular orientation.

tend to decrease the melting temperature, i.e. the larger the temperature interval where softening occurs, the larger the orientation. On the contrary, pyrolysis in a vacuum reduces molecular orientation extent to less than 50 Å.

Lignite shows a carbonization path (solid-line curve in Fig. 7.34) regularly tending toward the carbon pole with less tar yield than sporopollenin. Its maximum weight loss is much lower than for sporopollenin (60% as compared with 80%). Its DTA curve contains only diffuse endotherms, and the DWL curve shows only one maximum. It does not melt or show a marked orientation either, although it contains aromatic layer stacks as does sporopollenin. The orientation extent is reduced to about 50 Å and almost cannot be seen below 1 000° C.

Sporopollenin, when treated for 5 hrs at 200° C in air (reticulation), becomes absolutely comparable to lignite. Its carbonization path (Fig. 7.34, dotted-line curve) is slightly lower (smaller tar yield), and its WL is lower as well. The molecular orientation characteristic of sporopollenin (1 000 Å in size) is reduced to 50 Å, and the material does not melt.

These data can be correlated with IR spectra [42 to 45, 51]. For sporopollenin heated in a covered crucible (dashed line in Figs. 7.36a, b, c, d), the occurrence of molecular orientation coincides with the extremum of the aromatic CH groups (Fig. 7.36c), after the carbonyl and carboxyl groups have disappeared (Fig. 7.36d) and when the OH (Fig. 7.36a) and aliphatic CH groups (Fig. 7.36b) have reached a low value. This occurs slightly before the maximum spin concentration as measured by ESR. Lignite (solid line in Fig. 7.36) is quite different from sporopollenin, i.e. OH groups are more numerous and their number decreases very slowly during heat-treatment. Aliphatic CH groups are less numerous and their number decreases faster, while aromatic CH groups are both less numerous and more stable during heat-treatment. New C — OR groups not present in heat-treated sporopollenin occur in lignite and remain remarkably stable for an HTT as high as 600° C. Reticulation of sporopollenin (dotted line in Fig. 7.36) changes the amount of various groups so as to increase the OH and mainly the CO groups and to decrease the aliphatic CH groups.

Chars made from lignite and sporopollenin may be considered as built of molecules with a few aromatic rings (4 to 10 rings) piled up by twos or threes. They are connected by non-aromatic structures. If the latter are mainly aliphatic, tar yield is large. First melting occurs, then maximum tar departure (strong outgassing) is immediatly followed by molecular orientation which itself precedes the end of tar departure.

The molecular orientation is more developed when tar retention is favored. This results in an extension of tar departure throughout a larger temperature interval, i.e when the melting point is lower (for instance the melting point occurs at about 410° C for 5 bars instead of 435° C). The molecular orientation corresponds to a maximum of mobility of the aromatic molecule stacks characterized by an extremum of aromatic CH and favored by the disappearance of oxygenated groups. The molecular orientation is reduced or inhibited when the oxygen content increases and when the tar percentage decreases, which corresponds to a crosslinking between stacks of aromatic molecules. Finally, the reticulation mechanism should first be related to the combining of oxygen and aliphatic carbon groups [49] which produces water and is responsible for an OH increase and an aliphatic CH decrease in IR spectra. Then for a massive fixation of oxygen inside the material, the molecular orientation is inhibited by linkage developing after reticulation. This linkage is due to C — OR in lignite and to COOH (hydrogen bonds) in reticulated sporopollenin.

The improvement of molecular orientation (Fig. 7.10b) when HTT increases until 1 000° C corresponds to aromatic CH departure and probably causes a progressive delocalisation of the π electrons responsible for the decrease of the IR bands at 1 630 cm^{-1} and also responsible for the tendency of the ESR lines to become Lorentzian.

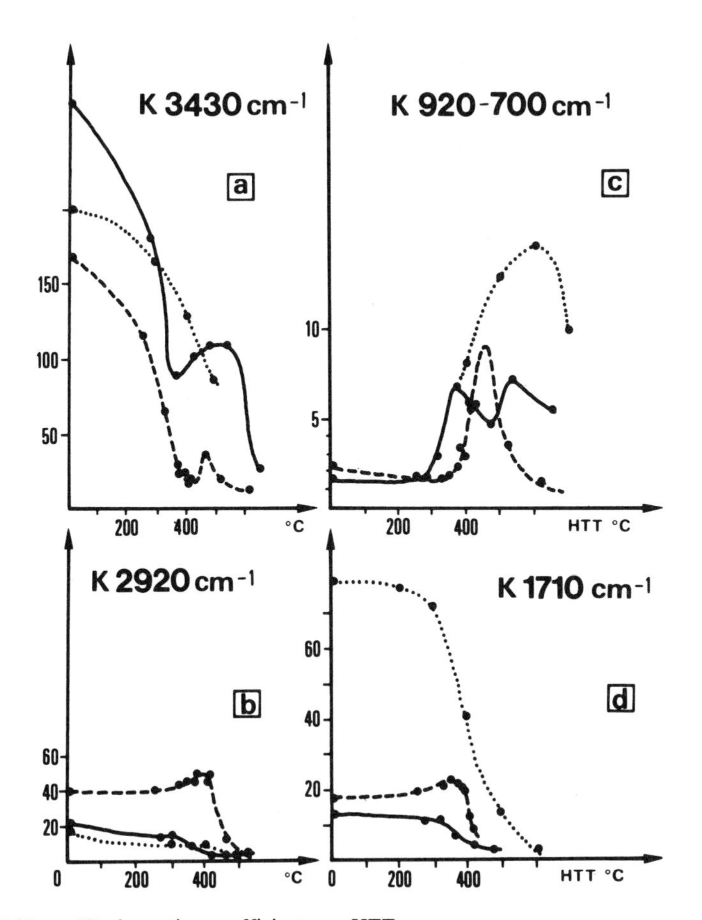

Fig. 7.36. — IR absorption coefficients vs. HTT.

a. OH groups. **b.** Aliphatic CH groups. **c.** Aromatic CH groups.
d. Carboxyl-carbonyl groups.
Solid line: lignite. Dashed line: sporopollenin. Dotted line: reticulated sporopollenin.

B. Industrial chars.

In graphitizing chars, the occurrence of mesophase spheres (part. 1, paragr. III.B.2) at about 400-450° C also corresponds to a molecular orientation of small mosaic elements (less than 10 Å). The parallel array of stacks occupies the whole liquid crystal sphere. Hence the molecular orientation extends over a few micrometres [6, 46]. In non-graphitizing chars the molecular orientation occurring in the pore walls (part. 1, paragr. III.B.2) is reduced to 50 A or less. There are apparently no structural differences between the molecular orientations of graphitizing and non-graphitizing chars except their size. Once again the reduced size of the orientation of non-graphitizing chars can be attributed to the occurrence of a high concentration of oxygenated functions, while graphitizing chars are almost oxygen free.

II. COMPARISON WITH KEROGENS

The data obtained by studying pyrolysis can be used to express the tar yield (or hydrocarbon yield) exclusively in terms of the molecular orientation (mesophase-like) extension. The final size of the orientation indeed depends on the oxygen content of the initial material, on the possible occurrence of oxygen in the environment and also on the possible tar retention; all of which are factors governing the tar yield.

In the same way, in the second part of the present paper, the extension of molecular orientation has been proposed as the criterion for evaluating the oil potential of a kerogen, since the composition of both the parent material and the environment is taken in account (1).

Such a proposal is fully justified if kerogens and pyrolyzed materials are not only similar in structure, as shown by CTEM, but are also similar in other properties. In natural and heat-treated kerogens the IR spectra show that molecular orientation corresponds to a low content of OH and aliphatic CH groups, to an extremum of aromatic CH and to the disappearance of carboxyl and carbonyl groups (Chapter 6). This orientation is produced slightly before maximum spin concentration (Chapter 8). In series I, OH and carboxyl carbonyl contents are low while aliphatic CH are numerous. In series III on the contrary, oxygenated groups are very numerous. These features exactly correspond to those determined during pyrolysis. The Van Krevelen diagrams themselves are similar in kerogens and pyrolyzed materials as seen by comparing Figs. 7.28 to 7.30 with Fig. 7.34.

To conclude, the molecular orientation extension can be considered as a criterion which includes the history of the material as a whole. This history includes the chemical composition of the organic parent material, its dispersion or local concentration in the parent rock, hydrocarbon retention or migration and finally oxidation if any.

Series II can be taken as an example. It is very sensitive to the physico-chemical variations of its environment since it contains hydrogenated and oxygenated functions is equal amounts. Its natural autoreticulation, which alone would lead to a reduced molecular orientation, is balanced by the pressure operating throughout burial which alone would lead a widely extended molecular orientation. In addition if the particles of the parent material such as spores are isolated in the parent rock, the orientation would be also reduced, while it would be developed if many particles are concentrated at the same point. On the contrary, in the absence of oxidation, the oil yield of series I is high since the hydrogen content is so great and the oxygen content so low that there is no effect of dispersion or concentration. At the other end, series III is relatively insensitive to all parameters since its oxygen content is so high that autoreticulation overshadows all other factors.

---

(1) At this point the remark already made on part. 2, paragr. II.D can be explained. All kerogens belonging to series III have usually been correctly classified since they contain such a high amount of oxygen that any additional oxidation cannot be detected easily. On the contrary series II or even I kerogens are sensitive to oxidation which eventually creates intermediate series or even brings them into series III.

GENERAL CONCLUSIONS

The contribution of CTEM to the study of kerogen is twofold. First this technique can be successfully applied to the characterization of kerogen (texture and structure) which contributes to the determination of the oil potential of an unknown sediment by evaluating the molecular orientation extent. Then it helps gain a better understanding of the catagenesis and metagenesis mechanism through a better understanding of the pyrolysis process which appear to be very similar to each other.

A. Characterization of kerogen.

(a) Since all the components of an unknown sample can be described, the degree of purity and homogeneity of the kerogen is known.

(b) The degree of catagenesis can be evaluated by considering both the interlayer spacing spreading and the occurrence of molecular orientation.

(c) The oil yield of the sediment during catagenesis can be evaluated by the extent of the molecular orientation. If the kerogen to be studied has too low a degree of catagenesis, the occurrence of molecular orientation is induced by a standard heat-treatment.

Table 7.5 summarizes the various steps to be undertaken to evaluate the characteristic features of an unknown sediment.

B. Mechanism of catagenesis.

1. Low and medium degrees of catagenesis.

DF electron microscope images of kerogen have been interpreted as entirely due to polyaromatic molecules whenever the H/C ratio is less than one. A structural model (Fig. 7.22) has thus been established which can be confirmed by more sophisticated reasoning taking into account more than two kinds of molecules. Fig. 7.37 represents the variation of the aromaticity factor f_a vs. the atomic ratio H/C calculated for structural units made of 20 carbons (thick lines) and of 30 carbons (thin lines). The composition of pericondensed aromatic molecules is represented in A and A'; B and B' correspond to catacondensed aromatic molecules. The aliphatic molecules characterized by $H/C \approx 2$ are indicated as D, while the alicyclic ones are noted C and C' ($H/C < 1.5$). The two series of straight lines limit the possible composition of any mixture of the four molecules. The dashed line represents exinites and the dotted line vitrinites as deduced from Table 7.4. If larger structural units (70 carbon atoms) are represented in the same way, the results completely disagree with experimental values. It can thus be concluded that kerogens are made of small structural units (less than 20-30 carbon atoms). Fig. 7.37 confirms what has previously been shown, i.e when $H/C \leq 1.48$, the percentage of aromatic molecules may be up to 40%; it may reach 75% when $H/C = 1$. Since

TABLE 7.5
PROCESS FOR EVALUATING THE OIL POTENTIAL OF A SEDIMENT BY ELECTRON MICROSCOPY (CTEM)

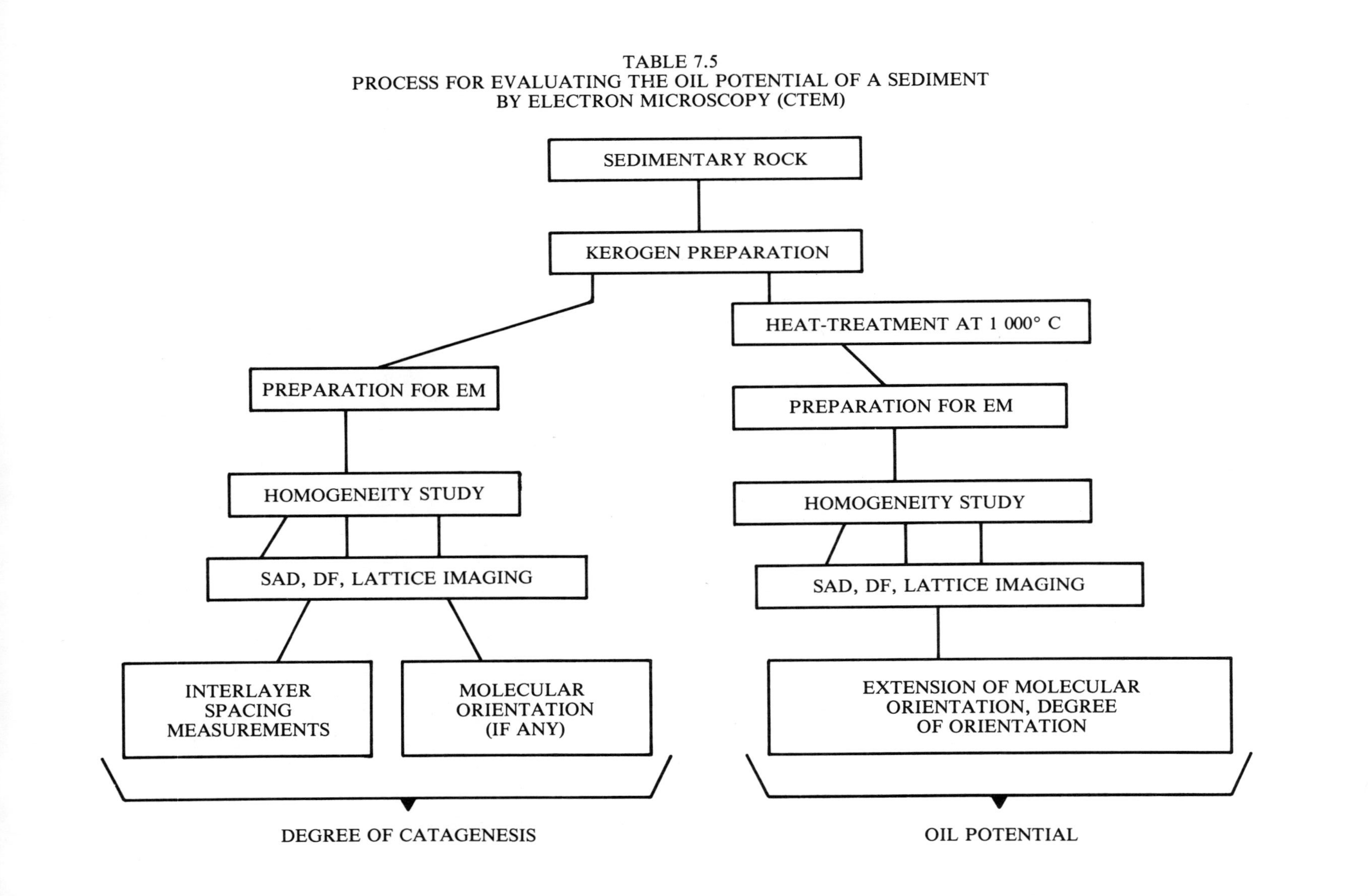

the percentage of non aromatic molecules is relatively low, the bright dots seen in DF images really represent aromatic ring clusters piled up by twos or threes and less than 10 Å in size. From IR studies [42 to 45, 51] the nature of the groups fixed around the graphitic layers may be inferred. Aliphatic CH, aromatic CH, hydroxyl, carbonyl, carboxyl, etc. have been found. The carbon layer stacks are thus connected by non-aromatic groups such as aliphatic, alicyclic and oxygenated ones (Fig. 7.22). In this range of catagenesis, the interlayer spacings are widely spread (3.4-8 Å). This is probably due to the fixation of non-aromatic groups on the graphitic layer edges which prevent them from getting closer. As catagenesis progresses the non-aromatic groups begin to disappear and the interlayer spacing spreading decreases.

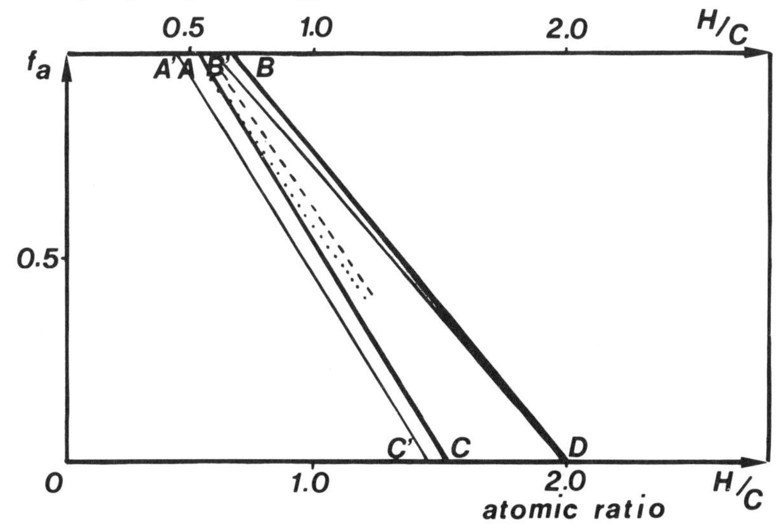

Fig. 7.37. — Phase diagram of a mixture of pericondensed aromatic molecules AA', catacondensed aromatic molecules BB', alicyclic molecules CC' and aliphatic molecules D.

2. *High degree of catagenesis (and metagenesis).*

As evolution progresses, a molecular orientation occurs slightly before the end of oil departure (slightly before the end of the release of the aliphatic CH groups). It corresponds to an extremum of aromatic CH groups, while carboxyl and carbonyl groups have disappeared. It also occurs slightly before the maximum spin concentration and corresponds to the 1.8-2% range of vitrinite mean reflectance. **The extension of the molecular orientation depends on the antagonistic action of both hydrogen and oxygen** and is more sensitive to the oxygen content. It is favored in materials which produce a large amount of tars (series I for instance) or by the factors which lead to tar retention. On the contrary it is reduced (series III for instance) in oxygen-rich or in oxidized materials. When the extent of the molecular orientation is small, IR spectra show the persistance of OH, C — OR and of carboxyl and carbonyl groups inside the material. These groups are responsible for the crosslinking of aromatic units which prevents the development of molecular orientation and at the same time reduces the tar yield. On the contrary, the molecular orientation is favored by the mobility of the aromatic stacks, i.e. by a high content of aromatic CH and of free radicals and by the lack of linkage (low content or lack of oxygenated groups).

At the highest degree of catagenesis, the material tends to be practically a pure carbon. Aromatic CH and free radicals tend to disappear. However there is no crystal growth. Each elementary domain remains biperiodic (turbostratic) and less than 10 Å in size. The extent of the molecular orientation remains the same, and the only change which occurs is an improvement in the degree of orientation. Materials having reached such a stage can be described by the structural model given in Fig. 7.10b. This model represents a molecular orientation about 50-100 Å in size and thus fits well with series III. For representing other series the number of elementary units associated in parallel has to be increased.

Molecular orientation extension integrates the entire history of a kerogen (chemical composition of the parent organic material, oil retention or migration and oxidation if any). Since it is the result of oil departure, it is also an absolute criterion for evaluating the oil potential of a sediment. When it is smaller than expected, the tar yield is also smaller. To sum up, if the orientation extension of a heat-treated kerogen is 500-1 000 Å, the oil yield is high; if it is 100 Å or smaller, the oil potential is low; for 200 Å it is medium.

REFERENCES

1. Ban, L.L., Vegvari, P.C. and Hess, W.M., (1973), *Norelco Reporter,* **20,** 1.
2. Blayden, H.E., Gibson, J., and Riley, H.L., (1944), *in: Proceedings of the Conference on Ultra Fine Structure of Coals and Cokes,* BCURA, London, 176.
3. Blyuman, B.A., Dyakonov, Y.S., and Krasavina, T.N. (1971), *Doklad. Akad. Nauk. SSSR.* **206,** 169.
4. Boulmier, J.L., (1976), Thesis, CNRS, Paris, AO 12748.
5. Boulmier, J.L., Oberlin, A., and Durand, B., (1977), *in: Advances in Organic Geochemistry, 1975,* ed. Campos R. and Goni J., Enadimsa, Madrid, 781.
6. Brooks, J.D., and Taylor, G.H., (1968), "Chemistry and Physics of Carbons" **4,** Marcel Dekker, New York, 243.
7. Cartz, L., Diamond, R., and Hirsch, P.B., (1955), *Nature,* **177,** 500.
8. Donnet, J.B., and Bouland, J.C., (1963), *in: Colloque National de Physico-Chimie du Noir de Carbone,* C.N.R.S., Paris, 43.
9. Durand, B., Nicaise, G., Roucaché, J., Vandenbroucke, M., and Haguemann, H.W., (1977), *in: Advances in Organic geochemistry, 1975,* ed. Campos R. and Goni J., Enadimsa, Madrid, 601.
10. Ergun, S., (1968), "Bureau of Mines Report 648", Washington.
11. Franklin, R.E., (1950), *Acta Cryst.,* **3,** 107.
12. Franklin, R.E., (1951), *Proceedings Roy. Soc.,* **209,** 196.
13. French, B.M., (1964), *Science,* **146,** 917.
14. Heidenreich, R.D., Hess, W.M. and Ban, L.L., (1968), *J. Appl. Cryst.,* **1,** 1.
15. Hirsch, P.B., Howie, A., Nicholson, R.B., Pashley, D.W., and Whelan, M.J., (1965), "Electron Microscopy of Thin Crystals", Butterworths, London.
16. Huc, A.Y., (1977), *Rev. Inst. Franç. du Pétrole,* **XXXII,** 703.
17. Inagaki, M., Oberlin, A., and Noda, T., (1965), *Tanso* (Japan), **81,** 68.
18. Landis, C.A. (1971), *Contrib. Mineral. and Petrol.,* **30,** 34.
19. Long, G., Neglia, S. and Favretto, L., (1968), *Geochim. et Cosmochim. Acta.* **32,** 647.
20. Maire, J., (1967), Thesis, CNRS, AO 1350, Paris.
21. Mentzer, M., O'Donnell, H.J., and Ergun, S., (1962), *Fuel,* **41,** 153.
22. Nicaise, G., (1977), "Institut Français du Pétrole, Rapport Géologie n° 21384".
23. Oberlin, A., (1977), *Analusis,* **5,** 85.
24. Oberlin, A., (1979), *Carbon,* **17,** 7.
25. Oberlin, A., "Orgon II", (1975), CNRS, Paris, 14.
26. Oberlin, A., Boulmier, J.L., and Durand, B., (1975), *in: Advances in Organic Geochemistry, 1973,* ed. Tissot B. and Bienner F., Editions Technip, Paris 15.
27. Oberlin, A., Boulmier, J.L., and Durand, B., (1976), *C.R. Acad. Sci. Paris,* C **280,** 501.
28. Oberlin, A., Boulmier, J.L., and Durand, B., (1974), *Geochim. et Cosmochim. Acta,* **38,** 647.

29. Oberlin, A., Terrière, G., Durand, B., and Clinard, C., (1972), *in: Advances in Organic Geochemistry, 1971* Von Gaertner and Wehner ed., Pergamon Press, 577.
30. Oberlin, A., Oberlin M., and Comte-Trotet, J.R., (1976), *J. Microsc. Spectrosc. Electron.*, **1**, 391.
31. Oberlin, A., Villey, M., and Combaz, A., (1978) *Carbon*, **16**, 73.
32. Oberlin, A., Endo, M., and Koyama, T., (1976), *Carbon*, **14**, 133.
33. Oberlin, A., Endo, M., and Koyama, T., (1976), *J. Cryst. Growth*, **32**, 335.
34. Oberlin, A., Oberlin, M., and Maubois, M., (1975), *Phil. Mag.*, **32**, 833.
35. Oberlin, A., Boulmier, J.L., and Terrière, G., (1975), *Bull. Inst. Chem. Res. Kyoto Univ.*, **53**, 81.
36. Oberlin, A., Terrière, G., and Boulmier J.L., (1974), *J. Microscopie*, **21**, 301.
37. Oberlin, A., and Terrière, G., (1972), *J. Microscopie*, **14**, 1.
38. Oberlin, A., and Terrière, G., (1973), *J. Microscopie*, **18**, 247.
39. Oberlin, A., and Terrière, G., (1975), *Carbon*, **13**, 367.
40. Oberlin, A., Terrière, G., and Boulmier, J.L., (1975), *Tanso* (Japan), Part I, **80**, 29; Part II, **83**, 153.
41. Ragot, J., (1977) Thesis, University of Toulouse.
42. Robin, P.L., (1975), Thesis, University of Louvain, Belgium.
43. Robin, P.L., and Rouxhet, P.G., (1976), *Fuel*, **55**, 177.
44. Robin, P.L., Rouxhet, P.G., and Durand, B., (1977), *in: Advances in Organic Geochemistry, 1975*, ed. Campos R. and Goni J., Enadimsa, Madrid, 693.
45. Rouxhet, P.G., Villey, M. and Oberlin, A., (1979), geochim. Cosmochim. Acta. **43**, 1705.
46. Shiraishi, M., Terrière, G., and Oberlin, A., (1978), *J. Mater. Sci.*, **13**, 702.
47. Tissot, B., and Espitalié, J., (1975), *Rev. Inst. Franç. du Pétrole*, **XXX**, 743.
48. Vandenbroucke, M., Rouzaud, J.N., and Oberlin, A., (1978), *in: Les réacteurs de fission naturels publ. IAEA*, Vienna, 307.
49. Van Krevelen, D.W., (1961), "Coal", Elsevier, Amsterdam.
50. Villey, M., Oberlin, A., and Combaz, A., (1976), *C.R. Acad. Sci. Paris*, D **282**, 1657.
51. Villey, M., Oberlin, A., and Combaz, A., (1979), *Carbon*, **17**, 77.
52. Villey, M., (1979), Thesis, University of Orléans.
53. Warren, B.E., (1941), *Phys. Rev.*, **59**, 693.
54. Yen, T.F., Erdman, G.J., and Pollack S.S., (1961), *Anal. Chem.* **33**,1587.
55. Yen, T.F., and Erdman, G.J., (1962), *in: Amer. Chem. Soc. Petroleum Chem. Atlantic City Meeting*, 99.

8

Electron paramagnetic resonance in kerogen studies

A. MARCHAND* and J. CONARD**

I. STABLE FREE RADICALS PRODUCED IN ORGANIC MATTER

Electron paramagnetic resonance (EPR or ESR) is a technique quite convenient for the detection and study of free radicals i.e. polyatomic structures having unpaired electrons. This Chapter deals with the application of EPR to kerogens.

In 1954, both Ingram [16] and Uebersfeld [12] used EPR to demonstrate the presence of stable free radicals in coals. During the following years this same technique has also shown the presence of free radicals in all types of carbons, either natural or artificial, and in the products of pyrolysis or irradiation of organic matter (OM). They were found in kerogens by Marchand *et al.*[18], and it can be said as a general fact that they appear in all degradation processes of organic or biological substances.

The free radicals formed in a solid may be very long-lived: their mobility is low, so they have a vanishing probability of getting together and recombining. Moreover theoretical chemistry considerations show that aromatic free radicals have a remarkably high stability.

The production of free radicals during carbonization has been extensively studied since 1954. For a given carbonization time, the free radical concentration increases with temperature of pyrolysis up to a maximum and then decreases. Under laboratory conditions this maximum depends upon the nature of the OM, the treatment time, and the nature of the surrounding atmosphere, but remains essentially between 500 and 700° C, and reaches values of 10^{19} to 10^{20} free radicals per gram [14].

But the same concentration of free radicals may be obtained using a short time pyrolysis with a higher temperature, or a long time treatment with a lower one. Considering the very long times typical for a geological process, it is natural to expect a high free radical content in sedimentary organic matter, even if treated at relatively low temperature. Such is the case of natural coals, and kerogens, which during their evolution have not been subsmitted to temperatures above 200° C.

---

* Centre de Recherche Paul Pascal (CNRS), Université de Bordeaux I. Talence, France.

** Centre de Recherches sur l'Organisation des Solides à Cristallisation Imparfaite (CRSOCI), CNRS, Orléans, France.

II. EPR AS A METHOD FOR STUDYING FREE RADICALS

A. Paramagnetism.

Only in ferromagnetic substances is a **permanent** magnetization found, which persists in the absence of any external magnetic field: but all materials acquire an **induced** magnetization when they are submitted to a magnetic field. If this magnetization is proportional to the applied magnetic field H, the substance is known as "paramagnetic" or "diamagnetic", with:

$$\vec{M} = \chi \vec{H} \tag{1}$$

where $\vec{M}$ is the induced magnetic moment per unit of mass and χ is the "magnetic susceptibility" per unit of mass (positive if paramagnetic, negative if diamagnetic).

In the international system of units $\vec{H}$ is measured in A/m and χ in m^3/kg. But these units are very seldom used. $\vec{H}$ is usually expressed in Gs or in T ([1]) and χ in CGS electromagnetic units (emu). Usual paramagnetic susceptibilities are in the order of 10^{-4} emu CGS/g at ambient temperature.

Paramagnetism originates from unpaired electrons. In quantum mechanics, each electron has a spin magnetic moment called the "Bohr magneton":

$$\beta = eh/4\,\pi m \tag{2}$$

with e, m, electrical charge and mass of the electron, and h the Planck's constant.

Pauli's exclusion rule requires that electronic spins belonging to an atom or a polyatomic system be paired off, with antiparallel spin magnetic moments. If the number of electrons is odd, one of them has to remain unpaired. It can be situated in:

(a) An atom (e.g. H, Na, Cl...).

(b) An ion (e.g. Cu^{++}...).

(c) A molecule (e.g. NO, NO_2...).

(d) A fragment of molecule, i.e. a free radical ($CH_3^{\cdot}$, $C_6H_5^{\cdot}$, $NH_2^{\cdot}$...).

If these atoms, ions, free radicals or molecules are submitted to a magnetic field $\vec{H}$, the majority of unpaired spins are oriented in the field direction, while the minority point in the opposite direction. The resulting magnetic moment is parallel to the direction of the magnetic field. This induced moment vanishes when the magnetic field is suppressed: the substance is paramagnetic. Its paramagnetic susceptibility χ_p is proportional to the concentration of "paramagnetic centers" (concentration of unpaired electrons).

([1]) We leave aside, in such a summary presentation, the distinction between magnetic field and induction.

If the unpaired electron is bound to an atom, ion, free radical or molecule, it is known as "localized" and its paramagnetism varies with temperature according to the Curie law:

$$\chi_p = \frac{C}{T} \tag{3}$$

where T is the absolute temperature, and constant C is characteristic of the substance under study ([1]).

Unpaired electrons in a solid can sometimes circulate freely from an atom or free radical to an other one: they are said to be "free". Their paramagnetism is almost temperature independant (Pauli's paramagnetism). Free electrons are found for instance in metals and in graphite.

An intermediate situation occurs when the unpaired electron is free to move within an extended molecule or radical, but cannot hop from one to the other. We may consider such an electron as partly delocalized, but it nevertheless obeys the Curie law.

Information about the extent of free motion of unpaired electrons can thus be obtained through a study of the thermal variation of χ_p.

The paramagnetic centers obtained through carbonization of an organic matter are at first localized. With increasing temperature or treatment time, or pressure, they grow more and more delocalized as the extension of the aromatic domains becomes larger and larger. Above 900° C (or when the pyrolysis time gets very long) the free electrons are increasingly numerous; the last step is the building of the graphite crystal lattice where electrons are completely free to move along the graphitic layers [14, 5].

B. Electron paramagnetic resonance.

The EPR phenomenon consists in changing the orientation of the magnetic moment of an unpaired electron in the magnetic field by means of an electromagnetic radiation of a suitable frequency.

Let us examine first the case of an isolated free electron: its magnetic moment originates only from its spin.

In a magnetic field $\vec{H}$, the spin moment may be oriented either in the field direction ("parallel") or in the opposite direction ("antiparallel"). The corresponding values of the magnetic energy are:

$-\ \beta H$ (parallel orientation)

$+\ \beta H$ (antiparallel orientation)

The energy difference between these two levels is:

$$\Delta E = 2\,\beta H \qquad \text{(Fig. 8.1).}$$

([1]) The Curie-Weiss law is not considered here, since kerogen susceptibilities always follow the simple Curie law.

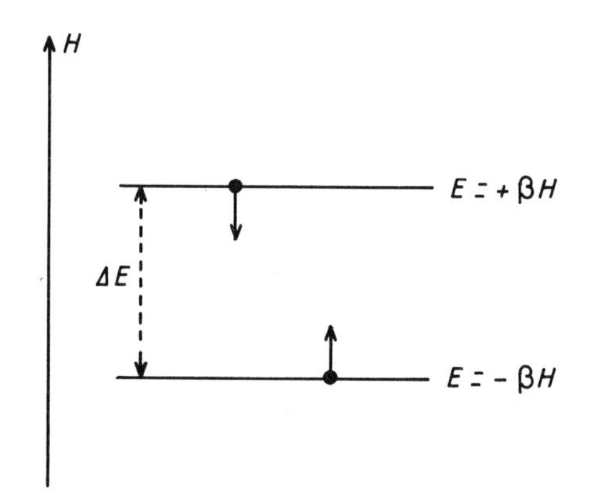

Fig. 8.1. — Magnetic energy levels of an unpaired isolated electron.

An electromagnetic radiation of frequency ν can induce a transition from the lower to the upper level if:

$$h\nu = \Delta E = 2\,\beta H$$

so that absorption of the photon energy $h\nu$ enables the spin moment to change its orientation in the magnetic field from parallel to antiparallel.

Considering now the more complex case of an unpaired electron in an atomic or molecular structure, a very similar phenomenon is observed.

The magnetic moment μ of the "paramagnetic center" originates from both the spin and the orbital moments of the electrons. In a magnetic field the magnetic energy of the whole structure depends upon the orientation of μ. But a suitable electromagnetic radiation can induce a transition between two energy levels corresponding to different orientations: the frequency ν must be such that the photon energy $h\nu$ is equal to the energy difference between the two levels:

$$h\nu = \Delta E$$

Quantum mechanics gives ΔE as:

$$\Delta E = g\beta H$$

where g is a dimensionless number called "spectroscopic splitting factor" or simply "g factor".

The frequency of the suitable radiation is:

$$\nu = \frac{\Delta E}{h} = \frac{g\beta H}{h}$$

For an isolated free electron, relativistic considerations lead to a g-value slightly different from that found above: $g = 2.0023$ (instead of $g = 2$).

In a paramagnetic structure (atom, ion, free radical...), the value of the g factor depends upon the extent of the interaction between the orbital moment and the spin moment of the electrons: the so-called "spin-orbit coupling" thus determines the g-value. In most cases this coupling is quite weak, so that g is very nearly 2. But g-values very different from 2 may be found; they may also depend on the orientation of the paramagnetic molecule in the field: such is the case, for instance, of the hemoglobin molecule, with a g factor which increases from 2 to 6 following the orientation of the molecule in the field.

From equation:

$$\boxed{h\nu = g\beta H}$$

it is clear that the radiation frequency is proportional to the magnetic field strength. Table 8.1 shows the order of magnitude of these frequencies ν (and wave-lengths λ). Theoretically EPR can be observed at any field strength, but the corresponding frequencies vary in such a wide range that techniques are extremely different at low fields and at high fields.

The resonance intensity increases (all other factors being kept constant) as the square of the radiation frequency, so that low frequency EPR (in the order of 10^6 Hz) requires very large quantities of paramagnetic substance (a few grams). Commercial equipments usually work between 300 and 35 000 MHz (ultra-high frequencies) with easily obtained magnetic fields (100 to 12 000 Gs) and paramagnetic samples of a few milligrams. Spectrometers making use of superconducting magnets (in liquid helium) can be built, but the corresponding frequencies are close to the far IR and necessitate techniques quite different from the usual ones.

TABLE 8.1

$g = 2$		
H (Gs)	ν (MHz)	λ
0.5 (earth field)	1.4	214 m
100 (simple coils)	280	1.07 m
3 500 (electromagnets)	9 850	3.05 cm
50 000 } superconducting magnets	1.4×10^5	2.14 mm
200 000 } superconducting magnets	5.6×10^5	535 μm

C. Instruments.

An EPR experiment can be devised in two different ways:

(a) The magnetic field is kept constant, and the radiation absorption is studied as a function of frequency.

(b) The radiation frequency is kept constant, and absorption is studied as a function of magnetic field.

The first method is the usual one in optical spectroscopy. But in the radio-frequency range the second method is more convenient: it is easier to keep a fixed frequency and vary the magnetic field.

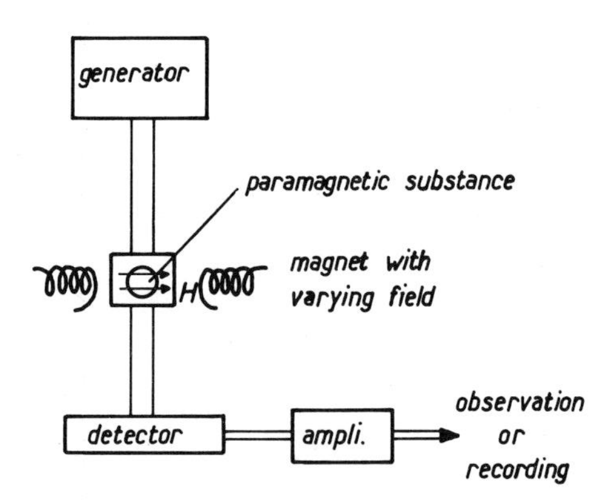

Fig. 8.2. — Schematic diagram of an EPR spectrometer.

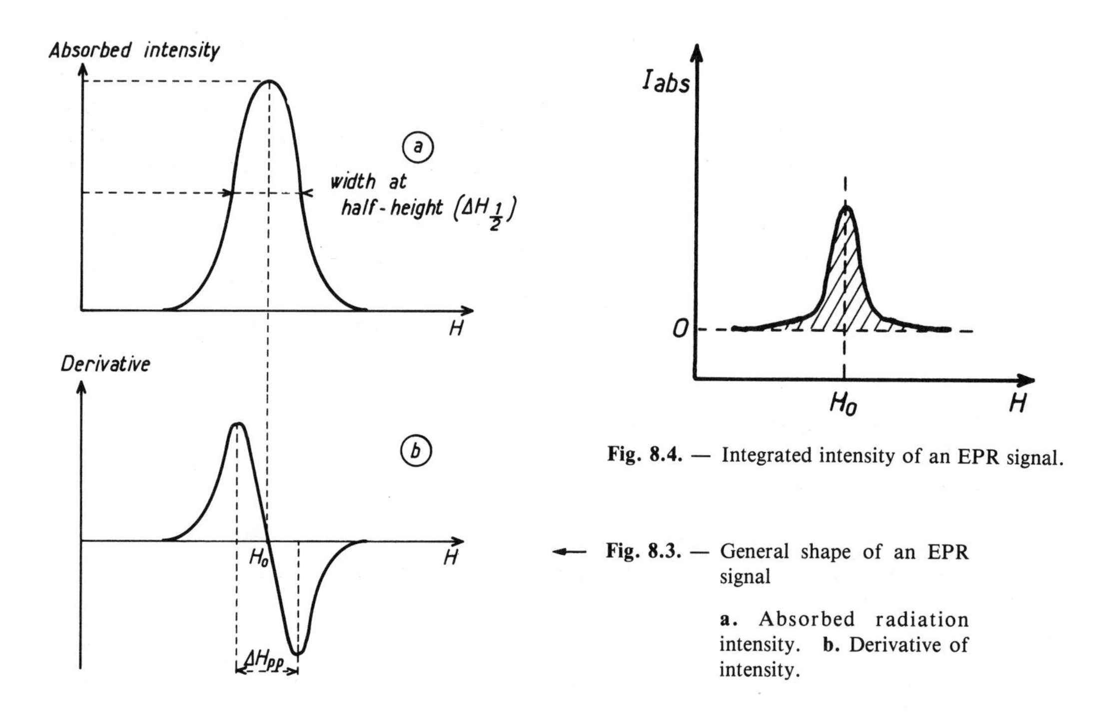

Fig. 8.4. — Integrated intensity of an EPR signal.

← **Fig. 8.3.** — General shape of an EPR signal

a. Absorbed radiation intensity. **b.** Derivative of intensity.

An EPR spectrometer is thus essentially made of (Fig. 8.2):

(a) A fixed frequency electromagnetic (EM) generator.
(b) An electromagnet (or a simple coil) giving a magnetic field H which can be varied.
(c) A "circuit" (coil or resonating cavity) situated in a region of homogeneous field, containing the paramagnetic substance. The role of the circuit is to concentrate the magnetic component of the EM wave upon the sample, with its vector direction perpendicular to the static magnetic field H.
(d) A detection system, sensitive to the power absorption by the paramagnetic substance, giving if possible a signal proportional to the absorption intensity.
(e) An amplifier for the signal, which is usually quite weak (in the order of a few microwatts).
(f) Devices for observing or recording the signal.

The intensity of the radiation absorbed by the paramagnetic sample increases rapidly when the field H reaches the value:

$$H_0 = h\nu/g\beta.$$

The observed or recorded signal shape is that illustrated by Fig. 8.3a.

A more sensitive detection ("synchronous detection") is actually obtained through low-frequency modulation of the main field H and recording of the derivative of the absorption intensity (Fig. 8.3b).

From the very elementary theoretical description presented here, one would expect the EPR signal to be an infinitely narrow line, since EM absorption would occur only when the condition $H = h\nu/g\beta$ is exactly satisfied. The absorbed intensity would vanish for any other field value.

Various theoretical considerations which will be examined later actually show that the signal width may vary greatly. It is possible to define a "width at half-height" $\Delta H_{1/2}$ (Fig. 8.3a) or a "peak-to-peak width" ΔH_{pp} between the peaks of the derivative curve (Fig. 8.3b). These widths may range from a few mGs to a few hundred or even a few thousand Gs.

D. Informations obtained through EPR.

It is possible to measure:

(a) The overall intensity of an EPR signal.
(b) Its position (as defined by the H_0 value of the field corresponding to maximum absorption, or by $g = h\nu/\beta H_0$).
(c) Its width.

It is also possible to study its shape or its variations as a function of environmental factors.

1. *Overall intensity of the signal.*

For a given EPR spectrometer, and all other factors being kept constant, the integrated intensity of the signal (i.e. the shaded area in Fig. 8.4. is proportional to $m\chi_p$ (m: mass of the substance; χ_p: its paramagnetic susceptibility per mass-unit), that is proportional to the number of paramagnetic centers or free radicals in the sample.

Thus measurement of the total integrated intensity of the EPR line is a method of determination of χ_p and of the free radical concentration. Generally an auxiliary sample of known susceptibility is used as a local standard for comparison.

There are other methods for measuring χ_p, namely the so-called "static methods" making use of the measurement of the magnetic force exerted on a sample in a magnetic field. But EPR has important advantages:

(a) EPR is at least 100 times more sensitive than the other methods: there is no major problem in measuring by EPR susceptibilities in the order of 10^{-10} emu CGS/g while the best usual "static" equipments reach their ultimate sensitivity at about 10^{-8} emu CGS/g, and cannot be used for kerogen studies.

(b) EPR allows the measurement of χ_p alone, while the other techniques give the global susceptibility:

$$\chi = \chi_p + \chi_d$$

sum of the para- *and* dia-magnetism. When χ_p is very small (free radicals or ions diluted in a diamagnetic substance – such as in the case of kerogens – or very paramagnetism of metals) the diamagnetic component χ_d may be much larger and the determination of χ_p by static methods is both difficult and inaccurate. EPR does yield directly the value of χ_p.

2. *Position of the EPR signal (measurement of g factor).*

The position of the signal is given by the H_0 value of the magnetic field corresponding to the maximum intensity of EM radiation absorption. It is equivalent to determine the g-value:

$$g = h\nu/\beta H_0$$

Since the g factor characterizes the "spin-orbit coupling" intensity, and may be measured with a high precision ($\pm 10^{-4}$), the g-value is often used as a characteristics of a paramagnetic substance.

It may even happen, when a mixture of several paramagnetic substances is studied, that their g-values are so far apart that the corresponding EPR signals are well separated: in such a case the constituents of the mixture can be identified, and their relative proportions determined from the relative intensities of their resonance lines.

3. *Linewidth and shape of the signal.*

The mathematical resonance equation shows that electromagnetic radiation can only be absorbed when the applied magnetic field has the value:

$$H_0 = \frac{h\nu}{g\beta}$$

and the "absorption line" should ideally be infinitely "thin" (e.g. a Dirac function).

There are however a number of factors which contribute to an increase of the linewidth of the signal, so that a study of the shape and width of the EPR line may yield interesting information.

The first factors are the "relaxation processes" which allow the spin magnetic moment to reverse its orientation and fall back to its lower energy level. These processes may correspond to interactions between paramagnetic species with different locations ("spin-spin" coupling) or to interactions between an unpaired electron and its surrounding medium ("spin-lattice" coupling). Both mechanisms have the effect of shortening the life-time of an unpaired spin in the excited state, and consequently increase the EPR linewidth (uncertainty principle).

Another factor is the interaction of the unpaired electronic spin with the magnetic moments (spins) of neighboring nuclei. Such a coupling, which also increases the linewidth in kerogens, can take place when H or N atoms are in the vicinity of the unpaired electron (^{12}C and ^{16}O have no nuclear spin) or if transition metal atoms or ions contribute to the EPR signal. In the latter case, when these paramagnetic metallic species are very diluted, a splitting of the EPR line is observed ("hyperfine structure"), with a number of components which is characteristic of the metal: 8 for ^{51}V, 6 for Mn (Fig. 8.5.), 2 for ^{57}Fe, etc...

An increase of the linewidth may also result from the inhomogeneity of the sample: different free radicals with slightly different *g*-values give different EPR lines very close to one another, which may be observed as a single broad line. Anisotropic paramagnetic particles randomly oriented in the magnetic field may also have a similar appearance: the anisotropy of *g*-values results in line-broadening.

4. Variations of the EPR signal with external causes.

The intensity, shape, and linewidth of the signal may be studied at various temperatures, yielding still more detailed information: the thermal variation of the intensity is related to the electron delocalization or degree of freedom, but the relaxation mechanisms efficiency is also a function of temperature.

III. EPR STUDY OF NATURAL KEROGENS

A. Purity. Homogeneity.

Purifying kerogens is the most difficult part of the investigation by EPR. Paramagnetic minerals are very often mixed with them and superimpose their own signals (sometimes with a hyperfine structure) to the one under study. Fig. 8.5 shows how complicated the EPR lines of non-purified kerogens may be. Even in less unfavorable cases, the impurity induced perturbation is such that it may be difficult to make an estimation of the intensity or linewidth (Fig. 8.6).

We shall not go here into the detail of purification methods (which are described elsewhere in this book). They make use of solvent-extractions, acid treatments (HCl, HF) and washing followed by centrifugation. They also quite often vary somewhat from one laboratory to the other, which may make comparisons difficult.

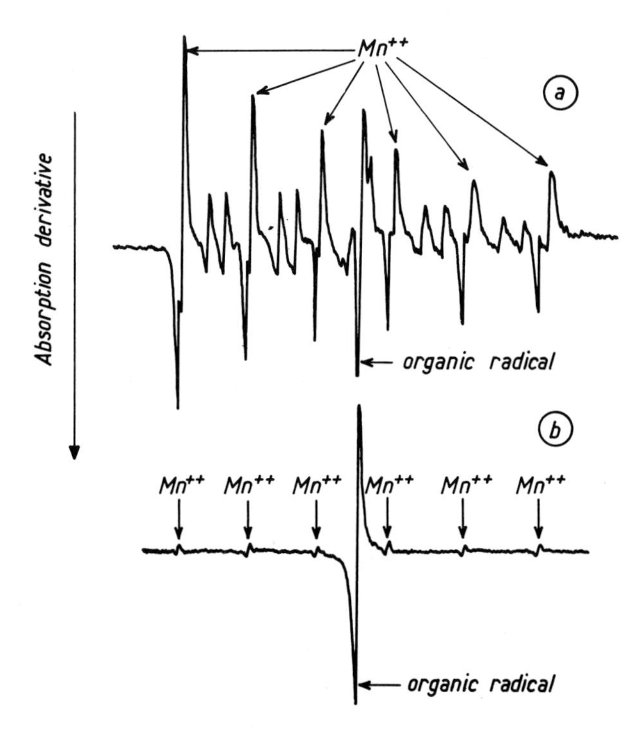

Fig. 8.5. — Examples of EPR signals of kerogen

a. Before purification. **b.** After purification.

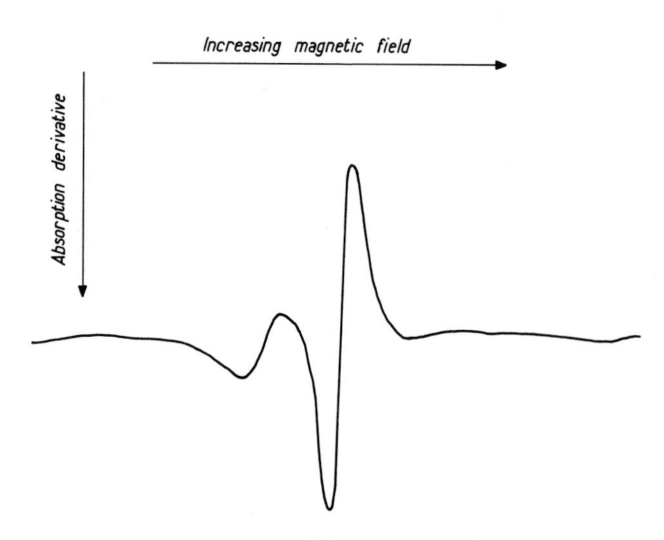

Fig. 8.6. — An EPR signal may be of little use, even after purification of the kerogen: example of a kerogen from Lybia (derivative curve).

Total purification is not always possible: pyrite especially can be so closely imbedded in kerogen that its separation is practically impossible. The resulting EPR signals have an intensity or a width which can indeed be measured (Fig. 8.7 shows two examples), but it is not possible to assign a clear meaning to the results of such measurements: they must be used with extreme caution.

However purification is generally quite effective, as illustrated by Figs. 8.5a (before purification) and 8.5b (after purification). But it must be emphasized that the treatment may change the free radicals of the kerogen. A systematic investigation of the influence of the purification treatment was undertaken by Durand *et al.* [7]. Their findings show that it is most desirable to keep the treatment at the minimum level, or to make use of various methods in order to compare their results. In any case it may be very dangerous to draw conclusions from the study of a single sample.

It should also be kept in mind that a kerogen sample, even one of very small dimensions (EPR utilizes only a few mg) is very seldom homogeneous: it is the product of numerous and diverse biological elements which may have had also different geological histories.

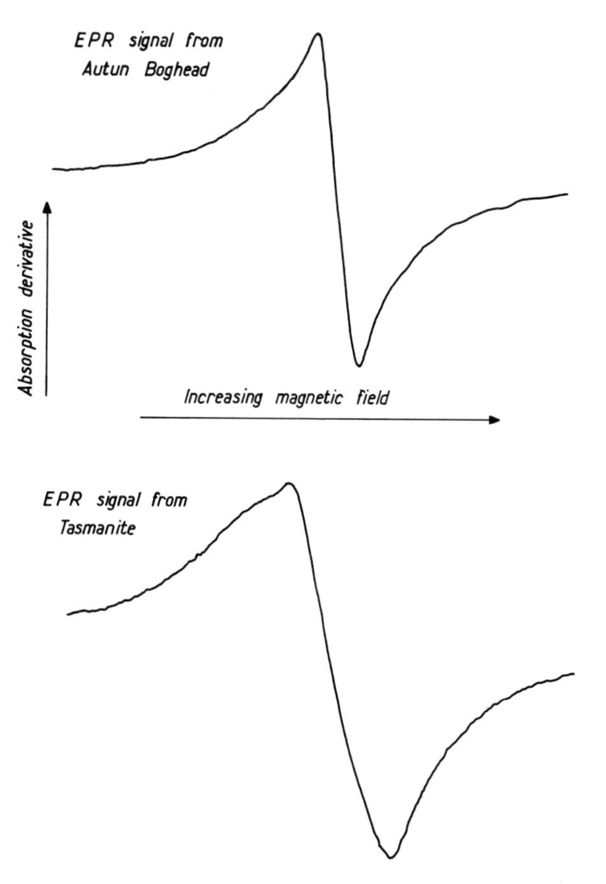

Fig. 8.7. — Two examples of usable, although rather unsatisfactory signals (derivative curves).

The experimental values consequently show some scatter even for samples taken at the same place and at the same depth. The ESR investigation yields only a general information, averaged over all the components of the sample under study: such a technique gives less detailed "pictures" than, for instance, the optical microscope examination or the reflectance measurement. But it can be used with any kind of kerogen, and its results are always valid provided the number of samples is not too small: only a large enough amount of data can cancel the effects of the scatter of experimental results.

B. Origin and nature of free radicals in kerogens.

Practically all kerogens, even those very recent or very little altered, give EPR signals. The paramagnetic susceptibility χ_p (per gram of organic carbon) varies approximately from 10^{-10} to 5×10^{-8} emu CGS at room temperature, which corresponds to free radical concentrations in the order of 5×10^{16} to 2.5×10^{19} [18, 7, 19, 17] per gram.

As for other physical properties of kerogens, the characteristic features of their EPR signal depends on the nature of the parent biological material and also on the "catagenetic process" this material has undergone. We will first examine the influence of catagenesis.

The free radicals of artificial carbons have been the subject of a large number of investigations, which have shed some light on the origin and nature of these radicals. It may be thought that those in kerogens are similar.

During the pyrolysis of an organic material, as well as during the natural degradation of a biological substance, volatile products of small molecular mass are generated (CO_2, H_2O, light hydrocarbons). These are torn off the skeleton of the parent substance; such a "peeling" leaves "scars": one unpaired electron for each broken chemical bond. New electron pairings and bond rearrangements of course do occur, and hydrogen atoms replace some of the torn functional groups of the aliphatic chains or aromatic rings of the parent material. But a number of free radicals remain: they cannot migrate through the solid substance, they cannot recombine, and they may be stable for a quasi-infinite time.

A traditional presentation of this degradation process is the "Van Krevelen" diagram, where the O/C and H/C ratios (determined from an elementary chemical analysis of the solid residual substance) are plotted as abscissae and ordinates respectively. The chemical evolution of any organic material is illustrated in this diagram by a "trajectory" or "catagenesis path" along which the O/C and H/C values are simultaneously decreasing (Fig. 8.12).

Studies of the carbonization process have shown that the production of volatile compounds, which starts below 200° C and can proceed up to 1 000° C, is always accompanied by a progressive aromatization of the structure of the residual solid, which can be followed, for instance, through the increasing intensity of the 1 600 cm^{-1} IR band [1].

Moreover, quantum chemistry calculations show that unpaired electrons generated in one ring of an aromatic polynuclear structure are delocalized throughout the whole structure, with a resulting increase of stability of the radical.

---

[1] It should be mentioned that Steelink [22] studied the free radicals of lignin degradation products and of humic acids and concluded that they originate mainly in semi-quinonic structures. In the more general case of kerogens, such structures may also account for a part of the total free radical concentration.

C. Intensity of the EPR signal.

Higher temperatures, or longer durations of the degradation process, will result in larger free radical concentrations and increasing aromatic character of these radicals: it can then be expected that the history of a kerogen "catagenesis" is reflected in the intensity of its EPR signal.

That it is indeed so will be illustrated by the following typical example.

A very complete investigation was performed on a series of kerogens from the upper cretaceous of the Douala basin (Logbaba): samples were studied down to a depth of 4 000 m.

The deeper a sediment is buried at a given location, the higher the temperature and the longer the duration of the heat-treatment it has undergone, so that depth of burial is an indication of the level of degradation ("catagenesis") of a kerogen. Fig. 8.8 shows the evolution of the paramagnetic susceptibility χ_p as a function of depth in the Logbaba series: after a steady increase, the paramagnetism reaches a maximum and then starts decreasing.

The maximum value corresponds to a concentration of 2.5×10^{19} free radical/g of organic carbon. At such high concentrations the mean distance between radicals is in the order of 30-40 Å, and small enough to allow direct recombinations: when the recombination rate grows larger than the creation rate of free radicals, their concentration starts decreasing. However the maximum concentration may be different for different types of organic or biological materials, since both the creation and the recombination rates depend on the chemical structure of the substance under study: it is well known for instance that oxygen-bridges between aromatic molecules inhibit the recombination of radicals and the further growth of the aromatic structures in artificial carbons [14].

The results illustrated by Fig. 8.8 show that the state of catagenesis of a kerogen can be characterized by the value of χ_p. It is important to compare this method with other

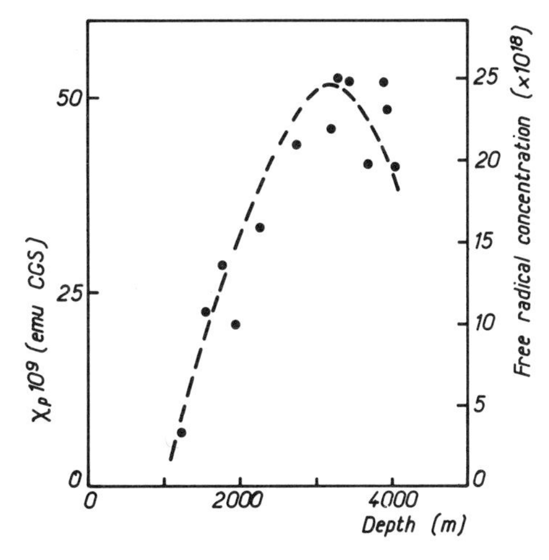

Fig. 8.8. — Paramagnetic susceptibility at room temperature and free radical concentration/g of organic carbon as a function of depth of burial. Unpurified kerogen from the Douala Basin [7].

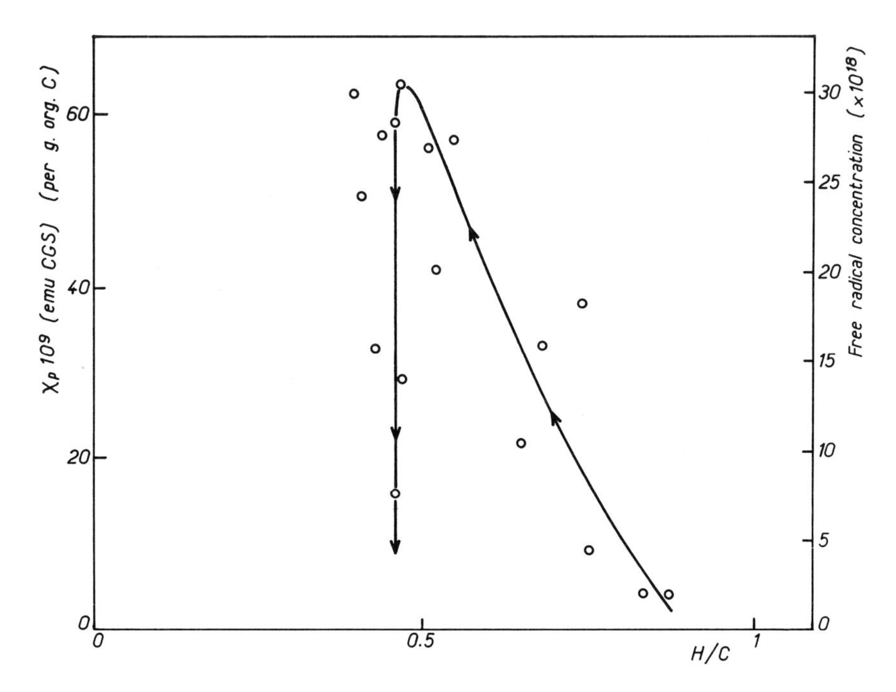

Fig. 8.9. — Correlation between atomic ratio H/C and paramagnetic susceptibility χ_p at room temperature (kerogen from Douala Basin) (Durand, *IFP* report, Feb. 1976).

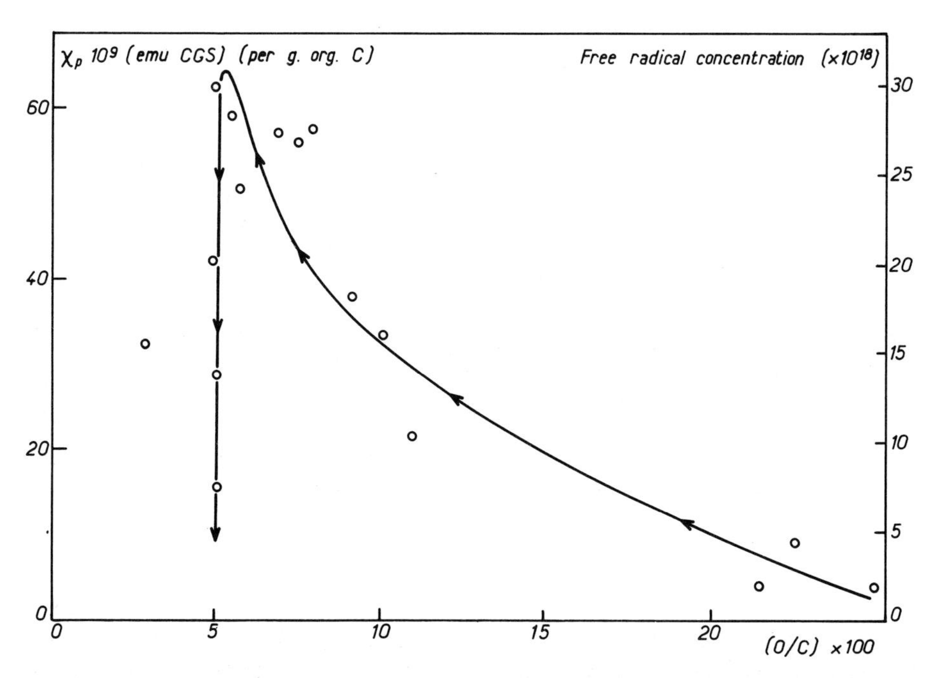

Fig. 8.10. — Correlation between atomic ratio O/C and paramagnetic susceptibility χ_p at room temperature (kerogen from Douala Basin) (Durand, *IPF* report, Feb. 1976).

techniques of characterization. The paramagnetic susceptibility of the Logbaba kerogens is plotted as a function of the H/C ratio (Fig. 8.9), O/C ratio (Fig. 8.10) and vitrinite reflectance (Fig. 8.11): arrows indicate increasing degradation of the material (increasing depth of burial). It is clear that there is no one-to-one correspondance between χ_p and the other criteria: the same paramagnetism may characterize two very different stages of catagenesis. For highly degraded kerogens χ_p is much more sensitive than the other parameters since it goes on decreasing while the H/C, O/C, and reflectance values remain nearly constant. At low catagenesis on the contrary all four criteria are equivalent since their variations are similar.

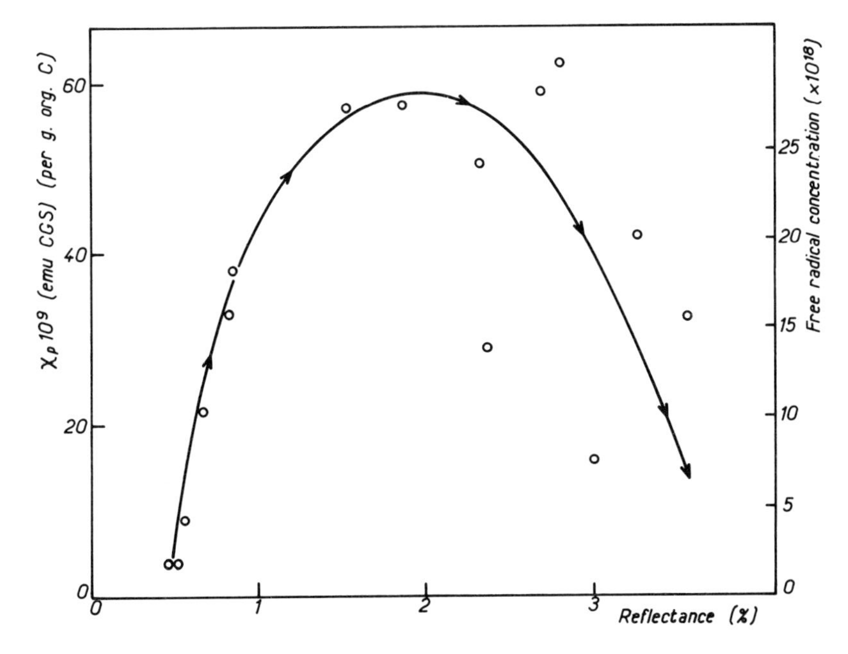

Fig. 8.11. — Correlation between vitrinite reflectance and paramagnetic susceptibility χ_p at room temperature (kerogen from Douala Basin).

IR absorption is another way of characterizing the evolution of a kerogen (*cf.* Chapter 6). It will be seen that a very close correlation can be observed between χ_p and the intensity of the vibration bands of the aromatic CH groups (Chapter 6, Figs. 6.8 and 6.12). There is also a good agreement of the variation of the 1 600 cm^{-1} aromatic band with the χ_p variation. These results were expected since the formation and stability of the free radicals is closely related to the aromatization of the kerogen structure.

D. The *g* factor.

The *g*-value is determined by measuring the magnetic field at maximum microwave absorption (zero-value of the derivative). The kerogens have *g* factors ranging from 2.0028 to 2.0037 (measured with a relative accuracy of 10^{-5}).

The higher g-values correspond to the first stages of the evolution of a kerogen, and a fast decrease is generally observed along the degradation process. Such a behavior is somewhat puzzling since it has been shown, both theoretically and experimentally, that the building of larger and larger aromatic structures corresponds to increasing g-values, up to the mean g factor of graphite (2.018 at room temperature) [14].

It is however a well established experimental fact that g decreases at the beginning of the carbonization or degradation processes of organic materials, and starts increasing only during much later stages, in very large aromatic systems [21].

Values of g higher than 2.0023 result from a contribution of the orbital magnetic moments of the electrons ("spin-orbit coupling"). Aromatization is not the only factor which may influence the spin-orbit coupling: atoms of oxygen or sulfur in the vicinity of the unpaired electron can also give a strong coupling [4]; this will result in rather high g-values when O or S atoms are included in the aromatic nuclei of free radicals, or closely linked to these nuclei [21].

E. Shape and line-width of the EPR signal.

The shape of the EPR signal of kerogens is generally ill-defined, even for well purified materials, because of sample-inhomogeneity. The peak-to-peak line-width ΔH_{pp} can however be measured usually without too much difficulty on the derivative curve (Fig. 8.3b); but it is rather strongly dependent on the experimental conditions (chemical nature of the surrounding atmosphere, electromagnetic power used for the measurement).

With 10 000 MHz radiation, the ΔH_{pp} values are generally in the 4-8 Gs range. The spin-lattice interactions have been investigated in a few cases [17, 20] and found to be rather weak, so that their contribution to the line-width is small in comparison to more important factors, namely:

(a) Residual iron, especially iron originating from pyrite, which laboratory treatments cannot completely eliminate.

(b) Nuclear spin coupling with protons located in the vicinity of the unpaired electrons.

(c) g factor anisotropy, which is difficult to measure, and results sometimes in a characteristic assymmetry of the line shape. By assuming a molecular structure with an axial symmetry (aromatic-likes structures) Sallé [20] was able to estimate the g factor anisotropy, and found its variations to be similar to those of the mean g-value: decreasing with increasing degradation of the kerogen.

Some changes in the line-width can be observed along the catagenetic process: there are indications of the existence of a maximum width. But variations are so small and the scatter of results so large that, contrary to Pusey's suggestions [15], it is not possible to use line-width measurements to characterize the catagenetic state of a kerogen. Using such criteria would anyway necessitate a most precise and rigorous definition of procedures for line-width measurements.

F. Influence of the parent biological material.

The general characteristics we have described are valid for all types of kerogens, and the evolution of χ_p and g during the degradation process are always similar. However some distinct features corresponding to the biological nature of the parent material can be shown by a more complete analysis of the results.

A detailed EPR investigation of typical series of kerogen was conducted by Durand *et al.*[7]. These are:

(a) **Series I:** issued from algae and vegetal waxes deposited in lacustral medium ("alginites"). It contains large amounts of amorphous OM.

(b) **Series II:** kerogens from the lower Toarcian of the Paris basin. They are well known from the studies of Tissot *et al.*[23], Alpern *et al.*[3], Durand *et al.*[8], Espitalié *et al.*[11]. Also kerogens from the german Toarcian. The parent OM corresponds to a marine environment. The amount of figured elements is 5-10%.

(c) **Series III:** kerogens from the upper Cretaceous of the Douala basin (Logbaba) previously studied by Albrecht *et al.*[1, 2] Dunoyer de Segonzac[6] and Durand *et al.*[9]. The parent material, mainly issued from higher plants, was deposited in a shallow coastal marine environment. They contain numerous samples of vitrinite. It includes also vitrinite containing coals described by Durand *et al.*[9].

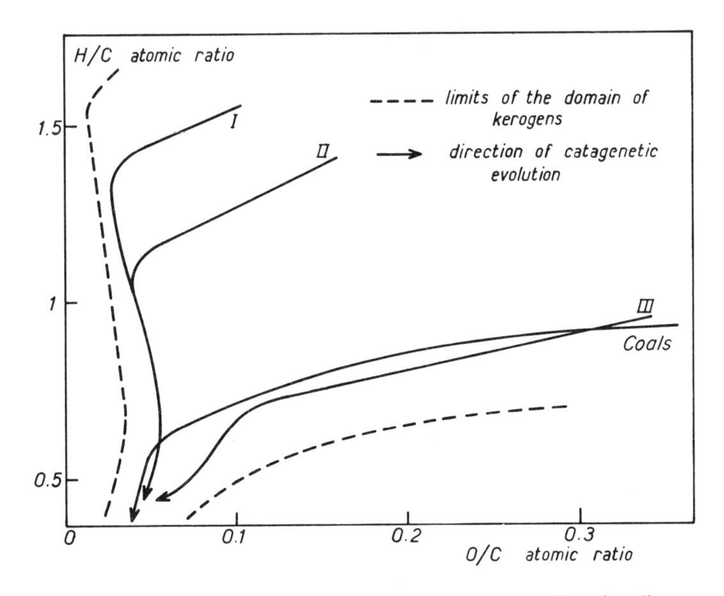

Fig. 8.12. — Evolution paths of kerogen series in the Van Krevelen diagram.

Figure 8.12 shows the evolution or catagenesis paths of all three series.

It must be understood of course that, in most real cases, the biological origin of kerogens is not so clearly determined: alterations, lithologic changes, deposition of varied organic materials, detritic contributions, all these factors play an important part, and real kerogens usually do not belong unequivocally to one of these standard "types" (Chapters 3 and 4).

Figures 8.13 and 8.14 show in each series, the variation of χ_p as a function of the vitrinite reflectance. The general evolution is of course in all cases similar to that of series III

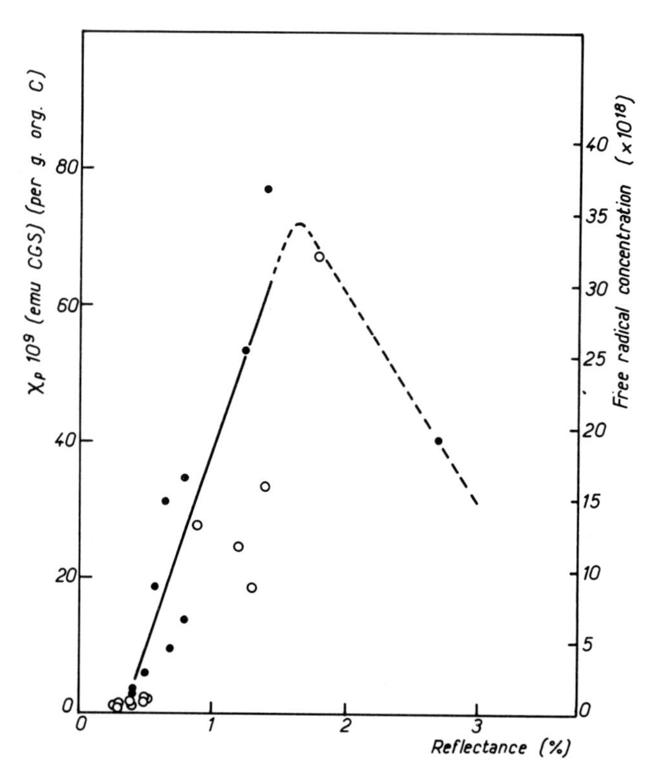

Fig. 8.13. — Evolution of paramagnetic susceptibility at room temperature as a function of vitrinite reflectance (kerogen of series I: ○ and II: ● [7].

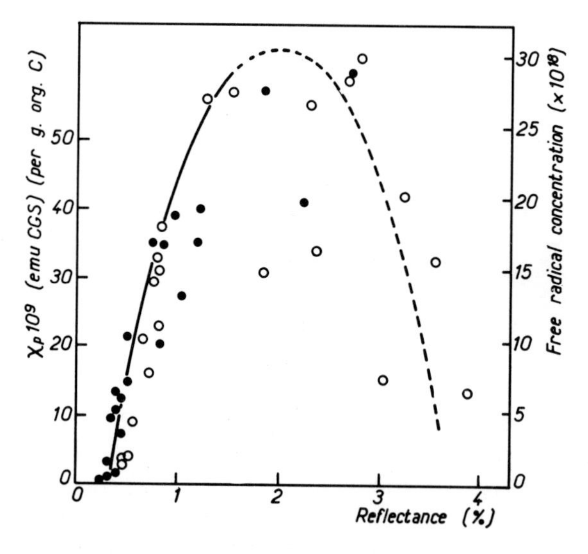

Fig. 8.14. — Evolution of paramagnetic susceptibility at room temperature as a function of vitrinite reflectance (kerogens from series III: ○ and coals: ●) [7].

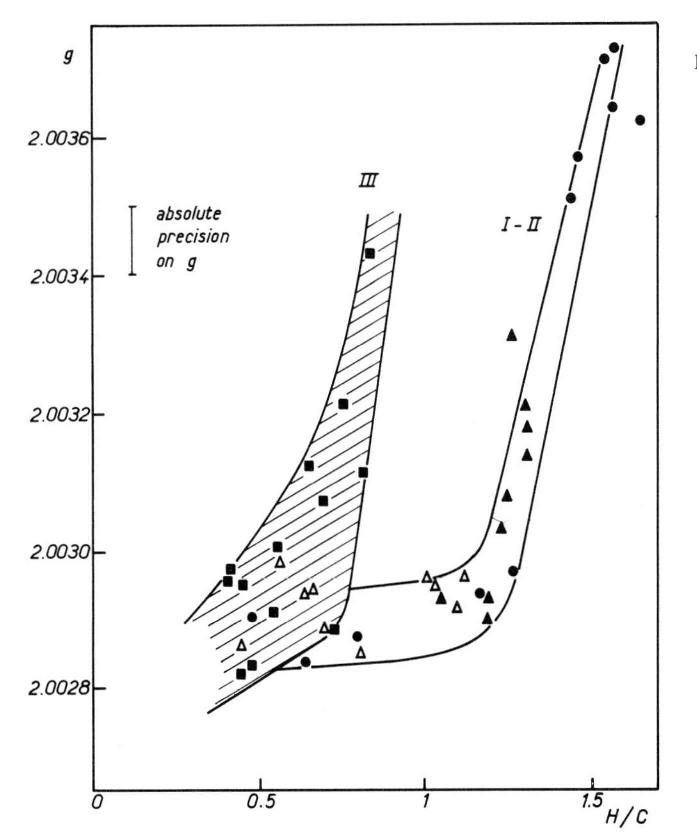

Fig. 8.15. — Variation of the *g* factor as a function of atomic ratio H/C:

Series **I**: ●
Series **II** (Paris Toarcian): ▲
Series **II** (German Toarcian): △
Series **III**: ■

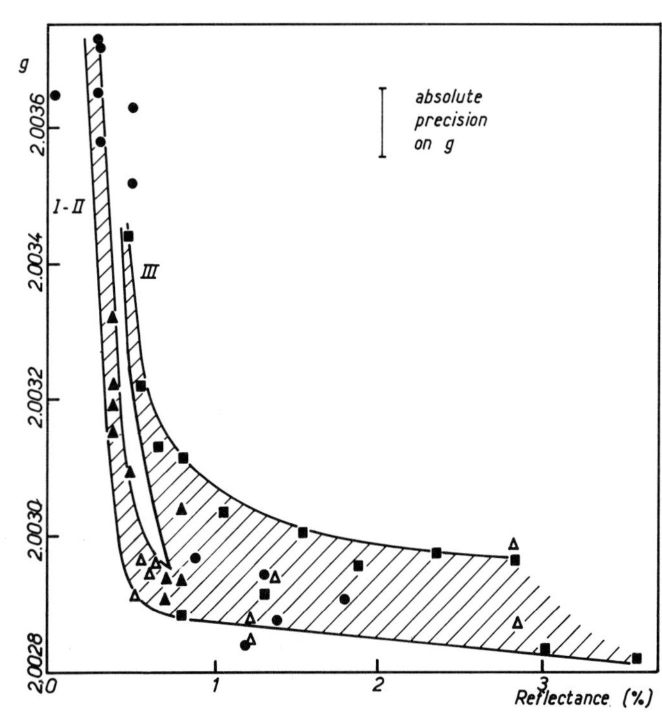

Fig. 8.16. — Variation of the *g* factor as a function of vitrinite reflectance:

Series **I**: ●
Series **II** (Paris Toarcian): ▲
Series **II** (German Toarcian): △
Series **III**: ■

(Fig. 8.11). The maximum in Fig. 8.14 is located very close to the maximum in Fig. 8.13. But points of series I-II are grouped along a curve with is somewhat different from the curve of series III. χ_p rises more slowly in Fig. 8.13, but later decreases faster than in Fig. 8.14.

It is interesting that the behavior of the EPR signal intensity is paralleled by that of the IR absorption intensity in the 700-930 cm^{-1} region (aromatic CH) (Chapter 6, Figs. 6.8 and 6.12). This is evidence that the production of free radicals and the fixation of hydrogen to aromatic rings occur as consequences of the same process, i.e. the "peeling" of the chemical skeleton of the parent material.

The evolution of the *g* factor is presented on Fig. 8.15 as a function of the H/C ratio and on Fig. 8.16 as a function of vitrinite reflectance. *g* decreases, as pointed out earlier, with progressive aromatization and increasing concentration of free radicals. But the *g* factor reaches a nearly constant value (2.0028-2.0030) very early in the aromatization process, and long before the maximum free radical concentration is observed. This rapid decrease of *g* shown in Fig. 8.16 is attributed to the loss of oxygen-containing groups, which is a preliminary condition for aromatization (paragr. III.C).

An important difference is also evident in the behavior of series III: the decrease of *g* occurs much later than in series I and II (Figs. 8.15 and 8.16): the larger values of *g* in series III for a given H/C or reflectance value are probably caused by the larger oxygen content of these kerogens (Fig. 8.12): as indicated earlier, the spin-orbit coupling is increased by the presence of oxygen atoms in aromatic nuclei or closely linked to them. This result gives the means of telling apart kerogens originating from cellulose-rich materials: there is some similarity between Fig. 8.15 and the Van Krevelen diagram (Fig. 8.12).

G. Influence of temperature.

The paramagnetic susceptibility of kerogens always follows rather closely the Curie law (formula No 3). This result shows that unpaired electrons are not "free" to move from one radical to another. But they may be "delocalized" throughout large aromatic systems the dimensions of which cannot be much in excess of 20 Å.

H. Oxygen effect.

Since the very first studies of free radicals in carbons, it was found by a number of investigators [14] that the EPR signal is strongly altered when a paramagnetic gas (O_2 or NO_2) surrounds the sample. This effect is mainly a physical one: an interaction between the magnetic moments of the gas molecules and those of the free radicals, which allows an easier relaxation of the unpaired spins (they fall back more easily to their lower energy orientation, parallel to the applied magnetic field). The result is a widening of the resonance signal, which may be so broad as to be undetectable.

This "oxygen-effect" is quite strong in artificial carbons treated above 600° C. It could be expected in kerogens having undergone a high enough catagenesis.

Durand *et al.* [7] did actually observe a narrowing of the line of some kerogens with a vitrinite reflectance higher than 2%, when EPR is studied in vacuum instead of air. In other cases a splitting of the signal may occur, when observed in vacuum: it could be an

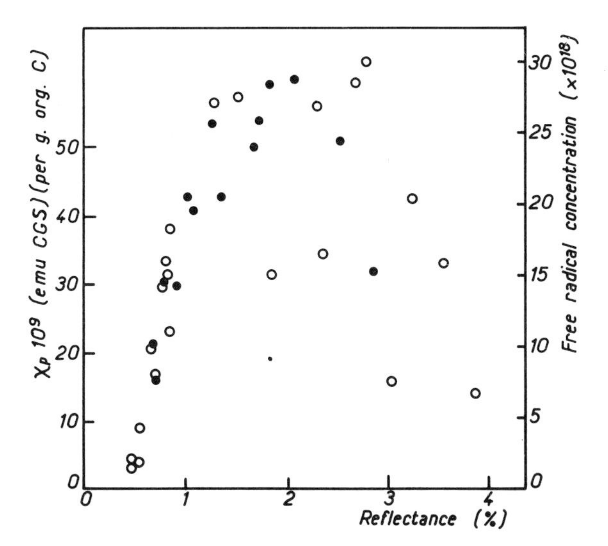

Fig. 8.17. — Paramagnetic susceptibility at room temperature as a function of vitrinite reflectance:

Natural kerogens of series **III**: **O** and artificial ones of series **III′**: ● [7].

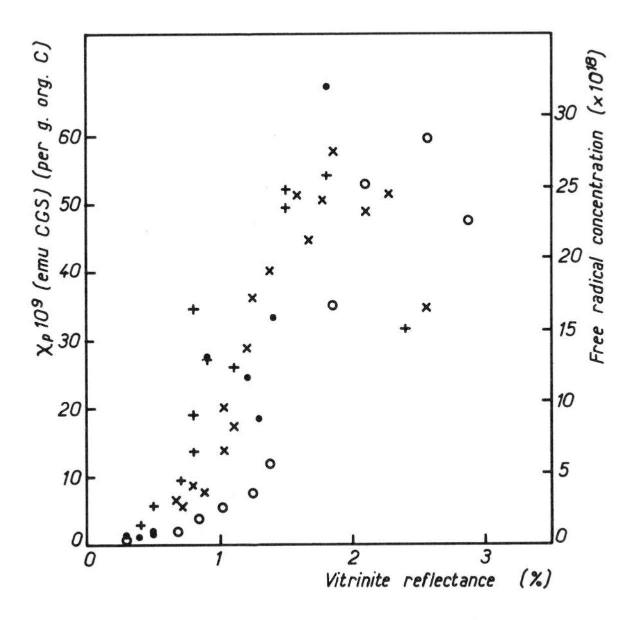

Fig. 8.18. — Paramagnetic susceptibility at room temperature as a function of vitrinite reflectance:

Natural kerogens of series **I**: ●. **II**: + and artificial kerogens of series **I'**: **O** and **II'**: × [7].

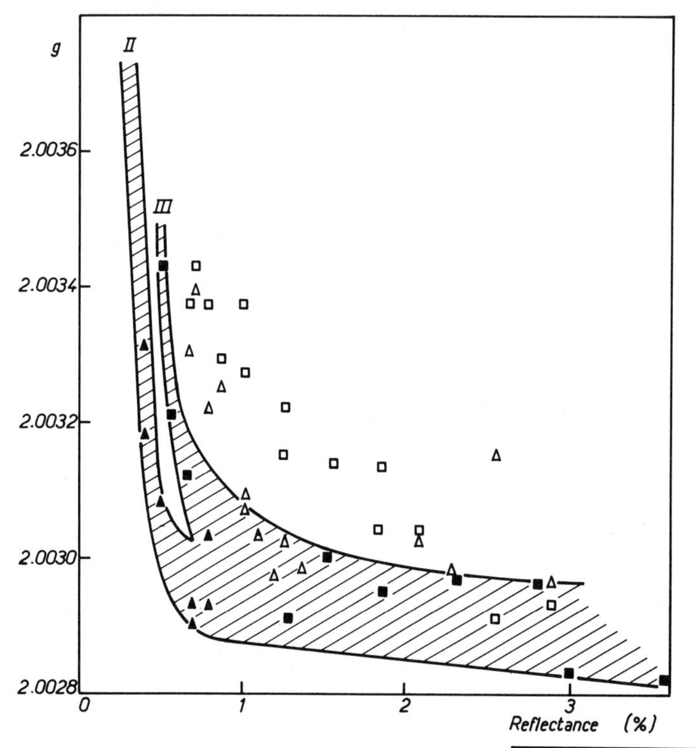

Fig. 8.19. — Values of the *g* factor as a function of vitrinite reflectance:

Kerogens of series **II** (natural: ▲ and artificial: △) and series **III** (natural: ■ and artificial: □). Compare with Fig. 8.16.

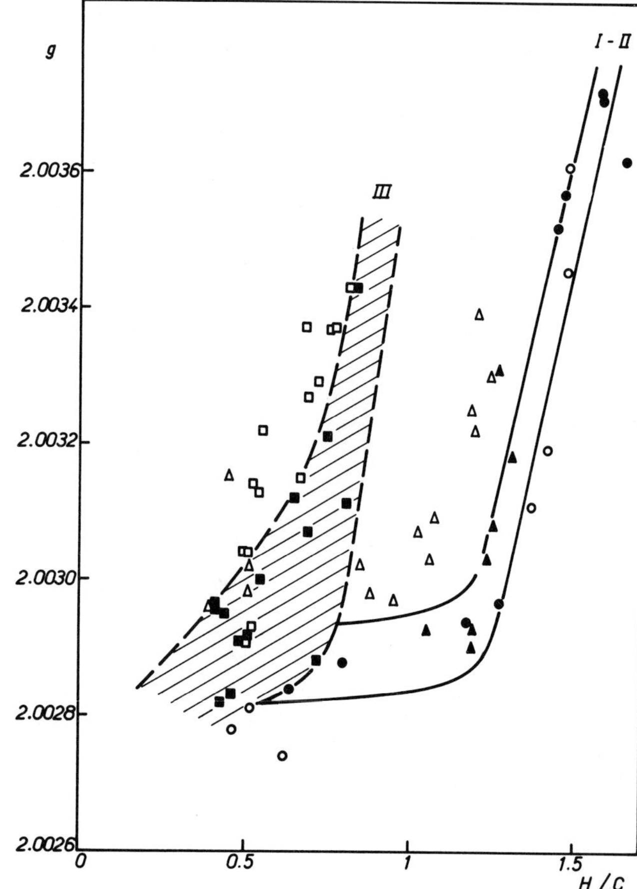

Fig. 8.20. — Values of the *g* factor as a function of atomic ratio H/C:

Kerogens of series **I** (natural: ● and artificial: ○). Series **II** (natural: ▲ and artificial: △) and series **III** (natural: ■ and artificial: □). Compare with Fig. 8.15.

oxygen-effect if the sample is not homogeneous. The general consequence will be an underestimation of χ_p whenever the measurement is performed in the presence of air.

The oxygen-effect is also linked to the kerogen porosity since the paramagnetic molecules must be very close to the free radicals in order to interact significantly with them.

It should be pointed out that in most oil-producing kerogens the vitrinite reflectance is lower than 2%, so that observation of the oxygen-effects is very unlikely in such kerogens.

IV. ARTIFICIAL KEROGENS

The EPR studies of free radicals in natural kerogens have been completed by similar studies on artificially degraded samples. Durand *et al.* [7] obtained artificial kerogens by heating shallow samples of series I, II, III in nitrogen atmosphere. The rate of temperature increase was 4° C min^{-1} and the maximum temperature was in the 300-600° C range. EPR studies of these I′, II′, III′ series were conducted, as well as measurements of the vitrinite reflectance in series III′.

Figure 8.17 shows the paramagnetism of series III′ samples as a function of their vitrinite reflectance: the evolution is identical to that of the natural series III. Similar results were obtained for series II and II′, but small differences may be observed between I and I′ (Fig. 8.18): the increase of χ_p in the artificial series I′ is slower and its maximum corresponds to a higher reflectance than in the natural series I.

It is also noteworthy that (as already pointed out for the natural series) the evolution of χ_p vs. reflectance in the artificial series is also paralleled by the evolution of the intensity of the 930-700 cm^{-1} IR bands (aromatic CH) (Chapter 6). Both effects result from the same "peeling" process of the chemical parent material.

However an examination of the *g* factor variations as a function of vitrinite reflectance (Fig. 8.19) or H/C ratio (Fig. 8.20) shows that, in this respect, artificial kerogens behave differently from one another according to their biological origin (Series I′, II′, III′) and that their behavior also differs from that of the corresponding natural kerogens of series II and III.

These results show that thermal treatments in the laboratory allow a fair simulation of the evolution of χ_p (if not of *g*) during the natural catagenesis of kerogens. Temperatures are of course quite different in the artificial and natural evolutions (natural kerogens have always been submitted to temperatures lower than 200° C), but time may to some extent compensate for temperature, and burial of an organic material, at a relatively low temperature for an extremely long geological time, may yield the same paramagnetic susceptibility as a higher temperature treatment of short duration.

V. APPLICATIONS TO OIL EXPLORATION

We emphasized previously the difficulties arising from the scattering and fluctuations of experimental EPR results on kerogens, due to impurities, sample inhomogeneities etc... and it was already pointed out that such data are significant and useful only if obtained through the study of a number of samples from an homogeneous series.

In the present state of our knowledge, only the EPR signal intensity, which has been the subject of extensive investigations, can be safely used in practical applications such as oil exploration. But various other data can be obtained from EPR, and it may very well be that other parameters, such as the *g* factor for instance, will turn out to be of practical value in the near future. We will examine these possible prospects later on, but will first discuss the use of the EPR signal intensity, i.e. the paramagnetic susceptibility of kerogens.

The evolution of the kerogen paramagnetism as a function of catagenesis suggests its utilization as a potential oil production index: the deeper an organic sediment is buried, the larger its hydrocarbon production, and the stronger its paramagnetism... up to a certain point, where the free radical concentration starts decreasing after reaching its maximum value (Fig. 8.8).

Since laboratory studies did also show that the EPR signal intensity changes in a similar way as a function of the heat-treatment temperature (HTT) of an artificial kerogen, a close correlation was suggested between the level of catagenetic evolution of a natural kerogen and the highest temperature it was submitted to. Laboratory HTT are of course much higher than those reached during the geological burial of an organic sediment, but it may be assumed that these higher temperatures compensate somehow for the very much shorter laboratory treatment times.

Paramagnetic susceptibility could then be considered as a "paleotemperature index": by using kerogens with a well known thermal history as standards, it would be possible to determine by EPR the "paleotemperature" of any natural kerogen.

It must be pointed out, however, than there is no one-to-one correspondence between the paramagnetic susceptibility χ_p of a kerogen and its paleotemperature. So that the concept of paleotemperature must actually be used with extreme caution.

First a given value of χ_p may correspond to two different paleotemperatures, since χ_p goes through a maximum during the catagenetic evolution: the lower χ_p, the larger the difference between the possible paleotemperature values.

Also the paleotemperature alone is not sufficient to determine the paramagnetic susceptibility of a kerogen. The duration of the kerogen burial is a factor which cannot be neglected: one million years at 150° C is not equivalent, in terms of free radical generation, to one hundred millions at the same temperature.

Other factors may also be effective in increasing the rate of catagenesis: pressure, catalyst... Neglecting them might lead to important errors.

Finally we have shown that significant differences in the evolution of χ_p may be observed in the various natural series of kerogens: at the beginning of the catagenetic process, a higher paleotemperature is for instance required in series I and II than in series III to reach a given paramagnetic susceptibility. In other words a single χ_p value does not allow the unambigous determination of a kerogen location in the Van Krevelen diagram. One has also to know the biological nature of the parent material.

This problem could be solved to some extent by using simultaneously the paramagnetic susceptibility (i.e. the EPR signal intensity or free radical concentration) **and** one other significant parameter, such as the O/C or H/C ratios, the vitrinite reflectance, or some characteristic intensity from the IR absorption spectrum (Chapter 6). These parameters are not equivalent since they originate from various components of the sample or from different features of the kerogen structure: by taking one of them **and** the paramagnetic suscep-

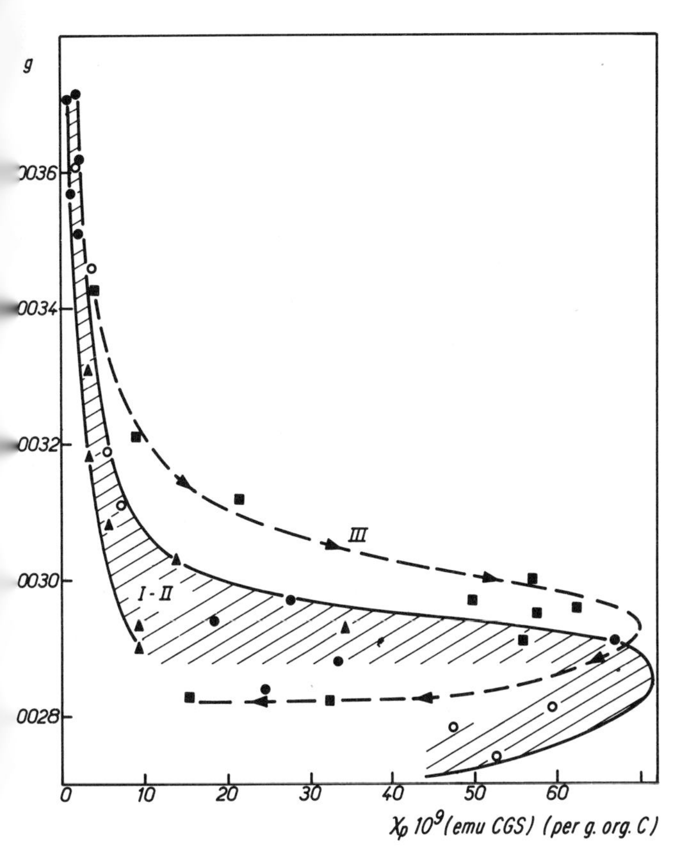

Fig. 8.21. — Variation of the g factor as a function of paramagnetic susceptibility at room temperature:

Kerogens of series **I** (natural: ● and artificial: ○). Series **II:** ▲ and series **III:** ■.
The direction of catagenetic evolution is shown by arrows.

Fig. 8.23. — Variation of the g factor as a function of paramagnetic susceptibility at room temperature: comparison of natural: ■ and artificial: □ kerogens of series **III** and **III′**.

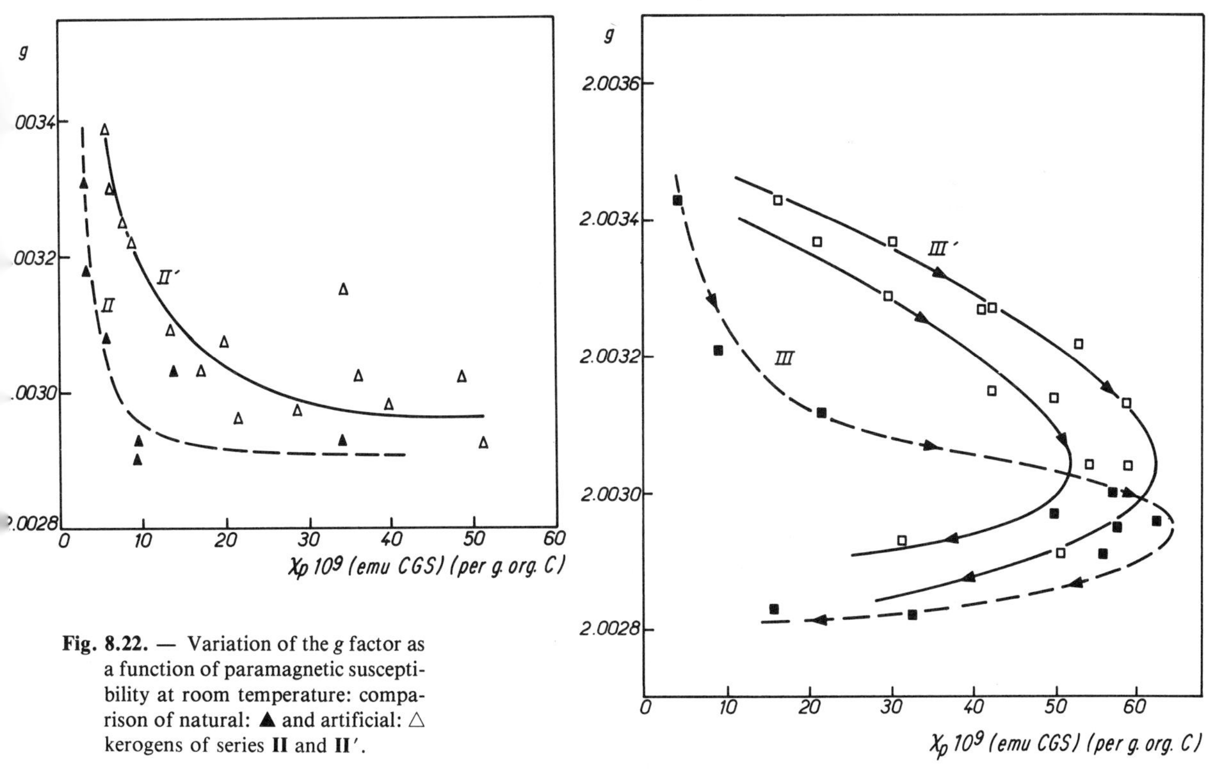

Fig. 8.22. — Variation of the g factor as a function of paramagnetic susceptibility at room temperature: comparison of natural: ▲ and artificial: △ kerogens of series **II** and **II′**.

tibility into consideration, information is obtained on both the biological origin and the catagenetic stage of the kerogen under study.

Another way of solving the question of biological origin, is to make use of both the χ_p and the g factor values. In such a method, only EPR measurements are necessary while other methods necessitate the use of another technique besides EPR (e.g. reflectance or chemical analysis). When the g vs. χ_p variations are represented, curves similar to those on Figs. 8.21, 8.22 and 8.23 are obtained. It can be seen on Fig. 8.21 that the decrease of g with increasing χ_p is the same for natural kerogens of series I and II, but different for series III. Moreover Figs. 8.21, 8.22 and 8.23 also show that the g vs. χ_p curves for natural and artificial kerogens are the same in series I, I′, slightly different in series II, II′ and quite different in series III, III′. Although we do not know yet the explanation of these different behaviors, they afford a convenient method for telling apart kerogens of type III from the others by using only EPR results.

From the point of view of oil exploration, anyway, the concept of paleotemperature is useless by itself. What is important is the knowledge of the evolution stage of a kerogen, which allows an estimation of its potential oil production. The paramagnetic susceptibility value gives direct information about this potential: the best oil producing kerogens have a paramagnetic susceptibility in the range 2.10^{-8}-3.10^{-8} emu CGS/g of organic carbon; lower values correspond to organic sediments where oil generation has hardly started, or to extremely degraded organic matter, well past the oil production stage; higher values on the contrary are indications of gas rather than oil production.

Since the same paramagnetism may correspond to two different catagenetic stages, the ambiguity can be resolved by making use of another parameter, such as the vitrinite reflectance; for instance values of χ_p close to 2.10^{-8} emu CGS/g of organic carbon characterize a good source-rock **provided the vitrinite reflectance does not exceed 1%.**

VI. FUTURE DEVELOPMENTS OF EPR IN KEROGEN STUDIES

Not only can EPR be used to detect OM in potential oil-bearing rocks, but also in any kind of sediments: for instance kerogens of low catagenetic evolution were recently found in phosphates[13].

Detection however is only a first step, and EPR studies may be expected, in not too distant a future, to yield more information on the composition or structure of a kerogen.

It is theoretically possible to calculate the g-value of a given molecular radical. But for large systems complete identification by EPR could only be expected in the liquid phase, where precise g-values are observed because of motional averaging of local dipolar interactions (paragr. II.D.3). Can a wonder-solvent or chemical treatment be found to solubilize at least some part of a kerogen, without too much alteration of the components? Identification by EPR and even quantitative analysis of functional groups by ^{13}C NMR (nuclear magnetic resonance) would then be available, as well as other familiar techniques.

In the solid state, the poor resolution of the signal results from the presence of local magnetic fields, one of them being due to residual iron. The question of why it is not possible to eliminate all the iron, and the relation between the presence of pyrite and the proportion of organic sulfur, are not at all clear. During the artificial pyrolysis of kerogens some enrichment in organic sulfur, probably originating from pyrite, has been noted, and further work in this field would be of interest. Sulfur is responsible for a large spin-orbit coupling and would result in g-values higher than does oxygen.

If all the iron could be eliminated, the EPR linewidth would be an efficient source of information about diagenesis, as it is in coals. The remaining width is due to the local field of protons: by irradiating simultaneously at the proton nuclear magnetic resonance frequency it is possible to measure the interactions between the protons and the unpaired electron, thus obtaining information on the free-radical localization. In another technique, a rapid reversal of proton spins is performed by saturation at the NMR frequency: the residual width and shape are then mainly due to g distribution and anisotropy; the signal resolution is proportional to the magnetic field strength, which shows the interest of high field EPR studies.

Even without these sophisticated methods, the changes of EPR signal width and shape of ironless samples are typical of the extension of aromatic systems.

Another approach consists in studying "modelized kerogens", obtained from very well defined precursors, chosen as typical for each biological family (e.g. sporopollenin, lignin, etc...). It may be expected to obtain a complete description of such an artificial diagenesis. From these studies it would then be possible to throw some light on the relation between geological time and temperature of pyrolysis, role of pressure, equilibrium with evolved gases etc... Research of this kind, which already gives interesting results in the domain of coals, is being started in kerogen studies, and further developments are expected in the near future. As an example of a practical rule obtained by this approach we can state: "A high g-value of an aromatic radical, at a sufficient level of catagenesis, proves the presence of oxygenated groups in the vicinity; it is typical of a type III kerogen. Now this family is known to be a poor oil source, due to inadequate reducing conditions". With such a rule we do not really identify the free-radicals present in the kerogen, but we recognize a character observed in the study of pure oxygenated aromatic radicals.

Finally there is another field of EPR applied to kerogens which has not been too much explored until now: it deals with vanadium and transition elements spectra. Since the VO_2^+ ion in porphyrins seems to be a typical geochemical witness in oil deposits, EPR studies in this field would very likely shed some light on the chemical filiation of kerogens.

VII. CONCLUSION

Although it has been only recently applied to oil exploration, EPR is now widely recognized as a useful technique. Future developments are promising, and it may be that this method will become self-sufficient to determine the oil-production potential of a sediment. For the time being it has to be used in association and as a complement to other techniques.

Some rules must be obeyed, of course, in chosing, purifying, and measuring the samples, but there is no difficulty in a routine utilization of EPR. Its most attractive features are that it is non-destructive and gives a general characterization of the whole OM in a sediment.

REFERENCES

1. Albrecht, P., (1969), Thèse, Strasbourg.
2. Albrecht, P. and Ourisson, G., (1969), *Geochim. et Cosmochim. Acta*, **33**, 138.
3. Alpern, B., Durand, G., Espitalié, J. and Tissot, B., (1972), *in*: *Advances in Organic Geochemistry* 1971, H. V. Gaertner and H. Werner ed., Pergamon Press, 1.
4. Bersohn, M. and Baird, J. C., (1966), "An Introduction to EPR", Benjamin.
5. Carmona, F. and Delhaes, P., (1974), *in*: *Proceedings of Internat. Conference Industrial Carbons and Graphites, London, 1974*, preprint n° **120**.
6. Dunoyer de Segonzac, (1969), *Thèse*, Strasbourg.
7. Durand, B., Marchand, A., Amiell, J. and Combaz, A., (1977), *in*: *Advances in Organic Geochemistry* 1975, R. Campos and J. Goñi ed, Enadimsa, Madrid, 753.
8. Durand, B., Espitalié, J., Nicaise, G. and Combaz, A., (1972), *Rev. Inst. Franç. du Pétrole,* **XXVII**, 865.
9. Durand, B., Nicaise, G., Roucaché, J., Vandenbrouke, M. and Hagemann, H., (1977), *in*: *Advances in Organic Geochemistry 1975*, R. Campos and J. Goñi ed., Enadimsa, Madrid, 601.
10. Erb. E., Thèse de doctorat d'Université, Paris.
11. Espitalié, J., Durand, B., Roussel, J.C. and Souron, C., (1973), *Rev. Inst. Franç. du Pétrole*, **XXVIII**, 37.
12. Etienne, A., Combrisson and Uebersfeld, J., (1954), *Nature*, **174**, 614.
13. Ferhat, M. and Conard, J., "Etude de la matière organique des phosphates par R.P.E.", à paraître dans la Rev. Inst. Franç. du Pétrole.
14. Groupe Français d'Etudes des Carbones (G.F.E.C.), (1965), *in*: "Les Carbones", **I**. 469, Masson éd., Paris.
15. Hwang, P. T. and Pusey, W. C., (1973), *U. S. Patent n° 3740*, June 19, 1973.
16. Ingram, D. J. E. and Bennett, J. E., (1954), *Phil. Mag.*, **45**, 545.
17. Libert. P., (1974), Thèse de Doctorat d'Etat, Bordeaux.
18. Marchand, A., Libert, P. and Combaz, A., (1968), *C.R. Acad. Sci. Paris*, D. **266**, 2316.
19. Marchand, A., Libert, P. and Combaz, A., (1969), *Rev. Inst. Franç. du Pétrole*, **XXIV**, 3.
20. Salle, M., (1977), Thèse de 3e Cycle, Université d'Orléans.
21. Simon, C., Estrade, H., Tchoubar, D. and Conard, J., (1977), *Carbon*, **15**.
22. Steelink., C., (1964), *Geochim. et Cosmochim. Acta*, **28**, 10, 1615.
23. Tissot, B., Califet-Debyser, Y., Deroo, G. and Oudin J. L., (1971), AAPG Bull. **55**, 2177.
24. Uebersfeld, J. and Erb, E., (1959), Proceedings 3rd Carbon Conference 1957, Pergamon, 107.

9

C^{13}/C^{12} in kerogen

E.M. GALIMOV*

I. INTRODUCTION

The carbon isotopic composition of a kerogen depends on the isotopic composition of its biological precursors as well as on the isotopic fractionation that has taken place in the course of the formation and chemical evolution of the kerogen. Regularities in carbon isotope distributions in kerogens must therefore be considered against a background of the general biogeochemical behaviour of carbon isotopes; such an approach has been attempted in this paper.

II. PRINCIPLES OF CARBON ISOTOPE DISTRIBUTIONS IN BIOLOGICAL PRECURSORS OF KEROGEN

Bioorganic carbon is known to be enriched in the C^{12}-isotope, in comparison with the carbon dioxide used for photosynthesis in plants. Different biochemical components of organisms have distinctive carbon isotope compositions. For example lipids are enriched in C^{12} while amino acids and carbohydrates are relatively depleted in the light isotope [1, 15, 42, 62, 63].

Although at one time the origin of the isotopic fractionations in biological systems remained vague, it was believed to be kinetic, i.e. the effects were dictated only by the greater mobility of the carbon-12 species compared to that of carbon-13 species and by the greater lability of C^{12}-X bonds compared to C^{13}-X bonds; briefly, by the preferential assimilation of $C^{12}O_2$ in comparison with $C^{13}O_2$ during photosynthesis.

But differences in the isotopic composition of individual biochemical compounds often proved to be regular and the observed naturally-occuring fractionations often duplicated those expected on a thermodynamic (not kinetic) basis. Investigations of this problem have led to the conclusion that in biological systems there is an ordered intra- and intermolecular carbon isotope distribution, which may in general be predicted by means of isotope thermodynamics [24]. A necessary first step in such studies was the development of convenient

* Institute of Geochemistry and Analytical Chemistry. Academy of Sciences. Moscow, USSR.

techniques for calculating the isotopic thermodynamic properties of complex organic substances.

The thermodynamic isotope exchange properties of compounds can be calculated using Urey's formula for the partition function ratio of isotopic forms of the compound, the so-called β-factor:

$$\beta = \prod_{i}^{3N-6} \frac{\nu_i^* \exp\left(-\frac{h\nu_i^*}{2\,kT}\right)\left[1 - \exp\left(-\frac{h\nu_i}{kT}\right)\right]}{\nu_i \exp\left(-\frac{h\nu_i}{2\,kT}\right)\left[1 - \exp\left(-\frac{h\nu_i^*}{kT}\right)\right]} \quad (1)$$

In this equation: ν_i and ν_i^* represent the oscillation frequencies of the isotopic species (here, the asterisk refers to a molecule containing a heavy isotope); h is Planck's constant; k is the Boltzmann constant; T is the absolute temperature, $K°$, and N is the number of atoms in the molecules.

The greater the value of the β-factor, the higher is tne concentration of the heavy isotope in the given compound provided that equilibrium takes place in the corresponding isotope-exchange system. For example, for the exchange reaction:

$$C^{12}O_2 + C^{13}H_4 \rightleftharpoons C^{12}H_4 + C^{13}O_2$$

we can write:

$$\delta C^{13}_{CH_4} - \delta C^{13}_{CO_2} = \left(\frac{\beta_{CH_4}}{\beta_{CO_2}} - 1\right) \times 10^3 \qquad (‰) \quad (2)$$

Carbon isotope compositions are given in δC^{13}-values and represent deviations per mil (‰) in the C^{13}/C^{12}-ratio of the sample from the C^{13}/C^{12}-ratio of the PDB standard.

The latter is equal to 0.0112372. Thus:

$$\delta C^{13} = \left(\frac{(C^{13}/C^{12})_{sample}}{0.0112372} - 1\right) \times 10^3 \qquad (‰) \quad (3)$$

Complete interpretation of the vibrational spectra of large molecules is extremely difficult so that precise calculations of β-factors for compounds such as complex biomolecules become impossible in practice. For such compounds the evaluation of β-factor values (with a good degree of approximation) can be done by means of a method based on the additive properties of these thermodynamic values [24, 29].

It has been shown that the β_ϵ-value, which characterizes the molecule as a whole, is the arithmetic mean of the β_i-factors characterizing monosubstitued isotopic species of the compound; in other words characterizing carbon atoms in different position in the molecule:

$$\beta_\epsilon = \frac{1}{n} \sum_{i=1}^{n} \beta_i \qquad \text{(the first rule of additivity)} \quad (4)$$

In turn, the magnitude of the β_i-factor can, to a certain degree, be expressed through the sum of scalar values which we called isotopic numbers of bonds L_j:

$$\beta_i = 1 + \sum_j L_j \qquad \text{(the second rule of additivity)} \quad (5)$$

where L_j refers to the bonds which the carbon atom forms directly.

For a better approximation one can use l_k values which allow to take into account the chemical surroundings of the adjacent carbon atoms [29].

The magnitudes of L_j and l_k values, at 300° K, for different types of carbon bonds are listed in Table 9.1A. Using these values, and the additivity rules, one can estimate the β-factor of any carbon compound.

TABLE 9.1A
ISOTOPIC NUMBERS OF BONDS AT T = 300 K

Type of the bond	L_j	l_k
C–H	0.0284	0
C–C	0.0464	0.0013
C–N	0.050	0.0016
C–O	0.055	0.0019
C–F	0.056	0.0020
C=C	0.0785	0.0016
C≡C	0.088	0.0003
C=N	0.090	0.0003
C=O	0.0958	0.0028

If we take, for example, a molecule of the amino acid alanine:

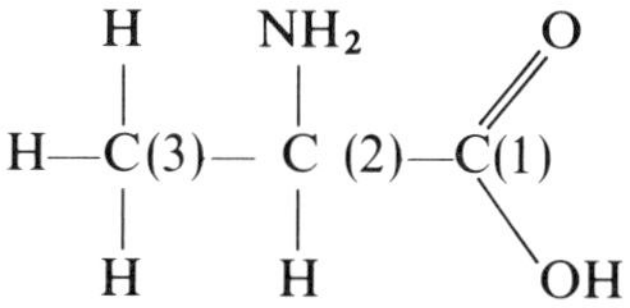

$$H-C(3)H_2-C(2)H(NH_2)-C(1)(=O)OH$$

then the values of the β_i-factors can be found as follows:

$$\beta_1 = 1 + (L_{C=O} + L_{C-O} + L_{C-C}) + (l_{C-N} + l_{C-H} + l_{C-C}) = 1.199$$

$$\beta_2 = 1 + (2L_{C-C} + L_{C-N} + L_{C-H}) + (l_{C=O} + l_{C-O} + 3l_{C-H}) = 1.180$$

$$\beta_3 = 1 + (3L_{C-H} + L_{C-C}) + (l_{C-N} + l_{C-C} + l_{C-H}) = 1.135$$

and hence the thermodynamic isotopic factor of the molecule as a whole:

$$\beta_\epsilon = \frac{1}{3}(1.199 + 1.180 + 1.135) = 1.171$$

One can see that the β_i-factors characterizing different carbon atoms in alanine are not equal. This implies intramolecular thermodynamic isotope effects.

Available data on the carbon isotope composition of biomolecules, and their calculated β-factors, suggests thermodynamically ordered isotope distribution in the biological systems: correspondance between experimentally measured δC^{13} and theoretically calculated β-values [24] (Fig. 9.1). In particular, such a correlation has proved to be characteristic of components in the lipid fractions of a number of the organisms [34] as shown on Fig. 9.2. It is noticeable that the same type of β_ϵ-δC^{13} relationship is found in the lipid fraction of organisms belonging to quite distinct taxonomic groups.

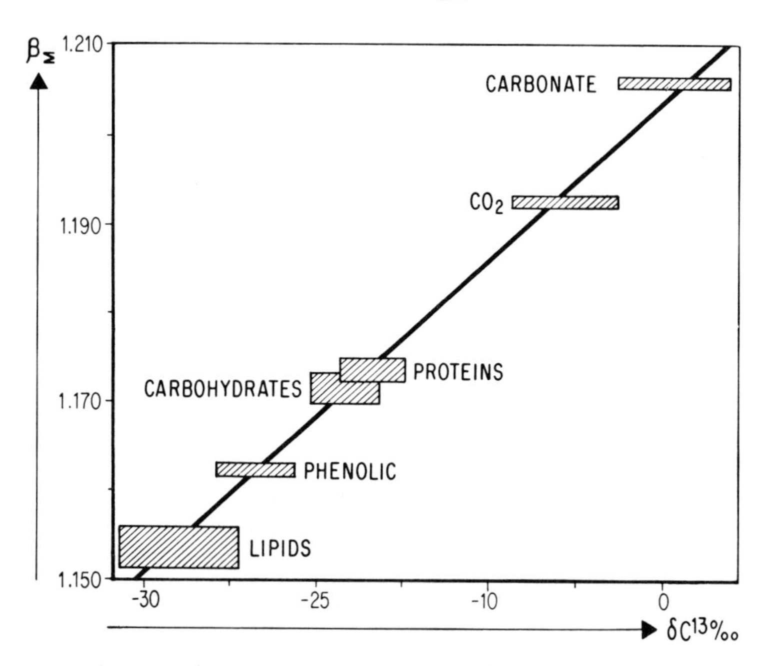

Fig. 9.1. — Correspondence between carbon isotope composition ranges of some components of biological system (including environmental CO_2 and biogenic calcite) and β-factor values of these components.

If the biological isotope effects are thermodynamically ordered the carbon isotopes will be regulated not only between biomolecules but also within them.

The β_i-factor values for carbon atoms in some functional groups, or structural positions, are listed in Table 9.1B.

TABLE 9.1B
THERMODYNAMIC ISOTOPE FACTORS (β_i-FACTORS) OF CARBON IN SOME STRUCTURAL POSITION IN ORGANIC COMPOUNDS

Structural group	Formula	β-factor
1. Methyl	$-CH_3$	1.131
2. Nitrile	$-C\equiv N$	1.137
3. Methoxy	$-OCH_3$	1.141
4. Methene	$-CH_2-$	1.149
5. Methine	$>CH-$	1.166
6. Aldehydic	$-CHO$	1.170
7. Amino	$>CH$ (with N bonded below)	1.172
8. Cetene	$C=C=O$	1.173
9. Phenolic	$\geqslant C-OH$	1.179
10. Carbonyl	$>C=O$	1.187
11. Carboxylic	$-COOH$	1.197

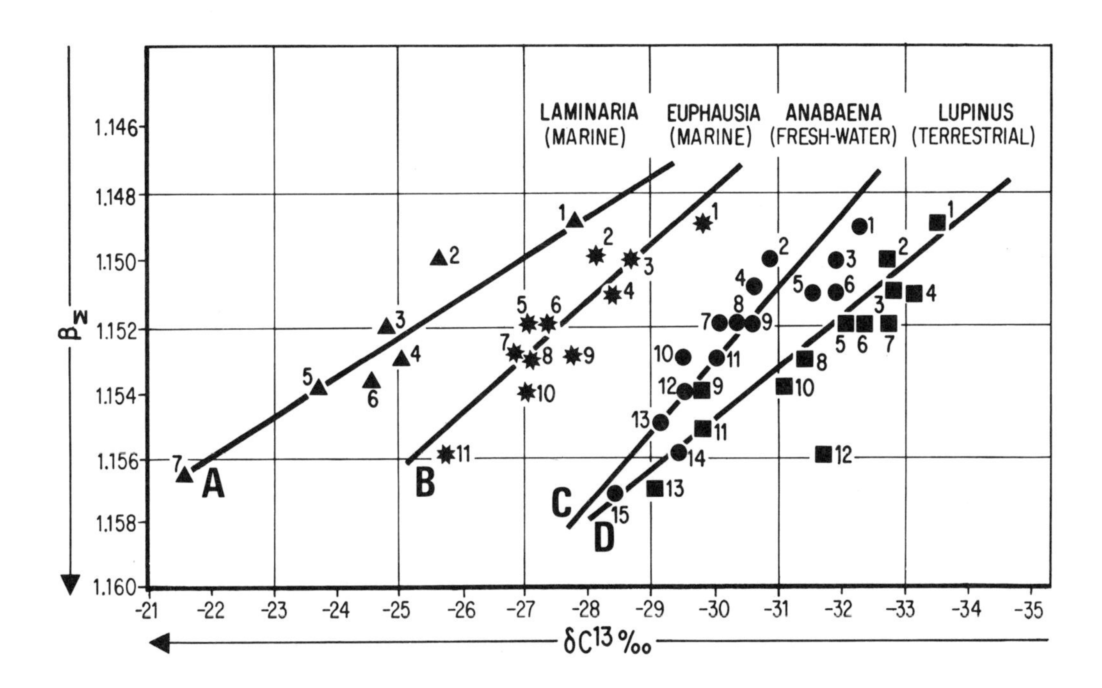

Fig. 9.2. — Carbon isotope composition of individual components of lipid fraction compared with the β-factor values of these components in some organisms [34].

A. Laminaria saccharina (seaweed).
1. Waxes, hydrocarbons. **2.** Fatty acids. **3.** Triglycerides. **4.** Monoglycerides. **5.** Fucoxanthin. **6.** Fucosterol. **7.** Chlorophyll.

B. Euphasia superba (marine).
1. Waxes, hydrocarbons. **2.** Fatty acids. **3.** Sphyngomyelin. **4.** Lecithin. **5.** Diglycerides. **6.** Triglycerides. **7.** Echinenon. **8.** Monoglycerides. **9.** Cardiolipine. **10.** Cholesterol. **11.** Astacene.

C. Anabaena variabilis (fresh-water blue-green algae).
1. Waxes, hydrocarbons. **2.** Sphyngomyelin. **3.** Fatty acids. **4.** Lecithin. **5.** β-Carotene. **6.** Phosphatidylserine. **7.** Diglycerides. **8.** Cephalin. **9.** Triglycerides. **10.** Echinenon. **11.** Monoglycerides. **12.** β-Sitosterol. **13.** Phosphatidylinositol. **14.** Myxoxanthophyll. **15.** Chlorophyll.

D. Lupinus luteus (terrestrial).
1. Waxes, hydrocarbons. **2.** Fatty acids. **3.** Lecithin. **4.** Phosphatidylserine. **5.** Diglycerides. **6.** Triglycerides. **7.** Cephalin. **8.** Monoglycerides. **9.** β-Sitosterol. **10.** Lutein. **11.** Chlorophyll, xantophyll. **12.** Monogalactosylglyceride. **13.** Chlorophyll.

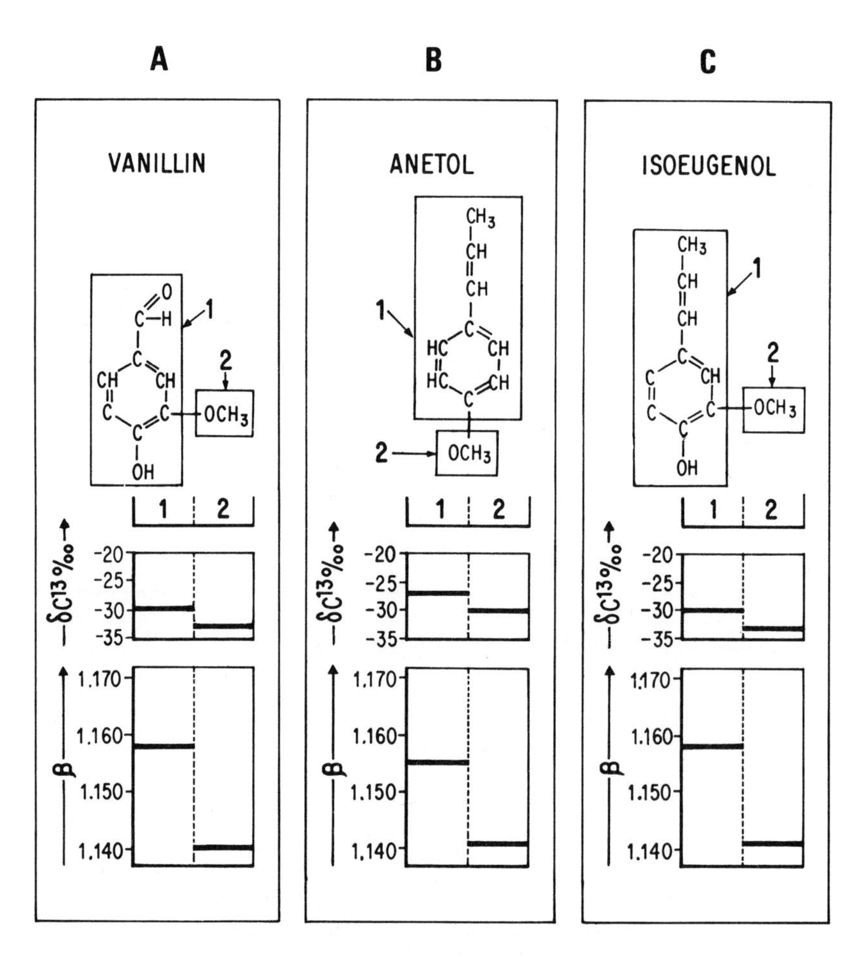

Fig. 9.3. — Thermodynamically ordered intramolecular carbon isotope distribution in some biological aromatic compounds :

A. Vanillin [30]. **B.** Anetol [84]. **C.** Isoeugenol [84]. Note that the measured isotope distribution is in correspondence with the calculated β-factor values of the same fragments.

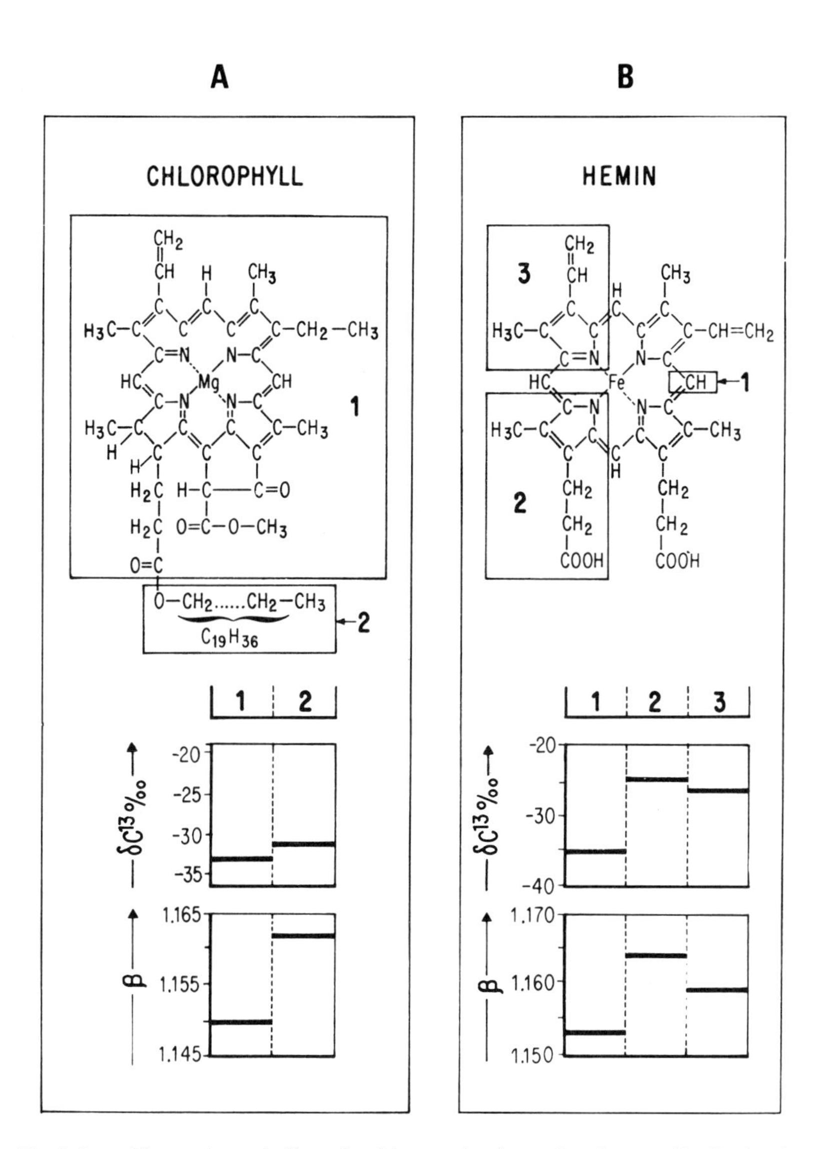

Fig. 9.4. — Thermodynamically ordered intramolecular carbon isotope distribution in:
A. Chlorophyll [31] of biological origin. **B.** Hemin.

Abelson and Hoering [1] showed that carboxyl groups of amino acids were enriched in the C^{13}-isotope compared to the molecule as a whole. Now this fact may be interpreted in terms of the difference of the corresponding β-values.

In our laboratory experiments were undertaken to study the intramolecular carbon isotope effects in biomolecules and to check the hypothesis of thermodynamically ordered intramolecular carbon isotope distributions. Thus, the relative concentration of the C^{13}-isotope in the methoxy groups of a number of aromatic lignin monomers was investigated [30]. In contrast to the carboxyl group, the carbon in a methoxy group has a low β_i-factor (Table 9.1B). Accordingly, the above experiment showed that the carbon in the methoxy groups of the aromatic compounds investigated was relatively depleted in the C^{13}-isotope. The data for the vanillin from a reed is shown in Fig. 9.3A. Similarly, it has been established that a distinction exists between the measured carbon isotopic compositions of certain fragments of hemin and chlorophyll molecules as expected from estimated β-factor values of these fragments (Fig. 9.4) [31]. Experimental results for acetic acid, and acetoin, obtained by Meinschein *et al.* [49] and Rinaldi *et al.* [66] are also consistent with the concept of a thermodynamically ordered carbon isotope distribution within these molecules (Fig. 9.5).

It has been supposed that the appropriate isotope fractionation takes place within enzyme-substrate complexes at each step of the enzyme-controlled formation of a biomolecule

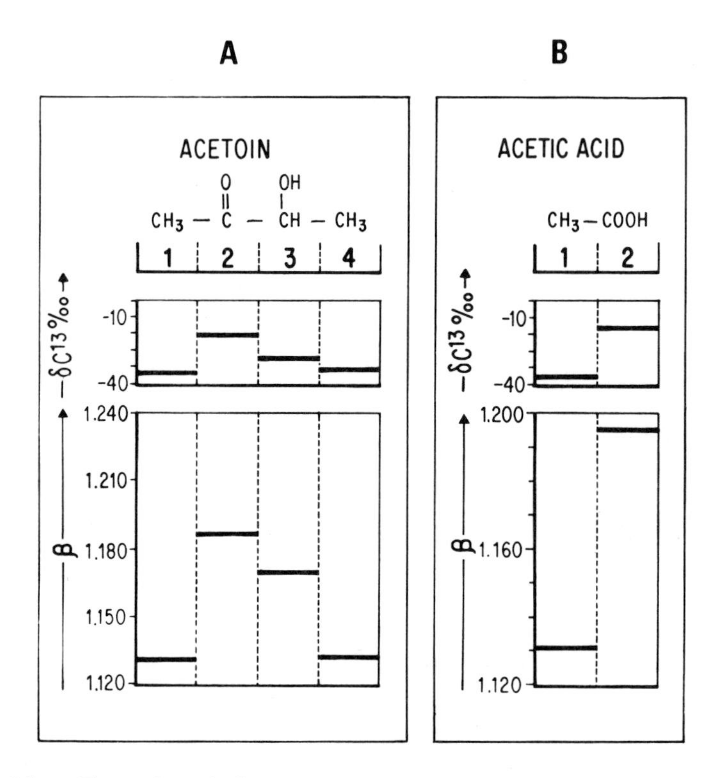

Fig. 9.5. — Thermodynamically ordered intramolecular carbon isotope distribution in :

A. Acetoin [66] of biological origin. **B.** Acetic acid [49].

[27, 29]. Analysis of this model indicates that the magnitude of the thermodynamic component of the biological isotope effect which occurs during enzymatic reaction is dependent on two quantities:

(a) β-factor ratios of the product and the substrate.

(b) The χ-value which depends on the kinetic constants (not confuse with kinetic isotope effects) of the reaction.

This gives the following relationship for biological isotope effects of enzymatic reaction in its simple version:

$$\delta C^{13}_{A} - \delta C^{13}_{B} = \chi\left(\frac{\beta_A}{\beta_B} - 1\right) \times 10^3 \qquad (‰) \tag{6}$$

χ-value may change from zero to unit. Under the normal conditions χ-value varies around 0.5. The changeability of this factor predetermines dependence of the biological isotope effects on the pathways of the biosynthesis and environment conditions. β_A- and β_B-values in expression (6) may represent both β_ϵ-and β_i-factors and expression (6) may correspondingly describe intermolecular or intramolecular carbon isotope effects.

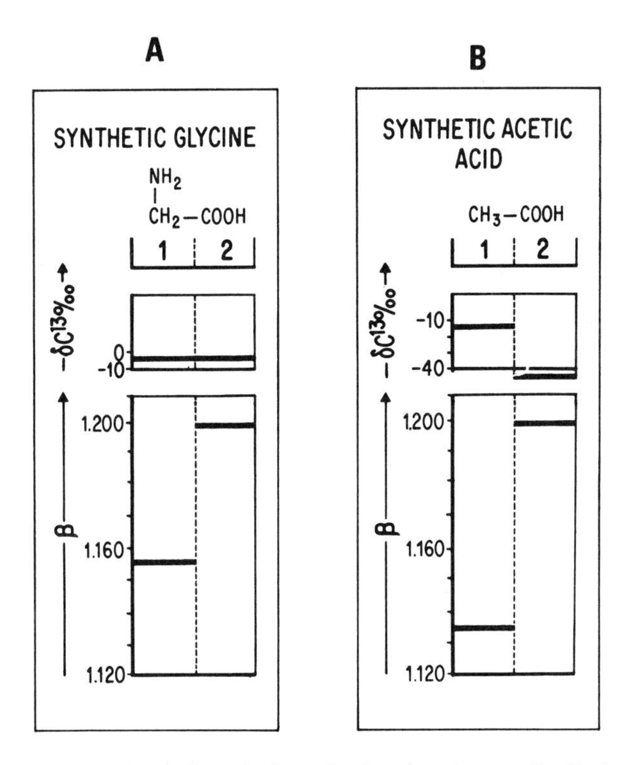

Fig. 9.6. — Lack of thermodynamically ordered carbon isotope distribution in compounds abiogeneously synthesized:

A. Glycine [84]. **B.** Acetic acid [49].

Thus, it is possible to evaluate carbon isotope distributions in different constituents of organisms, on some general basis, without resort to measurements in every case. Moreover, one can predict alterations in the carbon isotopic composition of organic matter (OM) during diagenesis. It is clear, for example, that elimination of functional groups with carbon enriched in C^{13} (high β_i-value) such as carboxyl, or formyl, results in the enrichment of the OM in the light carbon isotope. Alternatively, removal of isotopically light methyl or methoxy groups should produce OM richer in the heavy isotope.

Since the β-factor is temperature dependent, the carbon isotope composition of organisms will depend on the temperature of biosynthesis.

It should be emphasized also that the thermodynamically ordered isotope distribution appears to be a purely biological phenomenon. In other words the carbon isotope composition of **chemically synthesized** compounds should not be expected to correspond to β-factors of these compounds. For example, in contrast to biogenically produced carboxyl carbon, which is relatively enriched in the C^{13}-isotope, the carbon atoms of carboxyl groups formed by abiogenic oxidation will most likely be enriched in the C^{12}-isotope, due to kinetic isotope effects. Data presented in Fig. 9.6 confirm this proposition. Differences in intramolecular carbon isotope distributions in biogenic and abiogenic compounds may therefore be used to determine the biological or nonbiological origin of terrestrial and extraterrestrial compounds [24, 27, 34].

III. RELATIONSHIPS INHERITED BY FOSSIL ORGANIC MATTER FROM ITS BIOLOGICAL SOURCE

Terrestrial plants are enriched in the C^{12}-isotope by 5-8‰ when compared with marine ones (Fig. 9.7). This may be caused by more high degree of utilization of the ambient CO_2 by water plants. The another reason is difference between the isotopic composition of the carbon sources: the atmospheric CO_2 ($\delta C^{13} = -7‰$) and the seawater bicarbonate ($\delta C^{13} = -2‰$). Data in Fig. 9.2 also exemplify the influence of environmental carbon. Displacement of the lines along the horizontal axis, for various organisms, is also due to differences in the isotopic composition of the source carbon.

Difference in δC^{13}-values between OM of terrestrial and marine origin is preserved in fossil OM, but at a somewhat reduced level. Thus, if the difference between terrestrial and marine plants is about 8‰, for humic substances, and appropriate kerogens, it may amount to 3 to 5‰ only.

In transitional environments δC^{13}-values change gradually from about -20 to $-23‰$, characterizing open sea deposits to about -26 to $-28‰$, characterizing intracontinetal reservoirs [74]. The contribution of terrestrial OM to the total organic carbon of marine sediments, evaluated on this basis, becomes negligible at a distance of 60-100 km offshore [8]. However, due to undercurrents, continental material may be deposited as much as 1 000 km from its point of origin. The increased C^{12} content of deep sea sedimentary OM does not always appear to be related to the contribution of organic terrestrial matter. This might be due to more profound transformations of the OM within the massive water column. Temperature variations both throughout the water column, and during geological time also play a part.

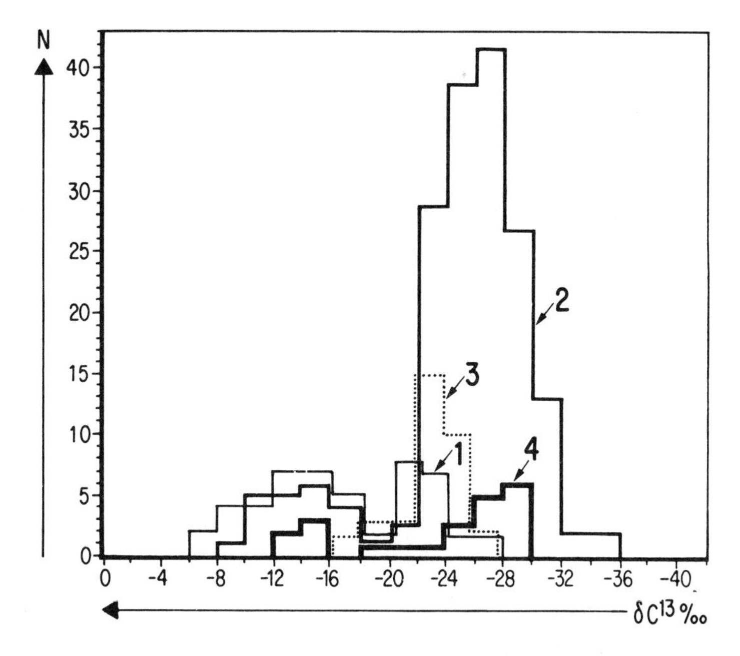

Fig. 9.7. — Carbon isotope composition of plants and humic acids from burial OM in marine and continental environments:

1. Marine plants. **2.** Land plants. **3.** Humic acids from marine sediments. **4.** Humic acids from soils and lake sediments.
δC^{13} – Values for humic acid samples are taken from: [65, 55, 7, 2, 57, 53]; marine and land plants data are taken from: [70, 79].

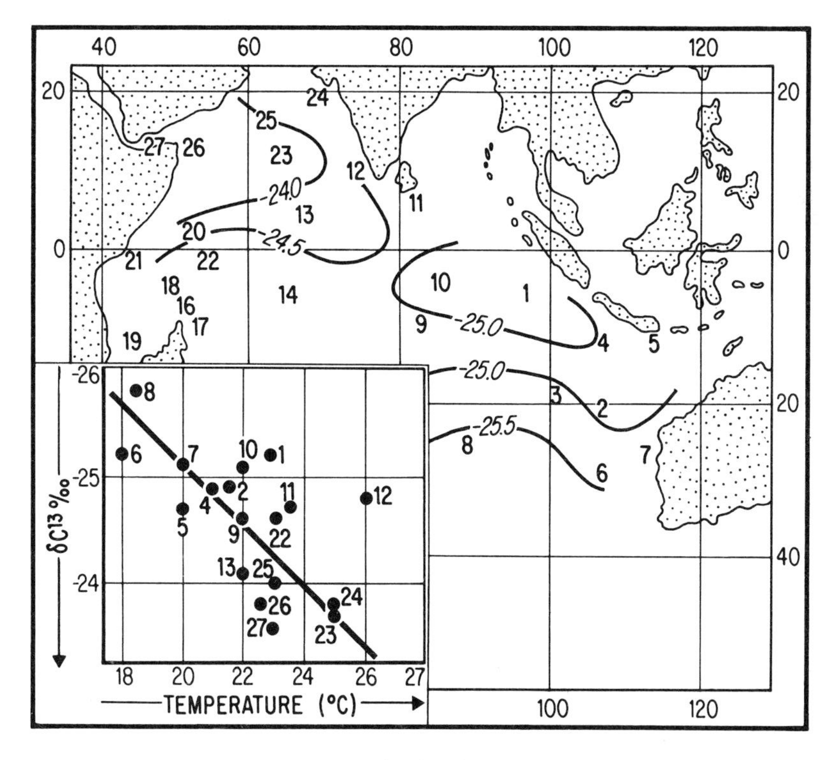

Fig. 9.8. — Temperature dependence of carbon isotope composition of the lipid fraction of plankton collected in Indian Ocean [35]. The figures mark the sites of sampling.

The theoretical temperature dependance of the thermodynamic carbon isotope effects for fossil OM is about 0.2-0.4‰ per 1° C; the same values are observed for naturally-occurring carbon compounds [29].

Attention has been paid before to correlations between the isotopic composition of organisms and their environmental temperature [15, 72] but correlations were ascribed to the influence of indirect factors. We have found [8] a definite correlation between the carbon isotope composition of the lipid fraction of plankton, collected at different latitudes in the Indian Ocean, and the corresponding average annual water temperature (Fig. 9.8).

The dependance of the isotopic composition of sedimentary OM on paleo-temperature, and its relationship with the history of glaciation was noted by Rogers and Koons [68] who showed that the organic carbon of periods of glaciation is 1-2‰ lighter. The ideas of Rogers *et al.* [68, 67] were criticized as being speculative [2] or as being valid in specific circumstances only [73]. These criticisms are invalid since the dependence of the isotopic composition of kerogen on paleotemperature may be a consequence of the thermodynamic nature of the biological isotope effects for which the temperature dependence is a fundamental property. Another thing is that the temperature dependence may be strongly masked by other factors. In some cases the dependence of the organic carbon isotope composition on temperature may be indirect [14, 15]; in several experiments this temperature dependence was not revealed [10]

It is known that the so-called C_4-plants, fixing CO_2 through phosphoenolpyruvate (Hatch-Slake's cycle) are enriched to a lesser extent in the C^{12}-isotope when compared with the usual C_3-plants fixing CO_2 through ribulose 1.5 diphosphate (Calvin cycle). If δC^{13}-values of the usual terrestrial plants are taken to be in the − 22 to − 28‰ range then the δC^{13}-values of the C_4-plants (and succulents) range from − 13 to − 19‰ [5, 79, 86]. The lower (left) maximum in the histogram for land plants (Fig. 9.7) corresponds to the C_4-plants. The low water and gas transfer and, in consequence, high degree of retention and utilization of CO_2 is a characteristic of these plants. Hence the χ-values is small during the biosynthesis of the primary products of CO_2-assimilation. Indeed, the isotope composition of the aspartic and malic acid carbon and other primary products of C_4-plant photosynthesis, was found to be close to the isotope composition of the initial carbon dioxide [86]. Carbon in the soil at sites where mainly C_4-plants grow (e.g. maize fields) is pronouncedly depleted in the light isotope as compared with the usual soil carbon (the left maximum in the histogram for soil humic substances, Fig. 9.7).

IV. DIAGENETIC ALTERATION OF ORGANIC CARBON ISOTOPE COMPOSITIONS DURING KEROGEN FORMATION

When an organism dies the mechanism providing the thermodynamically ordered isotope distribution ceases. Therefore, the isotopic distributions inherited from the various biological systems are preserved in the original molecules of the buried organic compounds. When subsequent bond formation, or bond breaking, occurs isotope fractionation may take place owing to the greater lability of C^{12}-C^{12} compared to C^{12}-C^{13}-linkages. This type of isotope fractionation is known as the kinetic effect. The more complete the chemical reconstruction of the original biomolecules the poorer is the relationship between the isotopic composition and the β-factor value of the compound.

Humic substances in recent sediments are considered to be early form of kerogen and are usually defined as dark-brown polymers that may be extracted from soils and sediments by dilute alkaline solutions. The alkaline extracts is usually subdivided into two fractions: alkali-soluble but acid-insoluble humic acids (HA), and fulvic acids (FA) which are alkali- and acid-soluble, the OM remaining in sediments after extraction by alkalis is called humin. Fulvic acids have lower molecular weights compared to humic acids. The percentage of native functional groups, as well as heteroatoms, in FA (C: 30-40%, H: 6-8%, O: 45-55%, N: 4.5-5.5%) is higher than in HA (C: 50-55%, H: 5.5.-6.5%, O: 30-35%, N: 1-4%) which themselves are less altered than humins (H). With increasing diagenetic transformation the fulvic acid content decreases at the cost of a rising humic acid content; that in turn drops as the humin content rises [7, 53, 54]. Indeed, the percentages of humic acids and humin in sediments are usually complementary.

The transformation of biomolecules begins in the water column and continues during sedimentation. As the first new polymers are formed there is an initial enrichment in the C^{12}-isotope relative to the plankton carbon [88].

Investigations of the carbon isotope composition of humic acids (Table 9.2) reveal, as a general rule, that humic acids are enriched in the C^{12}-isotope as compared with fulvic acids. Maturated humin carbon, in turn, is enriched in the C^{12}-isotope as compared with humic acids. These effects are more distinct for the organic carbon in sediments ($\Delta C^{13}_{HA-FA} = -1.9‰$) than for that in soils ($\Delta C^{13}_{HA-FA} = -0.7‰$). The formation of humic acids in soil is due to the transformation of lignin [45]. The formation of humic acids in marine sediments is supposed to be related to the reaction of proteins and carbohydrates. This is the so-called melanoidin reaction and involves the reaction of aldehyde and amino groups resulting in the formation of dark-brown substances similar to the humic acids of marine sediments [41, 44].

A relevant experiment on isotope fractionation during melanoidin formation was carried out in collaboration with Drozdova. A mixture of glucosamine and protein hydrolyzate was boiled under reflux (100° C) for 12, 18 and 30 hours. The melanoidin-forming character of this reaction has been demonstrated previously [17]. Data on experimental conditions, and results, are tabulated (Table 9.3) and show the following:

(a) Synthetic "fulvic" and "humic" acids are enriched in the C^{12}-isotope as compared to the starting carbon of glucosamine.

(b) Humic acids are richer in the C^{12}-isotope than the corresponding fulvic acids.

(c) Enrichement of humic acids in the C^{12}-isotope increases as the heating period increases, while enrichment of fulvic acid drops.

(d) Insoluble melanoidins ("humin") appeared to be the lightest fraction isotopically.

Thus the carbon isotope distribution in our experimental polymers coincides with the isotope distribution in the corresponding native polymers. This fact suggests that melanoidin formation is a sufficiently good model for the humidification of OM. The fact that fulvic acids are depleted in the C^{12}-isotope, as humic acids are enriched in it, indicates that fulvic acids are intermediate between a starting protein-carbohydrate complex and humic acids.

TABLE 9.2
CARBON ISOTOPE COMPOSITIONS OF HUMIC SUBSTANCES

Location, sample	δC^{13} (‰)			
	Fulvic Acid (FA)	Humic Acid (HA)	Humin (H)	ref.
MARINE SEDIMENTS				
Atlantic ocean, Mid-Atlantic Ridge, Hole 26 JOIDES, 5168 m water depth				
Depth 100 m		−24.7	−25.1	2
Depth 230 m		−25.8	−26.6	2
Depth 478 m		−24.4	−25.6	2
Average		−25.0	−25.8	
		$\Delta C^{13}_{H-HA} = -0.8$		
Pacific ocean				
Tanner Basin	− 20.3	− 22.0		1
Santa Barbara basin	− 19.1	− 22.7		1
Santa Cruz basin	− 20.7	− 21.8		1
San Pedro channel	− 21.7	− 22.7		1
Long Basin	− 20.3	− 22.3		1
Average	− 20.4	− 22.3		
	$\Delta C^{13}_{HA-FA} = -1.9$			
LITTORAL				
Hawaii, Kaneobe Bay,				
N° 1	− 23.3	− 24.2		1
N° 2	− 24.4	− 24.9		1
Florida, Humate cemented sand,				
N° 1	− 24.1	− 25.7		1
N° 2	− 24.1	− 25.7		1
California, Newport Marsh	− 17.4*	− 19.1*		1
Average	− 24.0	− 25.1		
	$\Delta C^{13}_{HA-FA} = -1.1$			
SOIL				
Nova Scotia, forest	− 26.3	− 26.2		1
Hawaii, Canefield	− 18.2*	− 14.8*		1
Israel				
Alluvial	− 26.8	− 28.1		4
Basaltic	− 27.8	− 28.4		4
Terra Rosa	− 28.6	− 29.4		4
Brown alluvial	− 26.1	− 27.0		3
Mountain rendzina	− 25.6	− 26.6		3
Valley rendzina	− 28.1	− 28.9		3
Arid	− 24.8	− 25.7		3
Peat	− 17.4	− 18.0		3
Hula peat	− 20.8	− 19.2		1
Canada, forest, Saanich Inlet	− 27.0	− 29.1		1
Average	− 26.8	− 27.5		
	$\Delta C^{13}_{HA-FA} = -0.7$			

(*) no taken into account while calculation of mean value
([1]) Nissenbaum and Kaplan [55]
([2]) Aizenshtat *et al.* [2]
([3]) Nissenbaum and Schallinger [57]
([4]) Nissenbaum [53]

Enrichment of the polymers in the C^{12}-isotope may be due both to the removal of the C^{13}-isotope-enriched functional groups (carboxyl,-formyl,-keto etc.) and to the kinetic isotope effect that accompanies the polymerization.

TABLE 9.3

δC^{13}-VALUES OF THE PRODUCTS OF MELANOIDINE REACTION

(Glucosamine 1 g, protein hydrolysate 0.012 g in the experiments 2 and 3, and 0.018 g in the experiment 1, H_2O – 15 ml, pH = 8.0)

Isotopic composition of the starting components:

glucosamine: – 26.14‰
protein hydrolysate: – 19.24‰

Experiment	Brown solution		Black precipitate			
	There is no precipitate with HCl. High-molecular substances are removed by dialysis		Dissolved in 1*N* NaON, precipitated in 3*N* HCl		Insoluble melanoidines	
	Fulvic Acids (FA)		Humic Acids (HA)		Humin (H)	
	δC^{13} (‰)	yield (%)	δC^{13} (‰)	yield (%)	δC^{13} (‰)	yield (%)
1. **Heating** for 12 h.	– 27.00	11.4	– 27.47	0.6	–	trace
2. **Heating** for 18 h.	– 26.68	10.7	– 29.19	2.0	–	0.33
3. **Heating** for 30 h.	– 26.32	3.6	–	trace	– 36.42	6.1

TABLE 9.4

ISOTOPE EFFECT OF POLYMERISATION REACTIONS

Reaction	δC^{13} (‰)		
	Monomer	Polymer	ΔC^{13}_{p-m}
Polymerisation of tetrahydrofurane...	– 34.15	– 37.30	– 3.15
Polymerisation of 1.3-dioxolane	– 28.97	– 33.20	– 3.23

When passing from fulvic acids to humic acids and then to higher molecular weight humic acids the percentage of functional groups decreases [48, 65, 75, 81]. Nissenbaum and Schalinger [57] indicated that the loss of carboxyl-groups was responsible for the increase in the C^{12}-isotope content of humic acids as compared to fulvic acids. Indeed, carbon atoms of carboxyl,-keto-groups and carbon atoms concerned with aldehyde- and hydroxy-groups are to be enriched in C^{13}-isotope. On the other hand, it is known that elimination of functional groups, specifically decarboxylation, results in the enrichment of the lost CO_2 in the C^{12}-isotope [6]. Thus, there is a competition of two trends and the net effect generally results although not always, in the elimination of CO_2, that is somewhat enriched in the C^{13}-isotope and consequently enrichment of the residual OM in C^{12}-isotope.

Polymerization seems to be one more possible source of isotopic fractionation during the diagenesis of OM. It follows, from general considerations, that the polymerization reactions should be accompanied by small isotope effects [50].

In cooperation with Berman and Klimov we have investigated two suitable polymerization reactions. The first of them involves the polymerisation of tetrahydrofuran:

$$\begin{array}{c} CH_2 - CH_2 \\ | \quad\quad | \\ CH_2 \quad CH_2 \\ \diagdown \ \diagup \\ O \end{array} \rightarrow \ldots - O - \begin{array}{c} CH_2 - CH_2 \\ | \quad\quad | \\ CH_2 \quad CH_2 \end{array} - \ldots \quad \ldots - O - \begin{array}{c} CH_2 - CH_2 \\ | \quad\quad | \\ CH_2 \quad CH_2 \end{array} - \ldots$$

the other, the formation of polydioxolane by polymerizing 1.3-dioxolane:

$$\begin{array}{c} CH_2 \quad CH_2 \\ | \quad\quad | \\ O \quad\quad O \\ \diagdown \ \diagup \\ CH_2 \end{array} \rightarrow \ldots - \begin{array}{c} CH_2 - CH_2 \\ | \quad\quad | \\ O \quad\quad O \end{array} - CH_2 - \ldots \quad \ldots - \begin{array}{c} CH_2 - CH_2 \\ | \quad\quad | \\ O \quad\quad O \end{array} - CH_2 - \ldots$$

Both reactions proceed with a catalyst at normal pressure and temperature (20° C). The yield of polymer (100-200 monomer units) is 20-25% and the reactions are not accompanied by the formation of artefacts. Therefore, a distinction between the isotope composition of the starting monomer and the resultant polymer may be ascribed only to the polymerization effect.

As seen in Table 9.4, the polymer is richer by nearly 3‰ in the light isotope. The extent of fractionation is comparable with that observed in the process of natural polymer formation in the series: fulvic acids-humic acids-humin.

The evolution of the isotopic composition of OM during diagenesis is illustrated schematically in Fig. 9.9. Obviously decomposition of biopolymers to monomers and repolymerization of the latter to resistant geopolymers is the leitmotiv of the early diagenetic transformation of sedimentary OM. The loss of the isotopically heavy CO_2, associated with carboxyl and other functional groups, as well as the polymerization isotope effect, specify the observed changes in the isotopic composition of the geopolymers, i.e. their enrichment in the C^{12}-isotope.

Humic acids are inclined to produce complexes with lipophilic compounds such as alkanes, fatty acids, pigments and other lipids [44, 59, 52, 64]. Therefore, the reaction of lipids and pigments with humic acids should result in a further enrichment of geopolymers in the C^{12}-isotope.

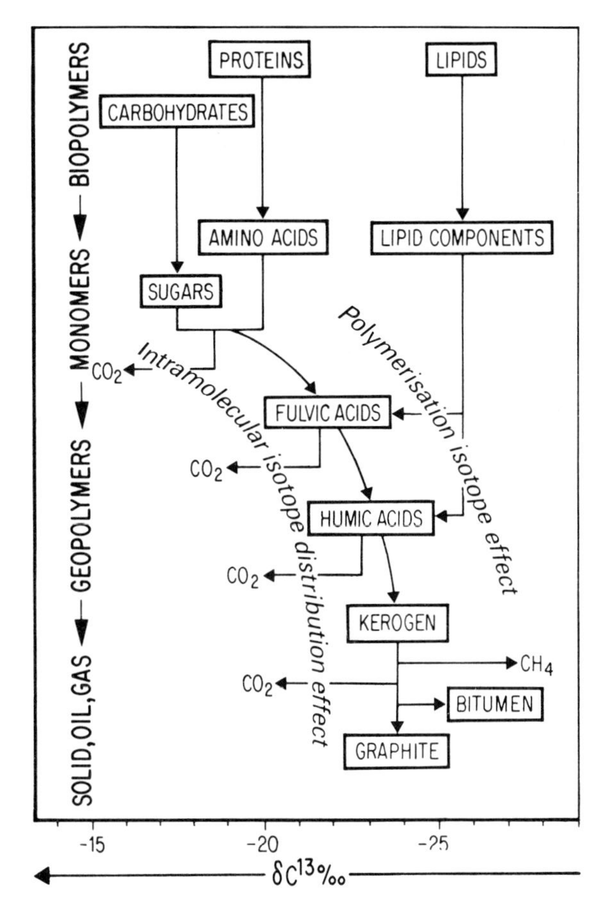

Fig. 9.9. — Sketch of diagenetical change of carbon isotope composition of OM on the pathway from biopolymers to kerogen.

V. RELATIONSHIP BETWEEN THE CARBON ISOTOPE COMPOSITION OF KEROGEN, CO_2 AND CH_4

Together with the degradation of polymeric biochemical compounds to monomers, followed by their repolymerization and stabilization in the form of kerogen, elimination of organic carbon occurs as a result of CO_2 and CH_4-formation. Diagenetic methane generation is usually related to reactions of the following two types [78, 11]:

$$\text{(a)}\ CH_3COOH \rightarrow CH_4 + CO_2$$

$$\text{(b)}\ CO_2 + 4H_2 \rightarrow CH_4 + 2H_2O$$

Methane-producing bacteria are obligate anaerobes and therefore methane forms after the oxygen reserve is completely exhausted, usually beneath the sulphate-reducing zone.

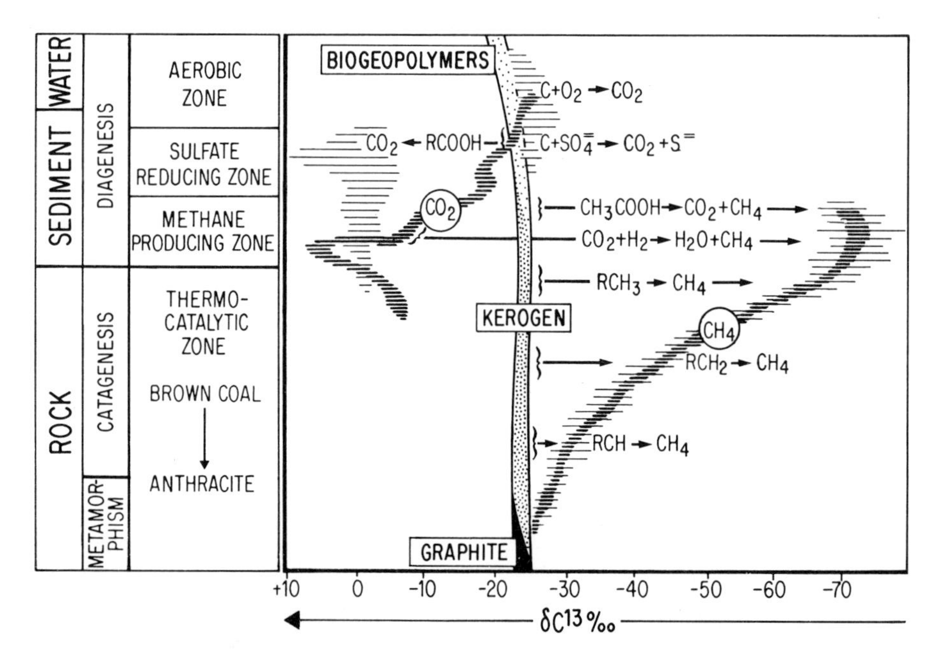

Fig. 9.10. — Relationship between carbon isotope composition of kerogen, CO_2 and CH_4 in different zones.

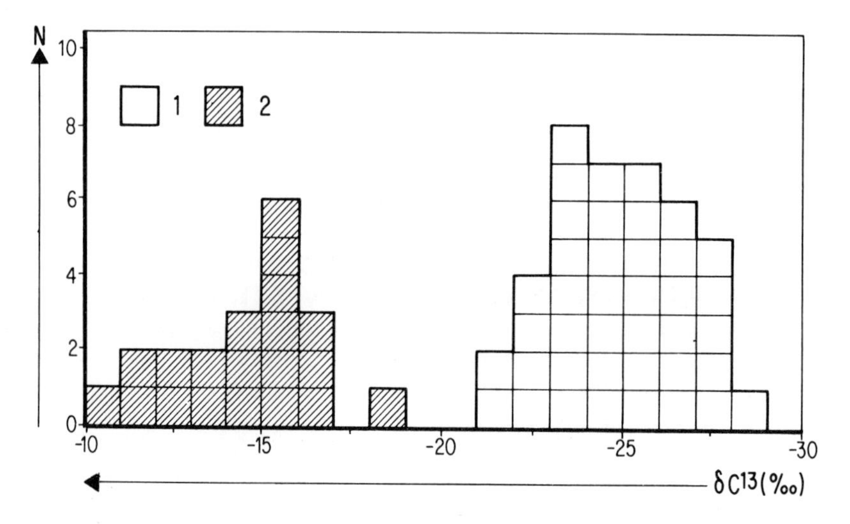

Fig. 9.11. — Distribution of isotope composition of carbon of soil carbon dioxide:

1. Location where predominantly grow C_3-plants [22]. **2.** Locations where predominantly grow C_4- and CAM-plants [45].

Measurements of the carbon isotope composition of methane in anaerobic environments [58, 11, 82], as well as in laboratory experiments [51, 69, 36], reveal that microbiologically-produced methane is, on average, 30-50‰ richer in the C^{12}-isotope as compared with other OM (Fig. 9.10).

Enrichment of methane in C^{12} is usually explained by the kinetic isotope effect. Nevertheless, it is logical to assume (keeping in mind the essential role of the thermodynamic isotope effects in biological systems) that the enrichment in C^{12} of the microbiological methane has a thermodynamic basis. In other words, the effect is due to isotope exchange in the CO_2—CH_4-system used by bacteria for methane formation. The calculated values for thermodynamic isotope effects in CO_2—CH_4-systems, at 20-40° C, are in good agreement with the isotope fractionation observed [24].

There are several CO_2 sources in sediments and rocks, two of which are of biological origin. The first results from the oxidation of OM while the second results from its decomposition:

$$\text{(a)}\ C_xH_yO_z + O_2 \rightarrow xCO_2 + \frac{y}{2}H_2O$$

$$\text{(b)}\ C_xH_yO_z \rightarrow C_{x-1}H_yO_{z-2} + CO_2$$

Carbon dioxide, resulting from the oxidation of OM, has an isotopic composition related to the average carbon isotopic composition of the OM. This is obviously true for both sediments and soils. δC^{13}-values for soil carbon dioxide range from -23 to -28‰ [22] with the exception of that which is isotopically heavier [47] from sites occupied by C_4-plants or CAM-type [1] plants (Fig. 9.11).

In anaerobic environments, carbon dioxide may be produced from oxygen-containing groups (e.g. carboxyl, formyl) depleted in the C^{12}-isotope. The balance of these types of carbon dioxide, and that of carbonate origin, determines the carbon isotope composition of the CO_2 in sediments (Fig. 9.10).

Carbon dioxide may be noticeably depleted in the C^{12}-isotope in the zone where methane-producing bacteria occur, as a result of the reaction:

$$CO_2 + H_2 \rightarrow CH_4 + H_2O$$

This was shown in experiments with the microorganism *Methanobacterium thermoautotrophicum* [36].

The isotopic composition of the CO_2 and CH_4 is determined by isotope redistribution in the CO_2—CH_4-system. Therefore carbon isotope fractionation in the course of gas-formation should not markedly affect the carbon isotope composition of kerogen.

VI. δC^{13} OF KEROGEN DURING CATAGENESIS

A. Sediments.

In the upper part of marine sediments a trend favouring an increase in the light carbon isotope, with depth of burial of the OM, is sometimes observed [46, 18, 77]. This may be due

(1) CAM-type plant = plant with the crassulacean acid metabolism.

to changes in the isotopic composition of the OM during diagenesis. However, in many cases the alteration is not systematic. To date, information is available on δC^{13}-variations in OM buried to several hundred metres in deep-sea sedimentary successions [9, 21, 67]. These variations seem to reflect alterations in the sedimentary environment, the temperature of the basin and the carbon source rather than process related to kerogen transformation.

Thus, on the strength of all the evidence, one may conclude that the isotopic composition of the kerogen is reached at an early stage of diagenesis, evidently after the humic carbon has entered the kerogen structure.

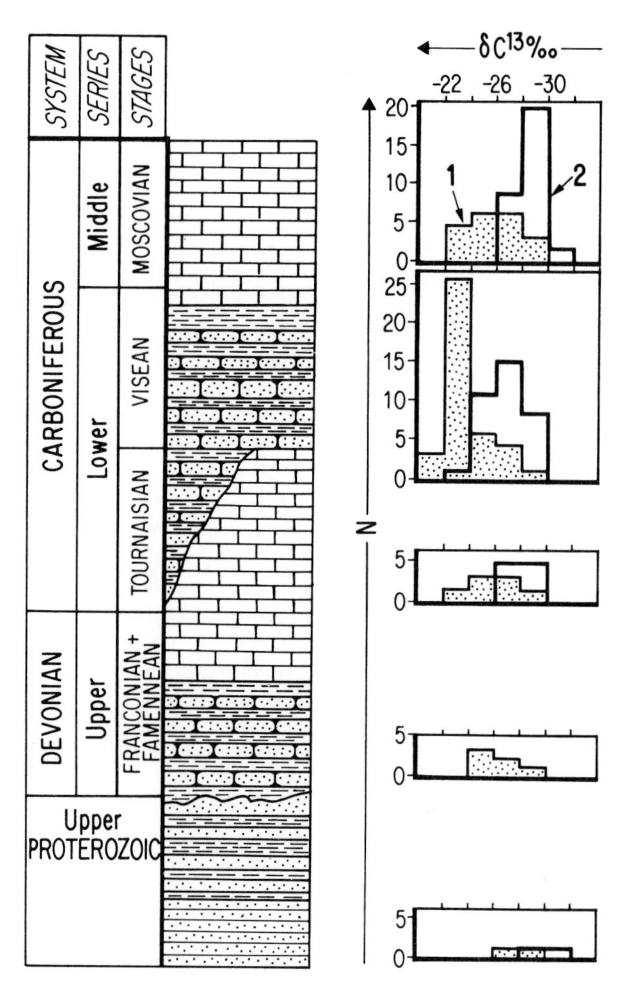

Fig. 9.12. — Distribution of isotope composition of carbon of kerogen [1] and chloroform-extracted bitumen [2] based on stratigraphic profiles of the Permian-Ural Area [24].

B. Ancient rocks.

Data on the carbon isotope composition of kerogens in Paleozoic deposits of the Volga-Ural area (the Russian Platform) are shown in Fig. 9.12. Some of the data are taken from previous work [24] but the rest are new and are listed in Table 9.5. As can be seen, the

TABLEAU 9.5
CARBON ISOTOPIC COMPOSITION OF KEROGENS AND BITUMENS(1)
FROM PALEOZOIC DEPOSITS OF THE PERMIAN URAL AREA

Stratigraphic complex	Area, borehole (No)	Sampling (depth:m)	Rock	Content in the rock (weight%)		δC^{13} (‰)	
				Kerogen	Bitumen	Kerogen	Bitumen
C_2vr Moscovian stage, Verejskyi horizon	Durinskaya, 12(2)	1 918	lm (3)	0.20	0.625	− 26.7	− 30.0
	Nozhovskaya, 36	1 117	lm	0.21	0.001	− 25.7	− 27.8
	Tartinskaya, 31	1 082	lm	0.54	0.313	− 23.8	− 27.5
	Andreevskaya, 23	1 163	sh (3)	0.36	0.003	− 25.9	− 26.8
	Tartinskaya, 31	1 115	sh	1.56	0.015	− 23.9	− 28.6
C_2b Bashkirskyi stage	Tazovskaya, 45	1 609	sh	0.71	0.080	− 23.5	− 27.0
	Tartinskaya, 19	1 151	lm	0.06	0.003	− 27.9	− 30.
C_1tl Visean stage, Tulskiy horizon	Tukachevskaya, 7	1 718	lm	0.43	0.010	− 26.3	− 29.1
	Tukachevskaya, 7	1 724	sh	0.56	0.010	− 24.8	− 26.8
	Tukachevskaya, 7	1 735	sh	1.35	0.010	− 27.3	− 29.6
	Romanshorskaya, 1	1 772	sh	0.57	0.060	− 24.5	− 27.9
	Vidrjanskaya, 39	1 998	sh	0.71	0.080	− 23.4	− 27.2
	Yayvinskaya, 4	2 327	lm	0.23	0.113	− 23.3	− 25.2
	Tazovskaya, 45	2 014	sh	0.85	0.080	− 22.9	− 24.7
	Tazovskaya, 45	2 014	sh	0.64	0.118	− 22.8	− 25.4
	Koltskovskaya, 1	1 463	sh	0.96	0.040	− 23.2	− 26.6
	Andreevskaya, 23	1 603	sh	1.09	0.010	− 22.8	− 24.9
	Gondyrevskay, 60	1 405	sh	1.04	0.020	− 24.1	− 26.2
	Gondyrevskaya, 60	1 409	sh	0.80	0.001	− 22.4	− 26.7
	Tartinskaya, 31	1 479	sh	1.60	0.015	− 22.9	− 27.3
	Tartinskaya, 31	1 512	sh	1.11	0.010	− 21.5	− 27.1
	Dmitrievskaya, 5	1 774	sh	13.23	0.080	− 22.5	− 24.0
C_1bb Visean stage, Bobrikovskyi horizon	Gondyrevskaya, 60	1 440	sh	2.19	0.080	− 25.6	− 29.2
	Olkhovskaya, 86	1 870	sh	3.00	0.080	− 23.5	− 24.7
	Yayvinskaya, 4	2 499	sh	0.82	0.030	− 23.7	− 27.5
	Nozhovskaya, 36	1 499	lm	0.23	0.010	− 28.2	− 29.2
	Tartinskaya, 19	1 531	sh	0.73	0.010	− 22.3	− 25.9
C_1mn Tournaisian stage	Dmitrievskaya	1 788	sh	1.09	0.040	− 24.5	− 27.2
	Dmitrievskaya, 5	1 809	sh	1.13	0.060	− 23.2	− 26.4
	Nozhovskaya, 36	1 527	lm	0.65	0.010	− 25.4	− 27.3
C_1t	Andreevskaya, 23	1 818	lm	0.50	0.010	− 26.6	− 28.0
	Koltsovskaya, 1	1 515	sh	0.64	0.040	− 27.4	− 29.1

(1) Chloroform extract
(2) Situation of the fields is pointed elsewhere [24]
(3) lm: limestone
sh: shales and sandstones

carbon isotope composition of kerogens in ancient deposits varies in the same range as in recent sediments. Obviously, no essential change in the isotope composition of kerogen occurs during catagenesis. This conclusion is confirmed from investigations of coals which show that their isotope composition is unrelated to their rank, or degree of metamorphism [13, 37, 83, 23].

Artificial coalification experiments reveal a small (0.5-1‰) enrichment in the heavy isotope as the degree of coalification increases [3, 38] but this effect has not been observed in nature.

C. Kerogens and bitumens.

During the catagenic maturation of OM, hydrocarbons and other compounds of low polarity are liberated. These compounds which can be extracted with organic solvents (bitumen fraction) are usually isotopically lighter than the kerogen carbon of the same sample. This difference is almost certainly from the biological source. Extractable carbon derives predominantly from kerogen moieties rich in C—H-linkages which contain relatively light carbon because of the thermodynamically ordered isotope distribution in biological precursors. Lipids are richest in C—H-linkages and are considered to be the most probable precursor of oil hydrocarbons [77].

The isotopic compositions of kerogens and associated bitumens have been investigated and the correlation shown (Fig. 9.13) indicates their genetic relationship.

The isotopic composition of oil does not always correlate with that of the kerogen from associated beds (Fig. 9.14). This is understandable, since, before its accumulation, oil

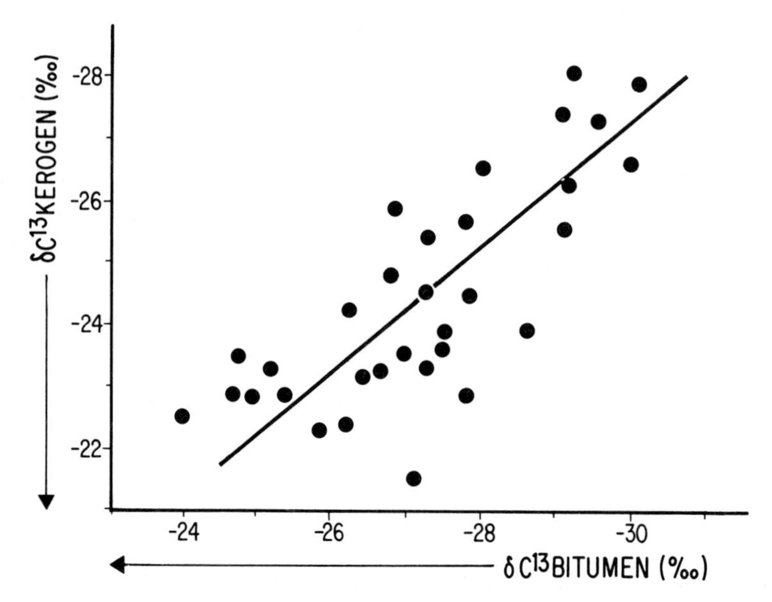

Fig. 9.13. — Relationship of carbon isotopic composition in kerogen and bitumen from the same core in Permian-Ural Area.

migrates, sometimes over long distances, both horizontally and vertically. For example, in the Permian region of the Volga-Ural Basin, the oil of the jasnopoljanskiy horizon (Lower Carboniferous) is one of the isotopically lightest in the section whereas the kerogen, in corresponding deposits, is isotopically heavier than in other intervals of the section [24]. On the other hand sometimes there is a positive correlation between the isotopic correlation of kerogen, bitumens and oil [87].

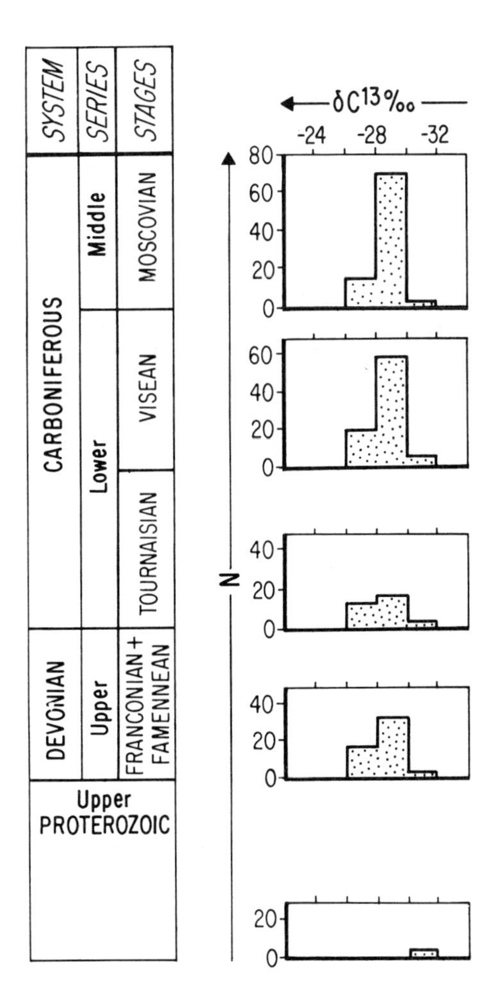

Fig. 9.14. — Distribution of isotopic composition of carbon of oils based on stratigraphic profiles of the Permian-Ural Area [24].

Obviously the correlation (or lack of it) in the isotopic composition of kerogen and oil, in certain sections, may serve as a basis for reconstructing the picture of oil migration within the formation.

D. Kerogen and natural gas.

Kerogen is the source of methane and other gaseous hydrocarbons in sedimentary rocks. The gases are formed by radical or thermocatalytic fragmentations which involve several types of isotope effects; these are reviewed in detail elsewhere [24]. Here all that we need to note is the relationship that exists between the isotopic composition of the gas and the intramolecular isotopic distribution in the kerogen. Methane forms by hydrogen disproportionation of the kerogen structure. The peripheral groups, rich in C—H-linkages such as CH_3-groups, appear to be cleaved first. As the CH_3-groups are exhausted structures less rich in C—H-linkages become the source of the methane carbon. Because of the thermodynamically ordered isotope distribution, methane formed from CH_3-groups ($\beta_{CH_3} = 1.131$) is isotopically lighter than methane formed from CH_2-moieties ($\beta_{CH_2} = 1.149$), CH-moieties ($\beta_{CH} = 1.168$) and so on. Thus, the intramolecular carbon isotope distribution in the kerogen determines carbon isotope composition of the methane (Fig. 9.10). A number of regularities in the isotopic distribution in gases result from this dependence, for instance the decrease in the C^{12} content of methane, with depth, [28, 71], the correlation between the carbon isotope composition of methane and vitrinite reflectance [80], and the occurrence of isotopically heavy methane (– 15 to – 25‰) in anthracites [12, 83]. A method identifying gas migration pathways was developed on this basis [24, 33].

VII. δC^{13}-VARIATIONS IN KEROGEN THROUGH GEOLOGICAL TIME

A gradual decrease in the enrichment in the C^{12}-isotope in passing from ancient to recent Phanerozoic kerogens has been noted [40,85]. Better data may result however by studying the alteration in the isotopic composition of kerogens in a regional Phanerozoic sequence, as was carried out for the Russian Platform [32]. Representative samples of each geological age of the sedimentary section were prepared by mixing several hundred specimens from different areas. The isotopic compositions of the bitumens and kerogens were determined separately. The δC^{13}-variation of organic carbon is shown in Fig. 9.15. Variations of certain other parameters, characterizing the sedimentary environment throughout the Phanerozoic of the Russian Platform, are shown in Fig. 9.16. There does not appear to be any distinct dependence of the organic carbon isotope composition on the lithofacies, sedimentation rate, concentration of organic carbon and so on. Variations in the isotopic composition of organic and carbonate carbon do not correlate, but the agreement between the carbonate values and other geochemical parameters is better than that of the organic carbon.

Two interesting features concerning the δC^{13}-variation of the organic carbon with geological time are:

(a) There is a sharp depletion in the C^{12}-isotope in the kerogens and bitumens in the Carboniferous and,

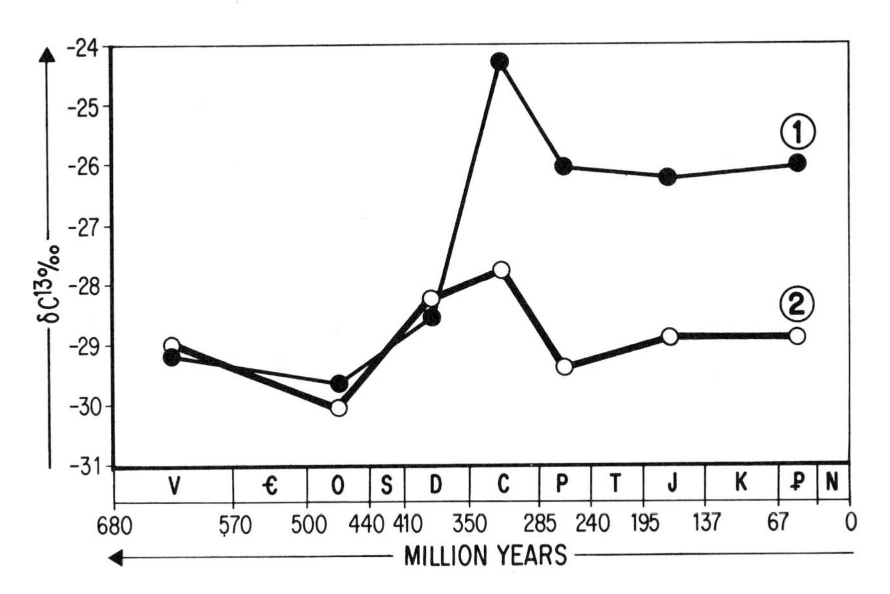

Fig. 9.15. — Dependence of the mean isotopic composition of carbon in:

1. Kerogen. **2.** Chloroform-extracted bitumen on geologic age of rocks in the Phanerozoic sequence of the Russian Platform [32].

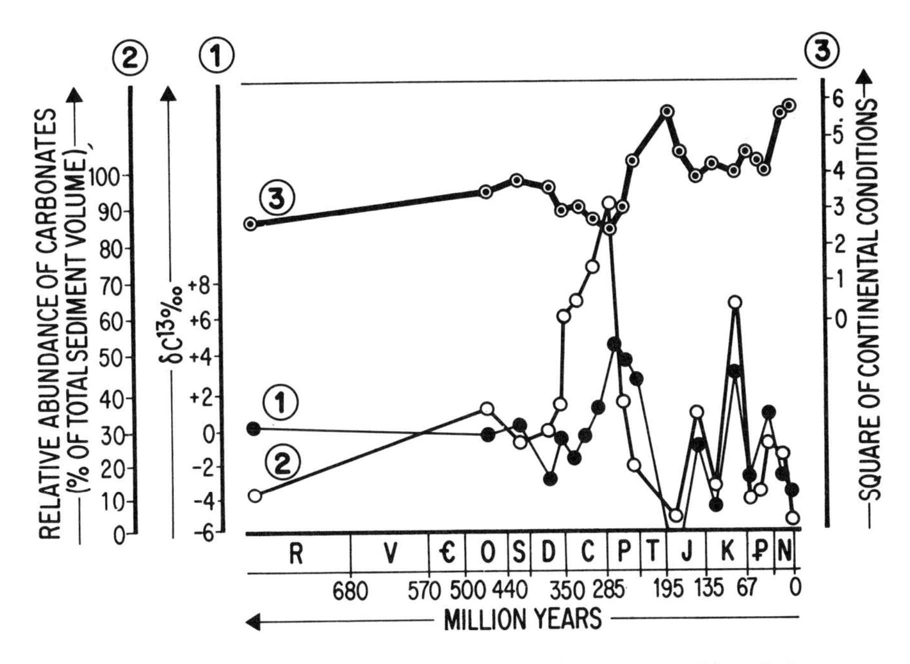

Fig. 9.16. — Comparison of variations of the carbon isotope composition of phanerozoic carbonates with the carbonate distribution in the deposits of each geological age and with the existence of terrestrial environment within the Russian Platform.

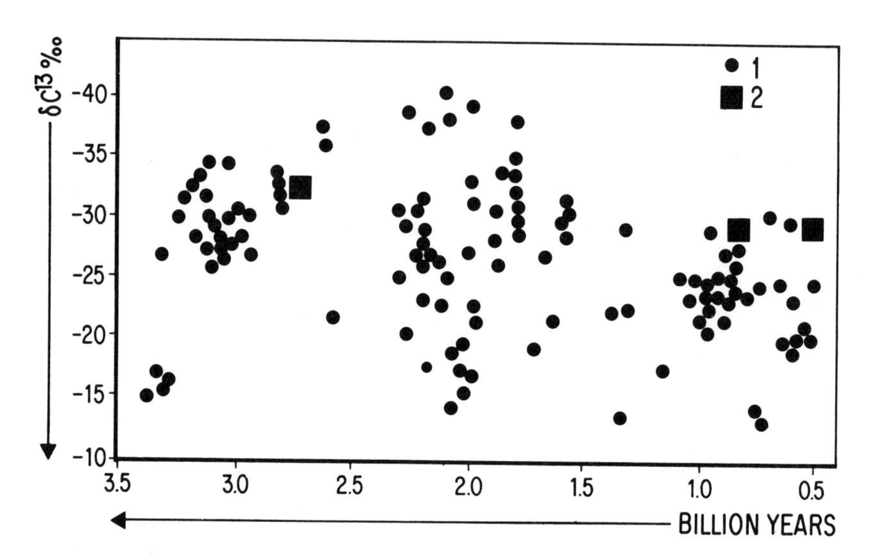

Fig. 9.17. — Carbon isotope composition of Precambrian kerogens:

●: denote results of separate measurements [40, 59, 19]. ■: mean-values [32].

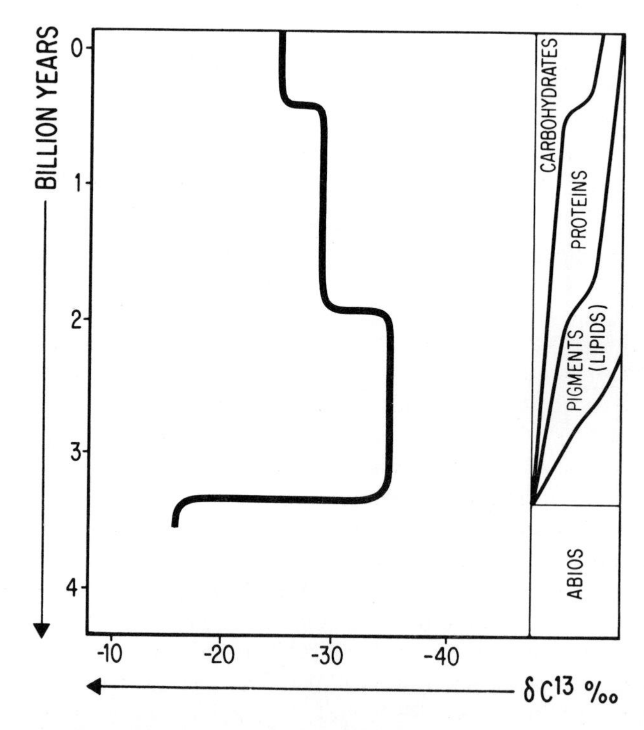

Fig. 9.18. — Generalized curve of alteration of isotope composition of organic carbon through geologic time on the ground of supposed evolution of the biochemical composition of living organisms.

(b) The differences between the isotopic composition of the kerogens and bitumens appears first in the Carboniferous.

These facts are apparently related, in some way, to the appearance of land vegetation.

Investigations of Precambrian kerogens have revealed that their carbon is enriched in the light isotope to nearly normal "biological" levels. This fact, together with the ubiquity of reduced carbon in the Precambrian [76] indicates an early formation of the biosphere. Data on the isotopic composition of Precambrian kerogens are summarized in Fig. 9.17. Isotope distributions in Precambrian organic carbon shows two characteristic features.

The first is the relative enrichment of Precambrian kerogen in the C^{12}-isotope. δC^{13}-values of kerogens older than 2 billion years range from -28 to $-38‰$, whereas for Phanerozoic kerogen these values range from -23 to $-30‰$; this was first noted by Hoering[40].

The second consists in the anomalous C^{12}-isotope depletion in the most ancient sediments. Shopf *et al.*[60] showed that kerogen from the Theespruit formation (Onverwacht, Swaziland, South Africa; $3.5 \times 10^9 yr$) was characterized by δC^{13}-values of about $-16‰$. These authors believe that this anomaly may record the transition from abiogenic to biogenic carbon, or the time at which the mechanism of biological isotope fractionation first appeared.

Several assumptions have been made regarding the C^{12}-isotope enrichment value of early Proterozoic kerogens. Epstein [20] assumed that it was related to the low abundance of biogenic carbon at that time which led to a corresponding displacement of the isotope balance in favour of a C^{12}-rich atmospheric CO_2. Degens [14] suggested that isotopically light organic carbon in the Precambrian might be related to a Precambrian sea characterized by lower pH-values and higher CO_2-concentrations. Pardue *et al.* showed, [61], from blue-green algal studies, that under certain conditions, (specifically a decreased cell concentration in the medium) the isotopic fractionation increased; this, they supposed, could account for isotopically light Precambrian carbon.

Schidlowski and cowokers [19, 43] note, and discuss, the enrichment in the C^{12}-isotopic composition of early Proterozoic kerogens as well as the C^{13}-isotopic enrichment of ancient Archean kerogens. They consider that the δC^{13}-variations of Precambrian organic carbon are, in general, in the range characteristic of Phanerozoic OM and that they are related to variations in the ratio of carbonate to organic carbon through geological time. However, this explanation requires synchronous δC^{13}-variations for carbonate and organic carbon; this does not seem to have been so. Variations in the isotopic composition of Precambrian carbonates have been found to correlate well with major tectonic events in the Earth's history [26, 32], but for organic carbon no such correlation has been found.

Nevertheless, in our view, variations in the δC^{13}-values of organic carbon throughout geological time are not random and reflect turning-points in the evolution of the biosphere. Variations in this value throughout the Precambrian and Phanerozoic are shown schematically in Fig. 9.18, with sudden changes at 3.5 billion, 2 billion and 0.4 billion years. The last corresponds to the appearance of land vegetation and a raising of the O_2-concentration in the atmosphere to present levels. The second abrupt change coincides with changes in the sulphur isotope composition and possibly corresponds to the appearance of sulphate-reducing bacteria. Fig. 9.18 also attempts to convey the idea that the changes may be related to milestones in the historical evolution of the principal biochemical constituents of organisms; discussion of this problem, however, is out with the framework of this paper.

REFERENCES

1. Abelson, P. H. and Hoering, T. C., (1960), *Carnegie Inst. of Washington*, **59**, 158.
2. Aizenshtat, Z., Baedecker, M. J. and Kaplan, I. R., (1973), *Geochim. et Cosmochim. Acta* **37**, 1881.
3. Bajor, M., Roquebert, M. and Van der Weide, B. M., (1969), *Bull. Centre Rech. Pausna*, **3**, 1, 113.
4. Behrens, E. W. and Frishman, S. A., (1971), *J. Geol.*, **79**, 1, 94.
5. Bender, M. M., (1968), *Amer. Jour. Sci. Radiocarbon Suppl.*, **10**, 468.
6. Bigelseisen, J. and Friedman, L. I., (1949), *Chem. Phys.*, **17**, 998.
7. Brown, F. S., Baedecker, M. J., Nissenbaum, A. and Kaplan, I. R., (1972), *Geochim. et Cosmochim. Acta*, **36**, 1185.
8. Calder, J. A. and Shultz, D. J., (1976), *Geochim. et Cosmochim. Acta*, **40**, 381.
9. Calder, J. A., Horvath, G. H. and Shultz, D. J., (1974), *in*: *Initial Reports of Deep Sea Drilling Project 1972*, US Government Printing Office, Washington, **26**, 613.
10. Calder, J. and Parker, P. L., (1973), *Geochim. et Cosmochim. Acta*, **37**, 133.
11. Claypool, G. E. and Kaplan, I. R., (1975), *in*: *Natural gases in marine sediments*, Kaplan, I. R., ed., New York, Plenum, 99.
12. Colombo, U., Gazzarini, F., Gonfiantini, R., Tongiorgi, E. and Caflish, L., (1969), *in*: *Advances in Organic Geochemistry, 1968*, Schenck, P. A., Havenaar, I., ed., Pergamon Press, 499.
13. Compston, W., (1960), *Geochim. et Cosmochim. Acta*, **18**, 1.
14. Degens, E. T., (1969), *in*: "Organic Geochemistry", Eglinton, G. and Murphy, M. T. J., ed., Springer Verlag, Berlin, 304.
15. Degens, E. T., Guilard, R. R. L., Sackett, W. M. and Helleburst, J. A., (1968), *Deep Sea Research*, **15**, 1.
16. Djuricic, M. V., Vitorovic, D., Andersen, B. D., Hertz, H. S., Murphy, R. C., Preti, G. and Bieman, K., (1972), *in*: *Advances in Organic Geochemistry, 1971*, von Gaertner, H. R. and Wehner, H. ed., Pergamon Press, 305.
17. Drozdova, T. V., (1957), *Biokhimiya* (in Russian), **22**, 3, 487.
18. Eckelman, W. R., Broecker, W. S., Whitlock, D. W. and Allsup, J. R., (1962), *AAPG Bull.*, **46**, 5, 699.
19. Eichman, R. and Schidlowski, M., (1975), *Geochim. et Cosmochim. Acta*, **39**, 585.
20. Epstein, S., (1969), *Calif. Inst. Techn., Contrib.* n° **1572**, 5.
21. Erdman, I. G., Schorno, K. S. and Scalan, R. S., (1974), *in*: *Initial Reports of Deep Sea Drilling Project, 1972*, US Government Printing Office, Washington, **24**, 1168.
22. Galimov, E. M., (1966). *Geokhimiya* (in Russian), **9**, 1110.
23. Galimov. E. M., (1968), "Geochimie des Isotopes du Carbone", Nedra Press, Moscow, translation CEA 2534-7.
24. Galimov, E. M., (1973), "Carbon Isotopes in Oil and Gas Geology". Nedra Press, Moscow, NASA translation TT F-682. Washington.
25. Galimov, E. M., (1974), *in*: *Advances in Organic Geochemystry, 1973*, Tissot, B., and Bienner, F., Editions Technip. Paris, 439.
26. Galimov, E. M., (1976), *in*: *Environmental biogeochemistry*, Nriagu, J. ed., Ann Arbor Sci., **1**, 3.
27. Galimov, E. M., (1976), *in*: "Science and Mankind", Nauka Press, Intern. Yearbook, 158.
28. Galimov, E. M., (1969), *Ziet. Angew. Geol.*, **15**, H.2, 63.
29. Galimov, E. M., "The origin of biological isotope effects", Nauka Press, Moscow, 280, in press.
30. Galimov, E. M., Kodina, L. A. and Generalova, V. N., (1976), *Geokhimiya* (in Russian), **1**, 11.
31. Galimov, E. M., Kodina, L. A., Generalova, V. N. and Bogacheva, M. P., (1977), *Proceeding 8-th Internat Congress Organic. Geochemistry, Moscow.*
32. Galimov, E. M., Migdisov, A. A. and Ronov, A. B., (1975), *Geokhimiya* (in Russian), **3**.
33. Galimov, E. M., Teplinsky, G. I., Tabassaransky, Z. A. and Gavrilov, E. Y., (1973), *Geokhimiya* (in Russian), **9**.
34. Galimov, E. M. and Shirinsky, V. G., (1975), *Geokhimiya* (in Russian), **3**, 503.
35. Galimov, E. M., Shirinsky, V. G., Bordovsky, O. K., and Zaikin, V. G., (1975), *Geokhimiya* (in Russian), **6**, 895.
36. Games, L. M. and Hayes, J. M., (1976), *in*: *Environmental Biogeochemistry*, Nriagu, J. ed., Ann Arbor Sci, **1**, 51.
37. Garcia-Loygorri, A., Bosch, B. and Marce, A., (1974), *in*: *Advances in Organic Geochemistry, 1973*, Tissot, B., et Bienner, F., ed., Technip, Paris, 859.
38. Geissler, C. and Belau, L. (1971), *Zeit. Angew. Geol.*, **17**, 1/2, 13.

39. Hoefs, J. and Schidlovski, M., (1967), *Science,* **155**, 1096.
40. Hoering, T. C., (1967), *in: Researches in Geochemistry*, Abelson. P. ed., **2**, Wiley.
41. Hoering, T. C., (1973), "Ann. Rep. of the Director Geophys. Lab. Carnegie Instit. Y. B.", **72**, 682.
42. Jacobson, B. S., Smith, B. N., Epstein, S. and Laties, G., (1970), *J. General Physiol.*, **55**, 1, 1.
43. Junge, C. E., Schidlowski, M., Eichmann, R. and Pietrek, (1975), *J. Geophys. Res.*, **80**, 4552.
44. Kalle, K., (1966), *in: Ocean Mar. Bio. Ann. Rev.*, Barnes, H., ed., **4**, 91.
45. Kononova, M. M., (1963), "Soil Organic Matter". Acad. Sci. Press, Moscow (in Russian).
46. Landergren, S., (1954), *Deep-Sea Research*, **1**, 98.
47. Lerman, J. C., (1972), *Proceedings 8th Internat. Conference Radio Carbon Dating, Willington, New Zealand, 1972*, **D93**.
48. Martin, F. Dubach, P., Mecta, N. C. and Deuel, H., (1963), *Z. Pflanzenernähr., Düng., Bodenkund*, **103**, 27.
49. Meinshein, W. G., Rinaldi, G. G. L., Hayes, J. M. and Schoeller, D. A., (1974), (preprint).
50. Melander, L. (1960), "Isotope Effects on Reaction Rates", Ronald Press, New York.
51. Nakai, N., (1961), *J. Earth Sci., Nagoya Univ.*, **9**, 59.
52. Neyroud, J. A. and Schnitzer, M., (1975), *Fuel*, **54**, 17.
53. Nissembaum, A., (1974), *in: Advances in Organic Geochemistry, 1973*, Tissot, B. and Bienner, F. Editions Technip, Paris.
54. Nissenbaum, A., Baedecker, M. J. and Kaplan, I. R., (1972), *in: Advances in Organic Geochemistry, 1971*, von Gartner, H. R., and Wehner, H., ed., Pergamon Press, Oxford, 427.
55. Nissenbaum, A. and Kaplan, I. R., (1972), *Limmology and Oceanography*, **17**, 4, 570.
56. Nissenbaum, A., Presley, B. J. and Kaplan, I. R., (1972), *Geochim. et Cosmochim. Acta*, **36**, 1007.
57. Nissenbaum, A., Schallinger, K. M., (1974), *Geoderma*, **11**, 137.
58. Oana, S. and Deevey, E. S., (1960), *Amer. J. Sci.*, **258-A**, 253.
59. Ogner, G. and Schnitzer, M., (1971), *Science*, **170**, 3955, 317.
60. Ohler, D. Z., Schopf, J. M. and Kvenvolden, K. A., (1972), *Science*, **175**, 1246.
61. Pardue, J. W., Scalan, R. S., van Baalen, Ch. and Parker, P. L., (1976), *Geochim. et Cosmochim. Acta*, **40**, 309.
62. Park, R. and Epstein, S., (1961), *Plant. Physiol.* **36**, 133.
63. Parker, P. L., (1964), *Geochim. et Cosmochim Acta*, **28**, 1115.
64. Philip, R. P. and Calvin, M., (1975), *in: Environmental Biogeochemistry*, Nriagu, J. O., Ann Arbor Sci., **1**, 131.
65. Rashid, M. A. and King, L. H., (1971), *Chemical Geol.*, **7**, 1, 37.
66. Rinaldi, G., Meinschein, W. G. and Hayes, J. M., (1975), preprint.
67. Rogers, M. A., van Hinte, J. E. and Sugden, J. G., (1972), *in: Initial Reports of Deep Sea Drilling Project*, **12**, U. S. Government Printing Office, Washington, 1115.
68. Rogers, N. A. and Koons, C. B., (1969), *Gulf Coast Assoc. Geol. Soc. Trans.*, **19**, 529.
69. Rosenfeld, W. D. and Silverman, S. R., (1959), *Science*, **130**, 1658.
70. Sackett, W. M., (1964), *Marine Geol.*, **2**, 173.
71. Sackett, W. M., (1968), *AAPG Bull.*, **52**, 853.
72. Sackett, W. M., Eckelman, W. R. and Bender, M. L., Be, A. W. H., (1965), *Science*, **148**, 235.
73. Sackett, W. M. and Rankin, J. G., (1970), *J. Geophys. Res.*, **75**, 4557.
74. Sackett, W. M., Thompson, R. R., (1963), *AAPG Bull.*, **47**, 525.
75. Schnitzer, M. and Desjardins, J. G., (1965), *Soil. Sci. Soc. Amer. Proc.*, **26**, 362.
76. Sidorenko, S. A. and Sidorenko, A. V., (1975), "Organic Matter in Sedimentary methamorphic Rocks of Precambrian", Nauka Press, Moscow.
77. Silverman, S. R. and Esptein, S., (1958), *AAPG Bull.*, **42**, 998.
78. Smith, P. H. and Mah, R. A., (1966), *Appl. Microbiol.*, **14**, 368.
79. Smith, B. N. and Epstein, S., (1971), *Plant Physiol.*, **47**, 380.
80. Stahl, W., Wollanke, G. and Boigk, H., (1975), *Proceedings 7th Internat. Congress Organic Geochemistry, Madrid.*
81. Stevenson, F. J. and Goh, K. M., (1971), *Geochim. et Cosmochim. Acta*, **35**, 471.
82. Takai, Y., (1970), *Soil. Sci. and Plant Nutrition*, **16**, 238.
83. Teichmüller, R., Teichmüller, M., Colombo, U., Gazzarrini, F., Gonfiantini, R. and Kneuper, G., (1970), *Geologische Mitteilungen*, **9**, 181.
84. Vinogradov, A. P., Galimov, E. M., Kodina, L. A., and Generalova, V. N., (1976), *Geokhimiya* (in Russian), **7**.
85. Welte, D. H., Hageman, H. W., Hollerbach, A., Leythaeuser, D. and Stahl, W., (1975), *9th World Petroleum Congress, Tokyo,* PD3, topic 5.
86. Whelan, T., Sackett, W. M. and Benedict, C. R., *Biochem. Biophys. Res. Commun.*, **41**, 1205.
87. Williams, J. A., (1974), *AAPG. Bull.*, **58**, 1243.
88. Williams, P. M., (1968), *Nature*, **219**, 152.

10

Structure elucidation of kerogen by chemical methods

D. VITOROVIĆ*

I. INTRODUCTION

The geochemical and economical significance of kerogen, which represents the major organic component of shales, and some other sediments, has in the past led to many attempts to unravel its chemical nature ("structure"). These investigations have been hampered not only by the insolubility and macromolecular nature of the substance but also by the fact that kerogens do not represent uniform molecules but assemblages of subunits of similar or varying molecular structures. Furthermore, in sediments, kerogens are most often intimately mixed with a large quantity of mineral matter which makes structural investigations even more difficult. Finally, kerogens from different geological sources differ in their general structure, and nature, since they vary depending on the type of precursors involved in their formation, the environmental conditions at the time of deposition and the diagenetic, catagenetic and metamorphic transformations of the original organic matter (OM). Hence, of all the organic material isolated from sediments of different ages, the determination of the chemical structure of kerogen appears to represent one of the most complex problems.

In attempts to solve this problem various approaches have been used such as microscopical investigations, pyrolytic degradation, the effect of various chemical reagents, and recently physical methods.

Since kerogen is insoluble in both organic and inorganic solvents, degradation to lower molecular weight products at high temperatures (pyrolysis, hydrogenolysis) or chemically at low temperatures (oxidation, reduction, hydrolysis) has often been used for its structural investigation. One aim of degradative studies is to make the degradation as specific as possible in order to obtain smaller identifiable compounds which still retain a structural relationship to the kerogen. In this respect chemical methods may appear advantageous when compared with thermal degradation methods. However, the reconstruction of degradation products into a structure representative of the original kerogen is difficult because the nature of the linkages between the various fragments is not clearly defined. Moreover, chemical reagents attack primarily at, or next to, carbon atoms bearing a functional group, and thus often

*Department of Chemistry and Physical Chemistry. Faculty of Science. University of Beograd, Yugoslavia.

do not give reliable information about the functional groups as they occur in the kerogen matrix. This may be illustrated using oxidations of low molecular weight products as examples:

$$RHC=CHR \longrightarrow RHC(OH)-C(OH)HR \longrightarrow 2\,RCHO \longrightarrow 2\,RCOOH$$

$$\text{cyclohexanone} \longrightarrow HOOC(CH_2)_4COOH$$

$$CH_3CH_2NH_2 \longrightarrow CH_3CHO \longrightarrow CH_3COOH$$

Therefore, functional group analysis of kerogen seems to be a useful complementary method which should help to provide a more complete insight into the kerogen structure. Thus chemical methods generally appear to be most useful for the examination of kerogen structures. Consequently structural investigations of kerogen by chemical methods will be reviewed in this Chapter, although physical and other methods are also very useful for the same purpose.

The definition of kerogen has changed much, since chemical methods were first used for its degradation. Consequently, for constitutional studies the starting material has not been prepared by identical methods. For a long time the term kerogen was used, as originally defined, to designate the total OM in shales, and during this period studies were mostly carried out with raw shales, or shale concentrates obtained by partial removal of mineral constituents only. Later, a distinction was drawn between kerogen and the soluble constituents in sediments; kerogen was then defined as the insoluble OM occurring in sedimentary rocks [74, 75, 192]. Kerogen may now be defined (*cf.* Chapter 1) as the organic residue which remains after exhaustive extraction of the soluble portion (commonly named bitumen) with organic solvents and acid treatment of the sediment to remove minerals.

Nevertheless, such definitions are still not sufficiently precise, particularly when the type of solvent used, the experimental conditions, and different isolation procedures are concerned. The mode of preparing the starting material is of particular significance in kerogen studies, if the results of different workers are to be correlated. In the future, efforts should be made to apply a generally agreed isolation procedure. Reviews on isolation procedures for kerogens and associated soluble organic materials, have been given by Forsman [75], Robinson [185], and Saxby [196, 197].

On this or other problems concerning kerogen, two oil shale bibliographies may also be consulted, one covering world literature [136], and the other covering publications from the USSR [141]. Several other smaller but very useful annotated or similar bibliographies in this field have also been published [10, 27, 28, 116, 195].

In this Chapter, methods and examples of structural studies of kerogens will be reviewed, and discussed, independently of the definition of kerogen used by individual authors.

II. METHODS, A REVIEW

Structural investigations of kerogen using various methods including oxidative degradation, hydrogenolysis, thermal degradation, degradation with hydrogen iodide and other agents, and functional group analysis, have been reviewed by Forsman [75]. Part of a review on oil shale and shale oil by Prien [156], was devoted to the organic constituents of shales and to the action of various chemical reagents on oil shale. A survey of methods for the structural determination of kerogens has been given by Philp [145].

Oxidative degradation seems to be one of the most useful chemical methods for the investigation of the structure of different kerogens. For this reason the largest part of the text in this Chapter will be concerned with methods of oxidative degradation. Investigation by other chemical reagents, and functional group analysis, will also be mentioned.

A. Oxidative degradation.

Oxidation converts most of the insoluble organic material (kerogen) to soluble degradation products. Oxidation should involve careful degradation of the kerogens into identifiable and structurally significant fragments followed by the structural identification of individual components in the generally complex mixture. The amount of product, and the molecular weights of the constituents depend upon the extent of oxidation. The separation, and identification, of such degradation products requires powerful analytical techniques, if they are to be accomplished within a reasonable period of time and with a reasonable amount of effort. The data obtained should allow one to attempt a reconstruction of the "structure" of the original kerogen.

Many reagents have been used in the past for oxidizing different kerogens, e. g. air, oxygen, ozone, hydrogen peroxide, potassium permanganate, chromic acid and nitric acid. These oxidants are active over different pH ranges and differ in their mechanism of degradation. Consequently, the degradation products may differ, to a certain extent, depending on the reagent used. However, in spite of the considerable differences amongst various oxidants, no substantial differences between oxidation products were observed, as might have been expected. On the other hand, the yields of oxidation products depended very much on the experimental conditions used.

Progress in the interpretation of kerogen structures has been assisted by the rapid development of powerful and sophisticated analytical techniques. By using these techniques it has become possible to isolate, and often identify, single components in very complex mixtures of kerogen degradation products. Examples of modern methods for isolating and identifying components, following the degradation of kerogen, will be included in this review. Analytical methods used in organic geochemical studies have been described in detail elsewhere [137].

1. Potassium permanganate.

Potassium permanganate is one of the oxidants most often used in structural studies of kerogens. Oxidation with alkaline potassium permanganate was first used many years ago for investigating the structure of coals. Thus, Bone *et al.* [15, 16] oxidized lignin, and coals of different rank, and found that coals were easily and totally, oxidized into various soluble products; this method has often been applied in kerogen studies.

Kogerman [103] and Down and Himus [47] were the first to use Bone's method for the oxidation of different oil shales. Two oxidation techniques were applied at that time: the "carbon-balance" and the "bulk-oxidation" technique.

The original "carbon-balance" technique [47] consisted of exhaustive oxidation of an amount of sample equivalent to 1 g of organic carbon with alkaline (1.6 % solution of KOH) potassium permanganate. When the oxidation was complete, the distribution of the original carbon was determined in the following oxidation products:

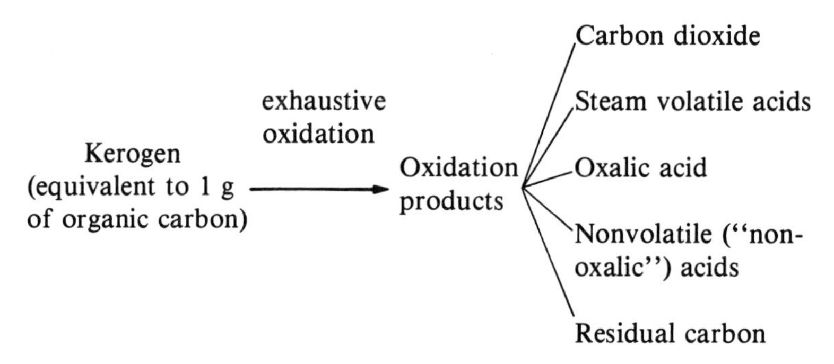

Since many organic materials, upon exhaustive oxidation, yield large amounts of carbon dioxide and oxalic acid, it was obvious that the ''carbon-balance'' oxidation products were not truly representative of the kerogen structures. Nevertheless, the yield of nonvolatile acids, consisting mainly of benzene-carboxylic acids (Fig. 10.1), reflected the aromaticity of the starting material. Comparison of the results of these oxidations with those obtained from known compounds as model substances [179, 194] and coals [15, 16] gave information about the structure of kerogens. In other words, the ''carbon-balance'' oxidation was a method to show structural similarities or differences among materials under investigation.

In the ''bulk-oxidation'' technique, larger (e. g. 0.5-1.0 kg) samples were degraded by the addition of solid potassium permanganate. The soluble oxidation products were separated, and identified as carbon dioxide and various volatile and nonvolatile acids including benzene-carboxylic acids [47, 192].

Somewhat modified ''carbon-balance'' and ''bulk-oxidation'' techniques were used by Robinson *et al.* [191, 192, 194] in structural studies of kerogens from Colorado and several other shales.

COOH

Benzoic acid

$(COOH)_2$

Benzene-dicarboxylic acids

$(COOH)_3$

Benzene-tricarboxylic acids

etc.

HOOC COOH HOOC COOH COOH COOH

Mellitic acid

Phthalic acid

Isophthalic acid

Terephthalic acid

Fig. 10.1. — Benzene-carboxylic acids.

Down and Himus [47] found that in the first stages of kerogen oxidation acids were frequently formed which were subsequently oxidized largely to carbon dioxide and oxalic acid. In order to prevent total oxidation, i.e. further appreciable degradation of the primary products, **stepwise oxidation** was introduced. The technique consisted in adding permanganate in small portions at a time and separating the oxidation products from the unreacted kerogen prior to each new addition of the oxidizing agent (Fig. 10.2). This approach led to high yields of degradation products [45, 46, 55, 62-65, 72, 79, 104, 122, 130, 149, 190, 192-194, 236-240] and also offered more information for the "reconstruction" of the original kerogen, as compared with "carbon-balance" or "bulk-oxidation" techniques.

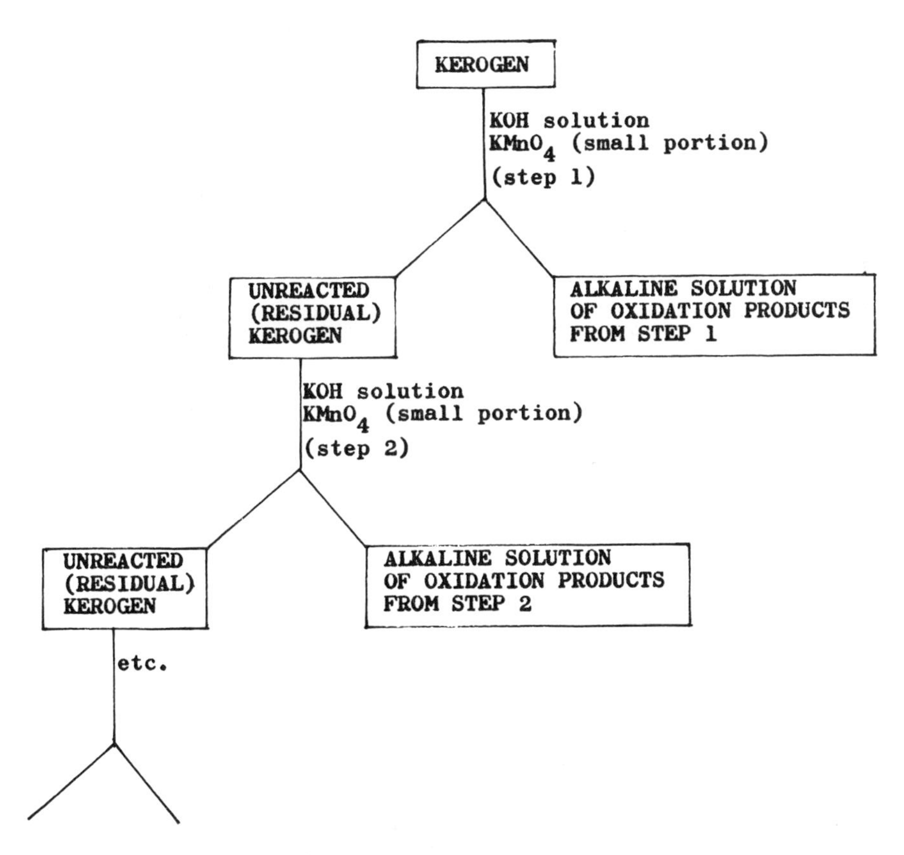

Fig. 10.2. — Scheme of stepwise alkaline permanganate oxidation.

The yields, and types of oxidation products, depended very much upon the conditions used (number of steps, temperatures, etc.); small [55, 62-65, 72, 104, 122, 149, 190, 192, 193, 237, 238] as well as larger [45, 46, 79, 130, 194, 236, 239, 240] numbers of steps were used at temperatures ranging from 50 to 90° C.

These stepwise degradative studies in which the oxidation products of all steps were combined and analyzed represents a useful method for the elucidation of the overall structure of kerogen subunits, and provides a good basis for the comparison of kerogens of various origins, as found in different shales. If these variations in structure could be correlated with the history and the environment of the shale, valuable information concerning the geochemical formation of kerogens should be obtainable.

However, since kerogens are not uniform organic materials but most likely conglomerates of a variety of compounds, Vitorović *et al.* [231] believed that more structural information on kerogens would be obtained, by separately investigating the products of each oxidation step of a stepwise degradation. The fact that significant qualitative and quantitative differences were found [231], rather than a uniform distribution of oxidation products throughout all the steps, indicated that such an approach might be advantageous. A similar technique was used by Yen [236]. Young and Yen [241] described a stepwise alkaline permanganate oxidation procedure at a constant temperature of 75° C, involving removal of the MnO_2 precipitate prior to each next oxidation step, as well as separate analysis of products from each oxidation step.

In order to achieve better selectivity of the site of permanganate oxidation and also to reduce even more the further degradation of primary oxidation products, a modification of the stepwise oxidation method was suggested [232]; this involved carrying out the first stages of the reaction at a lower temperature (20° C), adding permanganate in smaller portions, and frequent removal of oxidation products. This method consequently increased the number of steps. In such studies, an attempt should be made to correlate the rate of permanganate reduction at each step with the chemical nature of the degraded kerogen and the degradation products.

The rate of kerogen oxidation with permanganate allowed Robinson and Cook to suggest structural differences between different samples selected from cores of oil shale [187-189]. The method was essentially that of Erdman and Ramsey [50], with the exception that manual addition of the oxidant replaced the semi-automatic tirator. Each oxidation rate was corrected for the contribution from the pyrite present. The rate of oxidation indicated the type of structure present in the kerogens. Side chains in aromatic structures, partially unsaturated structures, and oxygen-containing groups or linkages are easily oxidized, whereas highly saturated hydrocarbon structures are oxidized quite slowly.

Khalifeh and Louis used controlled, mild alkaline [92] and acidic [93, 94] permanganate oxidation of sedimentary OM to distinguish petroleum bearing source rocks from those which had not furnished oil. Source rocks were characterized by an increased ability to reduce the permanganate. In addition to Khalifeh and Louis, several other authors have used oxidation methods for the identification of hydrocarbon source rocks (for references see the review of Cerchez and Anton [32]).

The kinetics of kerogen oxidation with alkaline permanganate have been investigated by Kogerman [104].

Alkaline permanganate degradative methods have been used by other authors to investigate the chemical nature of kerogens, and allied materials, in addition to those already mentioned, [14, 35, 66, 67, 81, 89, 106, 113, 123, 146, 148, 151, 153, 211, 213].

Stefanović and Vitorović [209] suggested that for structural determination of kerogen milder, and more specific, agents than alkaline permanganate might be advantageous. Such agents would oxidize the organic matrix at definite sites and thus enable primary degradation products to be isolated which would give more information about the structure of the original material. Their method consisted in portionwise addition of an excess of permanganate to shale concentrate dispersed in purified boiling acetone followed by refluxing for 16 hours.

This method was used only to a limited extent for the study of shales from Colorado, Aleksinac (Yugoslavia), and Kimmeridge (Dorset, England), as well as for the investigation of coorongite [29, 30, 210, 229, 230].

2. *Chromic acid.*

Hoering and Abelson [83] were the first to use chromic acid in degradative studies of kerogen, while Burlingame *et al.* [23-26, 181, 204, 205] used this method extensively in their investigations. The method described by Burlingame *et al.* consisted of stepwise chromic acid oxidation usually with different times being employed before separation of the products.

A similar chromic acid method was used for the determination of kerogen structures in recently deposited algal mats at Laguna Mormona, Baja California, by Philp and Calvin [147, 150]. Urov *et al.* [216] used chromic acid oxidation for the determination of the content of C-methyl groups in the OM of kukersite and Kashpir shales.

The stereochemistry of some acyclic isoprenoid compounds obtained in the chromic acid oxidation of Green River Formation oil shale kerogen was studied by Simoneit *et al.* [206].

3. *Nitric acid.*

Robertson [182] first used nitric acid for the oxidation of shale kerogens and compared the products with the corresponding oxidation products of cannel coals and peats. His method consisted of evaporating a mixture of the powdered sample and an excess of concentrated nitric acid to dryness.

Nitric acid has been used in various concentrations and under various reaction conditions (temperature and pressure) for kerogen oxidations. Nitric acid may react in different ways depending on the concentration and the reaction conditions (for example as an oxidant or as a nitrating agent). Since it may give nitro-derivatives with aromatic structures, it makes the oxidation products more complex. Hence it is considered to be more suitable for investigating aliphatic or alicyclic rather than aromatic structures, in spite of the fact that in some cases (e.g., phenolic structures) investigations with nitric acid may be complementary to studies with other oxidizing agents [222].

Fomina *et al.* [39, 40, 55-59, 66, 68, 177] showed that oxidation of Estonian shale kerogen with concentrated nitric acid at room temperature and atmospheric pressure gave a mildly-degraded nitro-kerogen, while at higher temperatures nitric acid, of any concentration, was able to oxidize the kerogen completely into soluble products in a relatively short time (2-8 hrs).

Stepwise oxidation of kukersite kerogen with nitric acid has been carried out by Veski *et al.* [226].

Nitric acid as an oxidant has been used in several other kerogen constitutional studies [14, 41, 102, 128, 160, 194, 195].

4. *Ozone.*

Ozonization at room temperature followed by analysis of the oxidation products was used in a study of Colorado shale by Robinson *et al.* [10].

Van den Berg *et al.* [218-220] selected ozone as an oxidizing reagent for kerogens in view of the selective electrophilic addition to double bonds known from reactions of ozone in homogeneous systems. The authors assumed that some of this selectivity would be retained in heterogeneous systems. They point out that from the experimental point of view reactions with ozone are simple, clean, and can be conducted easily on a microscale. Additional advantages are the variety of solvents which can be used and the wide temperature range in which the experiments can be carried out. The action of ozone on kerogen is essentially heterogeneous and therefore the nature of the kerogen surface which is directly in contact with attacking molecules has, together with experimental parameters such as solvent properties and temperature, a pronounced influence on the course of the reaction. Low temperature ozonization ensures maximal stabilization of the adsorption complex which is formed so that side reactions are minimized.

Ozonization followed by hydrogenolysis with $LiAlH_4$, alkaline hydrolysis or esterification, was used for the investigations of kerogen structures by Kutuev *et al.* [109-112]. The experiments were carried out in the temperature range of -70 to $-80°$ C in different reaction media such as hydrocarbons, chlorinated hydrocarbons, ethanol and low molecular-weight fatty acids. Glacial acetic acid was selected for the most experiments. According to Kutuev and Yakovlev [110] the advantage of the method is that it can be used at relatively low temperatures and that practically no carbon is lost in the form of volatile acids. Stepwise ozonization was used in kerogen structural studies by Egorkov *et al.* [48, 49].

Ozonolysis, combined with hydrogen peroxide oxidation, has been used for studying kerogen, coal and the insoluble OM in the Orgueil meteorite by Bitz and Nagy [11, 12].

Other kerogens have also been studied by ozonization [97].

5. *Air and oxygen.*

Proskuryakov *et al.* [20, 21, 36, 37, 80, 158, 159, 161, 162, 164-170, 172-176] have used atmospheric oxygen (air) for degrading kerogens of various shales. Their method consists in treating the kerogen concentrate dispersed in water, aqueous sodium hydroxide or sodium carbonate solution, with a continuous flow of air (1-7 MPa; 120-220° C). Acid [171] and non-aqueous media [163] have also been used for aerial oxidation of shale kerogens. The degree of degradation in non-aqueous media was low because high molecular weight products attached to the surface of the kerogen hindered its further oxidation. The yield and composition of oxidation products depended on reaction conditions as well as on the nature of investigated kerogens. Most of the studies of Proskuryakov *et al.* were aimed at the practical use of kerogen, however they gave much valuable information for the interpretation of the structure of oxidized kerogens.

Broi-Karre and Proskuryakov [20] and Mityurev [129] studied the kinetics of kerogen oxidation by atmospheric oxygen in aqueous alkali.

Air oxidation, similar to that developed by Mazumdar for coals [124], was used by Robinson *et al.* [194] in a study of oil shale kerogen.

Air [22, 98, 104, 138, 139] and molecular oxygen [8, 9, 35, 194, 195] have been used in other oxidation studies of shale kerogens.

6. *Other oxidants.*

Other oxidizing agents used in the structural investigation of kerogens include hydrogen peroxide [12, 35, 97], a mixture of potassium hydroxide and potassium nitrate at higher temperatures [95, 96], and periodate [134]. In addition to pure air or pure nitric acid, a combination of nitric acid and air has also been used [39, 54, 69, 71, 118-120, 178].

Degradation of kerogens with oxidizing agents, such as alkaline potassium permanganate, nitric acid, oxygen in aqueous alkali or a combination of nitric acid and atmospheric oxygen, have often been used to obtain dicarboxylic acids or other products, for industrial use or polyfunctional acids as plant growth stimulants [14, 20, 36-38, 42, 43, 54, 60, 61, 69-71, 73, 78, 99, 133, 142, 152, 155, 157, 161, 168, 171, 174, 178, 202, 208, 223-225, 239]. In spite of the fact that these studies were aimed at finding industrial applications for kerogen oxidation products, they also contributed to an understanding of the chemical structure of kerogen.

B. Reduction with hydrogen iodide.

Assuming that in some kerogens most of the oxygen is in the form of ether bonds, Raudsepp [180] proposed a degradative method based on treating kerogen with supersaturated hydriodic acid. The reagent was expected to cleave ether bonds and produce low molecular weight soluble products.

This procedure was seldom used in kerogen studies probably because it involved an uncommon reagent and an unusual technique. However, the effect of hydrogen iodide on shale kerogens of various ages was examined at the *U.S. Bureau of Mines* [195], by Hoering [81, 82], and by Jones and Dickert [88].

C. Other chemical reagents.

In addition to the above-mentioned chemical reagents (oxidants and hydrogen iodide), many others were tested in order to obtain some information on kerogen structure; these included chlorine [22, 100, 126, 127], bromine [100, 144, 232-235], sulphur [1, 105, 195], selenium [101, 195], phosphorus [82], metallic sodium in liquid ammonia [2, 3], lithium in ethylenediamine [88], lithium aluminum hydride [88, 115], sulphur dioxide [126, 127], sulphuryl chloride [126, 127], selenium oxychloride [126, 127], hydrogen bromide [22], fuming sulphuric acid [22], chlorosulphonic acid [22], a mixture of concentrated nitric and concentrated sulphuric acids [22], phosphoric acid [88], potassium iodide [88], aluminum chloride [2, 3, 88, 89], aluminum bromide [2, 3, 88], stannic chloride [2, 3, 215], boron tribromide [2, 3], mercuric acetate [2, 4], acetic anhydride [22], maleic anhydride [5, 22], paraformaldehyde in concentrated hydrochloric acid [2, 4], diazonium salts [2, 4], diazomethane [199], *p*-benzoquinone and *N*-bromosuccinimide [6], and some others [2].

In much of the work these reagents were not thoroughly investigated and some of the reactions gave disappointing results. A few reactions were carried out at too high temperatures so that pyrolysis of the OM may have occurred.

The purpose of testing the reactivity of kerogens with the reagents listed was principally to obtain information on some particular functional groups in the kerogen (based on the specificity of the reagents) or to cleave cross-linked or similar bonds and thus degrade and dissolve the kerogen. Since kerogen is insoluble and represents a macromolecular material of very complex structure containing various functional groups, and, since it is very difficult to completely separate kerogen from inorganic material, quantitative interpretations of the reactions carried out on such a heterogeneous mixture is generally extremely complicated.

However, some of the reagents were considered promising (for example, bromine) and, as such, were studied more extensively. Bromine was used in the form of various bromine-containing reagents for determining iodine numbers [144, 232-235]: as a bromine solution in carbon tetrachloride (McIlhiney, [154]), chloroform (Galpern and Vinogradova, [76, 77]), methanol saturated with sodium bromide (Kaufmann, [90]). Used under defined reaction conditions, bromine was expected to give quantitative information not only on the hypothetical unsaturation of kerogen but also on other reactive sites and functional groups of the kerogen structure.

D. Alkaline hydrolysis.

Alkaline hydrolysis is one of the methods which could indicate the presence of ester functions or other base labile functional groups.

Some such experiments have been primarily concerned with the quantitative aspects of kerogen analysis [51, 191, 203] (*cf.* Functional Group Analysis below). Kerogens from some shales have been hydrolyzed with alkali [2, 100, 107, 140, 201] but detailed identification of the products were not reported. However, when a detailed characterization of the hydrolysis products is carried out, additional information concerning the structure of kerogen can be obtained [26, 114, 130, 131, 149, 195].

Alkaline hydrolysis procedures consist in most cases in treating the kerogen concentrates with aqueous or alcoholic solutions of alkalies under various experimental conditions, for which see the references quoted.

E. Functional group analysis.

Functional group analysis may be considered as complementary to degradative methods, particularly to oxidative methods since oxidizing agents, as already mentioned, attack kerogen mostly at, or next to, carbon atoms bearing a functional group.

The functional group content of kerogens appears to be related to the nature of the precursor material, the environments of deposition, the age of the sediment and to diagenetic and catagenetic processes. Since most organic compounds possess one or more reactive functional groups, the major part of the original organic material deposited in the sediment would be a complex mixture having different functional groups. Interactions of these groups would lead to progressive condensation into kerogen. With increasing depth of burial kerogen may further lose functional groups, and thus change in composition and structure. Hence, an investigation of functional groups may help to a better understanding of the transformations of the organic material during its geological history. Moreover, it might have a practical interest, since the behavior of kerogen on pyrolysis may depend on the content and nature of some functional groups.

Most of the work in this fields has been devoted to determining oxygen-containing functional groups and therefore methods for this purpose will be reviewed. Problems arising from the heterogeneity of the reaction system, such as the specific reaction kinetics mentioned earlier, apply also to functional group analysis.

According to Aarna and Lippmaa [2] the first investigation of oxygen functional groups was carried out by Hüsse [87] on Baltic shales when he determined the carbonyl oxygen using the phenylhydrazine method.

To obtain data on the distribution of oxygen in the various functional groups in kerogens, Aarna and Lippmaa [2] determined most of them using the following methods:

(a) For **carboxyl groups** (–COOH): the barium acetate ion exchange method developed by Syskov and Kukharenko [212], as well as direct titration of kerogen with aqueous sodium hydroxide.

(b) For **hydroxyl groups** (–OH): acetylation with acetic anhydride in pyridine, followed by potentiometric titration of excess acid.

(c) For **ester groups** (–COOR): saponification with sodium hydroxide solution in alcohol.

(d) For **carbonyl groups** (>C=O): oximation with hydroxylamine as well as by treatment with formamide.

Similar methods were used by Siirde [203] and Yerusenko and Fomina [237] for the determination of oxygen functional groups in Dictyonema shale.

Hydroxyl groups in kerogen were determined by Aarna and Urov [7] by treating the kerogen concentrate with 3,5-dinitrobenzoyl chloride and/or hexamethyldisilazane. Periodate oxidation was used for the determination of vicinal OH groups in kerogen [134]. These groups are characteristic of carbohydrates and hence may serve as an index of the degree of preservation of the carbohydrate portion of the initial OM incorporated into the kerogen.

Functional groups in the OM of shales were also identified and determined by Semenov *et al.* [198, 200].

Fester and Robinson [51, 52, 53] have developed several methods for oxygen functional group analysis and used them mainly for the analysis of carboxyl, ester, amide (–$CONH_2$), hydroxyl, aldehyde (–CHO) and ketone (>C=O) groups in Green River shale kerogen. For the carboxyl content a modified Fuchs' calcium acetate ion exchange method was used [51, 52]. The ester function was determined by alkaline hydrolysis, followed by precipitation of the liberated acids as calcium salts, and from the increase in carboxyl function after hydrolytic cleavage [51]. Since it was considered possible that some of the amide structure, if present, would also react like an ester structure during the hydrolysis, a separate reaction was carried out to estimate the amide present. The hydroxylamine reaction was used to estimate ketone and aldehyde groups. Hydroxyl groups were determined by the acetylation method described by Blom *et al.* [13]. Ether groups (>C–O–C<) were estimated by difference.

Robinson and Dinneen determined the carboxyl, ester and hydroxyl groups in the kerogen concentrates of 12 shales which varied in geological age, environment, source material and elemental composition [191]. For these determinations, methods described by Fester and Robinson were used [53], except that the ester function was determined from the increase in carboxyl function after hydrolytic cleavage.

Carboxyl, carbonyl and hydroxyl groups were also determined in the kerogen of Aleksinac shale [233]. The carboxyl content was determined by the Fester-Robinson method [52], the carbonyl by the Bryant-Smith method [91], and the hydroxyl content by the Verley-Bölsing method [91].

III. EXAMPLES OF OXIDATIVE DEGRADATION STUDIES

A. "Carbon-balance" permanganate degradation of various shales.

As already mentioned, the "carbon-balance" alkaline permanganate oxidation technique developed for coal structure studies by Bone *et al.* [15, 16] represents a method which enables the general chemical nature of sedimentary kerogens and various types of coals to be compared.

Down and Himus [47] oxidized the kerogens of eight shales: Kimmeridge (Dorset, England), Ermelo (South Africa), Kukersite (Estonia, USSR), Amherst (Burma), Kohat (India), Broxburn Main (Scotland), Middle Dunnet (Scotland), and Pumpherston (Scotland) using the "carbon-balance" technique. Carbon dioxide, steam-volatile acids, oxalic acid (HOOC – COOH) and nonvolatile acids were formed in greatly varying proportions from different shales. The results obtained are shown in Table 10.1.

The distribution of organic carbon in the oxidation products of kerogens from Gdov (Estonia, USSR) and Volga (USSR) shales, as reported by Lanin and Pronina [113], and from St. Hilaire (France), and again from Ermelo (South Africa), and Kimmeridge (Dorset, England) shales, as reported by Dancy and Giedroyc [35], are also presented in Table 10.1. However, it should be noted that the oxidation periods for Gdov and Volga shales were much longer (500 hrs) than those used in the studies of other shales.

On the basis of their results Down and Himus suggested that the structure of some kerogens resembled that of coal as there was evidence of a benzenoid structure, although less well developed than in coal. Other kerogens did not show evidence of benzenoid structures.

TABLE 10.1
DISTRIBUTION OF CARBON IN OXIDATION PRODUCTS OF SHALE KEROGENS
(Weight %)

Shale	CO_2	Volatile acids	Oxalic acid	Nonvolatile "nonoxalic" acids	Unoxidized carbon	Oxidation period (hrs)
Kimmeridge	49.8	8.3	31.0	10.5	0.6	57
Ermelo	30.7	5.1	13.2	17.2	34.2	145
Kukersite	50.6	8.1	28.6	3.8	8.7	115
Amherst	40.9	2.9	27.2	4.0	25.1	132
Kohat	42.7	4.0	31.5	14.6	7.1	125
Broxburn Main	30.4	3.6	11.7	10.8	43.4	190
Middle Dunnet	28.1	5.1	14.9	13.7	38.9	120
Pumpherston	21.0	4.5	10.7	13.1	51.4	125
Gdov	98.0	2.2	3.8	0.5	0.0	500
Volga	88.1	3.9	5.2	4.1	0.0	500
St. Hilaire	35.3	0.2	6.6	10.1	48.5	100
Ermelo	30.7	5.1	13.2	17.9	34.2	120
Kimmeridge	49.7	8.3	30.9	10.5	0.6	30

Finally, certain kerogens contained two types of structure, one comparatively easily oxidized by alkaline permanganate and the other more resistant to attack. The kerogens were also differentiated according to the extent of degradation.

Gdov shale was rather stable to oxidation [113] and 500 hrs were required for its complete oxidation. Nonvolatile "nonoxalic" acids obtained by oxidation of this shale did not contain benzenoid acids. The results indicated that Gdov shale is free of humic type substances and is of "sapropelic" origin. The oxidation products from Volga shale [113] consisted of carbon dioxide, acetic (CH_3COOH), oxalic, succinic ($HOOC(CH_2)_2$ $COOH$), adipic ($HOOC$ $(CH_2)_4$ $COOH$), phthalic, benzene-tricarboxylic, benzene-tetracarboxylic and benzene-pentacarboxylic acids (Fig.10.1). Accordingly, Volga shale is said to be of mixed, "sapropelic-humic" origin.

After oxidation for 100 hrs, 48.5% of St. Hilaire shale kerogen, and after oxidation for 120 hrs, 34% of Ermelo shale kerogen remained unoxidized by alkaline permanganate [35]. The Ermelo torbanite contained algal material very similar to that found in the New South Wales torbanite; the remainder consisted of humic material basically benzenoid in structure. The St. Hilaire kerogen also consisted of two distinct types: the part oxidized more easily was similar to the Ermelo OM but probably more condensed, with fewer side chains. The other part is of algal type. The Kimmeridge kerogen is easily oxidized and has a high sulphur content. This kerogen was found to be very different from the kerogens in most sediments.

A similar "carbon-balance" technique was used by Robinson and Dinneen [191] in the exhaustive oxidation of 12 kerogen concentrates obtained from the following shales: Sao Paulo (Brasil), Shale City (Oregon, USA), Piceance Creek (Colorado, USA), Orepuki (New Zealand), Kiligwa River (Alaska, USA), San Juan (Argentina), St. Hilaire (France), Coolaway Mt. (Australia), Puertollano (Spain), Ermelo (South Africa), Dunnet (Scotland), and New Glasgow (Canada). Their results are shown in Table 10.2.

TABLE 10.2
DISTRIBUTION OF CARBON IN 100 hrs OXIDATION PRODUCTS OF SHALE KEROGENS
(Weight %)

Shale	Environment	Geological period	H/C (atomic)	CO_2	Volatile acids	Oxalic acid	Nonvolatile "nonoxalic" acids	Unoxidized carbon
Colorado	Lacustrine	Tertiary (Eocene)	1.58	62	2	33	1	1
New Zealand	Assoc. lignite	Tertiary	1.41	47	5	30	17	1
Argentina	Marine	Triassic-Permian	1.53	42	5	30	11	12
Oregon	Marine (?)	Tertiary (Oligocene)	1.52	38	8	28	13	13
Brasil	Lacustrine	Tertiary (Pliocene)	1.57	44	4	23	16	13
South Africa	Assoc. coal	Permian-Carboniferous	1.35	33	2	30	26	9
France	Assoc. coal	Permian	1.10	44	4	16	23	13
Spain	Assoc. coal	Permian-Carboniferous	1.38	33	4	25	27	11
Canada	Assoc. coal	Carboniferous	1.22	34	4	22	24	16
Scotland	Lagoon-Marine	Lower Carboniferous	1.35	26	3	21	22	28
Australia	Assoc. coal	Permian-Carboniferous	1.53	13	1	9	3	74
Alaska	Marine	Jurassic (?)	1.65	10	1	4	7	78

High atomic H/C ratios in several kerogen concentrates indicate a high degree of saturation and little aromatic or coal-like structures and vice versa.

Robinson and Dinneen's results suggested that the 12 kerogens could be classified into five different types as follows:

(a) Kerogens completely oxidized (represented by Colorado kerogen); this type of kerogen contains some highly condensed cyclic structure and a small quantity of "fatty" materials.

(b) Kerogens completely oxidized, producing significant amounts of nonvolatile "nonoxalic" acids (represented by New Zealand kerogen); the second type of kerogen contains little "fatty" materials but appreciable amounts of highly condensed cyclic structures.

(c) Kerogens not completely oxidized, producing 10 to 20% nonvolatile "nonoxalic" acids (represented by kerogens from Argentina, Brasil and Oregon); this type of kerogen contains significant amounts of highly condensed cyclic structures and "fatty" materials.

(d) Kerogens not completely oxidized, producing 20 to 30% nonvolatile "nonoxalic" acids (represented by Canada, France, Scotland, Spain and South Africa kerogens); the fourth type of kerogen contains as much as 50% highly condensed cyclic structures and "fatty" materials with some of the kerogens approaching coal-like structures and,

(e) Kerogens not completely oxidized, with large amounts of material resistant to oxidation (represented by kerogens from Alaska and Australia); this type of kerogen is predominantly "fatty". The degree of oxidation appeared to be partially age dependent, since Tertiary kerogens were more completely oxidized than the older kerogens.

In the opinion of Cane [31], the resistance of kerogen to oxidation is directly related to its structure and hence to the source material rather than to any diagenetic process. Vascular plant matter made some contribution to most oil shales, some of the shales being quite "coaly". According to Cane, three main types of kerogens in oil shales may be differentiated:

(a) Kerogen yielding benzenoid acids during permanganate oxidation which has a large contribution from vascular plant tissue and whose aromaticity is high but its oil yield is small.

(b) Kerogen yielding no benzenoid acids originating from algal fatty acids, the main oil-yielding component and,

(c) Kerogen yielding no benzenoid acids, arising from algal hydrocarbons, similar to kerogen (b).

Cane suggested that all oil shales are mainly made up of various tertiary mixtures of the three types of kerogen. The oil yield, which varies widely from deposit to deposit is probably caused by differences in the proportions of the three types of kerogen.

Several other similar classifications of kerogens based on oxidation behavior (or origin) have also been proposed [18, 47, 74, 125, 187-189, 214, 217, 231], Tissot *et al.* [214] have characterized kerogens as originating either from algal debris, from humic or plant debris, or from a mixture of both types of material. Algal kerogens have an aliphatic type structure and humic kerogens an aromatic type.

It is rather difficult to correlate these classifications reliably with kerogen classifications based on their elemental compositions [17, 19, 31, 33, 135, 183, 197, 207] because kerogens, frequently representing entire gradations of mixtures of various types of OM, may have a similar elemental composition but a different oxidation behavior. Forsman and Hunt [74] point out that kerogen, like coal, may be ranked according to its H/C ratio and fixed carbon, but that it is difficult to distinguish between the various types of kerogen by these measurements.

A detailed study of kerogen structure, i.e. the eventual determination of the ratio of various structural types in kerogen, is of great interest for solving several geochemical problems including source rock identification. Possible relationship of kerogen to petroleum was discussed by Forsman and Hunt [74] and by Hunt [84].

It has been suggested that the oil generating potential of sediments is greater for those with higher kerogen contents and more aliphatic and alicyclic structures in the kerogen (Vassoyevitch *et al.* [221]). The productive shales correspond to well defined organic type compositions such as those having a predominance of material of algal origin [34].

Kerogen varies considerably in its ability to yield gas or oil on heating and this depends mainly on its composition [85, 86]. The humic-coaly type, consisting of condensed functionalized aromatic and heterocyclic structures, yields mainly gas (methane), and the oily-sapropelic type of kerogen of aliphatic structure yields more oil and less gas.

Robinson and Cook [187-189] pointed out that in a geochemical study of an oil shale core one of the main goals is to prove maturational changes which are characteristic for oil yielding rocks, and which may be reflected in the ratio of aliphatic to aromatic components throughout the sequences being investigated. It is of interest to follow whether the aliphatic character of the soluble part increases with depth and whether it parallels an increase in the aromaticity of the kerogen. For this purpose, in addition to other methods (especially optical and some other physical methods), oxidative degradation may be very useful.

The amount of hydrocarbons that a kerogen can generate is a function of its original structure and hydrogen content and to a lesser extent of its oxygen content [125]. Hence, structural elucidation of kerogen leads to one of the important criteria for identification of potential source rocks.

B. Investigations of Green River Formation shale kerogen.

Green River Formation shale kerogen (Tertiary, Eocene) is one of the two extensively studied kerogens, the other being the Estonian kukersite kerogen. Several reviews on structural investigations of Green River shale kerogen have been published [44, 184, 186, 195]. A more complete description of the shale from the Green River Formation was given by Robinson [186].

Oxidative degradation has been widely used in studies of this kerogen; indeed, almost all the oxidizing agents proposed for kerogen structural studies have been used in its investigation.

Robinson *et al.* [192, 194, 195] oxidized Green River shale kerogen with alkaline potassium permanganate. The rate of oxidation was determined from the amount of oxidant reduced in a given interval of time, and the nature of the products obtained by exhaustive oxidation was investigated using the "carbon-balance" technique. The kerogen was almost completely oxidized in 100 hrs to nonbenzenoid water-soluble products (96-99%) and only 1% or less to nonvolatile "nonoxalic" acids. Samples from locations 50 miles apart within the Green River Formation in Colorado and Utah gave substantially the same distribution of final oxidation products. The behavior of the kerogen on oxidation appeared to be similar to that of Estonian algal limestone and Gdov shales, as similar oxidation products were obtained from each, but different from those of coals and most other oil shale kerogens. "Bulk-oxidation" of Colorado kerogen [192] also showed that no benzenoid acids were obtained. By contrast, coals produce large yields of nonvolatile "nonoxalic" acids, which are mostly benzene-carboxylic acids.

Oxidation rates showed that the initial rate was high, since approximately 70% of the total $KMnO_4$ was reduced in the first 10 hrs; the remaining 30% required an additional 90 hrs.

Based on these results, aromatic hydrocarbons, condensed aromatic hydrocarbons and condensed oxygen heterocyclic structures do not appear to be abundant in the Green River kerogen. However, noncondensed heterocyclic rings, noncondensed hydroxy- or methoxy-substituted benzene rings, aliphatic chains, noncondensed aromatic ethers and aldehydes, noncondensed cyclic ketones (saturated), monocyclic terpenes or carbohydrates may be present in the kerogen. In the same studies Robinson *et al.* [192, 194] used, for the first time, a stepwise alkaline permanganate oxidation procedure which converted approximately 80% of the kerogen into intermediate high molecular weight acids which they called "regenerated humic acids". They did not identify these oxidation products but indicated that they should be of value for further structural studies.

Robinson *et al.* [190, 194, 195] studied, in more detail, the structure of organic acids prepared by a two-step alkaline permanganate degradation of kerogen from Colorado oil shale. Approximately 58% of the total organic carbon in the kerogen was converted to the crude, nonvolatile acids, 15% was completely oxidized to carbon dioxide, 0.3% was converted to volatile acids and 26.7% of the initial carbon was unoxidized. The acids were converted to their *n*-butyl esters for fractionation and characterization. The degradation products were shown to consist of α, ω-dicarboxylic acids up to adipic acid (Fig. 10.3) together with dicarboxylic acids of higher molecular weight (up to about 800) that appeared to have saturated cyclic structures rather than aromatic or paraffinic structures. The predominant production of dicarboxylic acids indicated that the kerogen in this shale does not have a highly condensed structure, since the latter would yield multifunctional carboxylic acids on oxidation. Thus, the kerogen in Colorado shale appeared to be composed of high molecular weight substances with saturated rings which might be condensed or connected by short aliphatic chains. The high molecular weight acids contained approximately 1 atom of oxygen per molecule in excess of that present as carboxyl groups. This indicated the presence of carbonyl groups, hydroxyl groups, or ether linkages. Presumably substances of this type would be oxidized by alkaline permanganate to saturated dicarboxylic acids of progressively lower molecular weight until, finally, the ring structures would be ruptured to yield aliphatic dicarboxylic acids of six carbon atoms or less, and carbon dioxide.

The acids obtained in another two-step alkaline permanganate degradation of Green River shale kerogen [193, 194] were reduced to hydrocarbons and the resulting mixture was fractionated by various techniques and analyzed. Data from this study indicated that the oxidation mixture contained *n*-paraffinic (C_2-C_{38}), branched paraffinic, aromatic, naphthenic and heterocyclic acids (Fig. 10.3). The latter two types predominated, suggesting that Colorado oil shale kerogen contains mostly alicyclic and heterocyclic structures with smaller amounts of straight-chain and aromatic structures.

Chromic acid has been widely used in structural studies of Green River shale kerogen. For instance, Hoering and Abelson [83] obtained in the oxidation product *n*-acids (Fig. 10.3) in the C_{17}-C_{30} range in which the even numbered acids predominated, with *n*-C_{24} being the major component.

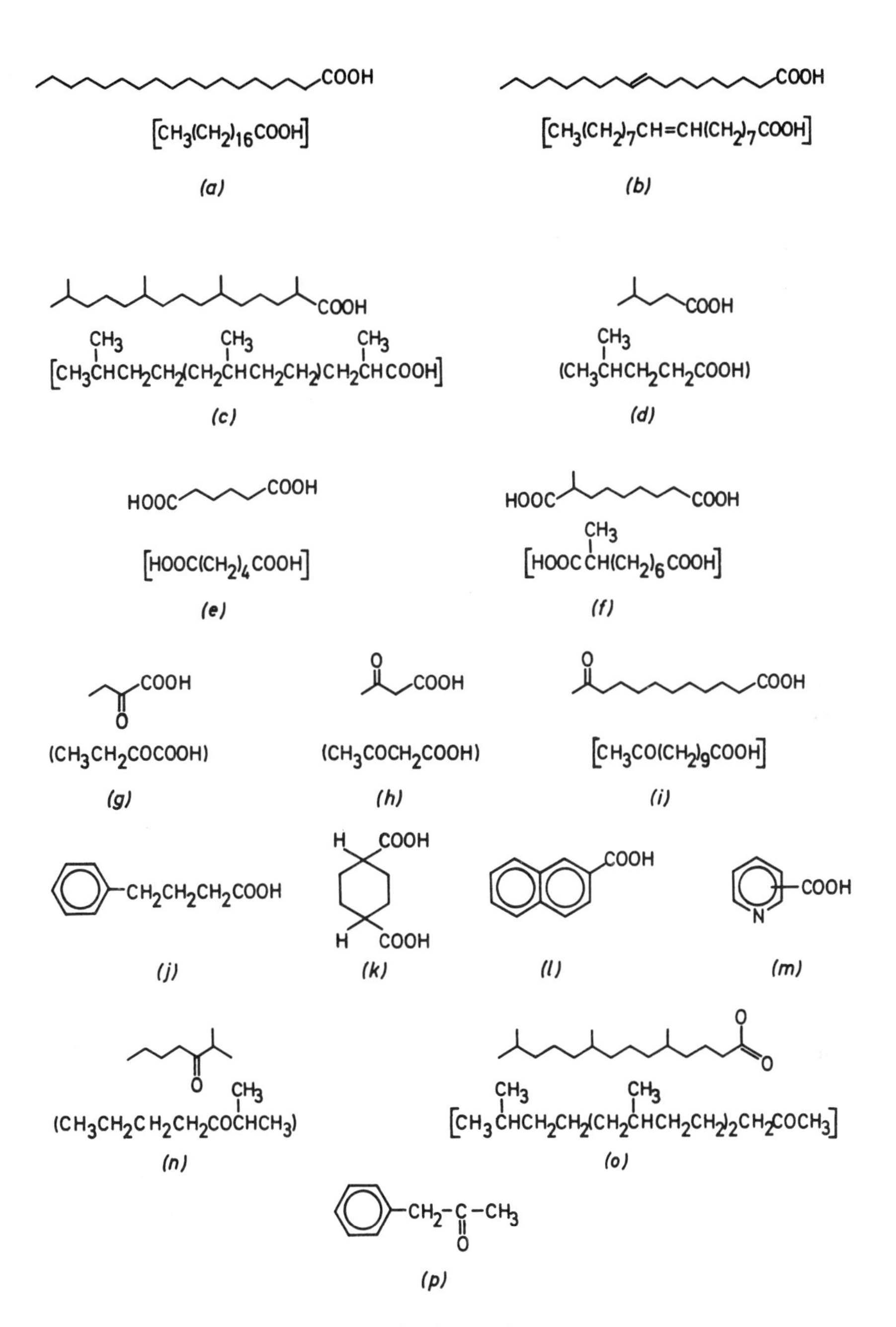

Fig. 10.3. — Examples of kerogen oxidation products:

a. Stearic acid (aliphatic saturated *n*-monocarboxylic acids). **b.** Oleic acid (monounsaturated *n*-monocarboxylic fatty acids). **c.** 2, 6, 10, 14-Tetramethylpentadecanoic-norphytanic acid (isoprenoid acids). **d.** *iso*-Caproic acid (*iso*-alkanoic acids). **e.** Adipic acid (aliphatic saturated α, ω-*n*-dicarboxylic acids). **f.** α-Methylazelaic acid (α-methyl-dicarboxylic acids). **g.** α-Ketobutyric acid (α-keto-acids). **h.** Aceto-acetic acid (β-keto-acids). **i.** (ω-1)-keto-lauric acid ((ω-1)-keto-acids or (ω-1)-oxo-acids). **j.** γ-Phenylbutyric acid (γ-phenyl-alkanoic acids). **k.** 1,4-Cyclohexane dicarboxylic acid (alicyclic acids). **l.** β-Naphthoic acid (aromatic acids). **m.** Pyridinecarboxylic acids (heterocyclic acids). **n.** *n*-Butyl isopropyl ketone (ketones). **o.** 6,10,14-Trimethylpentadecan-2-one (isoprenoid ketones). **p.** Benzyl methyl ketone (ketones).

Burlingame and Simoneit identified by high and low resolution mass spectrometry (MS), isoprenoid fatty acids (Fig. 10.3) ranging from C_{14} to C_{22} in the branched-chain acid fraction obtained by stepwise (3, 6, 15 hrs) oxidation of Green River Formation kerogen with chromic acid [23].The major components in this fraction were the C_{15} and C_{16} isoprenoid acids; the C_{17}, C_{19} and C_{20} acids were obtained in lower concentrations. The occurrence of isoprenoid acids in these oxidation products was taken as evidence that these acids are bound to the kerogen matrix.

In subsequent stepwise (3, 6, 15, 24 hrs) chromic acid oxidation experiments Burlingame *et al.* [24, 25] obtained fatty acids (0.60% of the shale) which were identified by gas chromatography (GC), low resolution and high resolution MS. The four oxidations, extending over 48 hrs, removed all of the organic carbon from the kerogen concentrate. The major homologous series found were C_5-C_{36} straight-chain acids, C_5-C_{26} branched-chain acids, (with a maximum concentration of 4,8,12-trimethyltridecanoic acid), and C_3-C_{25} α,ω-dicarboxylic acids (Fig. 10.3). The minor constituents were C_4-C_{20} methylketo-acids, C_5-C_{26} monounsaturated and/or cyclic acids, and tetracyclic- and pentacyclic monocarboxylic acids. Branched acids were major constituents (relative to normal acids) in the 3 hrs oxidation, decreasing in concentration as the oxidation time increased. Dicarboxylic acids, keto-acids, aromatic and cyclic acids, and polyfunctional acids, increased in concentration, and total yield, as the oxidation time increased.

According to Burlingame *et al.* [24] the acids obtained by oxidation could have arisen by several processes namely:

(a) "Loosening" of the kerogen matrix and removal of entrapped compounds which might subsequently be partially oxidized.
(b) Hydrolysis of ester linkages to give acids and alcohols, the latter being oxidized to carboxylic acids, and,
(c) Oxidative cleavage of carbon-carbon bonds.

All three processes probably contribute but their relative importance is difficult to assess. However, the authors advance arguments to support the view that a major portion of the acids is derived from carbon-carbon bond cleavage of side chains attached to the kerogen and cleavage of cross-linkages in the kerogen matrix itself. A suggested structure of kerogen based on the results obtained is illustrated in Fig. 10.4.

The kerogen seems to consist of a highly cross-linked polymer matrix of saturated hydrocarbons containing some aromatic nuclei and heteroatoms and with many long normal and isoprenoid peripheral chains. This means that early in the oxidation the long aliphatic side chains of the kerogen matrix are removed and, as the oxidation proceeds, breakdown of the cross-links in the matrix takes place. The "side-chain-poor" kerogen, judging from the products isolated, is highly cross-linked and not "very aromatized". The increase in concentration of the aromatic acids with increasing oxidation, especially the di- and tricarboxylic acids, is further evidence for carbon-carbon bond cleavage and indicates the presence of some aromatic nuclei within the kerogen. The very large increase in aliphatic dicarboxylic acids, with increasing oxidation, supports the conclusion regarding the aliphatic nature of the Green River Formation kerogen. The range and smooth distribution of the dicarboxylic acids suggest a random aliphatic cross-linked polymer matrix. The predominance of 4,8,12-trimethyltridecanoic acid in the branched fractions indicated cross-linking at the allyl rearrangement centre in phytol during kerogen formation.

In the mixture of acids obtained from Green River Formation kerogen by 24 hrs oxidation with chromic acid, Richter *et al.* [181] established the presence of a homologous series of (ω-1)-oxoacids ranging from C_4 to C_{12}. They developed a mass spectrometric method which allowed differentiation between the oxoacids and other acids such as dicarboxylic and aromatic acids. The method consisted in a borohydride reduction of the oxoacids followed by silylation of the hydroxy derivatives thus formed. The method also allowed the keto function to be located on the hydrocarbon chains of the oxoacid homologs.

In addition to acids (0.60% of the shale), successive (3, 6, 15 and 24 hrs) chromic acid oxidations yielded neutral fractions (amounting to 0.08% of the shale) which consisted of ketones (> 90%). These ketones were identified by Simoneit and Burlingame [205]. The major isoprenoidal ketone identified was 6,10,14-trimethylpentadecan-2-one, with minor amounts of 6,10-dimethylundecan-2-one and 6,10,14,18-tetramethylnonadecan-2-one (Fig. 10.3). A C_{30} triterpanone was also found in the neutral fractions. The results obtained confirmed, according to Simoneit and Burlingame, the conclusions that the Green River Formation shale kerogen is highly aliphatic. The identification of 6,10,14-trimethylpentadecan-2-one as a major oxidation product corroborates the authors' previous postulation that cross-linking occurred at the allyl rearrangement centre of phytol during kerogen formation [24, 25]. The isolation of C_{30} triterpanones as oxidation products indicated that triterpanols were bound to the kerogen matrix.

On the basis of chromic acid/sulphuric acid oxidation of Green River shale kerogen concentrates, Simoneit and Burlingame estimate that the relative overall proportion of isoprenoidal structures to straight-chain alkyl and polymethylene cross-links is about 38% for this kerogen.

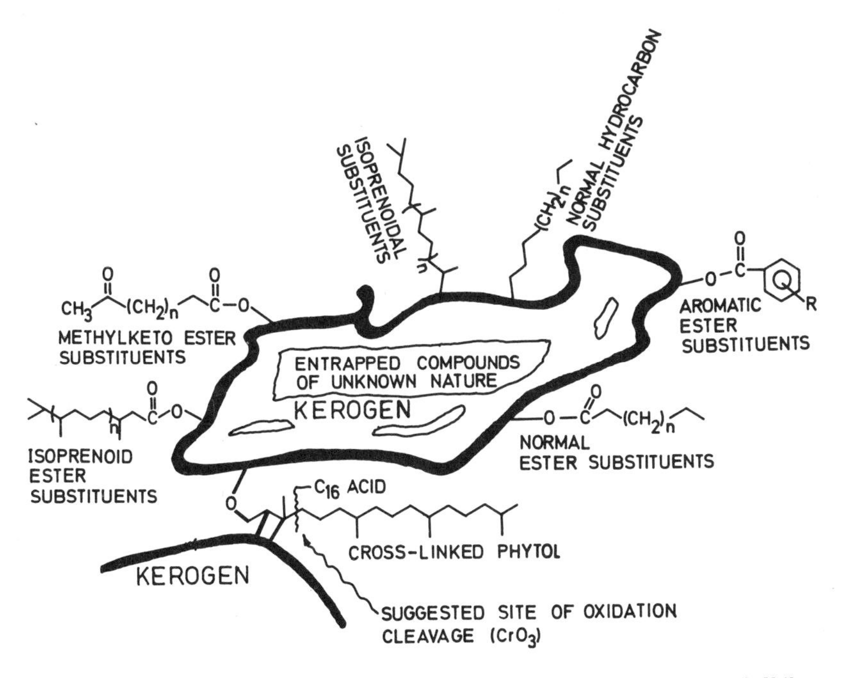

Fig. 10.4. — Structure suggested for Green River Formation kerogen network [24].

Djuričić *et al.* [45] degraded Green River shale kerogen by stepwise oxidation with alkaline permanganate to produce carboxylic acids in high yield (70% based on the original kerogen). The main products identified (as methyl esters) using a gas chromatograph-mass spectrometer-computer system, were unbranched aliphatic acids (C_8-C_{29}), saturated dicarboxilic acids (C_4-C_{17}) and isoprenoid acids (C_9, C_{12}, C_{14}-C_{17}, C_{19}-C_{22}) (Fig. 10.3). The products isolated in high yield were qualitatively the same as those found by Burlingame *et al.* [25, 26] in much lower yield using chromic acid. The major discrepancy between the results was in the abundance of keto-acids and aromatic acids. Keto- and aromatic acids were not significantly present in the permanganate oxidation products. On the basis of their results Djuričić *et al.* [45] suggested that the kerogen "nucleus" probably consists of interconnected, long aliphatic methylene bridges to which unbranched and isoprenoid chains are attached. Acidic products of alkaline hydrolysis of the Green River shale kerogen were also investigated (as methyl esters) [131]. The following acids were identified:

(a) C_9-C_{25} unbranched saturated aliphatic monocarboxylic acids.
(b) C_6-C_{15} saturated straight-chain dicarboxylic acids.
(c) C_{14}-C_{19} isoprenoid acids (except C_{18}).
(d) C_{11}-C_{16} *iso*-acids (except C_{15}) and,
(e) Two monounsaturated fatty acids (C_{16} and C_{18}) (Fig. 10.3).

On the basis of these results a suggested structure for kerogen was that shown in Fig. 10.5. The branching points (indicated by open circles) must be of a type that is susceptible to oxidative attack, alkaline hydrolysis, or both. Cleavage at a point of attachment of a saturated side chain would produce a monocarboxylic acid; cleavage at two carbon atoms connected by a methylene bridge would produce a dicarboxylic or even tricarboxylic acid.

Van den Berg *et al.* [218, 219], using ozonization for structural analysis of Green River and other kerogens, found that the uptake of ozone from methylene chloride solution was large compared to that using methanol and hexane. Despite the small initial surface area there was considerable ozone consumption. It appeared that methylene chloride was enlarging the effective solid/liquid interface for ozone adsorption. Much of the ozone forms an adsorption complex and degradation of the kerogen occurs mainly as a result of thermal decomposition of this adsorption complex. In the products obtained by decomposition of this adsorption complex, isoprenoid acids (C_{14}-C_{17} and C_{19}-C_{22}), ketones (C_{13} and C_{18}, in approximately equal quantity), mono-(C_7-C_{24}) and dicarboxylic (C_4-C_{17}) acids (Fig. 10.3) were identified by GC-MS analysis. Ozone does not appear to act specifically on double bonds in the case of Green River kerogen. The authors believe that most of the products were formed as a result of more or less random adsorption and reaction. According to these authors, the absence of any prominent isoprenoid compound in the kerogen ozonization products, confirms that this kerogen does not contain a phytyl unit retaining the double bond. According to the results obtained, Green River shale kerogen can not have a highly cross-linked structure since it exhibits gel-like properties (swelling in appropriate solvents). That is, Green River shale kerogen appears to contain more simple aliphatic chains.

Van den Berg [218] summarizing the findings of ozonization experiments with regard to the nature of typical structural elements of Green River shale kerogen noted that:

(a) The oxidized fragments were only attached to the kerogen matrix at one end.
(b) The oxidized fragments attached to the kerogen were predominantly functionalized on the other end, mainly by a carboxyl group.

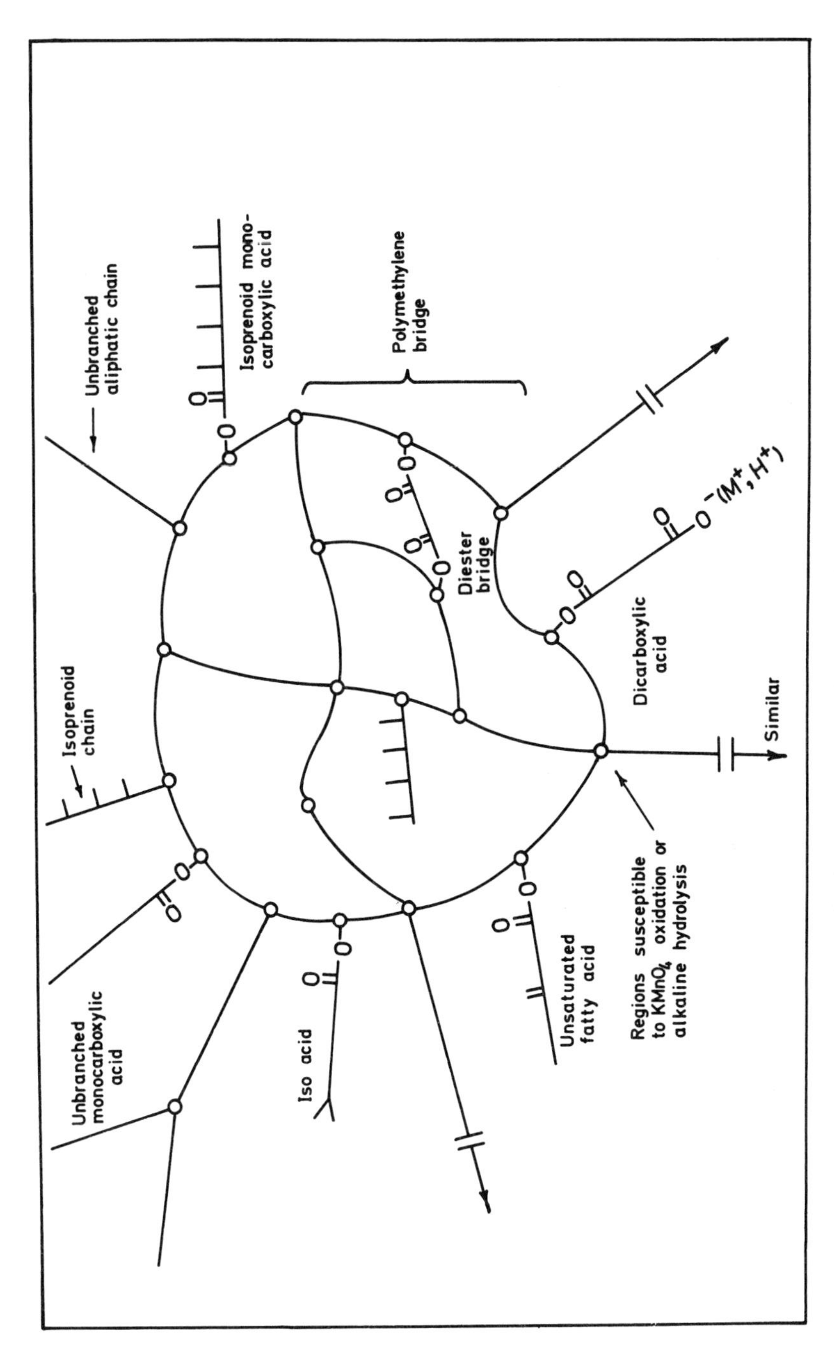

Fig. 10.5. — Diagram of a subunit of the Green River shale kerogen network [131].

(c) The oxidized fragments did not arise from cross-linked mono-unsaturated fatty acids but were probably mainly linearly attached to the rest of the kerogen by ether or ester linkages.

However, since only 2% of the kerogen was represented by the oxidation products, Van den Berg could not give a model concerning the overall structure of the kerogen. He believes that models by Burlingame *et al.* [24] and Djuričić *et al.* [45] are of limited value. The compounds that he identified as oxidation products are not necessarily randomly distributed in the kerogen matrix. In fact these compounds may well be derived from the oxidation of distinct parts of the kerogen that still reflect the molecular structure of plant and animal tissues. Van den Berg is of the opinion that Green River shale kerogen does not contain many long-chain compounds, but is formed predominantly by the condensation of non-lipid cell constituents such as proteins, polysaccharides, nucleic acids, etc. Such condensation processes may lead to the formation of condensed (heterocyclic) structures of a rather "structureless micrinitic-type" material. This picture is in essential agreement with that of Robinson *et al.* [186] who suggest that Green River shale kerogen is predominantly a linearly condensed cyclic structure in which straight-chain and aromatic structure are only minor components.

In another structural study of Green River kerogen stepwise permanganate oxidation was used by Yen *et al.* [236, 239]. The kerogen concentrate and the bioleached shale, provided over 90% by weight of acids. Products from ten oxidation steps were successively obtained and studied by various analytical methods such as GC, NMR (nuclear magnetic resonance), MS and IR. The kerogen concentrate and the shale samples, were also investigated by a number of physical methods. On the basis of combined chemical and physical studies, Yen *et al.* proposed a conceptual model of the Green River shale kerogen (Fig. 10.6), and came to the following conclusions concerning the structure:

(a) Aromatic systems are either absent or present in minute quantities in this kerogen. This does not exclude the possibility of the presence of isolated double bonds such as those present in hydroaromatic and heterocyclic systems.

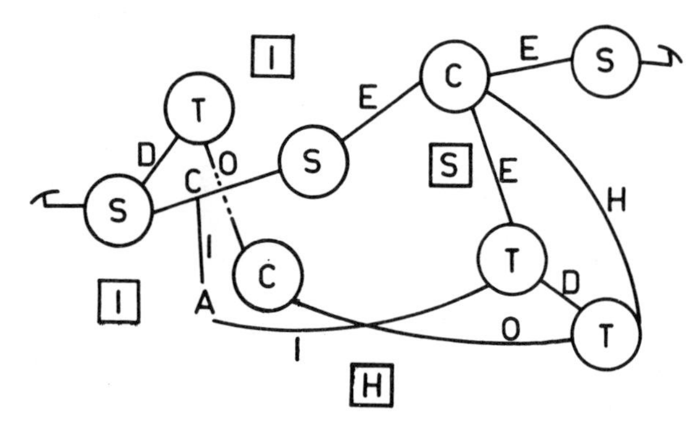

Fig. 10.6. — Hypothetical structural model of a multipolymer representing the organic component of Green River oil shale [236]. The circles represent essential components of kerogen; the squares represent molecules trapped in the kerogen network.

I. isoprenoids. **S.** steroids. **T.** terpenoids. **C.** carotenoids. **D.** disulfide. **O.** ether. **E.** ester. **H.** heterocyclic. **A.** alkadiene.

(b) The bulk of the carbon skeleton is naphthenic and contains 3-4 rings. It is possible that the skeleton consists of these naphthenic clusters which are linked by heterocyclic or randomly substituted branched hydrocarbon chains.
(c) Peripheral long-chain polymethylene structures are absent. The amount of long-chain alkane substituents (ranging from C_{17}-C_{31}) becomes greater toward the center of the kerogen core.
(d) Polar groups, as well as heterocyclic atoms, are located at the outermost shell region.
(e) The degree of cross-linking for the central core and the outermost shell is different: the outermost shell contains more cross-linking than the core region.
(f) Loosely held "monomers" may be entrapped in kerogen matrix.
(g) There is no regularity of cross-linking or branching in the kerogen.

In a mild stepwise alkaline permanganate oxidation study aimed at resolving contradictory conclusions about the size of the straight-chain aliphatic structures in the Green River kerogen, and hence carried out so that extensive degradation of kerogen-derived intermediates could be prevented, Young and Yen [241] came to the following conclusions:

(a) Green River kerogen contains a substantial portion (*ca.* 2-4 carbons out of every 10) of straight-chain aliphatic structures which are longer than C_4.
(b) The kerogen matrix forms a three-dimensional network of non-straight-chain clusters interconnected by long polymethylene cross-links and,
(c) The "core" in comparison with the "periphery" of the kerogen matrix contains a greater proportion of straight-chain and branched aliphatic structures which are attached to the kerogen matrix at one end.

Since it was shown that the structure of kerogen could vary according to the provenience of the sample, the controversy concerning the proportion of alicyclic to linear saturated structures in the Green River shale kerogen is probably much more a consequence of the differences in the shale samples that were investigated by different authors than of the differences in the methods used. Nevertheless, more work remains to be done to completely clarify this point.

C. Investigations of Estonian kukersite.

Estonian kukersite, one of the oldest shales in the world (Ordovician), is associated with Middle Ordovician marine sedimentary rocks, and Lower Ordovician Dictyonema shale. According to geological evidence, kukersite is shallow water, open sea sediment. It is referred to as a sapropelite. The average elemental composition of the kerogen of this shale is 77.1-77.8% C, 9.5-9.8% H, 1.7-1.9% S, and 9.7-10.2% O [55].

Kogerman may be considered the pioneer in the investigation of Estonian kukersite by oxidative degradation. On the basis of his early experiments with nitric acid [100] it appeared that the organic material in kukersite might be characterized as bituminous rather than humic. At that stage the OM of kukersite was regarded as a "highly polymerized resinic substance mixed with calcium salts of fatty acids", that can be depolymerized by heat and oxidation. In 1938, in his first application of alkaline permanganate oxidation, Kogerman [103] modified the "carbon-balance" method of Bone *et al.* [15, 16], oxidizing the shale at 30° C. The OM was almost completely oxidized yielding carbon dioxide (35.6%), oxalic acid

(43.8%), acetic acid (21.6%), and other mono- and dicarboxylic fatty acids. No benzene-carboxylic acids were found. Approximately 8% kerogen remained unoxidized. Using the "carbon-balance" technique, Down and Himus [47] obtained, in 115 hrs, carbon dioxide (50%), oxalic acid (29%), acetic acid (8%), nonvolatile "nonoxalic" acids (4%), and unoxidized organic carbon (9%). In the "bulk-oxidation" products of kukersite kerogen only carbon dioxide, acetic and oxalic acids were identified. Comparison of the data obtained by Kogerman and Down and Himus indicates that the balance of carbon in the oxidation of any shale may depend to a great extent on the reaction conditions, although in most cases there are no differences in the type of products formed.

Continuing his studies Kogerman [104] investigated the kinetics of the oxidation of kukersite kerogen with alkaline permanganate using again a modification of the method of Bone *et al.* [15, 16]. He studied the reaction rates over a temperature interval of 20 to 100° C, varying the concentration of base and permanganate. No benzene-carboxylic acids were found in the products but only aliphatic "ketonic and lactonic acids".

An extensive investigation of the chemical structure of kukersite kerogen was carried out by Fomina *et al.* [55, 62-65 69]. Fomina and Pobul [65] oxidized 50% of kukersite organic carbon with alkaline permanganate giving steam-volatile monocarboxylic (C_2-C_8) and non-volatile dicarboxylic acids such as oxalic (C_2), succinic (C_4), glutaric (C_5), adipic (C_6), pimelic (C_7), suberic (C_8), and azelaic (C_9) acids (in 10% yield). Unidentified higher molecular weight acids were also obtained. By further oxidation of these intermediate unidentified products, mono- and dicarboxylic acids and carbon dioxide were obtained. Again, no benzene-carboxylic acids were produced, a fact confirmed by Kokurin [106].

In similar studies about 55% of kerogen carbon was recovered as aliphatic compounds and about 40% in the form of C_4-C_{10} saturated dicarboxylic acids. A much smaller fraction of carbon was isolated in the form of monocarboxylic acids (1.5%): these authors assumed that kukersite kerogen may consist of approximately 70% aliphatic structures [55, 69].

Fomina *et al.* [39, 40, 55-57, 59, 68, 71] found that Estonian shale kerogen could be oxidized completely by concentrated (57-60%) nitric acid or by a combination of nitric acid and atmospheric oxygen. At room temperature and at normal pressure this acid produces the so called "nitro-kerogen" with a high content of nitrogen (up to 6%), which represents a partially degraded, insoluble product (78-92%, relative to starting OM). Prolonged reaction increases the solubility of the kerogen. The insoluble product may be decomposed by treatment with alcoholic potash yielding up to 40% of soluble and volatile products. At higher temperatures, nitric acid is able to degrade the kerogen completely, in a relatively short time (2 to 8 hrs) into a complex mixture containing saturated dicarboxylic acids (14-17% calculated relative to oxidized kerogen). Under certain conditions, the yield of C_4-C_{10} dicarboxylic acids may be as high as 55% [54].

The identification of a number of acidic products obtained by alkaline permanganate oxidation of kukersite kerogen, at 50° C, was reported by Pobul, Fomina *et al.* [67, 153]. The following acids were identified:

(a) α, ω-dicarboxylic C_4-C_{18} (up to 80%, among which the C_4-C_{10} acids represented over 70%).

(b) α-methyl dicarboxylic C_5-C_{18} (4 to 10%).

(c) Saturated tricarboxylic C_6-C_{17} (15-20%).

(d) *n*-monocarboxylic C_2-C_{26}.

(e) *iso*-monocarboxylic C_7-C_{19}.

(f) Phenyl-carboxylic C_8-C_{11} and,
(g) Terephthalic acids (Figs. 10.1 and 10.3).

Aromatic products represented less than 1% of the kerogen carbon. Hence, the main oxidation products were the C_4-C_{10} α, ω-dicarboxylic acids. The authors supposed the α-methyl dicarboxylic and tricarboxylic acids to be original structural units of the kerogen. Urov *et al.* [216] found that the content of C-methyl groups in the organic portion of kukersite was 3.3% relative to organic carbon.

Similar products were obtained using nitric acid as oxidant. On the basis of these results, the authors confirm previous hypotheses that the main part of the kukersite kerogen structure is aliphatic, and that about 40% of the hydrocarbon moieties are linear [14]. In addition, polymer structures of a lipocarbohydrate type should not be excluded, since in the hydrochloric acid hydrolysis products of kukersite, hexoses and pentoses were found [66, 132].

Mannik *et al.* [122] also used stepwise alkaline permanganate oxidation (20 and 50° C) for kukersite kerogen. At lower temperatures and with smaller permanganate concentrations, secondary reactions were suppressed, for instance the yield of tricarboxylic acids was considerably higher at 20° C. The major products were C_7-C_9 dicarboxylic acids (34-41%) and C_{10}-C_{12} tricarboxylic acids (13-17%). The authors believed that most of the C_4-C_6 dicarboxylic acids probably represented products of secondary reactions.

Ozonization (20° C/1.5 hrs) of a kukersite concentrate containing 82% OM, by Kutuev *et al.* [109, 110] was sufficient to convert 100% of the kerogen into a kerogen "ozonide" (80%) and low molecular weight products soluble in acetic acid (20%). Hydrogenolysis of this kerogen "ozonide" gave unbranched aliphatic compounds in the C_6 to C_{21} range, with components above C_{10} predominating. Among the low molecular weight ozonization products, α, ω-dicarboxylic acids ranging from oxalic to sebacic (C_{10}) were obtained in 8% yield. The rest of the products consisted of unbranched C_{11} to C_{25} aliphatic acids with a maximum at C_{16} and C_{18}. Alkyl-aromatic compounds were not found in any appreciable amount. On the basis of these results, the authors supposed that approximately 40% of the kerogen carbon was in the form of unbranched aliphatic structures containing C_2 to C_{25} chains, with chain lengths above C_{10} predominating.

On the basis of ozonization followed by alkaline hydrolysis of the solid product, Kutuev *et al.* [111] concluded that kukersite kerogen contained in its structure double bonds and ester groups. By GC analysis of methylated low molecular weight ozonization products [112] C_2-C_{10} dicarboxylic acids were identified as Baltic shale degradation products. Upon hydrogenation of the nonanalyzable portion of this ester mixture, C_6-C_{24} alkanes were identified.

By applying a stepwise ozonization at 20° C, Egorkov *et al.* [48, 49] were able to degrade the Gdov shale kerogen completely in a relatively short period of time, obtaining a large yield of products soluble in acetic acid. In the initial stages ozonolysis of double bonds seemed to play an important role. In the ozonization products C_6-C_{20} *n*-monocarboxylic acids, C_2-C_{16} α, ω-dicarboxylic acids and levulinic (CH_3COCH_2 CH_2 $COOH$) acid, were found.

Oxidation of kukersite kerogen with nitric acid (57%, 20° C, 24 hrs) by Veski *et al.* [222] gave a product consisting of C_7-C_{25} monocarboxylic and C_4-C_{18} dicarboxylic acids; tricarboxylic acids were also formed. Polyfunctional acids obtained in similar studies were difficult to analyze [177].

Fomina *et al.* [69] and Grassely *et al.* [79] found that the overall structure of kukersite kerogen does not change much during stepwise degradation by alkaline permanganate (the elemental composition of the residues and their properties were more or less constant), findings which suggest a uniformity of the kerogen.

In their nitric acid-air oxidation experiments, Mannik *et al.* [117-120] have found, in addition to α, ω-dicarboxylic acids, C_5-C_{14} α-methyldicarboxylic acids (Fig. 10.3), which represent 3-5% of the corresponding fraction of dicarboxylic acids. Since alkaline permanganate oxidation of kukersite kerogen also produced branched dicarboxylic acids [67, 118, 121, 153], the authors supposed that these acids had not been formed during the oxidative degradation process but that they rather represented a structural element of the kukersite kerogen. Propane- to nonane-tricarboxylic acids were also found (8-10% calculated relative to kerogen) as well as terephthalic acid in very low concentration. NMR analysis allowed the identification of 12 individual acids (as methyl esters) of the type $C_nH_{2n-1}(CO_2Me)_3$, where n = 3-8 [143]. The presence of tricarboxylic acids is also an indication of a branched kerogen structure. The presence of terephthalic acid in the oxidation products, although in small concentration, indicates that even aromatic rings may be incorporated into "aliphatic" kerogen structures.

By aerial oxidation of Gdov (Estonia) shale in alkali, Proskuryakov and Soloveichik *et al.* [80, 164-166, 168-170] obtained mainly saturated dicarboxylic acids. The composition of the oxidation products depended on the experimental conditions used. In the initial oxidation stages, high molecular weight acids were formed in high yield (up to 60 to 70%) while the yield of dicarboxylic acids was smaller, but with further oxidation the yield of dicarboxylic acids was increased to more than 40%. The degree of kerogen degradation was sometimes more than 90%. With increasing oxidation, both the yield and the composition of the dicarboxylic acids changed. Succinic, glutaric, adipic, pimelic, suberic, azelaic and sebacic acids were obtained, and phthalic acid was also found [169]. At the beginning of the oxidation, some volatile acids, mainly propionic and acetic, were formed from peripheral kerogen chains [20, 167].

D. Investigations of Aleksinac (Yugoslavia) shale kerogen.

Kerogen from Aleksinac shale, a lacustrine sediment of Miocene age, has been extensively studied by oxidative degradation. The composition of the OM of a typical Aleksinac shale sample was the following: 75.9% C, 9.3% H, 2.1% N, and 12.7% O + S (by difference). Since it was readily and completely oxidized by alkaline permanganate [211, 228] and since it did not give an appreciable amount of benzene-carboxylic acids, a predominantly non-aromatic structure was proposed.

Aleksinac shale kerogen was found to be more resistant towards permanganate in acetone than towards aqueous alkaline permanganate [209, 228, 230]. A major portion of the organic carbon remained insoluble (approximately 70%) after treatment with the former, and some ether-soluble products (10-30% of the total carbon), for the most part acids, were obtained. The acids were mainly dicarboxylic but some mono- and polycarboxylic types were also obtained. The residual insoluble carbon was readily oxidized with aqueous alkaline permanganate, mainly to carbon dioxide. These experiments suggested that the Aleksinac kerogen had principally an aliphatic and possibly hydroaromatic nature, containing some unsaturated and branched structures.

Despite the belief that the milder oxidation obtained with permanganate/acetone [209] should be advantageous in investigating kerogen, most studies have been carried out with aqueous alkaline permanganate. This choice was principally because it was shown that a carefully chosen stepwise alkaline permanganate degradation might produce high yields of structurally significant products.

Complete stepwise degradation of Aleksinac kerogen concentrate (9 steps, 80° C, 50 hrs) provided 60% of the original organic material as acidic products [46]. The acids were analyzed (as methyl esters) using a gas chromatograph-mass spectrometer-computer system. The compounds were identified by computer-comparison of the mass spectra from selected regions of the chromatogram, with the spectra of authentic compounds, plotting of the relative intensity of significant masses (mass chromatograms) or interpretation of individual spectra. Acidic products consisted mainly of C_2-C_{19} saturated, unbranched α, ω-dicarboxylic acids. Unbranched, C_8-C_{28} monocarboxylic acids and aromatic acids, such as benzoic, methyl-naphthoic, benzene-dicarboxylic, methyl-naphthalene-dicarboxylic, as well as benzene-tricarboxylic (3 isomers) and tetracarboxylic acids (Figs. 10.1 and 10.3) were also identified, but were present only in small amounts. No branched or isoprenoid mono- or dicarboxylic acids were detected in any appreciable amounts. From these oxidation products a preliminary structure for Aleksinac kerogen was constructed. The type of acidic products obtained indicated that the kerogen "nucleus" is mainly composed of polymethylene bridges with some aromatic structures. This was demonstrated by the low yield of aromatic acids as compared to long-chain dicarboxylic acids. The periphery consists of unbranched hydrocarbon chains as well as aromatic and heterocyclic structures.

Aleksinac kerogen was also degraded by a stepwise alkaline permanganate method involving the separate investigation of products at each step [231]. New structural information was obtained in this way, because significant qualitative and quantitative differences were found in the compositions of the acidic fractions obtained in the eleven degradation steps. It was estimated that most of the aromatic acids were obtained in the first relatively short steps, while the yield of saturated, unbranched, dicarboxylic acids gradually increased towards the later, more time consuming steps. The yield of unbranched monocarboxylic acids was found to be generally uniform throughout all steps. On the basis of these results, it was suggested that Aleksinac kerogen had neither a uniform nor a random structure, but that it is composed of at least two major types of OM differing in chemical nature and in origin. The more abundant type is more resistant towards alkaline permanganate and is composed mainly of aliphatic type material; the other type, which is more susceptible to alkaline permanganate is composed basically of aromatic type material. This conclusion was supported by micropetrographic investigations of the kerogen concentrate which indicated the presence of at least two types of OM. One is a humic (coaly) type material which is the most easily degraded and yields aromatic acids as the main product in the initial degradation steps; the other is a more resistant bituminous material which is progressively degraded to yield more and more unbranched dicarboxylic acids in the later degradation steps.

In order to obtain even more information on the structural nature and mechanism of degradation of Aleksinac kerogen, stepwise oxidation with alkaline permanganate (37° C) was supplemented by parallel petrographic and chemical investigations. This was done on partially degraded kerogen concentrates, isolated at each degradation step, and on some of the oxidation products, principally the higher molecular weight "precipitated acids" [108]. Ten identical portions of Aleksinac shale concentrate were oxidized by stepwise addition of alkaline permanganate solution. After each step, one reaction was terminated and a residue of partially degraded kerogen was recovered; thus ten kerogen concentrates with different degrees of degradation were obtained for parallel petrographic and chemical investigations. Degradation products (neutral products, precipitated acids, ether soluble acids and sulphate from pyrite) were isolated and analyzed separately. A detailed petrographic investigation of the initial kerogen concentrate, before oxidative degradation, indicated the presence of five different particle types. The amount of each type was also estimated in the partially degraded concentrates. It was apparent that particles of coaly type OM were the most easily degraded since they completely disappeared after the third degradation step. Particles representing a mixture of coaly and bitumen-like OM were also relatively easily degraded. The particles consisting of bitumen-like OM were more resistant towards alkaline permanganate.

On the basis of these results a heterogeneous structure for Aleksinac kerogen was strongly supported. That is, this kerogen is composed of two major types of organic material, a more susceptible aromatic (coaly) type material, and a more resistant aliphatic type material. In this way the usefulness of parallel petrographic and chemical investigations in kerogen studies was demonstrated.

IV. OXYGEN FUNCTIONAL GROUPS IN VARIOUS KEROGENS

Oxidative degradation limits itself mostly to the interpretation of kerogen skeletons and does not yield much information on functional groups. Functional sites in the kerogen structure are most easily attacked by oxidants. Therefore, to obtain a more complete insight into kerogen structures, functional group analysis should be considered a very useful complementary method.

Oxygen functional groups are the most abundant and diverse, and therefore they attracted maximum interest, although this does not mean that they have been extensively investigated. However, several examples of kerogen oxygen functional group analysis will be given.

Hüsse [87] was the first to determine the carbonyl group content in Baltic shale kerogens. Later, Semenov [198] found the following distribution of oxygen: 1.5% carboxylic, 1.3-3.7% hydroxyl- and 0.3-0.5% carbonyl-oxygen. He also indicated that ester- and methoxy groups were present. According to Semenov the functional groups are located on aliphatic side chains as terminal groups.

Aarna and Lippmaa [2] developed methods for oxygen functional group analysis, and applied them to a rich kerogen concentrate of Baltic shale containing 12.0% oxygen. The following distribution was determined by these authors:

Functional group	Percent of oxygen	
	Based on kerogen	Based on total oxygen
Carboxyl	0.16	1.33
Carbonyl	1.60	13.33
Ester	1.92	16.00
Phenol-ether	4.80	40.00
Hydroxyl	3.52	29.34

In another study Aarna and Urov [7] found that the true average value for the OH-content was approximately 42% of the kerogen oxygen; 10% of these OH groups were primary, 75% secondary and 15% tertiary. From the known content of methyl- and carboxyl groups and from the small content of primary hydroxyl groups, it was concluded that the kerogen is mainly cyclic and that 1 to 2 terminal hydroxyl groups occur per 100 carbon atoms. Nekrasov and Urov [134] found that vicinal hydroxyls in kukersite kerogen accounted for 4-5% of the total hydroxyl groups.

In Dictyonema shale, Siirde [203] determined the following distribution of oxygen in functional groups:

Functional group	Percent of oxygen	
	Based on kerogen	Based on total oxygen
Carboxyl	0.47	2.36
Carbonyl	6.56	33.05
Hydroxyl	1.92	9.63
Ester	10.93	54.96

Fester and Robinson [51, 52] estimated carboxyl, ester, amide, hydroxyl, aldehyde, and ketone groups for Green River shale kerogen, in a concentrate containing 86% organic material with 8.5% of oxygen (determined by difference). The following distribution of total oxygen in functional groups was found:

Functional group	Percent of oxygen	
	Based on kerogen	Based on total oxygen
Carboxyl	1.30	15.3
Ester	2.10	24.7
Amide	0.05	0.6
Carbonyl	0.10	1.2
Hydroxyl	0.40	4.7
Ether (by difference)	4.55	53.5

Hence, carboxyl and ester groups account for about half of the total oxygen while the ether group accounts for the other half. Minor amounts of hydroxyl, carbonyl, and amide groups were found.

Robinson and Dinneen [191] determined carboxyl, ester and hydroxyl groups for 12 kerogen concentrates (*cf.* "Carbon-balance" alkaline permanganate degradation of shales) and obtained the following data:

Source of kerogen Shale	Oxygen as –COOH (mg/g C)	Oxygen as –COOR (mg/g C)	Oxygen as –OH (mg/g C)
New Zealand	64	78	42
Brasil	39	61	33
Oregon	28	99	29
Argentina	19	109	17
Colorado	14	25	27
Canada	8	62	17
Spain	8	39	10
Scotland	7	31	12
South Africa	5	38	15
France	3	20	11
Alaska	2	37	4
Australia	1	1	13

The youngest (Tertiary) shales, had the highest content of carboxyl and hydroxyl oxygen, per gram of organic carbon, while the older shales had the lowest. Hence, the carboxyl and hydroxyl contents appear to be related to the age of the sediment. Robinson and Dinneen presume that the carboxyl groups are present as free acids as well as at the periphery of the kerogen. The hydroxyl groups may be present as alcohols or as groups attached directly to the kerogen nucleus. The controlling factor in the preservation of the ester function may have been the pH of the basin waters. In spite of the fact that no consistent relationship was apparent in the amount of ester oxygen for the 12 shales, the two with the largest amount of ester oxygen were of marine origin. The amount of ester preserved would be related to the environmental conditions and the amount of available precursor materials.

Investigations of Aleksinac shale kerogen also included the determination of oxygen functional groups [227, 228, 233].

In spite of the fact that much work has been devoted to the development of methods for the determination and estimation of oxygen functional groups in kerogen, no adequate effort has been made to interpret kerogen structures based on results obtained by different methods, e.g. functional group analysis and oxidative degradation. This is a requirement for the future.

V. CONCLUSIONS

Chemical methods, particularly oxidative degradation, have been very useful in kerogen structural studies. However, they have not yet given as good results as might have been expected. Many reagents have been tried but few with promising results. Very often, problems arising from the insolubility, inhomogeneity, complexity and diversity of kerogens, and even from the nonuniformity of the kerogen from one formation, have discouraged structural interpretations and further studies. Promising, oxidative degradation methods have been developed gradually, so too have identification techniques. Consequently, the interpretation of kerogen structures has progressed step by step. Unfortunately, although many complementary results have been obtained they have often been found to be controversial and contradictory, due sometimes to differences in the samples investigated and sometimes to differences in degradation methods and identification of the products. The above review, confirms the controversy that remains to date. In spite of that, chemical degradation continues to be an indispensable method for recognition of kerogen building blocks.

More advantage should be taken of quantitative data on functional group content to supplement structural and geochemical interpretations based on degradative methods.

Knowledge of the chemical nature of kerogen, of its composition and structure, certainly may help in a better understanding of:

(a) The source of the precursor organic material.
(b) The environmental conditions of deposition.
(c) The type of diagenetic, catagenetic and maturation processes.
(d) The gas and oil potential of kerogen and the source rock potential of the sediment.

However, the difficult task of structure elucidation of kerogen has not yet been fully accomplished and much work remains to be done using chemical methods or, what seems even more promising, a carefully chosen combination of chemical, physical and petrographic methods.

Acknowledgements. The author wishes to thank Professor M. Lj. Mihailović and Professor M. Gašić, for their stimulating comments and helpful suggestions.

REFERENCES

1. Aarna, A. Ya. and Urov, K.E., (1965), *Tr. Tallinsk. Politekhn. Inst.*, Ser. A, **228**, 9, 27 ; **230**, 3.
2. Aarna, A. Ya. and Lippmaa, E.T., (1955), *Tr. Tallinsk. Politekhn. Inst.*, Ser. A, **63**, 3.
3. Aarna, A. Ya. and Lippmaa, E.T., (1957), *Zh. Prikl. Khim.*, **30**, 312.
4. Aarna, A. Ya. and Lippmaa, E.T., (1957), *Zh. Prikl. Khim.*, **30**, 419.
5. Aarna, A. Ya. and Urov, K.E., (1965), *Tr. Tallinsk. Politekhn. Inst.*, Ser. A, **230**, 23.
6. Aarna, A. Ya. and Urov, K.E., (1965), *Tr. Tallinsk. Politekhn. Inst.*, Ser. A, **230**, 13.
7. Aarna, A. Ya. and Urov, K.E., (1965), *Tr. Tallinsk. Politekhn. Inst.*, Ser. A, **230**, 33.
8. Alumyae, T.E., (1954), Candidate Thesis, Tallin, USSR.
9. Alumyae, T.E., (1956), *Goryuch. Slantsy, Khim. i Tekhnol. Akad. Nauk Est. SSR, Inst. Khim.*, **2**, 17.
10. "Bibliography of Oil Shale and Shale Oil Bureau of Mines Publications, 1917-1973", (1973), compiled by M.P. Rogers, U.S. Bureau of Mines, Laramie Energy Research Center, Laramie, Wyo., USA.
11. Bitz, M.C. and Nagy, B., (1966), *Proc. Nat. Acad. Sci. U.S.*, **56**, 1383.
12. Bitz, M.C., (1967), "Dissertation Abstr. B 28", Univ. Microfilms, Ann Arbor, Mich., 808.

13. Blom, L., Edelhausen, L. and Van Krevelen, D.W., (1957), *Fuel,* **36,** 135.

14. Bondar, E., Veski, R.E. and Fomina, A.S., (1972), *Izv. Akad. Nauk Est. SSR, Ser. Khim. i Geol.,* **21,** 129.

15. Bone, W.A., Horton, L. and Ward, S.G., (1930), *Proc. Roy. Soc. London,* **127 A,** 480.

16. Bone, W.A., Parsons, L.G.B., Sapiro, R.H. and Groocock, C.M., (1935), *Proc. Roy. Soc. London,* **148 A,** 492.

17. Boulmier, J.L., Oberlin, A. and Durand, B., (1977), *in: Advances in Organic Geochemistry,* 1975, ed. Campos R. et Goni J., Enadisma, Madrid, 781.

18. Breger, I.A. and Brown, A., (1962), *Science,* **137,** 221.

19. Breger, I.A., (1963), "Origin and Classification of Naturally Occurring Carbonaceous Substances", *in: Organic Geochemistry,* ed. Breger I.A., chap. **3,** Pergamon Press, Oxford.

20. Broi-Karre, G.V. and Proskuryakov, V.A., (1965), *Zh. Prikl. Khim.,* **38,** 2779.

21. Broi-Karre, G.V. and Proskuryakov, V.A., (1966), *Zh. Prikl. Khim.,* **39,** 939.

22. Brower, F.M. and Graham, E.L., (1958), *Ind. Eng. Chem.,* **50,** 1059.

23. Burlingame, A.L. and Simoneit, B.R., (1968), *Science,* **160,** 531.

24. Burlingame, A.L., Haug, P.A., Schnoes, H.K. and Simoneit, B.R., (1969), *in: Advances in Organic Geochemistry,* 1968, ed. Schenck P.A. and Havenaar I., Pergamon Press, Oxford, 85.

25. Burlingame, A.L. and Simoneit, B.R., (1969), *Nature,* **222,** 741.

26. Burlingame, A.L., Wszolek, P.C. and Simoneit, B.R., (1969), *in: Advances in Organic Geochemistry,* 1968, ed. Schenck P.A. and Havenaar I., Pergamon Press, Oxford, 131.

27. Cane, R.F., (1973), "A Bibliography of Tasmanite with an Introduction and Annotations", *Papers and Proceedings Roy. Soc. Tasmania,* **108,** 211.

28. Cane, R.F., (1977), "Coorongite, Balkashite and Related Substances — an Annotated bibliography", *in: Trans. Roy. Soc. South Australia,* **101,** Pt 6, 153.

29. Cane, R.F., (1968), *Proceedings of Symposium on the Development and Utilization of Oil Shale Resources, Tallin, USSR,* 14 pp.

30. Cane, R.F., (1969), *Geochim. et Cosmochim. Acta,* **33,** 257.

31. Cane, R.F., (1976), "The Origin and Formation of Oil Shale", *in: Oil Shale,* ed. Yen T.F., Chilingarian G.V., Elsevier, Amsterdam, chap. **3,** 27.

32. Cerchez, V.T. and Anton, S., (1967), *Rev. Inst. Franç. du Pétrole,* **XXII,** 1818.

33. Combaz, A., (1974), *in: Advances in Organic Geochemistry,* 1973, ed. Tissot B. and Bienner F., Editions Technip, Paris, 423.

34. Correia, M. and Connan, J., (1974), *in: Advances in Organic Geochemistry,* 1973, ed. Tissot B. and Bienner F., Editions Technip, Paris, 153.

35. Dancy, T.E. and Giedroyc, V., (1950), *J. Inst. Petroleum,* **36,** 607.

36. Danilov, V.A., Yakovlev, V.I., Soloveichik, Z.V. and Proskuryakov, V.A., (1975-1976), *Okislenie Uglevodorodov i Kaustobiolitov, 45, 1975;* CA **84,** 76636a, 1976; Ref. Zh. Khim. 1975, Abstr. n° 20P46.

37. Danilov, V.A., Burlov, V.V., Yakovlev, V.I. and Proskuryakov, V.A., (1975-1976), *Okislenie Uglevodorodov i Kaustobiolitov,* **47,** 1975; CA **84,** 124246p, 1976; Ref. Zh. Khim. 1975, Abstr. n° 22P60.

38. Dashkovskii, I.D., Belotserkovskii, G.M., Yakovlev, V.I., Proskuryakov, V.A., Emelyanova, N.A., Chubarova, T.F. and Efremova, A.S., (1974), *Issled. Obl. Khim. Tekhnol. Prod. Pererab. Goryuch. Iskop.,* **1,** 26.

39. Degtereva, Z.A. and Fomina, A.S., (1959), *Izv. Akad. Nauk Est. SSR, Ser. Tekhn. i Fiz-Mat. Nauk,* **8,** 122.

40. Degtereva, Z.A. and Fomina, A.S., (1959), *Goryuch. Slantsy, Khim. i Tekhnol., Akad. Nauk Est. SSR, Inst. Khim.,* **3,** 5.

41. Degtereva, Z.A., (1962), *Candidate Thesis, Akad. Nauk Est. SSR, Tallin, USSR.*

42. Degtereva, Z.A., Fomina, A.S., Pobul, L.Ya. and Kyll, A.T., (1959), USSR Patent 115543, Nov. 29.

43. Degtereva, Z.A., Fomina, A.S. and Nutre, E.O., (1960), USSR Patent 127653, Apr. 12.

44. Dinneen, G.U., Smith, J.W., Tisot, P.R. and Robinson, W.E., (1968), *Proceedings of Symposium on the Development and Utilization of Oil Shale Resources, Tallin, USSR,* 22 p.

45. Djuričić, M., Murphy, R.C., Vitorović, D. and Biemann, K., (1971), *Geochim. et Cosmochim. Acta,* **35,** 1201.

46. Djuričić, M.V., Vitorović, D., Andresen, B.D., Hertz, H.S., Murphy, R.C., Preti, G. and Biemann, K., (1972), *in: Advances in Organic Geochemistry,* ed. von Gaertner H.R., Pergamon Press, Oxford, 305.

47. Down, A.L. and Himus, G.V., (1941), *J. Inst. Petroleum,* **27,** 426.

48. Egorkov, A.N., Mostetski, J., Yakovlev, V.I. and Proskuryakov, V.A., (1977), *Khim. Tverd. Topl.,* **2,** 53.

49. Egorkov, A.N., Mostetski, J., Yakovlev, V.I. and Proskuryakov, V.A., (1977), *Khim. Tverd. Topl.,* **2,** 57.

50. Erdman, J.G. and Ramsey, V.G., (1961), *Geochim. et Cosmochim. Acta,* **25,** 175.

51. Fester, J.I. and Robinson, W.E., (1966), "Coal Science", chap. **2,** *in: Advances Chem. Ser.,* **55,** 22.

52. Fester, J.I. and Robinson, W.E., (1964), *Anal. Chem.,* **36,** 1392.

53. Fester, J.I. and Robinson, W.E., (1964), American Conference on Coal Science, the Pennsylvanian State University, June 23.

54. Fomina, A.S., (1968), *Proceedings of Symposium on the Development and Utilization of Oil Shale Resources, Tallin, USSR,* 13 pp.

55. Fomina, A.S., (1968), *Proceedings of Symposium on the Development and Utilization of Oil Shale Resources, Tallin, USSR,* 28 pp.

56. Fomina, A.S., (1958), *Izv. Akad. Nauk SSR, Ser. Techn. i Fiz-Mat. Nauk,* **7,** 19; **7,** 91.

57. Fomina, A.S., (1959), *Genezis Tverd. Goryuch. Iskop., Akad. Nauk Est. SSR, Inst. Goryuch. Iskop.,* 77.

58. Fomina, A.S. and Degtereva, Z.A., (1956), *Izv. Akad. Nauk Est. SSR, Ser. Tekhn. i Fiz-Mat.,* **5,** 276.

59. Fomina, A.S. and Degtereva, Z.A., (1956), *Goryuch. Slantsy, Khim. i Tekhnol. Akad. Nauk Est. SSR, Inst. Khim.,* **2,** 7.

60. Fomina, A.S. and Degtereva, Z.A., (1958), *Tehn. ja Tootmine,* **5,** 14.

61. Fomina, A.S., Raig, H., Degtereva, Z.A., Veski, R. and Tiid, T., (1966), *Slants. Khim. Prom.,* **1,** 14.

62. Fomina, A.S. and Pobul, L. Ya., (1953), *Izv. Akad. Nauk Est. SSR,* **2,** 91.

63. Fomina, A.S. and Pobul, L. Ya., (1953), *Izv. Akad. Nauk Est. SSR,* **2,** 551.

64. Fomina, A.S. and Pobul, L. Ya., (1955), *Izv. Akad. Nauk Est. SSR,* **4,** 587.

65. Fomina, A.S. and Pobul, L. Ya., (1955), *Izv. Akad. Nauk Est. SSR,* **4,** 48.

66. Fomina, A.S., Pobul, L. Ya., Degtereva, Z.A. and Nappa, L., (1968), *Izv. Akad. Nauk Est. SSR, Ser. Tekhn. i Fiz-Mat.,* **17,** 139.

67. Fomina, A.S., Pobul, L. Ya., Mannik, A.O. and Veski, R.E., (1975), *First Republican Meeting on Oil Shales (Geochemistry and Lithology), Tallin, USSR, 1975, Book of Synopses,* 36.

68. Fomina, A.S., Pobul, L. Ya. and Degtereva, Z.A., (1959), *8th Mendeleev Meeting on General and Applied Chemistry and Chemical Technology of Fuels, USSR Academy of Sciences.*

69. Fomina, A.S., Pobul, L. Ya. and Degtereva, Z.A. (1965), *Izd. Akad. Nauk Est. SSR, Khim. Inst., Tallin, USSR, a review,* 215 pp.

70. Fomina, A.S., Veski, R.E. and Mannik, A.O., (1977), *Khim., Tverd. Topl.,* **3,** 170.

71. Fomina, A.S., Veski, R.E., Ilin, A.I., Kiviryak, S.V., Degtereva, Z.A., Tyanav, I.V. and Hanus, A.I., (1961), *Byul. Nauchn.-Tekhn. Inform. Goryuch. Slantsy,* **2,** 13.

72. Fomina, A.S. and Yerusenko, V., (1963), *Izv. Akad. Nauk Est. SSR, Ser. Tekhn. i Fiz-Mat.,* **12,** 189.

73. Fomina, A.S., Veski, R., Ilin, A.I. and Palvadre M., (1963), *Goryuch. Slantsy,* **6,** 24.

74. Forsman, J.P. and Hunt, J.M., (1958) *in: Habitat of Oil,* AAPG Symposium, ed. Weeks L.G.,747.

75. Forsman, J.P., (1963), "Geochemistry of Kerogen" *in: Organic Geochemistry,* ed. Breger I.A., Pergamon Press, New York, chap. **5.**

76. Galpern, G. and Vinogradova, J., (1936), *Neft. Khoz.,* **1,** 59.

77. Galpern, G. and Vinogradova, J., (1943), *Z. Anal. Chem.,* **125,** 47.

78. Gellis, Yu., Kozulin, N., Sokolov, V. and Shvortsman, Z., (1965), *Slants. Khim. Prom.,* **5,** 9; **6,** 13.

79. Grassely, Gy., Hetényi, M. and Agocs, M., (1973), *Acta Mineralogica-Petrographica, Szeged,* **21,** 55.

80. Gromova, V.V., Proskuryakov, V.A. and Yakovlev, V.I., (1967), *Zh. Prikl. Khim.,* **40,** 114.

81. Hoering, T.C., (1967), "The Organic Geochemistry of Precambrian Rocks", *in: Researches in Geochemistry,* **2,** ed. Abelson P.H., J. Wiley and Sons, New York, 87.

82. Hoering, T.C., (1964), "Annual Report of the Director of the Geophysical Laboratory 1963-1964", *Carnegie Inst. Washington Yearbook,* **63,** 258.

83. Hoering, T.C. and Abelson, P.H., (1965), *Carnegie Inst. Washington Yearbook,* **64,** 218.

84. Hunt, J.M., (1974), Preprint, *Soc. Petrol. Eng. AIME, Paper SPE* **5177.**

85. Hunt, J.M., "Geochemistry of Petroleum", *AAPG Continuing Education Lecture Series,* Woods Hole Oceanographic Institution, Woods Hole, Mass.

86. Hunt, J.M., (1977), *AAPG Bull.,* **61,** 100.

87. Hüsse, J., (1931), *Tehnika Ajakiri,* **10,** 115.

88. Jones, D.G. and Dickert, J.J., Jr., (1965), *Chem. Eng. Progr. Symp.,* Ser. **61,** 33.

89. Karavaev, N.M. and Vener, I.M., (1950), *Khimiya i Genezis Tverd. Goryuch. Iskop., Tr. Vses. Soveshch., Moskva,* 376.

90. Kaufmann, H.P. and Baltes, J., (1936), *Ber.,* **69,** 2679.

91. Kaufmann, H.P., (1958), "Analyse der Fette und Fettprodukte", Springer Verlag, Berlin.

92. Khalifeh, Y. and Louis, M., (1955), *Rev. Inst. Franç. du Pétrole,* **X,** 5, 340.

93. Khalifeh, Y. and Louis, M., (1958), *Rev. Inst. Franç. du Pétrole,* **XIII,** 1247.

94. Khalifeh, Y. and Louis, M., (1961), *Geochim. et Cosmochim. Acta,* **22,** 50.

95. Khristeva, L.A., (1938), *Zh. Prikl. Khim.,* **11,** 1506.

96. Khristeva, L.A., Kiosob, G.I., Dorofeev, I.M. and Veter, I.I., (1940), *Zh. Prikl. Khim.,* **13,** 132.

97. Kinney, C.R. and Leonard, J.T., (1961), *J. Chem. Eng. Data,* **6,** 474.

98. Kinney, C.R. and Schwartz, D., (1957), *Ind. Eng. Chem.,* **49,** 1125.

99. Kivirahk, S., Kercha, Yu. Yu. and Fomina, A.S., (1974), *Izv. Akad. Nauk Est. SSR, Ser. Khim. Geol.,* **23,** 31.

100. Kogerman, P., (1931), *Arch. Naturkunde Estlands, I Ser. Geol. Chem. et Phys.,* **10,** 6, 37, 66, 73, 82, Pt 2, 61.

101. Kogerman, P., (1935), *Keemia Teated,* **2,** 114.

102. Kogerman, P., (1949), *Issled. i Obz. Akad. Nauk Est. SSR,* **1,** 157.

103. Kogerman, P., (1938), "Oil Shale and Cannel Coal", *in: Proceedings of the Conference, Scotland, June 1938,* publ. by the Inst. of Petroleum, London, 115.

104. Kogerman, P., (1952), *Izv. Akad. Nauk Est. SSR,* **1,** 108.

105. Kogerman, P., (1927), *Sitzber. Naturforsch. Ges. Univ. Tartu,* **2,** 34, 166.

106. Kokurin, A.D., (1959), *Tr. Leningrad. Tekhnol. Inst. im. Lensoveta,* **51,** 58.

107. Kroepelin, H., (1967), *Beih. Geol. Jahrb.,* **58,** 499.

108. Krsmanović, V.D., Ercegovac, M. and Vitorović, D., (1976), The Microscopy of Organic Sediments. Coals and Cokes; Methods and Applications, April 1976, Oxford. Abstr. *in : Royal Microscopical Soc. Proceedings,* **11,** Pt 2, 88.

109. Kutuev, R.H., Yakovlev, V.I. and Proskuryakov, V.A., (1975), *First Republican Meeting on Oil Shales (Geochemistry and Lithology), Tallin, USSR, 1975. Book of Synopses,* 55.

110. Kutuev, R.H. and Yakovlev, V.I., (1975), *Okislenie Uglevodorodov i Kaustobiolitov,* 37.

111. Kutuev, R.H., Egudina, O.G., Yakovlev, V.I. and Proskuryakov, V.A., (1976), *Khim. Tverd. Topl.,* **2,** 108.

112. Kutuev, R.H., Yakovlev, V.I. and Proskuryakov, V.A., (1976), *Khim. Tverd. Topl.,* **4,** 3.

113. Lanin, V.A. and Pronina, M.V., (1944), *Izv. Akad. Nauk SSSR, Otd. Tekhn.* Nauk, **10-11,** 745.

114. Lawlor, D.L. and Robinson, W.E., (1965), *Am. Chem. Soc., Div. Petrol. Chem.,* Preprints, **10,** 5.

115. Lawlor, D.L., Fester, J.I. and Robinson, W.E., (1963), *Fuel,* **42,** 239.

116. "List of Bureau of Mines Publications on Oil Shale and Shale Oil, 1917-1968", (1969), compiled by M.P. Rogers, *US Bureau of Mines Inform. Circ. 8429.*

117. Mannik, E.P., Fomina, A.S., Mannik, A.O., Ikonopistseva, O.A. and Erm, A. Yu., (1972), *Khim. Tverd. Topl.,* **6,** 23.

118. Mannik, E.I., Fomina, A.S., Pehk, T.I. and Mannik, A.O., (1972), *Khim. Tverd. Topl.,* **3,** 142.

119. Mannik, E.I., Fomina, A.S., Kann, Yu. and Ikonopistseva, O., (1968), *Izv. Akad. Nauk Est. SSR, Ser. Khim. Geol.,* **17,** 118.

120. Mannik, E.I., Fomina, A.S. and Mannik, A.O., (1971), *Khim. Tverd. Topl.,* **3,** 106.

121. Mannik, E.I., Fomina, A.S. and Mannik, A.O., (1971), *Khim. Tverd. Topl.,* **4.**

122. Mannik, A.O., Pobul, L.Ya. and Saluste, S. Ya., (1975), *First Republican Meeting on Oil Shales (Geochemistry and Lithology), Tallin, USSR, 1975, Book of Synopses,* 77.

123. Manskaya, S.M., Kodina, L.A. and Generalova, V.N., (1974), *in: Advances in Organic Geochemistry,* 1973, ed. Tissot B. and Bienner F., Edition Technip, Paris, 97.

124. Mazumdar, B.K., Anaud, K.S., Roy, S.N. and Lahiri, A., (1957), *Brennstoff Chem.,* **38,** 305.

125. McIver, R.D., (1967), *Proceedings 7th World Petroleum Congress, Mexico City,* **2,** 25.

126. McKee, R.H. and Goodwin, R.T., (1923), *Colorado School of Mines Quart.*, **18**, Suppl. A, 1.

127. McKee, R.H. and Goodwin, R.T., (1923), *Ind. Eng. Chem.*, **15**, 343.

128. McKinney, J.W., (1924), *J. Amer. Chem. Soc.*, **46**, 968.

129. Mityurev, A.K., (1956), *Tr. Vses. Nauchn.-Issled. Inst. po Pererabotki Slantsev*, **6**, 79.

130. Murphy, R.C., (1970), Ph. D. Thesis, Dept. of Chem., Massachusetts Inst. of Technol., Cambridge, Mass.

131. Murphy, R.C., Biemann, K., Djuričić, M.V. and Vitorović, D., (1971), *Bull. Soc. Chim. Beograd*, **36**, 281.

132. Nappa, L. and Fomina, A.S., (1965), *Izv. Akad. Nauk Est. SSR, Ser. Fiz-Mat.i Tekhn. Nauk*, **14**, 163.

133. Needham, R.B. and Peacock, D.W., (1970), US Patent 3.499.490, March 10.

134. Nekrasov, V.I. and Urov, K.E., (1971), *Tr. Tallinsk. Politekhn. Inst.*, Ser. A, **311**, 79.

135. Oberlin, A., Boulmier, J.L. and Durand, B., (1974), *in: Advances in Organic Geochemistry*, 1973, ed. Tissot B. and Bienner F., Editions Technip, Paris, 15.

136. "Oil Shale Bibliography", Instituto de Quimica, Universidade Federal, Rio de Janeiro, Inst. Bras. Bibl. Doc. 1971, 5232 entries.

137. "Organic Geochemistry — Methods and Results", ed. Eglinton G. and Murphy M.T.J., Springer Verlag, Berlin, 1969.

138. Orlov, N.A., (1936), *Khim. Tverd. Topl.*, **7**, 419.

139. Orlov, N.A. and Radchenko, O.A., (1934), *Zh. Prikl. Khim.*, **7**, 1476.

140. Orlov, N.A. and Radchenko, O.A., (1934), *Khim. Tverd. Topl.*, **5**, 506.

141. Pata, E., (1968-1970), "Bibliography of Baltic Oil Shales, 1777-1968", 5 books, 7843 entries, Estonian Acad. Sci., Tallin, USSR. (in Russian).

142. Pehk, T., Veski, R. and Fomina, A.S., (1965), *Slants. Khim. Prom.*, **6**, 21.

143. Pehk, T., Mannik, E., Fomina, A.S., Laht, A. and Mannik, A., (1974), *Khim. Tverd. Topl.*, **1**, 102.

144. Pfendt, P. and Vitorović, D., (1973), *Erdöl, Kohle, Erdgas, Petrochem., Brennst. Chem.*, **26**, 143.

145. Philp, R.P., in press, *Scient. Amer.*, quoted according to the preprint.

146. Philp, R.P., Calvin, M., Brown, S. and Yang, E., in press, 25th Internat. Geological Congress, Sydney, Australia, August 16-23, 1976.

147. Philp, R.P. and Calvin, M., (1976), *Environmental Biogeochemistry, Proceedings Internat. Symposium*, ed. Nriagu J.O., Ann Arbor Sci., Ann Arbor, Mich., **1**, chap. **10**, 131.

148. Philp, R.P. and Calvin, M., (1976), *Nature*, **262**, 134.

149. Philp, R.P. and Yang, E., (1976), *Energy Sources*, **3**, 149.

150. Philp, R.P. and Calvin, M., (1977), *in: Advances in Organic Geochemistry*, 1975, ed. Campos R. and Goni J., Enadisma, Madrid, 735.

151. Pobul, L.Ya., (1957), Candidate Thesis, Akad. Nauk Est. SSR, Tallin.

152. Pobul, L.Ya. and Fomina, A.S., (1957), *Izv. Akad. Nauk Est. SSR, Ser. Tekhn. i Fiz-Mat. Nauk*, **6**, 190.

153. Pobul, L.Ya., Mannik, A.O., Fomina, A.S., Ikonopistseva, O.A. and Bondar, E.B., (1974), *Khim. Tverd. Topl.*, **3**, 115.

154. Polgar, A. and Jungnickel, J.L., (1963), "Olefinic Unsaturation", *in: Organic Analysis*, Interscience Publ., New York, vol. **III**, 234.

155. Polozov, V.F., (1954), *Tr. Vses. Nauchn.-Issled. Inst. po Pererabotki Slantsev*, **2**, 5.

156. Prien, C.H., (1951), "Oil Shale and Cannel Coal", *in: Proceedings 2nd Oil Shale and Cannel Coal Conference*, Inst. Petroleum, London, 76.

157. Proskuryakov, V., (1964), *Slants. Khim. Prom.*, **6**, 18.

158. Proskuryakov, V.A., Belotserkovskij, G.M. and Emelyanova, N.A., (1966), *Zh. Prikl. Khim.*, **39**, 2731.

159. Proskuryakov, V.A., Belotserkovskij, G.M. and Emelyanova, N.A., (1967), *Zh. Prikl. Khim.*, **40**, 109.

160. Proskuryakov, V.A. and Broi-Karre, G.V., (1963), *Khim. i Tekhnol. Topl. i Prod. Ego Pererab.*, **12**, 5.

161. Proskuryakov, V.A., Emelyanova, N.A. and Novoshilov, E.N., (1964), *Tr. Leningrad. Tekhnol. Inst. im. Lensoveta*, **63**, 45.

162. Proskuryakov, V.A. and Novoshilov, E.N., (1964), *Tr. Vses. Nauchn.-Issled. Inst. Pererabotki i Ispol'z. Topliva*, **13**, 5.

163. Proskuryakov, V.A., Soloveichik, Z.V., Shuvalov, V.I. and Yakovlev, V.I., (1968), *Khim. Tekhnol. Goryuch. Slantsev Prod. Ikh Pererab.*, 184.

164. Proskuryakov, V.A. and Soloveichik, Z.V., (1961), *Tr. Vses. Nauchn.-Issled. Inst. Pererabotki i Ispol'z. Topliva,* **10**, 64.

165. Proskuryakov, V.A. and Soloveichik, Z.V., (1961), *Tr. Vses. Nauchn.-Issled. Inst. Pererabotki i Ispol'z. Topliva,* **10**, 81.

166. Proskuryakov, V.A. and Soloveichik, Z.V., (1964), *Tr. Leningrad. Tekhnol. Inst. im. Lensoveta,* **63**, 68.

167. Proskuryakov, V.A. and Soloveichik, Z.V., (1964), *Tr. Leningrad. Tekhnol. Inst. im. Lensoveta,* **63**, 62.

168. Proskuryakov, V.A. and Soloveichik, Z.V., (1965), *Zh. Prikl. Khim.,* **38**, 389; **38**, 632.

169. Proskuryakov, V.A, Petrov, A.A. and Soloveichik, Z.V., (1966), *Zh. Prikl. Khim.,* **39**, 144.

170. Proskuryakov, V.A., Soloveichik, Z.V. and Gromova, V.V., (1965), *Zh. Prikl. Khim.,* **38**, 936.

171. Proskuryakov, V.A., Soloveichik, Z.V. and Shuvalov, V.I., (1968), *Zh. Prikl. Khim.,* **41**, 225.

172. Proskuryakov, V.A., Soloveichik, Z.V. and Yakolev, V.I., (1971), *Okislenie Uglevodorodov, Ikh Proizvod. Bitumov,* **9**, 76.

173. Proskuryakov, V.A. and Yakovlev, V.I., (1964), *Tr. Leningrad. Tekhnol. Inst. im. Lensoveta,* **63**, 39.

174. Proskuryakov, V.A., Yakovlev, V.I., Otrodnykh, T.F. and Lukin, A.J., (1975), *Okislenie Uglevodorodov i Kaustobiolitov,* 43.

175. Proskuryakov, V.A., Yakovlev, V.I. and Potekhin, V.M., (1963), *Tr. Vses. Nauchn. - Issled. Inst. Pererabotki i Ispol'z. Topliva,* **12**, 11.

176. Proskuryakov, V.A., Yakovlev, V.I. and Kurdyukov, O.I., (1962), *Tr. Vses. Nauchn. - Issled. Inst. Pererabotki i Ispol'z. Topliva,* **11**, 20.

177. Punga, V., Fomina, A.S., Degtereva, Z.A. and Palu, V., (1973), *Izv. Akad. Nauk Est. SSR, Ser. Khim. i Geol.,* **22**, 230.

178. Punga, V., Fomina, A.S., Degtereva, Z.A. and Palu, V., (1970), *Izv. Akad. Nauk Est. SSR, Ser. Khim. i Geol.,* **19**, 195.

179. Randall, R.B., Benger, M. and Groocock, C.M., (1938), *Proceedings Roy. Soc. London,* **165 A**, 432.

180. Raudsepp, K.T., (1954), *Izv. Akad. Nauk SSSR, Otd. Tekhn. Nauk,* **3**, 130.

181. Richter, W.J., Simoneit, B.R., Smith, D.H. and Burlingame, A.L., (1969), *Anal. Chem.,* **41**, 1392.

182. Robertson, J.B., (1914), *Proceedings Roy. Soc. Edinburgh,* **34**, 190.

183. Robin, P.L., Rouxhet, P.G. and Durand B., (1977), *in: Advances in Organic Geochemistry,* 1975, ed. Campos R. and Goni J., Enadisma, Madrid, 693.

184. Robinson, W.E., (1976), "Origin and Characteristics of Green River Oil Shale", *in: Oil Shale,* ed. Yen T.F. and Chilingarian G.V., Elsevier, Amsterdam, chap. **4**, 61.

185. Robinson, W.E., (1969), "Isolation Procedures for Kerogens and Associated Soluble Organic Materials", *in: Organic Geochemistry - Methods and Results,* ed. Eglinton G. and Murphy M.T.J., Springer Verlag, Berlin, chap. **6**.

186. Robinson, W.E., (1969), "Kerogen of the Green River Formation", *in: Organic Geochemistry - Methods and Results,* ed. Eglinton G. and Murphy M.T.J., Springer Verlag, Berlin, chap. **26**, 619.

187. Robinson, W.E. and Cook, G.L. (1971), *U.S. Bureau of Mines, Rep. Invest. 7492,* 32 pp.

188. Robinson, W.E. and Cook, G.L., (1975), *U.S. Bureau of Mines, Rep. Invest. 8017,* 40 pp.

189. Robinson, W.E. and Cook, G.L., (1973), *U.S. Bureau of Mines, Rep. Invest., 7820,* 32 pp.

190. Robinson, W.E., Cummins, J.J. and Stanfield, K.E., (1956), *Ind. Eng. Chem.,* **48**, 1134.

191. Robinson, W.E. and Dinneen, G.U., (1967), *Proceedings 7th World Petroleum Congress, Mexico City, 1967,* **3**, 669.

192. Robinson, W.E., Heady, H.H. and Hubbard, A.B., (1953), *Ind. Eng. Chem.,* **45**, 788.

193. Robinson, W.E. and Lawlor, D.L., (1961), *Fuel,* **40**, 375.

194. Robinson, W.E., Lawlor, D.L., Cummins, J.J. and Fester, J.I., (1963), *U.S. Bureau of Mines Rep. Invest. 6166,* 33 p.

195. Robinson, W.E. and Stanfield, K.E., (1960), "Constitution of Oil Shale Kerogen, Bibliography and Notes on Bureau of Mines Research", 237 entries, *U.S. Bureau of Mines Inform. Circ. 7968.*

196. Saxby, J.D., (1970), *Chem. Geol.,* **6**, 173.

197. Saxby, J.D., (1976), "Chemical Separation and Characterization of Kerogen from Oil Shale" *in: Oil Shale,* ed. Yen T.F. and Chilingarian G.V., Elsevier, Amsterdam, chap. **6**.

198. Semenov, S.S., (1954), *Izv. Akad. Nauk Est. SSR,* **3**, 391.

199. Semenov, S.S., Kornilova, Y.I. and Dokshina, N.D., (1959), *Tr. Vses. Nauchn.-Issled. Inst. Pererabotki i Ispol'z. Topliva,* **8**, 28.

200. Semenov, S.S., Kornilova, Y.I. Gurevich, B.E. and Orlova, N.S. (1955), *Tr. Vses. Nauchn. - Issled. Inst. po Pererabotki Slantsev,* **3**, 11.

201. Semenov, S.S. and Kornilova, Y.I., (1955), *Tr. Vses. Nauchn. - Issled. Inst. po Pererabotki Slantsev,* **3**, 5.

202. Shadrina, N.E., Gorobotsov, A.S., Shulman, A.J. and Proskuryakov, V.A., (1974), *Issled. Obl. Khim. Tekhnol. Prod. Pererab. Goryuch. Iskop.,* **1**, 52.

203. Siirde, A.K., (1956), *Tr. Tallinsk. Politekhn. Inst.,* Ser. A, **73**, 3.

204 Simoneit, B.R. and Burlingame, A.L., (1973), *Geochim. et Cosmochim. Acta,* **37**, 595.

205. Simoneit, B.R. and Burlingame, A.L., (1974), *in: Advances in Organic Geochemistry,* 1973, ed. Tissot B. and Bienner F., Editions Technip, Paris, 191.

206. Simoneit, B.R., Clews, L.A., Watts, C.D. and Maxwell, J.R., (1975), *Geochim. et Cosmochim. Acta,* **39**, 1143.

207. Souron, C., Boulet, R. and Espitalié, J., (1977), *in: Advances in Organic Geochemistry,* 1975, ed. Campos R. and Goni J., Enadisma, Madrid, 797.

208. Standard Oil Development Co., (1950), British Patent 636.033, April 19.

209. Stefanovic, Gj. and Vitorović, D., (1959), *J. Chem. Eng. Data,* **4**, 162.

210. Stefanović, Gj., Vitorović, D. and Djuričić, M., (1960-1961), *Bull. Soc. Chim. Beograd,* **25/26**, 411.

211. Stefanović, Gj., Vitorović, D. and Djuričić, M., (1960-1961), *Bull. Soc. Chim. Beograd,* **25/26**, 425.

212. Syskov, K.I. and Kukharenko, T.A., (1947), *Zavodsk. Lab.,* **13**, 25.

213. Teichmüller, M., Khalifeh, Y., Roucaché, J. and Louis, M., (1960), *Rev. Inst. Franç. du Pétrole,* **XV**, 1567.

214. Tissot, B., Durand, B., Espitalié, J. and Combaz, A., (1974), *AAPG Bull.,* **58**, 499.

215. Urov, K.E., (1975), *Khim. Tverd. Topl.,* **4**, 134.

216. Urov, K.E., Litvinovskaya, V.I. and Stoler, E.I., (1969), *Tr. Tallinsk. Politekhn. Inst., Ser. A,* **270**, 137.

217. Uspenski, V.A. and Radchenko, O.A., (1974), *in: Advances in Organic Geochemistry,* 1973, ed. Tissot B. and Bienner F., Editions Technip, Paris, 481.

218. Van den Berg, M.L.J., (1975), Ph. D. Thesis, Technische Hogeschool, Delft, Drukkerij Princo, Culemborg.

219. Van den Berg, M.L.J., Leeuw, J.W. de, and Schenck, P.A., (1974), *in: Advances in Organic Geochemistry,* 1973, ed. Tissot B. and Bienner F., Editions Technip, Paris, 163.

220. Van den Berg, M.L.J., Mulder, G.J., Leeuw, J.W. de, and Schenck, P.A., (1977), *Geochim. et Cosmochim. Acta,* **41**, 903.

221. Vassoyevitch, N.B., Akromkhodzhaev, A.M. and Geodekyan, A.A., (1974), *in: Advances in Organic Geochemistry,* 1973, ed. Tissot B. and Bienner F., Editions Technip, Paris, 309.

222. Veski, R.E., Bondar, E.B. and Fomina, A.S., (1975), *First Republican Meeting on Oil Shales (Geochemistry and Lithology), Tallin, USSR; Book of Synopses,* 80.

223. Veski, R.E., Fomina, A.S., Ilin, A.I. and Palvadre, M., (1964), *Slants. i Khim. Prom. Byul., Nauchn. Tekhn. Inform.,* **1-2**, 31.

224. Veski, R.E. and Fomina, A.S., (1968), *Goryuch. Slantsy,* **2**, 11.

225. Veski, R.E., Fomina, A.S., Room, A., Degtereva, Z.A., Pobul, L.Ya., Mannik, A.O. and Purn, A., (1971), *Khim. Tverd. Topl.,* **5**, 90.

226. Veski, R.E., Bondar, E.V. and Fomina, A.S., (1977), *Khim. Tverd. Topl.,* **4**, 93.

227. Vitorović, D., (1968), *Proceedings of Symposium on the Development and Utilization of Oil Shale Resources, Tallin, USSR,* 12 pp.

228. Vitorović, D., (1974), *Proceedings of Symposium on Science and Technology of Shale, Curitiba, Brasil, 1971,* ed. Brasilian Acad. Sci., 111-136.

229. Vitorović, D. and Djuričić, M., (1963), *Bull. Soc. Chim. Beograd,* **28**, 291.

230. Vitorović, D. and Djuričić, M., (1963), *Bull. Soc. Chim. Beograd,* **28**, 543.

231. Vitorović, D., Djuričić, M.V. and Ilić, B., (1974), *in: Advances in Organic Geochemistry,* 1973, ed. Tissot B. and Bienner F., Editions Technip, Paris, 179.

232. Vitorović, D., Krsmanović, V.D. and Pfendt, P., (1977), *in: Advances in Organic Geochemistry,* 1975, ed. Campos R. and Goni J., Enadisma, Madrid, 717.

233. Vitorović, D. and Pfendt, P., (1967), *Proceedings 7th World Petroleum Congress, Mexico City, 1967,* **3**, 691.

234. Vitorović, D. and Pfendt, P., (1968), *Bull. Soc. Chim. Beograd,* **33**, 557.

235. Vitorović, D. and Pfendt, P., (1974), *Anais Acad. Brasil. Cienc.,* **46**, 49.

236. Yen, T.F., (1976), "Structural Aspects of Organic Components in Oil Shales", *in: Oil Shale,* ed. Yen T.F. and Chilingarian G.V., Elsevier, Amsterdam, chap. **7**, 129.

237. Yerusenko, V. and Fomina, A.S., (1964), *Izv. Akad. Nauk Est. SSR, Ser. Fiz.-Mat. i Tekhn. Nauk,* **13**, 319.

238. Yerusenko, V. and Fomina, A.S., (1966), *Izv. Akad. Nauk Est. SSR, Ser. Fiz. - Mat. i Tekhn. Nauk,* **15**, 106.

239. Young, D.K., Shih, S. and Yen, T.F., (1974), *Amer. Chem. Soc., Div. Fuel Chem.,* Preprint 19, **2**, 169.

240. Young, D.K., Shih, S. and Yen, T.F., (1976), "Mild Oxidation of Bioleached Oil Shale", *in: Science and Technology of Oil Shale,* ed. Yen T.F., Ann Arbor Sci. Publ., Ann Arbor, 65.

241. Young, D.K. and Yen, T.F., (1977), *Geochim. et Cosmochim. Acta,* **41**, 1411.

11

Pétrographie du kérogène

B. ALPERN *

I. INTRODUCTION. CONSIDÉRATIONS CRITIQUES SUR LE CONCEPT DE KÉROGÈNE

Le concept traditionnel de **kérogène** tel qu'il ressort de la littérature (cf. Foreword) est malheureusement ambigu car il ne sépare pas ce qui revient, d'une part, au **rang** (catagenèse), d'autre part au **type** (nature).

La houillification, lors de sa progression, est nécessairement productrice de kérogène puisqu'au stade terminal de l'évolution toutes les matières organiques (MO) sont insolubles. Il en résulte que, réciproquement, la solubilité est un caractère essentiellement fugitif et transitoire.

D'un autre côté, certains constituants sont insolubles dès l'origine (lignine par exemple) et certains processus primaires précoces, tels la fusinisation, renforcent encore le cheminement vers l'insolubilité.

Au total, le même vocable de kérogène recouvrira par exemple des substances aussi fondamentalement différentes que :

1. *les anthracites,* produit final de la catagenèse, ayant fourni en général surtout du gaz. Il ne s'agit plus alors de *MO dispersée* dans une matrice minérale, mais de *MO concentrée* à ciment organique, donc d'une roche différente des roches à kérogène dispersé;
2. *la fusinite,* constituant oxydé dès le départ et qui ne produira pas, de ce fait, des hydrocarbures;
3. le kérogène insoluble dispersé classique.

La notion de kérogène est également critiquable du fait qu'elle est liée à une pratique, *non standardisée,* et pas à une définition scientifique.

Le fait d'ailleurs de placer la pétrographie organique dans la géochimie témoigne également de positions scientifiquement critiquables sur ces problèmes. La chimie est une des méthodes qui peuvent aider le naturaliste et s'il peut y avoir réciprocité, il y a cependant un ordre logique dans les investigations qui ne doit pas être oublié sous peine de pénalité finale.

---

* Chef du Laboratoire de Pétrographie Organique et Conseiller Scientifique (jusqu'au 1.5.78). CERCHAR. Verneuil-en-Halatte, France.
Directeur du Groupe d'Etude des Combustibles Fossiles. Université d'Orléans, 45045 Cedex France.

C'est le cas par exemple des matières volatiles pour les charbons lorsqu'on veut leur faire traduire la seule houillification en oubliant qu'elles traduisent aussi le type pétrographique, d'où la crise actuelle qui frappe la classification internationale des combustibles (B. Alpern, publication en cours).

C'est la démarche naturaliste qui doit être, chronologiquement bien sûr, la première. Elle doit créer les concepts que la chimie, la physique, les mathématiques, lui permettront d'affiner ou de modifier. La démarche inverse est souvent une impasse.

C'est pourquoi nous proposons de compléter la notion chimique de kérogène, basée sur la seule insolubilité, par un concept génétique recouvrant « l'ensemble des constituants organiques des sédiments avec tous les produits de dégradation et de néoformation qui leur correspondent, quel que soit leur stade d'évolution donc leur degré de solubilité, et quels que soient leur âge géologique et le type de roche dans lequel ils se trouvent ». Nous avons proposé pour cela, faute d'avoir trouvé un meilleur terme, le mot d'**organophase** (voir ci-dessous). Sur cette base et à partir des considérations précédentes, le projet suivant pourrait constituer une solution sur le plan de la nomenclature :

A. Concepts pétrographiques et génétiques	
Phase organique totale des sédiments : solides, liquides, gaz	**Organophase**
Constituants organiques reconnaissables en microscopie	**Organoclastes**
Produits décomposés ou néoformés, solubles ou insolubles, autochtones (protobitumes) ou migrés (migrabitumes)	**Bitumes**
Note : Dans notre projet, les Bitumes font partie des organoclastes	
B. Concepts chimiques	
Phase organique soluble des sédiments	**Bitumoides**
Phase organique insoluble des sédiments	**Kérogène**

Cette solution lève la contradiction qui provenait de l'existence de nombreux « bitumes », au sens pétrographique et au sens général du terme, totalement insolubles (anthraxolite, shungite, etc.), donc contraires à la définition chimique.

Le terme « Bitumoïde », repris de la littérature russe (O. Radchenko) (1), permet de garder au terme de Bitume son sens le plus général. Par ailleurs, si le kérogène conserve un sens chimique strict, rien ne s'oppose alors à ce qu'il s'applique aussi bien à la MO dispersée que concentrée (charbons).

II. PÉTROGRAPHIE ORGANIQUE ET KÉROGÈNE

Il convient maintenant, sur la base de ce qui est dit dans l'introduction ci-dessus, de bien définir l'objet de la pétrographie organique et de comparer cet objet au kérogène du chimiste d'une part, à la MO totale réellement présente d'autre part.

(1) Lettre personnelle en réponse à une enquête sur les Bitumes (Commission MO dispersée du *Comité International de Pétrographie des Charbons).*

1. Si nous conservons la conception chimique classique et définissons le bitume comme la fraction extractible (chloroforme, benzène, etc.) et le kérogène comme la fraction insoluble de la MO fossile, il apparaît que, dans une certaine mesure, le pétrographe voit plus que le seul kérogène puisqu'il observe parfois des produits fluorescents qui se solubilisent dans l'huile d'immersion et la résine d'enrobage et qui seraient donc des « bitumes » au sens chimique du terme.

2. D'un autre côté, travaillant au microscope, le pétrographe ne voit pas, par définition, ce qui est *submicroscopique* et qui peut ne pas être négligeable. En lumière réfléchie notamment, le polissage requiert une dimension minimale des particules d'au moins 5 μ pour permettre une photométrie correcte, même si le champ mesuré n'est que de 1 μ de diamètre (effets de bordure et de bombement). En fluorescence, par contre, des objets de l'ordre du micron sont aisément détectés et localisés. En outre, des inclusions ou imprégnations organiques dans les cristaux, notamment carbonatés, sont visibles par cette technique alors qu'ils ne sont probablement pas extractibles (pl. 11.1).

3. Les méthodes de préparation peuvent introduire une distorsion modifiant les proportions initiales des constituants organiques. Les séparations par voie chimique, lorsqu'elles sont bien faites, permettent de concentrer de la matière organique colloïdale souvent abondante mais perdue lors de la concentration par voie physique. Les organismes figurés ont la proportion originale qu'ils avaient dans la roche et qui peut être très faible. Inversement, la voie physique suivie au *Centre d'Etudes et de Recherches des Charbonnages de France (CERCHAR)* (broyage $< 200\ \mu$, dispersion par ultra-sons, centrifugation en milieu dense $d = 1{,}7$) permet une concentration remarquable des particules et un renforcement, d'ailleurs recherché, du nombre des organismes figurés. Cette procédure permet l'exploitation (palynologie, réflectométrie) de roches très pauvres qui seraient inutilisables par d'autres techniques. Ajoutons que les techniques chimiques modifient la fluorescence des palynomorphes et qu'on cherche de ce fait à les éviter. Pour illustrer l'influence du mode de préparation sur la composition de l'ensemble organique observé au microscope, nous prendrons l'exemple suivant, établi sur cinq schistes bitumineux du Bassin de Paris :

	Roche brute (n = 5) (%)	Concentré organique (n = 5)	
		chimique (kérogène IFP) (%)	physique (concentré CERCHAR) (%)
Algues et particules fluorescentes	12,01	25,40	58,14
Particules gélifiées	4,71	6,05	5,41
Particules fusinisées	0,66	1,03	10,49
	5,37	7.08	15,90
Fond fluorescent	76,12	60,68	0

On constate très clairement que la méthode physique concentre de préférence les micro-organismes figurés qui augmentent de 12 à 58 % et l'ensemble des particules, tant gélifiées que fusinisées (augmentation de 5 à 16 %). Par contre, la masse organique amorphe fondamentale est perdue alors qu'elle représente en fait le constituant prédominant du cortège organique (76 %).

La diminution de la proportion de fond fluorescent entre roche brute et kérogène provient du fait qu'une fraction minérale des schistes est fluorescente comme il a été établi après action thermique [6]. Cette fraction disparaît lors de la déminéralisation chimique et le fond fluorescent du kérogène représente bien alors le total de la seule MO amorphe présente dans la roche. Notons à ce propos que le rapport MO fluorescente/MO réfléchissante égal à 86/7 (sur kérogène) aura des conséquences lors du choix du meilleur « paléothermomètre » (voir ci-dessous), dont on aura intérêt à ce qu'il soit le plus abondant.

En définitive, l'observation directe des roches brutes, lorsqu'elle est possible, constitue une étape importante de la démarche du pétrographe dans le cas des roches suffisamment riches en matière organique.

4. Précisons d'emblée qu'une partie non négligeable de la MO dite « amorphe » correspond probablement à la composante nanoplanctonique, donc organisée, de la MO fossile. Même en microscopie optique, de nombreuses microlamelles se révèlent être circulaires, à contours définis, dans les sections polies // au plan de stratification et sont probablement des microalgues : Schizophytes, Nostococaccées [39]. Avec le microscope électronique, de très belles structures de Coccolithophoridées ont été mises en évidence. De ce fait, le qualificatif d'amorphe doit être considéré avec prudence.

5. Il faut noter aussi que la réflectométrie se pratique sur un matériel d'origine lignocellulosique, souvent négligeable dans les kérogènes du type I [49] où il a toutes les chances d'être remanié. La « paléothermométrie » (1) porterait alors sur des constituants à la fois quantitativement accessoires et douteux quant à leur syngénétisme. Il existe heureusement une parade à cet inconvénient : la fluorescence (voir ci-dessous) qui porte sur le matériel pollinique et algaire majoritaire dans les kérogènes I et II [48,7].

6. Enfin, il faut reconnaître que la diagnose de microdébris isolés de leur contexte lithologique est parfois très malaisée. Dans les roches inorganiques, un microfragment amorphe peut provenir de familles très différentes sans qu'il soit toujours possible de trancher avec certitude (voir ci-dessous). C'est encore plus difficile lorsque ces microdébris sont isolés de la roche de départ. En contexte charbonneux, la vitrinite de base est là pour faire référence et étalonner les réflectances des autres macéraux et donc aider à leur diagnose.

III. CLASSIFICATION PÉTROGRAPHIQUE DE LA MATIÈRE ORGANIQUE DISPERSÉE DANS LES SÉDIMENTS

La matière organique dispersée des sédiments, que nous noterons dans ce qui suit MOD, est le résidu de la fossilisation de nombreuses familles naturelles d'origine botanique et zoologique.

---

(1) Nous verrons ci-dessous les restrictions qu'il faut apporter à cette notion.

A. Diagenèse et catagenèse.

La matière organique fossilisée subit, pratiquement dès l'origine, un processus de décomposition biochimique (diagenèse) puis de maturation thermique (catagenèse) que nous subdiviserons en utilisant la terminologie suivante qui ne représente qu'une possibilité parmi d'autres :

Phase biochimique	Phase géothermique		
Diagenèse		Catagenèse	
	Epigenèse	Mesogenèse	Métagenèse
Tourbe	Lignite	Houille	Anthracite
Réflectance (%)	0,2	0,5	2,5

Nous avons repris dans ce tableau les termes de diagenèse et de catagenèse en partie au sens de la littérature russe, mais leurs limites sont encore discutées et ces notions sont comprises très différemment d'un pays à l'autre; il faut donc les utiliser avec prudence.

L'avantage de notre choix — outre la quantification par la réflectance et la précision de langage qui en découle — est de faire correspondre l'ensemble de la houillification : des lignites aux anthracites inclus, à la seule catagenèse. Nous n'avons pas encore pour l'instant de limite à proposer entre la catagenèse et l'anchimétamorphisme, mais il est certain qu'une limite objective pourrait être trouvée.

La houillification se traduit, non seulement par une transformation profonde et convergente des matières diversifiées du départ, mais par l'apparition de produits de néoformation, l'expulsion éventuelle de certains d'entre eux sous forme de liquide ou de gaz et, réciproquement, l'incorporation de certains autres ayant subi une migration.

Devant un complexe aussi vaste, il paraît normal de suivre la démarche classique du naturaliste consistant à inventorier, caractériser (diagnose), nommer (nomenclature) et enfin classer (systématique) les objets qu'il étudie.

B. Pétrographie organique et pétrographie des charbons.

Dans un premier temps, les pétrographes de la MOD ont tout naturellement utilisé les termes et la classification (tableau 11.1) des pétrographes du charbon, qui avaient le mérite d'exister et de fournir une base pour la réflectométrie.

Une grande partie de la littérature, aussi bien passée qu'actuelle, porte, d'une manière globale, sur le terme de **vitrinite** sans qu'il soit connu des non-spécialistes que la définition, comme les subdivisions de ce concept, ont été l'objet d'un long et difficile travail d'équipe du groupe de l'*International Committee for Coal Petrology (ICCP),* qui avait à intégrer, en un compromis acceptable, à la fois les données scientifiques modernes et le passé bibliographique relatif à ce terme. Le stade actuel de la nomenclature est donc autant historique que purement logique.

TABLEAU 11.1
LISTE COMPARÉE DES MACÉRAUX DES LIGNITES ET DES HOUILLES

MACERAUX DES LIGNITES				MACERAUX DES HOUILLES		
SUBMACERAL	MACERAL	SOUS-GROUPE	GROUPE	GROUPE	MACERAL	SUBMACERAL
	TEXTINITE	HUMOTELINITE	HUMINITE	VITRINITE	TELINITE	Telinite 1 Telinite 2
Texto-ulminite Eu-ulminite	ULMINITE					
Porigelinite Levigelinite	GELINITE	HUMOCOLLINITE			COLLINITE	Telocollinite Desmocollinite
Phlobaphinite Pseudophlobaphinite	CORPOHUMINITE					Gelocollinite Corpocollinite
	ATTRINITE DENSINITE	HUMODETRINITE			VITRODETRINITE	
	SPORINITE CUTINITE RESINITE SUBERINITE ALGINITE LIPTODETRINITE CHLOROPHILLINITE BITUMINITE (FLUORINITE)* (EXSUDATINITE)*		EXINITE ou LIPTINITE	EXINITE ou LIPTINITE	SPORINITE CUTINITE RESINITE ALGINITE LIPTODETRINITE (BITUMINITE)* (FLUORINITE)* (EXSUDATINITE)*	
	FUSINITE SEMI-FUSINITE MACRINITE SCLEROTINITE INERTODETRINITE		INERTINITE	INERTINITE	FUSINITE SEMI-FUSINITE MACRINITE MICRINITE SCLEROTINITE INERTODETRINITE	Pyrofusinite Degradofusinite

* terme en cours de discussion

A l'origine, la vitrinite, définie à partir des Houilles ne se subdivisait qu'en deux macéraux :

— *télinite,* tissu gélifié mais avec structure botanique encore visible;
— *collinite,* gel vrai sans structure.

Il est apparu, au cours du temps, que la collinite n'était pas formée que de gels, mais comprenait aussi de nombreux *tissus totalement gélifiés* et reconnaissables seulement par attaque chimique (ou d'autres procédés), d'où le sub-macéral « télocollinite » qu'on trouvera dans la littérature sous les noms de « vitrinite A » [15] et de « homocollinite » [1] qui l'on précédé.

Par ailleurs, les gels vrais pouvaient être, soit purs : gélocollinite, soit de véritables mixtes submicroscopiques : la desmocollinite (vitrinite B et hétérocollinite, des mêmes auteurs) se traduisant par une réflectance légèrement plus faible que la télocollinite au stade de l'épigenèse.

Pour les lignites, dont la nomenclature a été fixée plus tardivement, l'**huminite,** équivalent et précurseur de la vitrinite, comporte des divisions similaires, mais aussi des catégories complémentaires marquant, notamment, l'achèvement de la *compaction* (tableau 11.1) :

— porigélinite puis lévigélinite (compacte);
— attrinite (poreuse) puis densinite (compacte).

La série qui intéresse le pétrographe pour les mesures de houillification est celle des gels vrais et des tissus totalement gélifiés qui ont une réflectance à peu près égale :

— tissus : eu-ulminite → télocollinite,
— gels : lévigélinite → gélocollinite et desmocollinite.

C. Difficultés de la pétrographie organique.

Au cours du temps, diverses difficultés sont apparues lorsqu'on a voulu étudier les sédiments et non plus seulement les charbons.

1. Toutes les particules amorphes, grises en lumière réfléchie, ne sont pas pour autant de la gélinite-collinite; on peut en effet retrouver les mêmes aspects, surtout dans le cas de très petites particules, à partir (pl. 11.2) :

— de cryptotissus dont la structure, comme nous l'avons vu ci-dessus, n'apparaît qu'après attaque chimique (pl. 11.2, photo 1);
— de bitumes primaires ou migrés (pl. 11.2, photos 2 et 3);
— de fragments de mégaspores et cuticules évoluées (méso et métagenèse) (pl. 11.2 et photos 4 et 5);
— d'inertinite massive de basse réflectance (macrinite grise);
— de Graptolithes (pl. 11.2 photo 7) et de Chitinozoaires (pl. 11.2, photo 8);
— de restes indéterminés et provenant probablement de zoo-épidermes (pl. 11.2, photo 9).

Au total donc, de nombreuses interprétations sont possibles, en particulier lorsque le constituant est microfragmenté et séparé de son contexte lithologique, sans parler du *remaniement* qui est très fréquent et de la *contamination* (retombées et additifs tourbeux aux boues de forage) qui sont de règle dans le cas des cuttings. Nous ne mentionnons pas, dans cette liste.

des confusions possibles, la **pseudovitrinite** des auteurs américains [11] et la **semivitrinite** reconnue par la norme russe, dont les réflectances sont légèrement plus élevées que celle de la vitrinite correspondante et qui ne se décèlent que si cette dernière est présente. Ces termes démontrent l'existence de stades transitoires continus entre vitrinite et semifusinite (pl. 11.2, photo 6).

2. La pétrographie des charbons constitue, comme nous l'avons vu, un ensemble historique bien fixé, fondé sur les concepts de **macéral** (voir tableau 11.1) constituant élémentaire et de **microlithotype,** mélange de macéraux (> 5 %) en lits de plus de 50 μ.

Ce cadre, relativement rigide, bien adapté au cas des charbons formés à partir de l'accumulation de restes de plantes supérieures sur les continents à partir du Dévonien, n'est pas adapté au classement de nouveaux constituants, notamment ceux provenant du règne animal et de ses produits de décomposition, ni au classement des hydrocarbures et bitumes solides, *dans tous les types de roches et pour tous les âges,* du Précambrien à l'actuel.

Tout nouveau macéral doit effectivement, soit entrer dans l'un des trois groupes existants : vitrinite, exinite (liptinite), inertinite, et en modifier ainsi la définition initiale, soit entrer dans un ou plusieurs groupes de macéraux supplémentaires qui n'existent pas pour l'instant.

Rappelons très brièvement que la *vitrinite* (voir ci-dessus) correspond en principe à des produits gélifiés (cryptotissus et gels vrais), que l'*inertinite* regroupe des produits fusinisés et devenus de ce fait inertes du point de vue des propriétés cokéfiantes (pas de phase plastique lors de la pyrolyse), et que l'*exinite* (liptinite) enfin rassemble : les exines des spores fossiles, les cuticules, les produits de sécrétion des plantes supérieures (huiles végétales, résines et tanins). A ce groupe appartiennent aussi les *Algues* en raison de l'analogie optique avec les spores (basse réflectance, forte fluorescence), mais ce regroupement est critiquable du fait du chimisme différent (1) ainsi que de la paléogéographie distincte. Les Algues, organismes autonomes, ont d'ailleurs des propriétés optiques nettement différentes des spores.

La récente inclusion [47] dans le groupe de l'exinite de nouveaux macéraux fluorescents (bituminite, fluorinite, exudatinite) correspondant à des bitumes primaires, des secrétions cellulaires et des exsudats néoformés, outre qu'elle implique un changement dans le concept d'exinite, conduit à en faire un groupe encore plus hétérogène et risquant à la longue de devenir démesuré si l'on incluait, toujours pour des raisons d'analogie optique, des restes zoo- et phyto- planctoniques. Le fait de remplacer exinite par liptinite afin d'élargir le concept est une fausse solution, les deux termes étant et devant rester rigoureusement synonymes [36]. S'ils ne l'étaient pas, on aurait deux groupes différents, ce qui modifierait fondamentalement les bases de la nomenclature macérale.

Inversement, si l'on met les nouveaux constituants hors des trois groupes classiques et que l'on crée un nouveau groupe (par exemple bituminite), on déclenche automatiquement la création d'une série nouvelle de *microlithotypes* (pour les mélanges qui le contiennent), ce dernier concept devenant alors pratiquement inutilisable.

---

(1) Lignée évolutive de l'alginite [34] et des kérogènes I [49].

D. Essai de classification des constituants du kérogène.

Pour tenter de surmonter ces difficultés, j'ai essayé (*29e réunion ICCP, Newcastle 1976*) d'élaborer une classification totalement ouverte et souple, indépendante de la terminologie charbonnière classique. Cette classification naturelle (voir tableau 11.2) permet :

— de redonner à la nature originelle de l'objet la priorité sur le comportement à la carbonisation et de regrouper, par exemple, tous les tissus botaniques de plantes supérieures alors que dans la nomenclature macérale classique un tissu végétal change de nom de groupe selon qu'il a été gélifié (vitrinite) ou fusinisé (inertinite);
— de traiter les bitumes dans une catégorie séparée;
— d'ouvrir des possibilités pour classer chaque type d'organisme animal ou végétal;
— de séparer les spores des algues, donc les lignées évolutives distinguées déjà par Van Krevelen, ainsi que les divers types de kérogènes [49];
— de ne pas imposer le changement de nom lorsqu'un même constituant franchit la réflectance 0,5 % (gélinite-collinite), ce qui se produit parfois dans une même population.

J'ai tenté, en outre, de mettre en évidence les meilleurs indicateurs thermiques et de caractériser les zones évolutives dans lesquelles ils se trouvent au moyen d'un préfixe : épi, méso, méta, identique pour tous les constituants.

J'ai proposé enfin [5] de désigner ces constituants de la MOD par le terme « d'organoclaste » plus restreint qu'organolite [3] mais plus large que macéral et que phytoclaste [3]. Ce concept recouvre la particule dans son ensemble, indépendamment de sa taille, de la proportion respective de ses constituants, s'il s'agit d'un mixte, et de la position du réticule oculaire.

En effet, dans la pétrographie des charbons classique, le nom d'une particule est donné par le macéral qui est au point de croisement du réticule oculaire. Ainsi, pour une particule mixte à ciment de collinite contenant un peu de micrinite, d'inertodétrinite, une microspore et de la pyrite finement dispersée, on pourra être amené à utiliser cinq noms différents selon les positions respectives du grain et du réticule.

Cette procédure est statistiquement défendable avec des analyses macérales portant sur 500 points. Avec des organoclastes, souvent beaucoup moins nombreux, une dispersion importante risquera d'apparaître si on aborde un jour l'étape des analyses quantitatives des constituants. Avec la conception proposée une seule désignation est possible au lieu de cinq; de plus, le préfixe indique la zone de maturation et donne de ce fait une information plus complète.

On pourra critiquer la racine « claste » qui s'applique mal aux gels et aux bitumes mais, faute de mieux, nous donnons ici à organoclaste le sens plus général de *microconstituant organique,* aussi bien détritique que néoformé.

L'organoclaste intègre donc les notions de macéral et de microlithotype; il est moins marqué par les propriétés cokéfiantes et plus orienté vers les applications géologiques vers lesquelles notre système est essentiellement orienté. Les macéraux de la pétrographie des charbons classique ne couvrent qu'une partie seulement du tableau proposé.

Les catégories que la notion d'organoclaste (pl. 11.3) permet d'introduire sont illimitées. Nous ne reviendrons pas sur les groupes traités par la pétrographie des charbons mais placés différemment dans notre système : tissus, gels, détrites (mixtes), épidermes, sécrétions, spores, etc.

TABLEAU 11.2

PROJET DE CLASSIFICATION DES ORGANOCLASTES (B. ALPERN, 1978)

Groupe	Division	Sous-division	Termes
I PHYTOCLASTES faciès continental plantes supérieures		Xylites (tissus)	Fusitextites Gélitextites : Epitextites, Mésotextites, Métatextites
		Gélites (gels lignocellulosiques)	Epigélites Mésogélites Métagélites
		Détrites (mixtures)	Epidétrites Mésodétrites Métadétrites
		Epidermes : cuticules, suber Inclusions : huiles, cires, résines, tanins	
II FUNGI		Pseudoxylites (hyphes) Sclérotes Chitinomycètes et champignons marins	
III SPORES transgression du faciès		Sporites (plantes vascul.)	Episporites Mésosporites Métasporites
IV ORGANISMES autonomes, lacustres et marins	Phyto	Algues benthiques : Phéophycées, Rhodophycées	
	Phyto	Algues planctoniques	Chlorophytes : Botryococcales, Tasmanales Chromophytes : Diatomées, Silicoflagellés, Coccolithophoracées Schizophytes, Nostococcacées
	Phyto	Sapropélites (gels algaires et bactériens)	
	Incertae Sedis	Acritarches Chitinozoaires	
	Zoo	Zooplancton	Foraminifères (p.p.) Radiolaires, Ciliés Crustacés (Copepodes, Ostracodes) Graptolithes
	Zoo	Zoobenthos	Vers (Scolécodontes) Mollusques (p.p.) Cœlentérés Foraminifères (p.p.) Spongiaires Arthropodes (Trilobites)
	Zoo	Necton : Poissons, Céphalopodes, etc	
V _ BACTERIES		Eubactéries, Sidérobactéries, Thiobactéries, etc	
VI BITUMES		Protobitumes (dispersés, en place)	Epibitumes Mésobitumes Métabitumes
		Migrabitumes (déplacés) : terminologie en cours d'unification	
		Pyrobitumes (cracking)	sphérolites anisotropes, enduits des fissures inclusions cristallines sédimentaires
		Abiobitumes Cokes naturels	inclusions cristallines ignées coacervats, météorites, etc métamorphisme de contact des charbons

Pour les Algues, deux groupes commencent à être connus et utilisés : Botryococcacées et Tasmanacées. Comme il s'agit d'organismes autonomes complets, que leur position dans le diagramme H/C-O/C est différente de celle des spores, que la fluorescence entre spores et algues est aussi très différente (le cas est particulièrement net pour le Toarcien), nous pensons qu'il est préférable de disjoindre les algues de l'exinite et de considérer ces deux catégories séparément, certains phénomènes géologiques ne pouvant être interprétés correctement lorsque l'on prend le groupe de l'exinite dans la conception globale de l'*ICCP.* Des exemples de courbes évolutives de Tasmanacées sont donnés ci-dessous (fig. 11.5).

Pour les autres Algues, notamment celles de très petite taille, décrites sous le nom de Nostocopsis [39], les mesures sont encore à faire, mais elles seront plus difficiles étant donné la taille. Il faudra utiliser des sections // à la stratification. Pour les Zoorestes et le Zooplancton, tout le travail de diagnose et de mesure est encore à faire, il est à peine amorcé pour le groupe des Chitinozoaires (ou peut-être Chitinomycetes [37]) dont la réflectance est proche de celle des vitrinites. Le groupe des Bactéries, qui a dû jouer un rôle important avec les Schizophytes dans le Précambrien, est encore totalement à explorer.

Pour les bitumes nous avons adopté la solution suivante :

a) les *protobitumes* regroupent tous les bitumes primaires *autochtones* qui peuvent donc servir d'indicateurs paléothermiques. Ils sont divisés en trois en fonction de la fluorescence (stade épi) et de l'anisotropie entre nicols croisés (stade méta);

b) les *migrabitumes,* correspondent à l'ensemble des produits filoniens ou de remplissage des vides, donc migrés, et qui ont reçu, lorsqu'ils sont en grande masse, des noms locaux : albertite, gilsonite, grahamite, impsonite, wurtzlite, etc. Cette nomenclature est de ce fait très désordonnée; elle serait à unifier sur une base qui reste encore à déterminer, mais qui ne peut pas être uniquement optique. De toute manière ces constituants, migrés parfois tardivement, ne peuvent être considérés comme des paléothermomètres valables;

c) enfin la catégorie des *pyrrobitumes* peut être réservée aux sphérolites de carbone pyrolitique, formés probablement par cracking thermique gazeux, et que l'on retrouve sur les parois de certains fours industriels ainsi que dans des charbons de divers rangs;

d) le problème des *cokes naturels* resterait encore à traiter, probablement dans un groupe tout à fait à part. Les *abio-bitumes* (1) (roches ignées, météorites carbonées) vrais ou hypothétiques sont également à étudier.

IV. SÉLECTION DES INDICATEURS OPTIQUES DE LA HOUILLIFICATION

Le pétrographe de la MOD cherche à exploiter au mieux la relation irréversible qui lie l'histoire d'une formation sédimentaire et les propriétés optiques des inclusions organiques syngénétiques que celle-ci renferme. Dans une large mesure (voir ci-dessous) les transformations sont liées aux variations de température et c'est en définitive l'histoire thermique des sédiments que le pétrographe tente de reconstituer.

Dans la masse des organoclastes que le pétrographe peut rencontrer (voir chap. 3) certains sont mieux adaptés que d'autres à l'enregistrement des effets thermiques.

---

(1) Terme proposé par J. Connan en réponse à une enquête sur les bitumes.

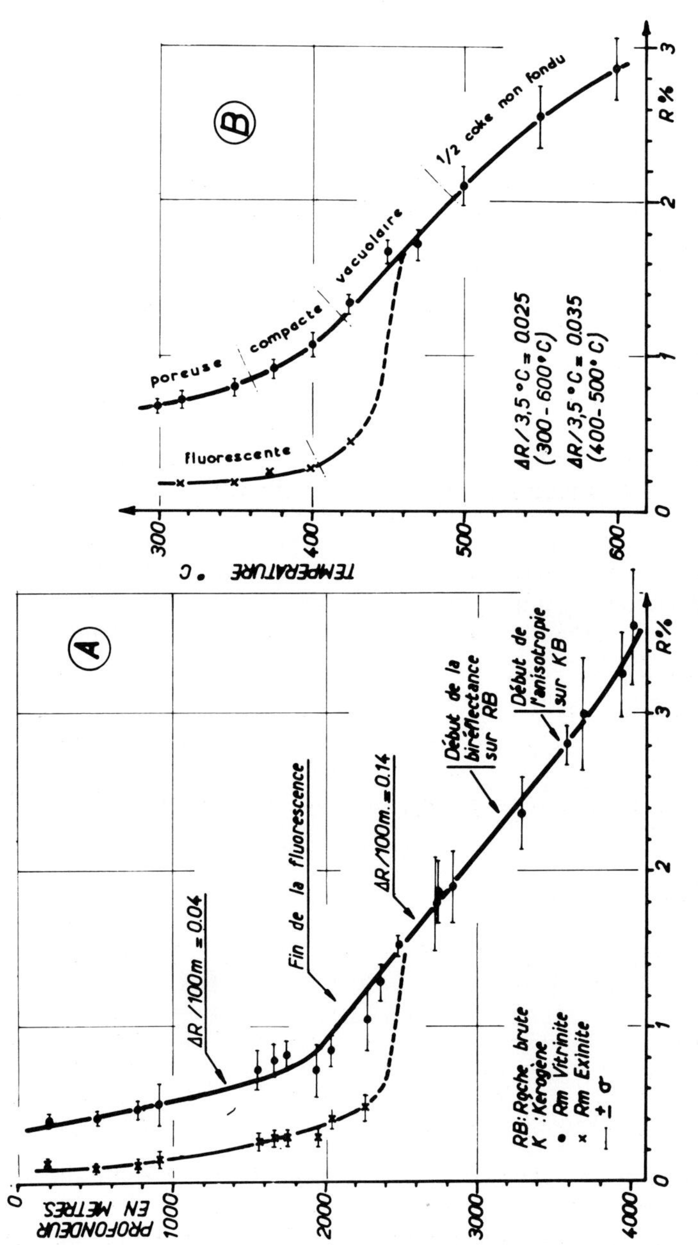

Fig. 11.1. — Variation de la réflectance (*R*) avec la profondeur (**A**) et la température de traitement (**B**). Logbaba (Cameroun).

Avant d'examiner les critères qui peuvent permettre, soit de sélectionner, soit de rejeter une particule en tant qu'indicateur de maturation valable, précisons cette notion de « paléothermomètre » optique actuellement très utilisée mais parfois mal comprise.

A. Notion de « paléothermomètre ».

Il a été amplement démontré par les géochimistes de l'*Institut Français du Pétrole (IFP)* [50] que l'évolution de la MO dans les sédiments se décrit mieux en termes de cinétique chimique qu'en termes de thermodynamique, l'équilibre thermodynamique n'étant pas atteint à une profondeur donnée des sédiments.

Il n'y a donc pas, entre une propriété optique déterminée et une certaine température, une relation *universelle, constante et nécessaire,* permettant de dire que cette particule, après étalonnage, constitue un thermomètre absolu.

Pour des niveaux de maturation ayant dépassé le stade épigénétique, temps et température peuvent se compenser, un même stade final d'évolution, repéré par un paramètre optique, peut donc correspondre à des histoires thermiques *très différentes.*

Cela devient particulièrement évident si l'on compare l'échelle géologique à l'échelle du laboratoire. Ainsi, le kérogène des forages de Logbaba (Bassin de Douala, Cameroun) atteint la même réflectance de 1 % (fig. 11.1) :

— dans les conditions naturelles du forage à la profondeur de 2 000 m et à une température de 80 °C d'après le degré géothermique actuel;
— en laboratoire, après chauffage à 4 °C/min. jusque 400 °C.

Il faut cependant remarquer qu'en cinétique chimique (loi d'Arrhenius par exemple) l'avancement des réactions est plus sensible à la température qu'au temps. La température reste donc le facteur déterminant pour l'évolution physico-chimique et optique des particules.

La notion de paléothermomètre est donc toute relative et n'est valable, après décryptage correct, que pour un ensemble sédimentaire déterminé et d'histoire géologique connue.

B. Causes de distorsion.

Les circonstances qui font qu'une particule enregistre mal les effets thermiques sont diverses :

1. La particule était déjà marquée par un phénomène thermique et/ou chimique antérieur annulant toutes ses capacités ultérieures de maturation. C'est le cas de la **pyrofusinite** marquée par la combustion, et plus généralement de toutes les particules ayant subi une forte action aérobie avant enfouissement (fusinisation).

2. La particule a subi une histoire antérieure dans d'autres sédiments (particules remaniées) et a été soumise dans le passé à des actions thermiques plus fortes que celles qu'elle subira ensuite. Dans certains cas la population remaniée peut être très largement prédominante.

3. La particule migre tardivement dans des sédiments dont elle n'a pas partagé l'histoire (bitumes secondaires migrés).

C. Influence des familles naturelles.

Toutes les familles botaniques ou zoologiques naturelles n'évoluent pas de la même façon pour une même tranche de la maturation. Il en résulte, d'une part, que certains paramètres optiques sont mieux adaptés que d'autres pour suivre la maturation, d'autre part, que certaines familles naturelles présentent une sensibilité plus marquée aux effets thermiques à tel ou tel stade de la houillification.

Ainsi, spores et algues ont une fluorescence qui évolue vite aux stades *épi* et *méso* mais qui est nulle au stade *méta.* Inversement, leur réflectance, nulle à faible au stade épi, évolue très vite au stade méso (où elle rejoint la réflectance de la vitrinite) puis continue d'augmenter fortement jusqu'à la fin du stade méta. A ce stade, dans les houilles, l'exinite, sauf exception (polarisation) ne se distingue plus de la vitrinite.

Il faudra donc examiner dans chaque cas quelle est la meilleure famille présente possible et quel est l'indice optique le plus approprié.

Enfin de nombreuses familles naturelles ont des propriétés optiques qui sont encore mal connues (Algues) ou totalement inconnues en fonction de l'évolution (Chitinozaires, Scolecodontes, Graptolithes, Zooplancton, etc.). Un important travail d'étalonnage par rapport à la vitrinite est à entreprendre.

D. Sélection des indicateurs optiques.

Pour toutes les raisons exposées ci-dessus les indicateurs optiques actuellement les plus usités pour suivre la maturation sont :

— *les gels et tissus gélifiés* lignocellulosiques appartenant à ce que les pétrographes du charbon appellent la série huminite-vitrinite :
. gels : **gelinite** (lignites, réflectance $< 0,50$); **collinite** (houille, réflectance $> 0,50$);
. tissus gélifiés : **humotelinite** (lignites); **telinite** (houilles);

— *les spores* dont on utilise tant la fluorescence au stade épi que la réflectance au stade méso.

Nous avons vu (pl. 11.2) que la diagnose correcte des gels et substances amorphes présente de nombreux pièges du fait du peu de caractères morphologiques utilisables.

V. MÉTHODES OPTIQUES

A. Réflectance.

1. Appareillage et méthode.

La réflectance est mesurée depuis longtemps par les pétrographes du charbon qui ont défini, standardisé puis publié des conditions opératoires dans le lexique international de pétrographie des charbons [36].

A l'heure actuelle cette méthode, déjà normalisée dans divers pays (Allemagne, URSS, Etats-Unis, etc.) est en voie de standardisation à l'échelle internationale (norme ISO en cours de discussion).

Il est donc inutile de revenir en détail sur une procédure largement décrite; nous nous bornerons à discuter des aspects spécifiques de cette technique lorsqu'elle est utilisée à de la MO finement dispersée. D'une manière générale, la procédure est rigoureusement la même, elle consiste à mesurer l'intensité d'un faisceau réfléchi et ceci relativement à un étalon dont la réflectance théorique est connue d'après ses indices de réfraction n et d'absorption k :

$$R = \frac{(n - N)^2 + n^2k^2}{(n + N)^2 + n^2k^2} \quad (1)$$

N étant l'indice du milieu d'immersion, en général une huile non fluorescente d'indice $n = 1{,}517$ à 24 °C. En fait, nous avons établi [4] que l'huile peut atteindre des températures plus élevées sous l'objectif.

L'étalon le plus classique est un saphir synthétique d'indices $n = 1{,}769$, $k = 0$ dont la réflectance mesurée à $\lambda = 546$ nm est $R = 0{,}604\ \%$ dans une huile d'indice $N = 1{,}517$.

Plus récemment divers autres étalons, chimiquement stables, de dureté élevée (8 1/2) et de réflectance supérieure ont été proposés [32] :

- grenat (Yttrium, Aluminium) YAG; $n = 1{,}84$; $R = 0{,}92\ \%$;
- grenat (Gadolinium, Gallium) 3G; $n = 1{,}98$; $R = 1{,}73\ \%$.

Dans le cas de substances homogènes la connaissance de R dans des milieux différents, par exemple : air, eau, huile, permet de déterminer n et k par le calcul [18] :

$$n = \frac{1/2\,(N^2 - 1)}{N\left(\dfrac{1+R_h}{1-R_h} - \dfrac{1+R_a}{1-R_a}\right)} \quad (2)$$

$$K = n\sqrt{\frac{2}{n}\left(\frac{1+R_a}{1-R_a} - \frac{n^2+1}{n^2}\right)} \quad (3)$$

$$K = k_n \quad (4)$$

K : indice d'extinction

n : indice de réfraction

k : indice d'absorption

N : indice de l'huile

R_a : réflectance dans l'air

R_h : réflectance dans l'huile

Notons ici que les mesures d'absorption lumineuse au travers de l'exine des spores, utilisées par divers palynologues pour estimer la diagenèse [23] sont liées également à n et k, mais font intervenir en outre l'épaisseur de l'objet (équation n° 5), valeur généralement non connue, non constante d'une espèce à l'autre ou d'un point à l'autre du même individu et peu facile à mesurer.

$$\frac{I}{I_0} = \exp\left(-4\pi\frac{nd}{\lambda}k\right) \quad (5)$$

I : intensité via l'objet
I_0 : intensité via le milieu sans objet
d épaisseur de l'objet

Du fait de la petite taille des particules, les pétrographes pétroliers ont tendance à prendre des objectifs de plus fort grandissement que le classique 25 × des charbonniers. S'il est possible, dans cette perspective, d'utiliser un objectif 50 ou un 60 ×, il est préférable de s'abstenir de dépasser le grandissement 85 ×, ceci pour des raisons d'orthogonalité du faisceau incident. En effet, pour des angles d'incidence de plus de 25°, Caye [17] a montré que les réflexions selon des plans de vibration parallèles ou perpendiculaires au plan d'incidence n'étaient plus symétriques. La loi normale (équation n° 1) ne s'applique plus, ce qui peut entraîner une erreur sur la première décimale.

Dans les photomètres modernes il n'est plus nécessaire d'augmenter le grossissement pour diminuer le diamètre du champ de mesure, on utilise des diaphragmes circulaires permettant de mesurer des grains de 1 µ seulement de diamètre, ou des diaphragmes rectangulaires adaptés à la forme allongée des particules. Il est à noter que, dans de tels cas et compte tenu des très faibles réflectances mesurées, le niveau de stabilisation des alimentations, la fiabilité et les performances du photo-multiplicateur doivent être au-dessus de toute critique. De plus, il faut considérer que le polissage de microparticules aussi fines, incluses dans un milieu inorganique de dureté généralement très différente, conduit à une mise en relief et à un bombement de la surface incompatibles avec une mesure optique correcte. Il sera donc préférable de mesurer des plages choisies sur des grains nettement plus gros que le strict minimum du champ de mesure afin de s'assurer d'une bonne planéité.

Sur le plan statistique, la richesse des sédiments ne permet pas toujours de mesurer 100 particules comme le veut la norme charbons. On se contente alors nécessairement du nombre de particules photométrables recensées.

La photométrie se complique fortement lorsque, dans les stades finaux de la métagenèse, la MO devient anisotrope et présente, lorsqu'elle reçoit un faisceau de lumière polarisée (un seul nicol sur le faisceau incident), une valeur maximale et une valeur minimale de réflectance dont la différence constitue la **biréflectance**. Ces valeurs ne peuvent être correctement établies que sur des sections polies perpendiculairement à la stratification.

Le charbon se comportant généralement comme un uniaxe négatif [1], il en résulte que $R_a = R_b > R_c$. La réflectance maximale s'obtient sur des plans parallèles à la stratification et la réflectance minimale à 90° de ces plans.

Sur des sections de grains coupés au hasard on ne peut établir, sans polariseur, qu'une valeur statistique moyenne $\bar{R}$ et, avec polariseur, qu'une valeur statistique approchée

[1] Parfois aussi comme un biaxe.

$\bar{R}_{max}$ et $\bar{R}_{min}$ de R_a et de R_c. Les relations liant ces divers indices ont été calculées par Hevia *et al.* [27].

$$\bar{R}_{max} = R_a$$

$$\bar{R}_{moy} = \frac{2\,R_a + R_c}{3}$$

$$\bar{R}_{min} = \frac{R_a + 2\,R_c}{3}$$

$$\bar{R}_{max} - \bar{R}_{min} = 2/3\,(R_a - R_c)$$

Il suffira donc de deux de ces paramètres pour déterminer le troisième et la valeur de la biréflectance vraie. Il ne faut pas confondre celle-ci avec l'anisotropie optique entre nicols croisés qui caractérise : les stades finaux de la métagenèse (anthracites), certains bitumes et pyrocarbones (produits de cracking) ou certains cokes naturels produits par un intense métamorphisme de contact.

On conviendra finalement de mesurer le rang à partir de la réflectance statistique moyenne sans polariseur et sans rotation de la platine, cette procédure étant plus courte et par ailleurs théoriquement justifiée [26].

2. *Définition des objets mesurés et évaluation de la dispersion.*

Nous avons vu ci-dessus que la meilleure famille naturelle pour l'enregistrement (par la réflectance) des effets thermiques était celle des gels et tissus gélifiés lignocellulosiques formant la série huminite-vitrinite.

Dans les charbons, cette famille se caractérise, au microscope, par un aspect amorphe (par définition), une couleur gris foncé devenant blanc jaunâtre à la fin de l'évolution, une réflectance dans l'huile variant de 0,2 à plus de 8 %.

Du fait de son caractère amorphe, cette série évolutive humique se distingue mal d'autres familles naturelles, parfois très différentes, et nous avons donné ci-dessus la liste partielle des confusions possibles.

Les risques d'erreur de diagnose sont d'autant plus grands que la particule est plus petite. Les difficultés d'identification augmentent donc après broyage et extraction du milieu lithologique originel.

Certaines erreurs ne sont pas trop graves du point de vue de la mesure du rang; ainsi gels, cryptotissus et certains bitumes primaires peuvent avoir des courbes de réflectance assez analogues; de même pour l'exinite après son point de convergence avec la vitrinite.

Par contre, la confusion vitrinite-exinite avant convergence (fig. 11.1) vitrinite-macrinite, vitrinite-bitume secondaire, vitrinite-chitinozoaire et plus généralement vitrinite-zooclaste, peut conduire à des erreurs plus graves pour la mesure de l'évolution.

A titre d'exemple, les dispersions de réflectance mesurées sur une série de charbons du Donetz par des pétrographes de l'*ICCP* appartenant à des laboratoires spécialisés de neuf pays différents ont été les suivantes (11e série d'analyses comparées) :

Charbon du Donetz n°	Réflectance moyenne	Dispersion entre laboratoires $n = 9$	
		Ecart-type	Ecart-type %
1	0,49	0,03	6,1
2	0,92	0,03	3,3
3	1,02	0,05	4,9
4	1,53	0,10	6,5
5	2,04	0,12	5,9
6	1,81	0,08	4,4
7	0,91	0,04	4,9
8	1,13	0,04	3,5
9	1,63	0,07	4,3
10	2,29	0,11	4,8
			Moyenne : 4,86

On constate que la dispersion entre laboratoires est, en moyenne, inférieure à 5 % de la valeur de la réflectance et que, pour des charbons de réflectance inférieure à 1,5 %, l'écart-type est toujours inférieur à 0,10.

On peut donc considérer que, pour des charbons peu ou moyennement évolués, la première décimale de *R* est significative dans tous les cas où aucun problème de diagnose de la vitrinite ne se pose.

Considérons maintenant (tableau ci-dessous) des échantillons plus composites ou plus délicats, tels que ceux qui ont fait l'objet d'analyses comparées (dites analyses MOD) de la *Commission Internationale de Pétrographie de la Matière Organique Dispersée des Sédiments*. Pour dix analyses comparées, la moyenne des écarts-types est de 25 %, soit une dispersion 5 fois plus élevée qu'avec un faciès charbonneux classique.

Echantillon	Nombre d'analyses	*R* moyen	Ecart-type	Ecart-type %	Valeurs limites
MOD 1	24	0,42	0,05	11,90	0,34-0,47
MOD 2	24	0,69	0,05	7,25	0,62-0,82
MOD 11	13	0,51	0,15	29,66	0,19-0,79
MOD 12	14	0,41	0,12	29,27	0,28-0,78
MOD 13	13	1,53	1,36	23,71	1,19-1,84
MOD 14	13	1,14	0,31	26,88	0,69-1,69
MOD 15	14	0,82	0,20	24,97	0,56-1,24
MOD 16	27	0,50	0,06	13,15	0,38-0,62
MOD 17	25	0,65	0,15	24,12	0,39-1,07
MOD 18	21	0,36	0,21	59,33	0,10-0,94
				Moyenne : 25,03	

B. La fluorescence.

1. Définitions.

On peut dire, d'une manière générale, que la fluorescence varie à l'inverse de la réflectance. La fluorescence n'est marquée que lorsque la réflectance est faible et, par conséquent, elle concerne principalement le groupe de l'exinite et les stades initiaux de la houillification (épigenèse). De nombreux zooclastes sont également fluorescents, ils évoluent d'une manière encore mal connue mais qui semble différente, à l'échelle du laboratoire, de celle du groupe de l'exinite.

La fluorescence primaire (autofluorescence) est la propriété que possèdent certaines substances d'émettre, lorsqu'elles sont excitées par une radiation de courte longueur d'onde, un rayonnement de longueur d'onde plus élevée. En microscopie par réflexion appliquée à la MO sédimentaire, on utilise l'excitation par l'UV proche (360 nm). L'utilisation de l'UV plus court impliquerait l'emploi d'une optique en quartz, ce qui est généralement exclu.

La fluorescence se distingue de la phosphorescence dans la mesure où l'effet cesse avec l'irradiation. La phosphorescence ne jouant parfois que sur des fractions de seconde, il est difficile de départager les deux phénomènes et on les regroupe alors sous le terme de **luminescence.**

Dans les sédiments la fluorescence n'est pas seulement le fait de la matière organique, mais également de nombreux minéraux (zircon, monazite, apatite, fluorite, sheelite, etc.). Cependant, dans certains cas, la fluorescence des minéraux est due à des inclusions organiques (calcite, quartz, etc.).

La fluorescence de la MO a en outre ceci de particulier que sa couleur et son intensité peuvent changer au cours du temps pendant l'excitation. Ce phénomène a reçu le nom de « *fading* » [21] ou « d'*altération* » [40].

La fluorescence des substances organiques serait due aux électrons π des systèmes conjugués, moins liés à la molécule que les électrons σ et capables, de ce fait, de passer sur une orbite plus externe du noyau par absorption d'une radiation électromagnétique de relativement basse énergie, reçue lors de l'excitation lumineuse et obéissant à loi de Stockes ($E = h\nu$). L'électron revient ensuite à sa position initiale d'équilibre en restituant une partie de l'énergie reçue sous la forme d'une radiation émise dans une longueur d'onde du spectre plus élevée que celle d'excitation et qui est caractéristique, à la fois des conditions optiques et de la structure chimique de l'objet étudié.

On considère généralement que les structures chimiques responsables de la fluorescence appartiennent à des molécules rigides, planes, présentant des doubles liaisons aromatiques conjuguées. Cependant, Stach [45] considère cette condition comme nécessaire mais pas suffisante et cite les groupements perhydrogénés : $> CH_2$, $-OCH_3$, $-CH_2OH$ comme des renforçateurs de fluorescence (auxoflor).

Cette opinion est également exprimée par Wehry [52] à propos des groupes : OH, OCH_3 et OC_2H_5 qui renforcent l'intensité de fluorescence (mais abaissent la fréquence, donc augmentent la longueur d'onde) des cycles aromatiques sur lesquels ils sont substitués.

Les composés aliphatiques, par ailleurs, sont considérés comme non fluorescents à l'exception des aldéhydes et des cétones. De toute manière, puisque la fluorescence diminue puis disparaît lorsque la houillification augmente et que celle-ci est marquée par une évolution vers la polycondensation, il faut admettre que les structures aromatiques impliquées dans la fluorescence sont nécessairement peu condensées (probablement 1 à 2 cycles).

Pour Stach [45] cette baisse de fluorescence est due, en partie, à la perte des groupements hydrogénés au cours de la maturation.

En définitive, il paraît probable que la fluorescence est plus un problème de structure que de composition chimique élémentaire, certains isomères se comportant différemment selon qu'ils sont en posisition *cis* ou *trans* [52].

2. *Appareillage et méthode.*

La fluorescence se pratique en lumière transmise (biologie) et en lumière réfléchie; dans ce dernier cas, on n'est pas gêné par les lames porte- et couvre-objet.

Le microscope peut être de type classique avec une optique en verre normal; cependant, l'huile d'immersion et la résine d'enrobage devront, de préférence, être non fluorescentes. Au *CERCHAR,* la MO est collectée sur un tamis soluble et polie directement sur une lame support sans autre enrobage. Pour la simple observation, on utilise un illuminateur vertical normal permettant des examens successifs en réflexion et en fluorescence. Pour la mesure on utilise de préférence l'illuminateur de Ploëm (voir ci-dessous)..

La source d'excitation est une lampe à vapeur de mercure ou une lampe au Xénon. La seconde est plus stable que la première dans les courtes longueurs d'ondes, mais elle est nettement moins intense. On isole de toute manière une bande précise d'excitation, soit à 390 nm (BG 12) pour l'observation, soit à 360 nm (UG_1) pour la mesure. Ce n'est que dans ce dernier cas, malheureusement moins favorable sous l'angle des intensités, que le spectre émis couvre correctement tout le visible (430 à 700 nm). Ce spectre est envoyé vers le photomultiplicateur au travers :

a) de l'illuminateur de Ploëm comportant un filtre d'arrêt (K 430) et un miroir dichroïque (TK 400) qui ont pour but d'arrêter toute réflexion directe du faisceau incident. Dans ces conditions, l'analyse du spectre ne commence qu'au-dessus de 430 nm.

(b) d'un filtre interférentiel mobile à déplacement continu, Véryl B 60, qui analyse le spectre au cours d'une rotation de 25 s commandée électriquement. La courbe brute est enregistrée sur un dispositif mobile Servogor ou sur une table traçante. La courbe de réponse du photomultiplicateur étant très affaiblie dans le rouge une courbe de correction doit être calculée qui tient compte également du bruit de fond et de l'influence de l'optique. Cette courbe corrigée est établie au moyen d'une lampe à ruban de tungstène dont le spectre théorique est connu en fonction de la température de couleur [42].

On dispose finalement d'une courbe spectrale corrigée (Fig. 11.2) établie à partir de la moyenne de *n* courbes, généralement 10, fournissant un spectre moyen caractéristique :

- de la nature de la particule;
- de son degré d'évolution.

A partir de ce spectre, on établit les caractéristiques suivantes :

- position du maximum : λ_{max}
- quotient rouge/vert : $Q = \dfrac{\text{intensité à 650 nm}}{\text{intensité à 500 nm}}$
- signe et intensité du fading.

Pour cette dernière propriété, on distingue :

– la variation d'intensité ΔI au cours du temps (30′) dans une même longueur d'onde (546 nm)

– le déplacement de λ_{max} d'une même particule après 30′ d'irradiation. On mesure en nm le décalage $\lambda'_{max} - \lambda_{max}$ et sa variation d'intensité en %.

La première méthode donne le fading apparent, la seconde le fading réel. Nous proposons de différencier ces deux mesures et d'appeler la première **fading** d'après Van Gijzel [20, 21] et la seconde **altération** (ou fading spectral) d'après Ottenjann *et al.* [40].

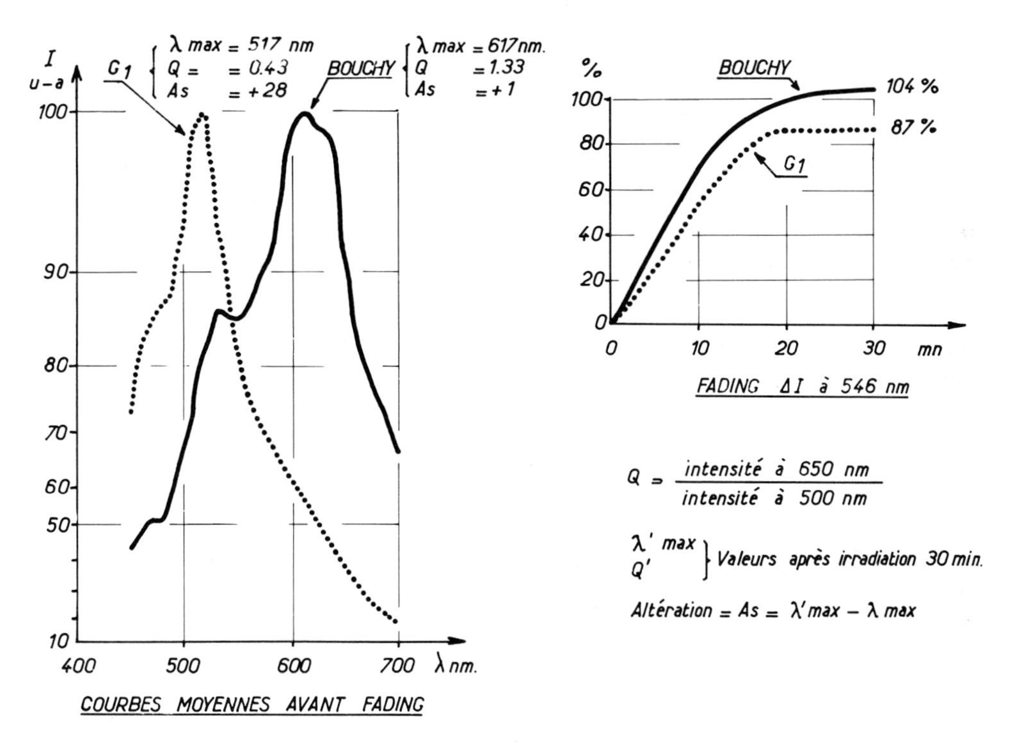

Fig. 11.2. — Fluorescence des Tasmanites et catagenèse dans le Bassin de Paris (Toarcien).

Toutes les opérations d'enregistrement et de calcul peuvent faire l'objet d'un programme de mini-ordinateur de laboratoire automatisant toutes les corrections. Dans ce cas, la durée de l'ensemble des opérations est tout à fait compatible avec le travail statistique de routine.

Au total cependant, et pour les laboratoires à équipement classique, le nombre des paramètres retenus paraît trop élevé et la procédure trop lourde. Nous pensons que l'augmentation du nombre de mesures d'un seul ou de deux paramètres, par exemple *Q et* λ, est préférable au maintien d'un grand nombre d'indices, certains d'entre eux pouvant avoir *grosso modo* la même signification catagénétique, d'où une duplication un peu inutile du travail analytique et une difficulté, pour la fluorescence, à passer du stade de la recherche fondamentale au stade opérationel. Pour le Bassin de Paris il a été établi que l'augmentation de 10 à 100 des mesures de Q rendait ce paramètre significatif pour la catagenèse (σ_{10} = 0,20 à σ_{100} = 0,05) tout en simplifiant le processus opératoire, le tracé complet du spectre pouvant être supprimé et remplacé par deux mesures seulement : à 500 et 650 nm.

VI. MESURE OPTIQUE DE LA CATAGENÈSE

Considérons séparément réflectance et fluorescence puisque nous avons vu ci-dessus qu'elles ne s'appliquaient pas aux mêmes familles ni aux mêmes zones de maturation.

A. Mesure par réflectométrie.

1. Dans toute série sédimentaire continue normale, la réflectance du matériel organique gélifié augmente avec la profondeur, en fonction du gradient géothermique, d'une manière non linéaire. Dans le cas le plus banal on obtiendra une courbe évolutive permettant de préciser dans quelle zone de maturation on se trouve en prenant comme référence l'échelle classique d'évolution des charbons graduée en réflectance. Dans le cas d'un sondage suffisamment profond comme, par exemple, celui de Gironville [1], fait à la base du houiller lorrain, ou celui de Munsterland [24] , on recoupera une série de stades d'évolution. Ainsi en Lorraine-Gironville on passe des charbons flambants aux anthracites et on rencontre successivement tous les phénomènes optiques caractérisant qualitativement les diverses étapes de la houillification (fig. 11.3) :

— convergence des courbes de réflectance exinite-vitrinite;
— fin de la macérabilité (extraction des spores);
— fin de la fluorescence et de la transparence des spores;
— début et développement de la biréflectance;
— apparition de l'anisotropie entre nicols croisés, etc.

Les zones successives recoupées montrent des $\Delta R/100$ m de sédiments croissants, non que les paléogradients aient augmenté avec la profondeur, mais du fait de l'action exponentielle de la température et du caractère non linéaire de la variation de réflectance pour un gradient géothermique donné.

2. Le fait que la variation de la réflectance avec la température et la profondeur semble, pour une partie du moins de la catagenèse, de type exponentiel, explique le souci de certains de rechercher une variation de type linéaire. C'est le cas des pétrographes de la Shell [28] avec leur échelle LOM (Level of organic metamorphism) divisée en 20 parties égales.

On peut tout aussi bien prendre les logarithmes des réflectances [3, 20]. Ce procédé, appliqué aux forages de Gironville, de Logbaba et de la Mer du Nord (fig. 11.4) donne des relations quasi-linéaires permettant de comparer les ΔR quelle que soit la zone de maturation où ils ont été calculés. Il est généralement plus facile par cette méthode d'évaluer l'épaisseur des terrains éventuellement érodés en prolongeant les droites obtenues jusqu'à leur point de croisement avec la réflectance 0,15 %, valeur moyenne de formations tourbeuses superficielles. La partie manquante des terrains est immédiatement lisible sur la figure.

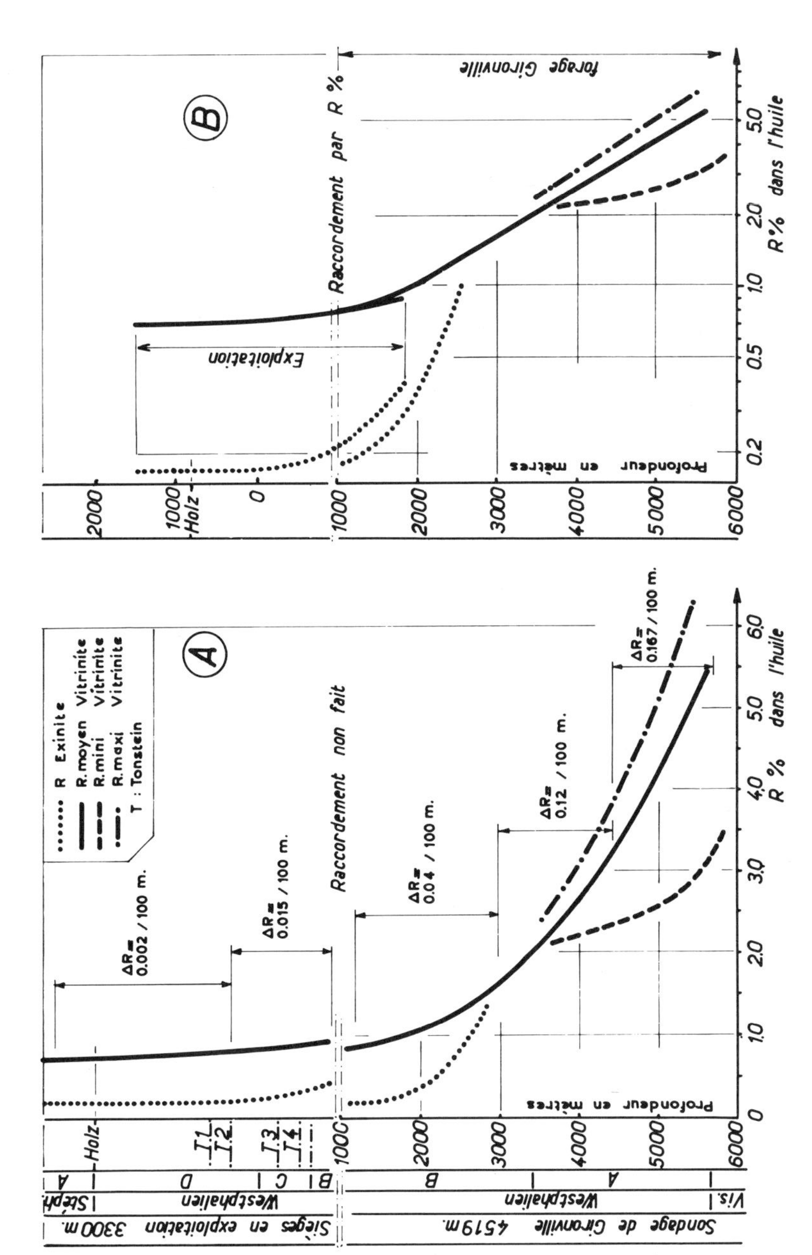

Fig. 11.3. — Evolution de la réflectance (R et log R) dans le Carbonifère lorrain.

3. En fait on peut supposer, comme l'ont fait Tissot *et al.* [50] pour le Bassin de Paris, que le gradient géothermique est resté linéaire avec la profondeur et constant sur les plates-formes stables. Cette supposition ne peut être faite que si l'orogenèse est ancienne. Le gradient serait alors en moyenne de 25 à 35 °C/km, ce qui se traduirait par une augmentation de 0,5 à 1 °C par million d'années dans des conditions moyennes de sédimentation.

4. La relation maturation-réflectance subit l'influence de divers facteurs qui viennent expliquer les dispersions plus ou moins importantes observées dans une même série sédimentaire.

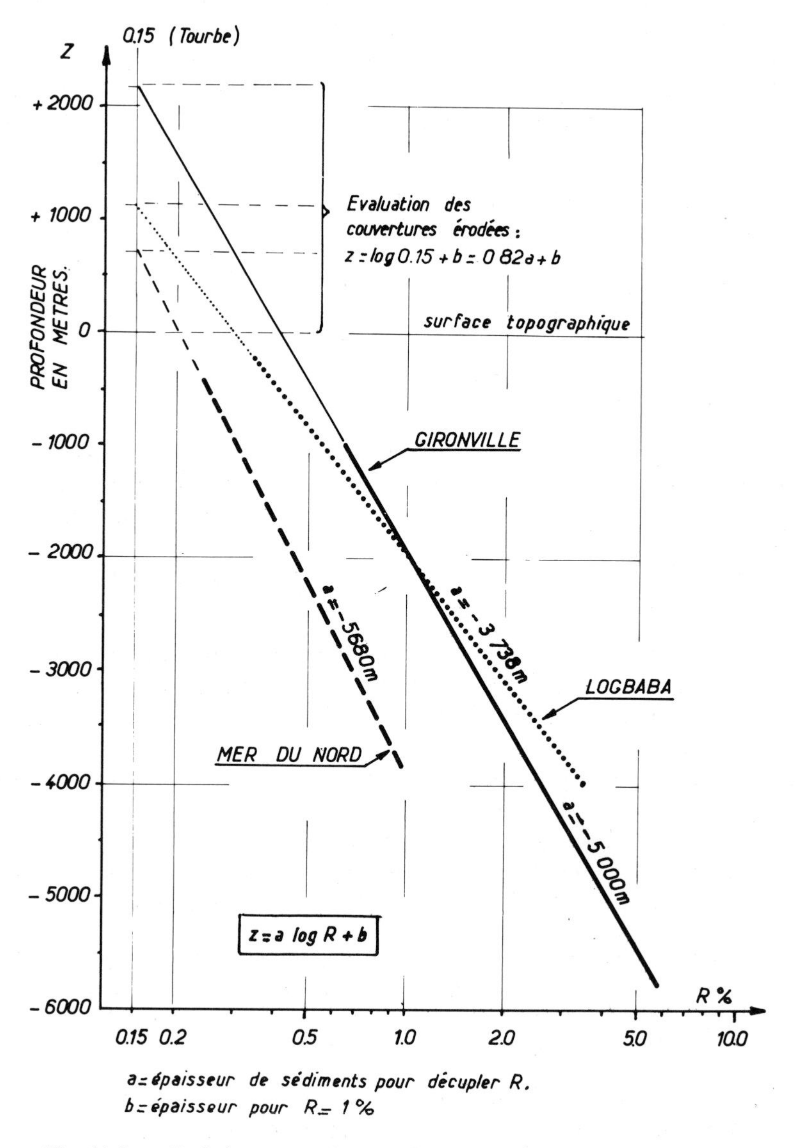

Fig. 11.4. — Variation comparée des réflectances (log *R*) dans divers bassins.

Une même couche de charbon, correspondant donc à un même faciès et à une tranche de temps relativement brève, présente généralement, pour $R = 1$ %, un écart-type de 0,05 environ autour de la moyenne (100 mesures par couche). Cela signifie que 99 % des valeurs individuelles sont comprises dans un intervalle de $\pm 2\,\sigma$, soit une dispersion maximale individuelle de 0,20, ce qui est considérable. D'où l'obligation pour le pétrographe de procéder à des études statistiques faisant intervenir au moins 100 valeurs par niveau et portant sur un nombre suffisant d'échantillons répartis sur la colonne stratigraphique inventoriée et échantillonnée en fonction de sa lithologie.

Même si l'on suppose en effet que la vitrinite constitue un objet élémentaire bien individualisé et constant, ce qui n'est pas le cas comme nous l'avons vu ci-dessus : tissus gélifiés ou gels vrais, imprégnations submicroscopiques bitumineuses, résineuses ou argileuses, légères variations du stade de gélification, oxydation météorique primaire ou secondaire, etc.; la particule de vitrinite est incluse dans un milieu organique ou minéral qui peut conditionner son évolution.

5. L'étude statistique de l'influence de la lithologie sur la réflectance a intéressé de nombreux auteurs. D'une manière générale, il semble que la réflectance de la MO des veines de charbon soit plus élevée que celle des autres sédiments. Ainsi, Bostick [14] trouve des valeurs, rapportées à la vitrinite du charbon, allant :

— de + 2 à − 31 % pour les grès;
— de − 5 à − 18 % pour les schistes;
— de + 2 à − 17 % pour les calcaires.

Kunstner [35] trouve une réflectance moyenne plus élevée dans les couches de charbon que sur les microparticules dispersées. Blanquart *et al* [12] trouvent également des valeurs augmentant systématiquement avec l'épaisseur des niveaux charbonneux, mais plus élevées dans les grès que dans les schistes (+ 14 %).

Ces fluctuations peuvent être interprétées de différentes manières, la vitrinite étant supposée correctement reconnue :

— variations de l'histoire thermique elle-même, par exemple par des circulations préférentielles de fluides météoriques froids ou chauds (thermalisme) en fonction de la lithologie [33]. La circulation est évidemment plus active dans les grès grossiers que dans les sédiments fins argileux et charbonneux;
— différences de conductivité thermique : elle est beaucoup plus forte dans les grès et conglomérats que dans les schistes et les charbons [19];
— effet catalytique variable des minéraux, notamment argileux;
— altération différente ou oxydation précoce des particules organiques en liaison avec les conditions sédimentologiques et la porosité du milieu (cas des pseudovitrinites?).

6. Les auteurs russes [10] tiennent également compte de la concordance ou de la discordance des formations et des rapports angulaires entre les strates géologiques et la surface topographique qui peuvent jouer, en l'absence de couverture imperméable, sur l'évacuation plus ou moins aisée des flux thermiques.

7. L'une des limitations de la réflectance pour la mesure de la maturation est particulièrement sensible dans les roches-mères à kérogène I ou II ne comportant que peu ou pas de vitrinite syngénétique.

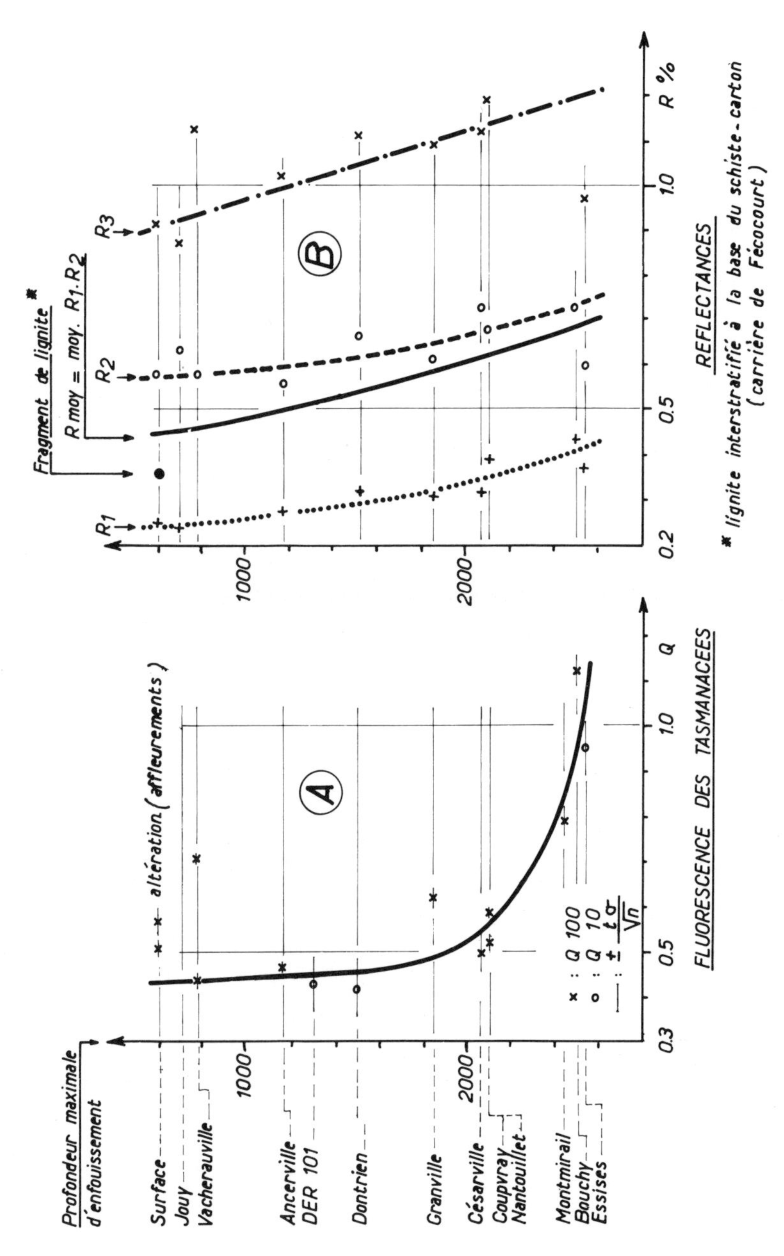

Fig. 11.5. — Evolution des indices optiques avec la catagenèse. Toarcien. Bassin de Paris.

Le Bassin de Paris est particulièrement significatif à cet égard. On trouve en effet dans les schistes bitumineux du Toarcien, pour un même stade de maturation, des populations de réflectances différentes [7] :

— des microlamelles interstratifiées noir-rougeâtre, de réflectance 0,10-0,15 % de fluorescence et de structure souvent composites (matrice) parfois homogènes mais de contours souvent définis. Il s'agit probablement de ce que Teichmüller *et al* [48] ont appelé bituminite 1 et 2;

— des zooclastes de réflectance 0,10-0,20 % et de fluorescence brune homogène;

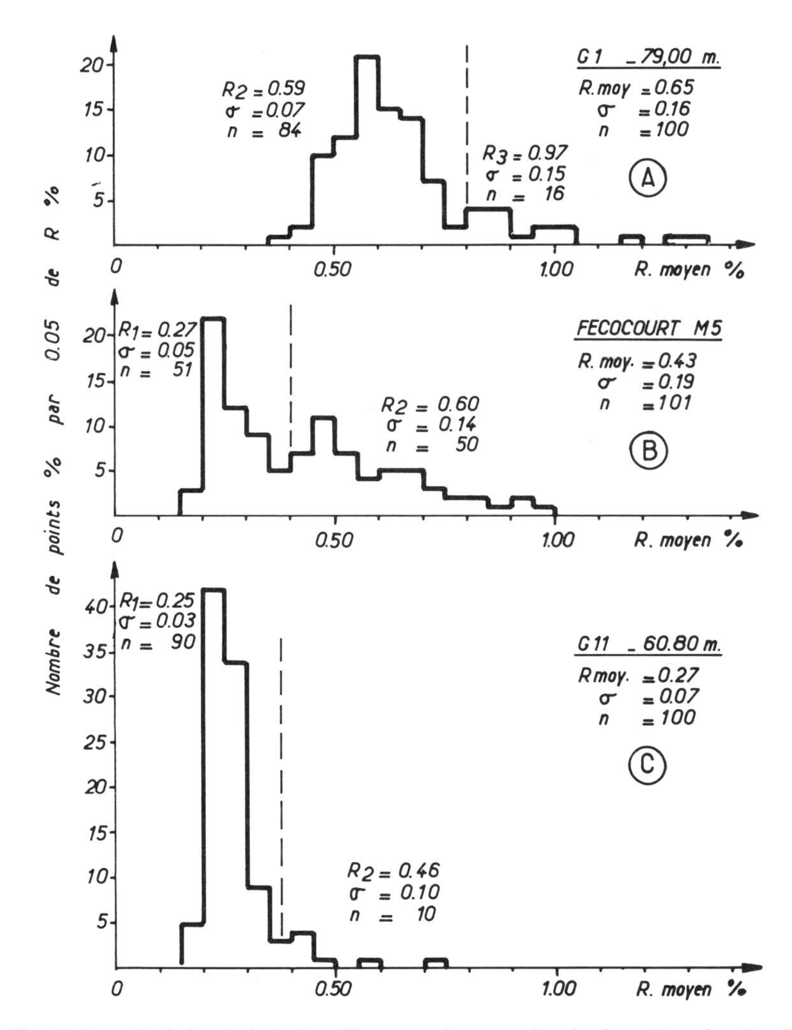

Fig. 11.6. — Variation latérale des réflectances dans un même horizon. Toarcien. Bassin de Paris:

A. dominante remaniée. **B.** composition mixte. **C.** dominante autochtone.

— des gels granulaires (R_1) plus ou moins poreux de réflectance 0,20-0,30 % (moyenne 0,26 – n = 794). C'est peut-être la bituminite 3 des mêmes auteurs;
– de la vitrodétrinite (R_2) de réflectance 0,46-0,67 % (moyenne 0,58 – n = 908);
– de la vitrinite remaniée (R_3) de réflectance 0,80-1,30 %.

La réflectance moyenne R_1-R_2 (fig. 11.5) traduit assez bien (lorsque R_1 et R_2 sont présents) le niveau de maturation réel tel qu'il est établi par les données géochimiques, mais il faut reconnaître que c'est la population de réflectance 0,10-0,15 % (matrice) qui est de loin la plus abondante, tandis que R_1 (Bituminite 3) est probablement l'indicateur optique le plus correct (0,30 % à l'affleurement). Par ailleurs, les proportions relatives R_1-R_2 varient considérablement avec les lieux de prélèvement, l'une des populations pouvant, à la limite, être quasi-absente. De ce fait, la valeur de la moyenne est très douteuse (fig. 11.6).

Si l'on devait utiliser les bituminites comme indicateurs de maturation, l'entrée dans la fenêtre pétrolière se ferait à la réflectance 0,30 % environ (fig. 11.5).

L'origine de la matrice n'est pas claire, l'aspect est souvent celui d'un microaggloméra t composite très fin à ciment fluorescent. Les relations optiques avec la fenêtre pétrolière ne sont évidemment pas celles de la vitrinite. Cependant, les kérogènes préparés par l'*IFP* sont essentiellement formés de ce type de constituant qui est le type du kérogène II et la vitrinite classique en est pratiquement absente (pl. 11.4, photo 7).

Pour mesurer correctement la maturation de telles séries, il est alors préférable de recourir aux organismes figurés, notamment les Tasmanites, fréquentes dans tous les échantillons (voir ci-dessous) ou à la bituminite primaire intraformationnelle.

8. Indépendamment des causes naturelles de variation de la réflectance, une dispersion inévitable existe entre laboratoires différents du fait des différences entre les appareillages, les processus de concentration et le coefficient personnel lors de la sélection des particules (voir tableaux 11.1 et 11.2).

B. Mesure par fluorescence.

1. C'est Mme Teichmüller qui, dans un remarquable article paru dans *Inkohlung u. Erdöl* (1974), a fait la première, sur des séries charbonneuses, les observations et les hypothèses décisives. Elle a remarqué en effet :

— qu'une partie des constituants noirs, pris antérieurement pour de l'argile ou de la liptodétrinite, était formée en réalité de matières fluorescentes brunes qu'elle a nommées « bituminite »;
— que l'évolution de ce constituant s'accompagnait souvent de l'apparition du macéral « micrinite » (groupe de l'inertinite) dont l'origine était jusqu'alors très controversée;
— qu'il existait, dans les fissures de nombreux charbons, des produits secondaires fluorescents (exsudatinite) ne se formant qu'à l'entrée de la fenêtre pétrolière (pl. 11.5, photo 1);
— qu'au moment où apparaissait la micrinite, la vitrinite devenait elle-même parfois fluorescente, probablement par imprégnation d'hydrocarbures, le polissage devenant alors quelquefois difficile du fait de la formation de films d'huile.

2. La fluorescence quantitative est plus compliquée et plus récente que la réflectométrie, elle n'est pas normalisée à l'échelle internationale. Dans une série continue normale, la fluorescence du matériel algaire et sporo-pollinique (groupe de l'exinite) diminue avec la profondeur (pl. 11.6); la couleur, de verte devient jaune, puis brune. Toute fluorescence disparaît ensuite lorsque la réflectance du matériel gélifié correspondant atteint la valeur 1,30 % environ. Ces observations ont été quantifiées par Van Gijzel [22] sur du matériel pollinique extrait chimiquement des roches inorganiques, par Ottenjann *et al.* [41] sur du matériel pollinique *in situ* en section polie, puis par Teichmüller *et al.* [48] sur le ciment bitumineux-minéral.

Selon ces auteurs (1975), la longueur d'onde de l'intensité maximum de fluorescence des spores varie de :

— 490 nm (vert) au stade tourbe (réflectance 0,20 %) à
— 670 nm (rouge) au stade du charbon gras à coke (réflectance 1,34 %).

Le rapport *Q* rouge/vert (650 nm/500 nm) passe parallélement de 1 à 6,3.

Ainsi le spectre de fluorescence se déplace très nettement vers le rouge tandis que l'intensité diminue jusqu'à s'annuler totalement.

Pour les algues, les phénomènes sont différents. Nous avons pu [7] suivre l'évolution des Tasmanacées dans le Bassin de Paris depuis Fécocourt (affleurement) jusque Essises (2 600 m) et nous avons obtenu les résultats de la figure 11.5.

On constate qu'un approfondissement de 2 000 m se traduit *grosso modo* par un déplacement de λ_{max} de 519 à 595 nm tandis que le quotient *Q* 100 (650/500) passe de 0,45 à 1,04. L'augmentation correspondante des réflectances est de 0,25 % à 0,40 % pour la population de réflectance minimale (bituminite) et de 0,46 à 0,70 % pour la réflectance moyenne.

L'évolution statistique du quotient *Q* (fig. 11.5) paraît traduire la maturation d'une manière correcte. La fenêtre pétrolière serait atteinte pour la valeur $Q = 0,5$. L'allure de la courbe est proche de celle des extraits chloroformiques fournis par les mêmes séries [8]; elle montre que les fortes transformations se font à partir de cette valeur (1 800 m environ).

Sur le ciment bitumineux-minéral, c'est la variation d'intensité pendant 30' (fading à 546 nm) qui passe du domaine positif au domaine négatif lorsqu'on va de l'immature au mature [48].

3. La fluorescence, outre cet aspect quantitatif, fournit le moyen d'observer la génération de certains hydrocarbures et permet, qualitativement cette fois, d'augmenter la liste des critères de maturation.

Ainsi, en dehors des repères d'évolution basés sur la réflectance, toute une série d'observations nouvelles, tirées de la fluorescence, confirment ou même enrichissent la liste des critères et des étapes de la maturation décelables au microscope, du moins dans les séries charbonneuses classiques. Dans des séries stratigraphiques sans veines de charbon, tous ces critères n'apparaissent pas nécessairement. Inversement, la formation et la migration des exsudats, qui jouent dans les charbons un rôle secondaire, revêtent un aspect primordial dans les séries pétrolières.

VII. CONCLUSIONS : INDICES OPTIQUES ET FENÊTRE PÉTROLIÈRE

Le tableau 11.3 tente de synthétiser l'ensemble des observations et critères optiques recensés à ce jour et de les corréler avec les propriétés du combustible fossile, charbon ou pétrole, correspondant.

On remarquera à ce propos :

— que l'entrée dans le tableau se fait par un indice de réflectance, celui-ci paraissant à l'heure actuelle le plus universellement admis pour mesurer la catagenèse;

— que la position de la fenêtre pétrolière varie avec les auteurs. Restreinte à l'origine [53] avec une dead-line placée à 0,7 % puis à 1 % [46,9] , elle varie actuellement de :

. 0,7 à 1,35 % [28], elle inclut alors la possibilité de l'expulsion et implique une concentration minima;

. 0,5 à 1,35 % [51,47];

. 0,53 à 1,57 % [38].

Ces différences sont moins graves qu'il ne paraît *a priori.* En effet, il est bien difficile de mettre une ligne de démarcation très fine dans un processus évolutif naturel continu, surtout si le paléogradient géothermique est faible. Où se trouve avec précision le début d'un phénomène exponentiel? Il y a là une place pour une certaine marge d'incertitude, nullement négligeable, mais dans laquelle les mesures de réflectance ne sont pas nécessairement impliquées.

Pour réduire cette incertitude, il suffirait :

a) de préciser la notion de fenêtre pétrolière par des indications numériques basées, par exemple, sur le taux d'extractibles (ou sur tout autre critère géochimique). Malheureusement, celui-ci varie avec le type de kérogène et la définition manquera d'universalité.

En fait il a été démontré [50] que la fenêtre liquide du kérogène I est plus profonde que celle des kérogènes II et III, à conditions géologiques égales, les énergies d'activation pour rompre les molécules correspondantes étant centrées autour de 70 Kcal/mole pour le kérogène I et 50 seulement pour les kérogènes II et III. En série normale (35 °C/km) la différence peut se traduire par 1 000 m d'écart entre les zones maximales de production liquide. Il devient alors parfaitement clair que la notion de fenêtre pétrolière doit être considérée avec nuance et précaution, un kérogène complexe pouvant donner une succession de maximums des quantités de produits liquides formés, chaque famille progénitrice atteignant tour à tour sa zone optimale de production.

b) d'accroître nos informations statistiques et d'établir empiriquement, avec une meilleure précision, les relations entre indices optiques et potentiel pétrolier, quantifié cette fois par un paramètre géochimique à déterminer comme il est souhaité ci-dessus.

Il y a place dans ce domaine pour un travail commun des géochimistes et des pétrographes qui, bien qu'amorcé, mérite d'être amplifié et poursuivi.

Au total, les indices optiques traduisent un *état final* d'évolution sans que les actions respectives de la température et du temps puissent être départagées par le pétrographe. Ceci, encore une fois, n'est pas une objection quant à l'application de ces indices à la prospection

TABLEAU 11.3
CATAGENÈSE DES CHARBONS ET POTENTIEL PÉTROLIER. SYNTHÈSE

STADE		RANG DU CHARBON			REFLECTANCE %	REPERES OPTIQUES: transparence	REPERES OPTIQUES: réflexion	REPERES OPTIQUES: fluorescence	POTENTIEL PETROLIER
Diagénèse			TOURBE	H_2O 75	0.20	incolore		bleu - vert	CH_4 biogénique
CATAGENESE	Epigénèse	LIGNITE hypo	brun poreux	35		jaune clair		vert - jaune	
		LIGNITE meso	brun compact	25		jaune	Attrinite ↓ Densinite	jaune	Zone immature
		LIGNITE méta	noir brillant	12	0.50	jaune rouge	vitrinisation	jaune brun	0.50
	Mésogénèse	HOUILLE hypo	début fusion, fluidité ↗			jaune brun	– début micrinite –	exsudats et V. fluorescente	0.50 generation huile K_2 K_3*; 0.60 generation huile K_1; 0.70 migration → K_2 K_3; 0.80 production maxi → K_1
		HOUILLE meso	gonflement maximal		1.00	brune		spores brunes	1.00
		HOUILLE méta	fluidité ↘, fin de gonflement	M.V. 8	1.50	opaque (Limite de	convergence V – E (macérabilité)	fin de fluorescence	1.35 Zone de destruction; condensats + gaz humides; 2.00
	Metagénèse	ANTHRACITE hypo	élasticité**	5	2.50		biréflectance (1 polariseur)		gaz secs
		ANTHRACITE meso	polarisation	3	3.00; 5.00		polarisation (2 pol. ⊥)		3.00; méthane catagénique
		ANTHRACITE méta	début graphitisation		8 à 10		inversion des réflectances E. > V. > I.		
Métamorphisme			GRAPHITE		15.00				

** au microduromètre

*, K_1 . K_2 . K_3 . : Types de kérogène

pétrolière, puisqu'il est admis *grosso modo* qu'au-delà des phénomènes biochimiques temps et température se compensent. Il reste encore à déterminer si un même stade optique final correspond toujours à la même histoire thermique et a engendré (à faciès organique constant) la même quantité d'hydrocarbures. Cela n'est pas certain et reste encore du domaine des hypothèses, notamment dans le cas où, après avoir subi une température élevée, les sédiments sont relevés dans une zone moins chaude, mais où certains processus chimiques pourraient encore se poursuivre alors que la réflectance, ayant atteint son maximum, ne varierait peut-être plus.

Il reste encore à ajouter que la position de la fenêtre pétrolière dans un bassin dépend de l'histoire de son paléogradient thermique. Elle peut se situer [43] :

— entre 700 et 1 500 m pour les paléogradients forts (80 °C/km);
— de 2 500 à 6 000 m pour les paléogradients faibles (20 °C/km).

De telles évaluations ne peuvent être qu'approximatives puisque Pusey ne tient compte que des gradients actuels et les suppose constants dans le passé, qu'il suppose aussi une vitesse de subsidence constante et qu'il ignore les divers types de kérogènes.

L'avantage final du microscope est sa double capacité de préciser simultanément et sans extraction préalable le **niveau** de maturation et le **faciès** de l'assemblage organique présent.

Les indices optiques permettent également de préciser la position de la fenêtre pétrolière dans les séries où les roches mères ont subi des phénomènes d'accumulation ou de drainage et pour lesquelles, de ce fait, le taux d'extractibles ne traduit plus fidèlement le niveau de maturation.

Les photographies en couleurs des six planches insérées dans ce chapitre ont été prises au CERCHAR avec un microscope Orthoplan MPV2 Wild-Leitz.

BIBLIOGRAPHIE

1. Alpern, B., (1966) *in : Advances in Organic Geochemistry, 1964,* Pergamon Press, Oxford, 129-145.
2. Alpern, B., (1969), *Ann. Soc. Géol. du Nord,* **89**, 143.
3. Alpern, B., (1970), *Rev. Inst. Franç. du Pétrole,* **25**, 1233.
4. Alpern, B., (1971), *Bull. Soc. Fr. Min. Crist.* **94**, p. 179-180.
5. Alpern, B., (1976), *Bull. Centre Rech. Pau, SNPA,* **10**, 1, 201.
6. Alpern, B., Durand, B., Espitalie, J. et Tissot, B., (1972), *in : Advances in Organic Geochemistry,* 1971, 1, Pergamon Press, Oxford, 1.
7. Alpern, B. et Cheymol, (1978), *Rev. Inst. Franç. du Pétrole,* **XXXIII, 3**, 515.
8. Alpern, B. Durand, B. et Durand-Souron, (1978), *Rev. Inst. Franç. du Pétrole,* **XXXIII, 6.**
9. Ammosov, I., (1961), *Sovet. Geol.* **4**, 7.
10. Ammosov, I., (1967), *Akad. Nauk, SSSR. Inst. Géol. Moscou,* **5.**
11. Benedict, L.G., Thompson, R.R., Shigo III, J.J. et Aikman, R.P., (1968), *Fuel,* **47**, 125.
12. Blanquart, P. et Meriaux, E., (1975), *in :* Coll. Int. « Pétrographie de la matière organique des sédiments », 1973, Ed. B. Alpern, CNRS, Paris, 27-39.
13. Bostick, N., (1973), *7e Congr. Internat. Géologie. Stratigraphie. Carbonifère, Krefeld, 1971,* **2**, p. 183-192.
14. Bostick, N., (1975), *in :* Coll. Int. « Pétrographie de la matière organique des sédiments » 1973, Ed. B. Alpern CNRS, Paris, 13.
15. Brown, H.R., Cook, A.C. et Taylor, G.H., (1964), *Fuel,* **43**, 111.

16. Caye, R., (1970), *Bull. Soc. Fr. Mineral. Cristallogr.*, **93**, 249.

17. Caye, R. et Ragot, J.P., (1972), *in : Advances in Organic Geochemistry*, 1971, Pergamon Press, Oxford, 591.

18. Cervelle, B., (1971), *Bull. BRGM* (2), **11**, 5, 9.

19. Damberger, H., (1968), *Brennst. Chem.*, **49**, 73.

20. Dow, (1977), *J. Geochem. Explor.* **7**, 79.

21. Gijzel, P. van, (1971), *in : Sporopollenin*. Ed. J. Brooks *et al.*, London, Acad. Press, 659.

22. Gijzel, P. van, (1975), *in :* Coll. Int. « Pétrographie de la matière organique des sédiments », 1973, Ed. B. Alpern, CNRS, Paris, 57.

23. Grayson, J.F., (1975), *in :* Coll. Int. « Pétrographie de la matière organique des sédiments », 1973, Ed. B. Alpern, CNRS, Paris, 261.

24. Hedemann, H.A. et Teichmüller, R., (1966), *Z. Deutsch. Ges.*, **115**, 787.

25. Hevia, V., (1970), Thèse, Oviédo.

26. Hevia, V., (1977), *in :* Advances in Organic Geochemistry, 1975, Ed. R. Campos et J. Goni, Enadisma, Madrid, 655.

27. Hevia, V. et Virgos, J.M., (1976), *Meeting of the Microscopy of Organic Sediments, coals and cokes : Methods and applications,* Oxford.

28. Hood, A., Gutjahr, C. et Heacock, R., (1975), *AAPG Bull.*, **59**, 6, 986.

29. Jacob, H., (1967), *Erdöl und Kohle,* **20**, 393.

30. Johns, W.D. et Shimoyama, A., (1972), *AAPG Bull.*, **56**, 11.

31. Jongsama, A.P.M., Hijmans, W. et Ploem, J.S., (1971), *Histochemie,* **25**, 329.

32. Jukes, L.M., (1974), *Mineralogy and Materials News Bulletin for Quantitative Microscopic Methods,* **4**, 12.

33. Karweil J., (1975), *in :* Coll. Int. « Pétrographie de la matière organique des sédiments », sept. 1973, Ed. B. Alpern, CNRS, Paris, 195.

34. Van Krevelen, D.W., (1961), « Coal », Elsevier, Amsterdam.

35. Kunstner, E., (1971), Thèse, Freiberg, 1971.

36. « Lexique International de Pétrographie des Charbons ». 2e édition (1963), 1er supplément à la 2e édition (1971), 2e supplément à la 2e édition, (1975), CNRS, Paris.

37. Locquin, (1977), « Relations chronophénétiques entre taxons fossiles ». Ecole Pratique des Hautes Etudes, Paris.

38. Lopatin, N. et Bostick, N. (1973), Trad. anglaise : Illinois State Geol. Surv. Reprint series 1974, Q. Art. original : Nauka Press, Moscou 1973, 79.

39. Madler, (1963), *Beih. Geol.* JB, **58**, Hanover.

40. Ottenjann, K., Teichmüller, M. et Wolf, M., (1974), *Forstschr. Geol. u. Westf.*, **24**, 1.

41. Ottenjann, K., Teichmüller, M. et Wolf, M., (1975), *in :* Coll. Int. « Pétrographie de la matière organique des sédiments », 1973, Ed. B. Alpern, CNRS, Paris, 49.

42. Ploem, J., Sterke, J., Bonnet, J. et Wasmund, H. (1974), *J. Histochem. Cytochem.* **22** 7 668.

43. Pusey, III W.C., (1973), *Petroleum Times,* **12**, 21.

44. Raynaud, J.F. et Robert, P., (1976), *Bull. Centre Rech. Pau, SNPA,* **10**, 1, 109.

45. Stach, E., (1969), *Freiberger Forschungshefte H.C.*, **242**, 35.

46. Teichmüller, M., (1958), *Rev. Industrie Minérale,* n° spécial 15 juillet 1958, 99.

47. Teichmüller, M., (1974), *Fortsch. Geol. Rheinl. u. Westf.* **24**, 37.

48. Teichmüller, M. et Ottenjann, K., (1977), *Erdöl u. Kohle,* **30**, 387.

49. Tissot, B., Durand, B., Espitalié, J. et Combaz, A., (1974), *AAPG. Bull.*, **58**, 499.

50. Tissot, B. et Espitalié, J., (1975), *Rev. Inst. Franç. du Pétrole* **XXX**, **5**, 743.

51. Vassoyevitch, N., Kortchagina, J., Lopatin, N. et Tchernichev, W., (1970), *Moscou Univ. Vestnik,* **6**, 3 (1969). Trad. anglaise : *Intern. Geol. Rev.*, **12**, 1276.

52. Wehry, (1967), *in : « Fluorescence »* Ed. G. Guilbault, New York, 37.

53. White, D., (1915), *J. Wash. Acad. Sci.*, **6**, 189.

PLANCHE 11.1
COMPARAISON DES MÉTHODES OPTIQUES

Photos 1 à 3. — Lignite de Moscou.
Lame mince polie non recouverte vue successivement en transparence, réflexion et fluorescence. Les spores sont visibles par les trois procédés mais seule la réflexion montre des produits gris (vitrinite) et blancs (inertinite).

Photos 4 à 6. — Concentré d'organoclastes par voie physique.
Seule la fluorescence montre la présence d'une population algaire majoritaire formant le fond de la préparation.

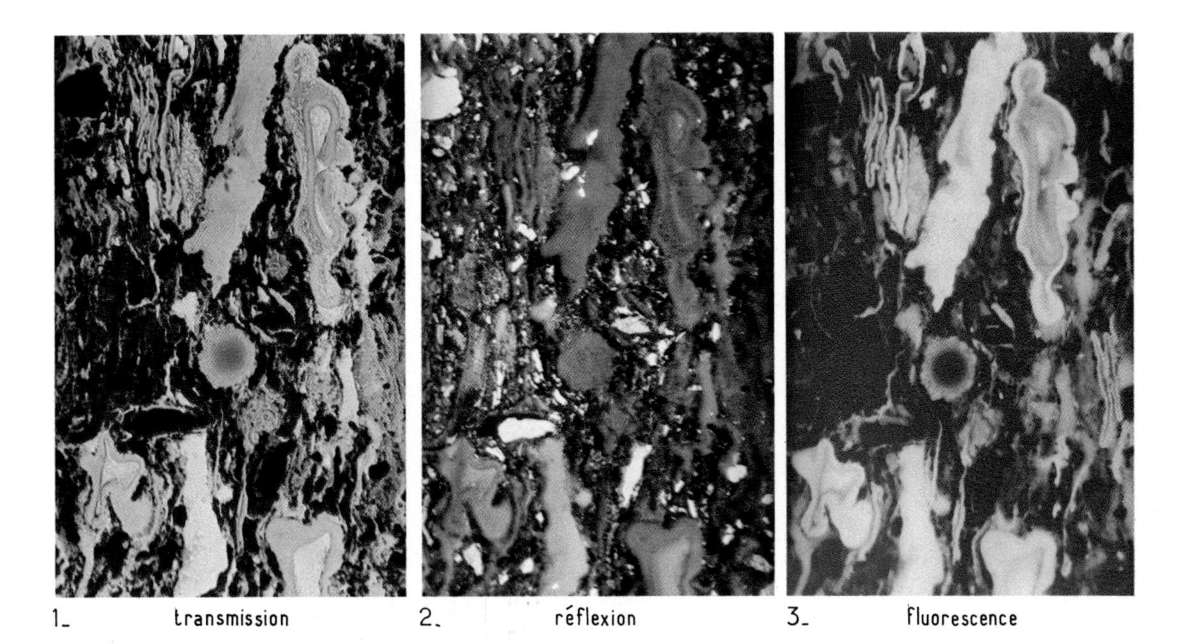

1_ transmission 2_ réflexion 3_ fluorescence

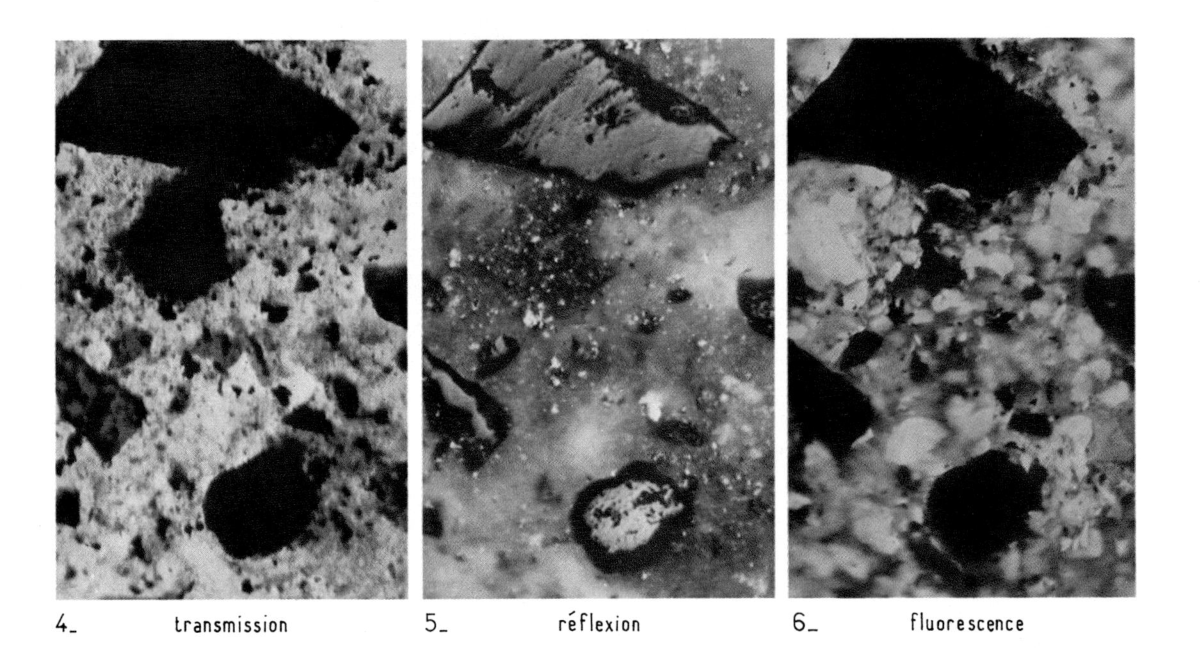

4_ transmission 5_ réflexion 6_ fluorescence

PLANCHE 11.2
VRAIE ET FAUSSES VITRINITES

Photo 1. — **Charbon de Ste-Fontaine (Lorraine).** (G. × 137). Les deux types principaux de vitrinite. Au centre, un cryptotissu, la télocollinite. Autour des microspores noires, un ciment gélifié : la desmocollinite.

Photo 2. — **Asphaltite de March Bank (Colorado).** (G. × 137). Fragmentée et hors de son contexte lithologique d'origine, une telle particule (au centre) peut être confondue avec la collinite.

Photo 3. — **Sondage N à 2634 m (Algérie).** (G. × 137). Bitume révélé par sa forme anguleuse aciculaire (remplissage intercristallin) et l'attaque au permanganate de *K* qui a corrodé la partie centrale. La partie claire, prise isolément, peut être confondue avec la collinite.

Photos 4 et 5. — **Sondage R à 870 m (Spitzberg).** (G. × 137 et G. × 330). Toutes les particules gris blanc, homogènes, que l'on peut prendre pour de la vitrinite, sont en réalité des ornements de mégaspores comme on le voit nettement sur la photo 5. La membrane des mégaspores est corrodée mais les ornements arrondis sont massifs.

Photo 6. — **Charbon de Ste-Fontaine, veine G (Lorraine).** (G. × 126). Avant carbonisation (gauche), après carbonisation à 600 ° (droite). Cette photographie montre les difficultés de diagnose dues à la transition continue vitrinite → inertinite. La télocollinite (R = 0,88 à 0,92 ‰) passe à la semifusinite très graduellement tandis que le gonflement du coke diminue fortement (dès R = 0,94 ‰).

Photo 7. — **Sondage K à 2516′ (Canada).** (G. × 330). Fragment de zooclaste : Graptolithe. Des particules isolées plus petites peuvent être confondues avec de la gélinite interstratifiée avec de la résinite.

Photo 8. — **Gothland.** (G. × 275). Partie basale de Chitinozoaire avec son opercule. Les fragments isolés peuvent être confondus avec la vitrinite.

Photo 9. — **Sondage H à 8300′ (Kenya).** (G. × 275). Zooclaste? La partie gris clair amorphe, ressemble à la vitrinite. La forte fluorescence à droite montre que c'est inexact. Il s'agit probablement d'un bitume d'origine animale.

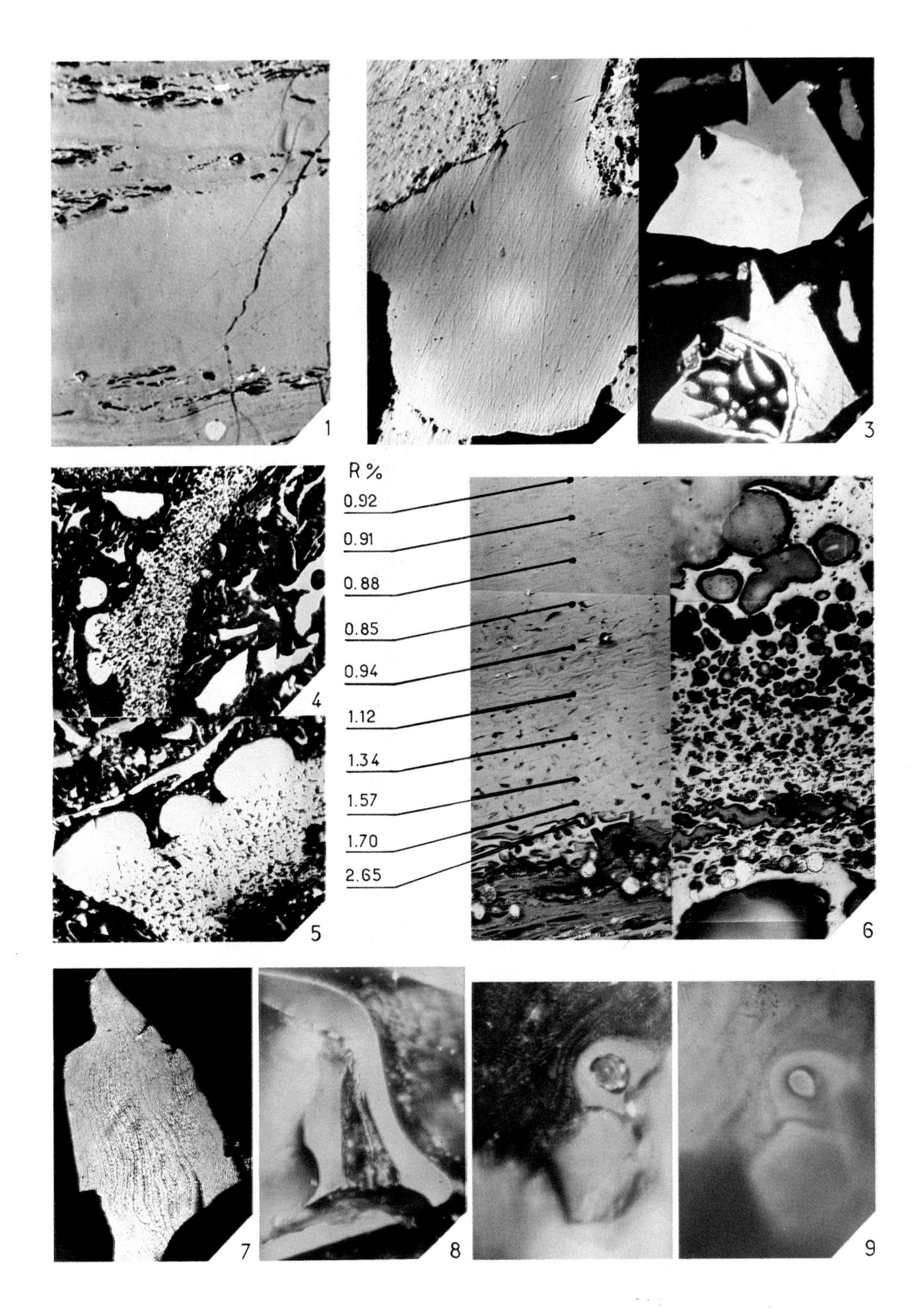
1
3
4
5
R %
0.92
0.91
0.88
0.85
0.94
1.12
1.34
1.57
1.70
2.65
6
7
8
9

PLANCHE 11.3
LES ORGANOCLASTES FIGURÉS

Photo 1. — Charbon de l'Aumance (France). (G. × 275).
a. Fusinite à structure botanique nette. **b.** Les cavités sont remplies de bitumes fluorescents qui se dissolvent dans l'huile d'immersion, d'où le voile verdâtre.

Photo 2. — Sédiment marin récent (vase). (G. × 275). Tissu cuticulaire à membranes fluorescentes, en réflexion (**2a**) et en fluorescence (**2b**).

Photo 3. — Charbon de la Houve, veine Henri (Lorraine). (G. × 275). Mégaspore (fragment) et microspores vues en fluorescence. On notera que les microspores massives (*Torispora* pour la plupart) fluorescent en jaune plus vif que les microspores fines (*Lycospora* pour la plupart).

Photo 4. — Charbon de Faulquemont, veine 4ter (Lorraine). (G. × 275). Microspore fortement ornementée. Les ornements ont une fluorescence plus vive.

Photo 5. — Charbon de Faulquemont, veine 4ter (Lorraine). (G. × 275). La microspore centrale présente des protubérances de teinte jaune vert tandis que la spore est brunâtre. Ces excroissances ne sont pas des ornements réguliers sur toute la surface mais peut-être des produits bitumineux néoformés que l'on retrouve parfois en amas isolés (Fluorinite?).

Photo 6. — Lignite du Bassin de Moscou. (G. × 275). Microspores cingulées (*Anulatisporites*) et Algues (*Botryococcus*).

Photo 7. — Boghead de Fréjus (France). (G. × 275). Accumulation d'algues du genre *Botryococcus.*

Photo 8. — Sondage H à 7800′ (Kenya). (G. × 275). Épiderme de zooclaste probable, noir-rougeâtre, en réflexion (**8a**) et en fluorescence (**8b**).

Photo 9. — Schistes de Decazeville (France). (G. × 176). Zooclaste. Reste de poisson.

Photo 10. — Sondage D à 5940′ (Danemark). (G. × 330). Tasmanacée vue en fluorescence.

Photo 11. — Toarcien du Bassin de Paris. (G. × 330). Tasmanacée vue en fluorescence.

Photo 12. — Sondage N à 622 m (Sahara). (G. × 330). Acritarches (*Veryachium?*).

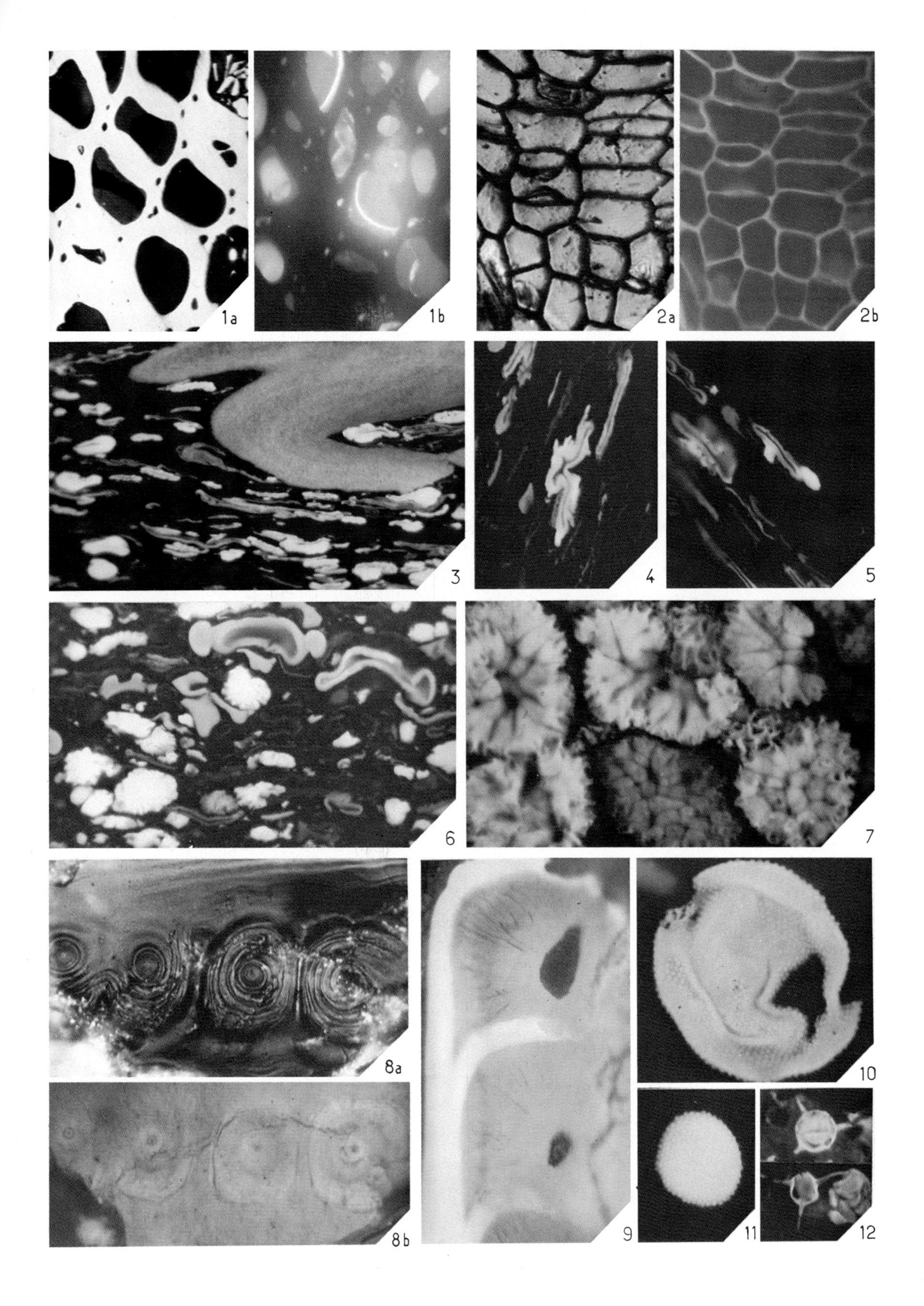
1a
1b
2a
2b
3
4
5
6
7
8a
8b
9
10
11
12

PLANCHE 11.4
SCHISTES BITUMINEUX ET KÉROGÈNE

Photo 1. — **Green River Shales (Colorado/USA).** (G. × 275). Section perpendiculaire à la stratification. Lumière réfléchie. Lamelles bitumineuses grises interstratifiées avec des lits carbonatés.

Photo 2. — **Green River Shales (Colorado/USA).** (G. × 275). Vue en fluorescence. Les lits carbonatés sont imprégnés. (V. notamment à droite). Au centre le lit vertical est composé de lamelles accumulées, aux limites indistinctes.

Photo 3. — **Green River Shales (Colorado/USA).** (G. × 275). Section parallèle à la stratification. Lumière fluorescente. Certaines zones sont imprégnées d'une matière diffuse mais dans de nombreux cas on devine des organismes figurés de petite taille (algues).

Photo 4. — **Sondage à 6680′ (Afrique).** (G. × 137). Roche imprégnée d'hydrocarbures (réservoir).

Photo 5. — **Sondage G11 à 60,80 m. Toarcien du Bassin de Paris.** (G. × 275). Concentré par voie physique *CERCHAR*. Lumière réfléchie. On distingue quatre types de particules réfléchissantes : de l'inertinite blanche, des particules poreuses grises piquetées de micrinite (bituminite?), deux particules rougeâtres homogènes (centre), une particule noire (inférieur gauche).

Photo 6. — **Sondage G11 à 60,80 m. Toarcien du Bassin de Paris.** (G. × 275). Idem en fluorescence. De nombreuses algues apparaissent, une Tasmanacée de grande taille et des petites algues (*Nostocopsis*) ainsi que de nombreux fragments amorphes. Les particules poreuses piquetées ne sont pas fluorescentes mais les particules rougeâtres le sont d'autant plus qu'elles sont plus foncées en lumière réfléchie normale.

Photo 7. — **Sondage de Coupvray. Toarcien du Bassin de Paris.** (G. × 825). Kérogène concentré par voie chimique (IFP). Lumière fluorescente. Tasmanacée et accumulation de microlamelles jaunes (algues).

Photo 8. — **Green River Shales (Colorado/USA)** (G. × 825). Kérogène concentré par voie chimique (IFP). Lumière fluorescente. Accumulation de microlamelles fluorescentes jaunes parfois encore stratifiées.

Photo 9. — **Sondage W à 789 m (Australie).** (G. × 275). Montage du concentré physique (*CERCHAR*). Lumière réfléchie. Spores rouges, certaines membranes sont en partie vitrinisées, fragment de spore bisaccate, plus particules grises et blanches.

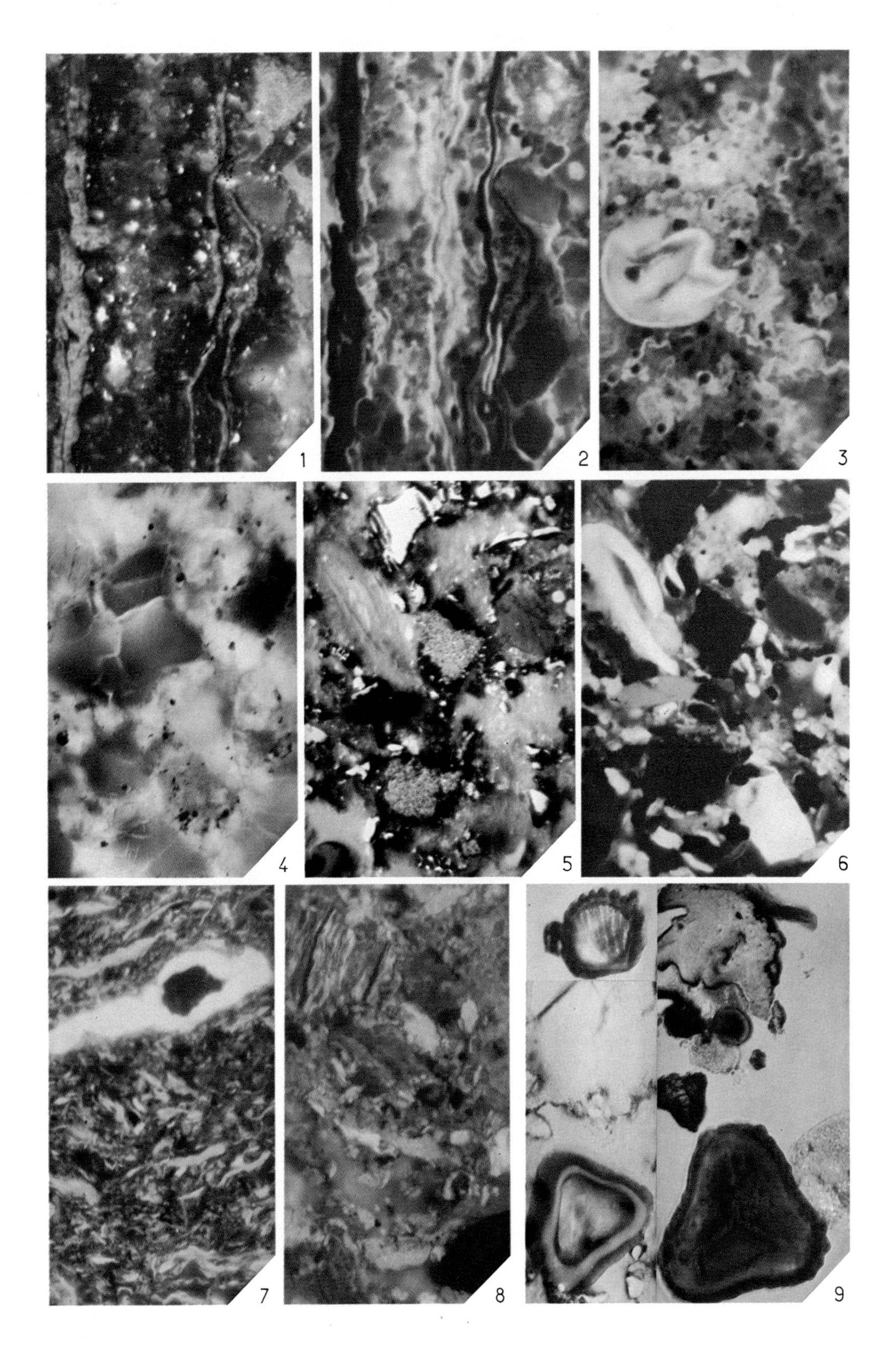
1
2
3
4
5
6
7
8
9

PLANCHE 11.5
EXSUDATS ET BITUMES

Photo 1. — **Lignite de Port Christmas (Kerguelen).** (G. × 275).
a. Lumière réfléchie. Vitrinite grise, résinite et micrinite en bordure. **b.** Fluorescence. On constate que les fractures sont remplies de bitumes fluorescents (exsudatinite).

Photo 2. — **Charbon de la Houve, veine Henri (Lorraine).** (G. × 275). La photo en fluorescence (**2b**) montre que les fractures sont remplies de bitumes migrés (exsudatinite).

Photo 3. — **Sondage H à 2 236 m (Afrique du Sud).** (G. × 55). Dégagement de gouttelettes fluorescentes (hydrocarbures) au cours de l'examen.

Photo 4. — **Lignite de Port Christmas (Kerguelen).** (G. × 137). Inclusion bitumineuse stratifiée ayant généré des exsudats transverses.

Photo 5. — **Schiste radioactif de Lodève (France).** (G. × 192). Diverses générations de bitumes en réflexion (**a**) puis fluorescence (**b**). La fluorescence est annulée par des inclusions radioactives, notamment dans la fracture du haut.

Photo 6. — **Sondage W à 4 278 m (Australie).** (G. × 275). Vitrinite (télocollinite) nettement fluorescente (**6b**).

Photo 7. — **Sondage B à 10 020′ (Australie).** (G. × 275). Sphère résineuse à inclusion radioactive, pourtour fluorescent (**7b**) et centre noirci par les radiations alpha.

Photo 8. — **Sondage G à 1 424 m (Sénégal).** (G. × 275). Globule à deux phases fluorescentes (résinite ou bitume).

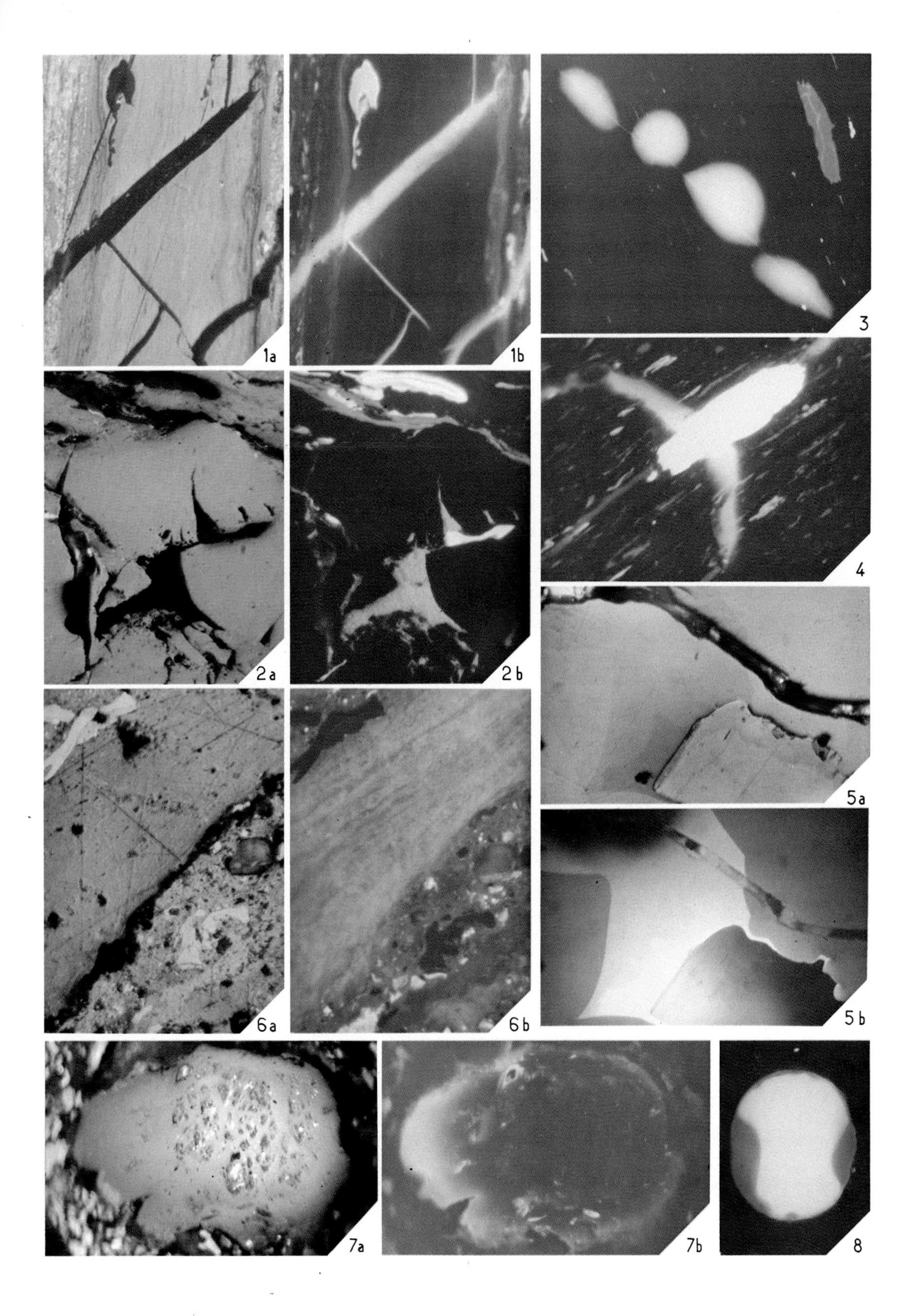

1a
1b
3
2a
2b
4
5a
6a
6b
5b
7a
7b
8

PLANCHE 11.6
ACTION DE LA DIAGENÈSE ET ACTION THERMIQUE

Photo 1. — Schistes bitumineux de Fécocourt. Toarcien du Bassin de Paris. (G. × 137). Affleurement. Fluorescence par excitation dans le bleu (390 nm). La Tasmanacée et les microalgues sont franchement visibles en jaune. La diagenèse est faible.

Photo 2. — Sondage Bouchy. Toarcien du Bassin de Paris. Profondeur 2 500 m. (G. × 137). Fluorescence par excitation dans le bleu (390 nm). La Tasmanacée est passée au brun, les microlamelles du fond sont devenues presque invisibles.

Photo 3. — Schistes bitumineux de Fécocourt. Toarcien du Bassin de Paris. (G. × 137). Affleurement. Fluorescence par excitation dans le violet (360 nm). Aspect d'une Tasmanacée à faible degré de diagenèse (affleurement).
λ_{max} = 514 nm Q = 0,68.

Photo 4. — Sondage Bouchy. Toarcien du Bassin de Paris. (G. × 137). Fluorescence par excitation dans le violet (360 nm). La couleur est devenue franchement brune.
λ_{max} = 617 nm Q = 1,33.

Photo 5. — Green River Shales (Colorado/USA). Kérogène IFP. (G. × 275). Kérogène chauffé à 350 °C. L'aspect est pratiquement celui d'un schiste cru.

Photo 6. — Green River Shales (Colorado/USA). Kérogène IFP. (G. × 275). Kérogène chauffé à 480 °C. Le kérogène a fondu et donné des particules ressemblant à de la vitrinite. Les organoclastes initiaux ne se distinguent plus, on ne voit que des inclusions de pyrite. La différence de réflectance est peut-être due à des différences locales de température?

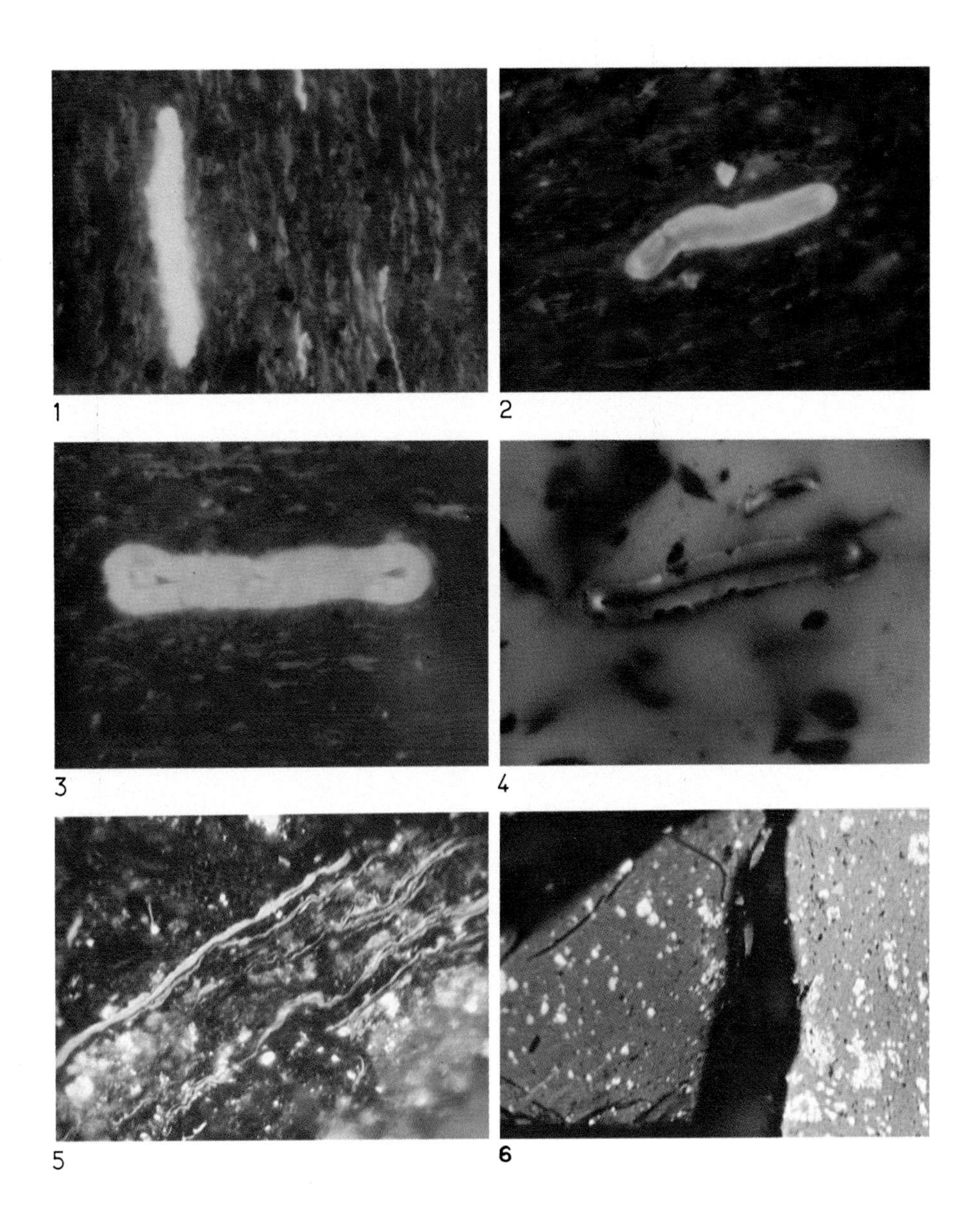
1
2
3
4
5
6

12

The optical evolution of kerogen and geothermal histories applied to oil and gas exploration

P. ROBERT*

I. INTRODUCTION

In the area of petroleum exploration opticals methods are much used to identify kerogen types. When this data is complemented by measuring the state of catagenesis of the kerogen, it may be possible to provide useful information on:

(a) The nature of the fluids generated from kerogen during the course of its thermal evolution.

(b) Their generation history, and also possibly their migration characteristics.

(c) The use of thermal history as an aid in exploring for favourable zones of petroleum generation on a regional basis.

The aim, here, will be to examine:

1. How various optical procedures have been adapted to exploration requirements.

2. The way in which the thermal history of kerogen and hydrocarbons develops, and the role of optical studies in the knowledge of hydrocarbon generation.

3. The geothermal history of series by means of vitrinite reflectance to reveal the succession of various periods in this history. This helps in finding hyperthermal causes of the hot periods and leads to a better understanding of sedimentary basins.

*Société Nationale Elf Aquitaine (Production), Direction Exploration. Département Laboratoire de Géologie. Boussens, 31360 Saint-Martory, France.

II. OPTICAL METHODS FOR DETERMINING ORGANIC CATAGENESIS AND THEIR APPLICATION TO PETROLEUM EXPLORATION

A. Concentration processes of organic matter for optical studies.

Oil and gas research involves the examination of all kinds of sedimentary rocks. Hence, the kerogen, which frequently represents no more than 1.0 to 5.0‰ of the rock in weight, has to be concentrated. The resulting concentrates often differ widely in quality: they are generally less rich in carbonates than in sand-shale rocks, and, outcrops from hot, dry areas are especially poor in organic particles. It is more advisable, in all cases, to select the richest beds or levels, i.e. the ones that are usually darkest in colour.

The main methods of concentration processing are the following.

1. Conventional processes.

These processes are based on the destruction of the mineral phase with acids (ClH, FH), liberating the organic fraction; this fraction is then concentrated through gravity by centrifuging in heavy liquid, a technique used in palynology.

The great advantage of this procedure is that the main part of the organic fraction is reçovered; on the other hand, there is the disadvantage that particles are finely comminuted, so that constituent coal macerals may dissociate, resulting in greater problems of diagnosis.

Acid attack, generally performed by heating, is considered to undermine fluorescence to a greater or lesser extent.

2. Physical processes.

Processes based on the following principles have been developed to avoid these difficulties:

(a) Gravity separation in heavy liquids by centrifugal application.

(b) "*Froth flotation*", i.e. separation in water, using a selective wetting agent: bubbles formed by passing air through and stirring in the grinded sample separate particles that are partially or entirely organic, due to favourable surface tension characteristics [61].

In both cases, part of the mineral matrix is present, which eases diagnosis, the concentrated fragments obtained are larger in size and the fluorescence remains intact; this can be confirmed by comparing the concentrates with non-concentrated rocks under the microscope (if the rock is rich enough in kerogen).

These methods, on the other hand, may be slightly selective, as the resulting product is not a pure kerogen, and, they do not provide a *total* concentration of organic matter (OM).

B. Parameters for rank measurements (Table 12.1).

1. Physical and chemical parameters.

Physical and chemical analyses of *coals* have provided parameters that enabled them to be classified into types that reflected their thermal histories. This is embodied in Hilt's Law [32] which relates *coalification* to depth of burial, coalification being taken as an irreversible concentration in carbon, together with a progressive loss of volatile matter, including water, nitrogen, carbon dioxide and methane.

TABLE 12.1. — MAIN PARAMETERS IN ORGANIC CATAGENESIS
After Stach's textbook [65], modified Soviet scale after Kouznetsova *et al.* [45]

COAL RANK USA	USSR	GERMAN	C %	H_2O %	V.M. %	Ro %	TAI (E.C.)	Spor. FLUO max nm	LOM	Hydrocarbon generation
									0	Coals / Source Rocks
Peat		Peat			70	.20	1	500		
			60	75				600		
					60	.30				
							(2)		2	Early "diagenetic" Gas
Lignite	B_1	Browncoal: soft								
				35	56					
									4	
	B_2	mat								
Sub-bitum. C						.40	2			
					50				6	
Sub-bitum. B	B_3		71	25						
					46	.50	2.5	580		
		brilliant								
Bituminous High Volatile C	D								8	Gas / Oil — Oil
						.60				
			77	8-10	43					
Bituminous High Volatile B		Hardcoal: flaming			40	·70	(3)	590		
								630		
Bituminous High Volatile A	G	gas flaming			37	80	3		10	
	J	gas			33	1.		660	11	
Bituminous M. Vol			87		28	1.25	4	675		Wet Gas
	K	fat			22	1.50			12	
Bituminous L. Vol		half fat			19	1.60	(4)			(+condensaté)
	OS				14	1.90		no more visible fluorescence		
						2.				
Semi		lean			12	2.1			14	Catagenic Dry- Natural gas
Anthracite	T				10	2.3				
							5			
			91		8	2.6			16	
		Anthracite			6	2.9				
						3.				
Anthracite	PA				5				18	No Economical Gas
			93.5		4	3.5				
		Meta				4.				
					2					
Meta Anthracite	A	Anthracite							20	

C = Carbon
H_2O = Moisture content
VM = Volatile Matter
Ro = Vitrinite Reflectance in oil
TAI = Thermal Alteration Index
(E.C.) = "Etats de conservation"
FLUO = λ max fluorescence sporinites
LOM = Level of Organic Metamorphism after HOOD et al(34)

The following parameters, then, provide a means of determining the successive degrees of the evolution of coals or "**rank**" and they form the basis of our knowledge on the thermal evolution of OM yielding either coal or petroleum, i.e., kerogen. They can be used over a part or the total range of coal rank, depending upon the range to which they express such evolution:

(a) VM% (volatile matter) covers the whole range.

(b) Moisture content parameter is used for coals of low rank such as peat, lignite, subbituminous and high volatile bituminous varieties.

(c) Carbon and hydrogen contents are only used for the higher ranks, i.e. medium and low-volatile bituminous coals and anthracites; however, they are not commonly measured for classification purposes, but rather as indicators of coal composition.

These parameters (see Table 12.1 for VM%, C%, H_2O%) have been developed for use on coals. An empirical correlation has been used for rocks with dispersed kerogen, by means of such indices as:

$$\text{The carbon-ratio, i.e. } \frac{\text{fixed carbon}}{\text{total organic carbon}}$$

the fixed carbon being measured by pyrolysis in an atmosphere of nitrogen at 900° C [84, 28, 47].

2. *Vitrinite reflectance* (R_o).

Microscopic study of the OM in reflected light, or coal petrology, as developed widely by coal petrologists over the past 50 years, is nowadays as reliable an analytical method as chemical methods. A further feature is that it enables coal constituents, or "*macerals*" (Alpern, *in:* Chapter 11) to be identified and hence mixtures of coals to be differenciated. *Vitrinite reflectance* [46], using an oil immersion technique, R_o, has proved to be the most suitable physical parameter for checking the rank of OM in sediments.

Some advantages of this technique are:

(a) Measurements are carried out by photometry which is an objective and easily reproduced means of data recording; the value obtained is the ratio of the amount of reflected light to light received by the measured surface area.

(b) This method has been standardized for about 20 years and has been correlated with physical and chemical parameters of coals [81, 1, 51, 70]. Data on an international scale have been cross-referenced to provide a high degree of accuracy and research support material [MOD (Dispersed Organic Matter) Commission of *ICCP* (*International Committee of Coal Petrology)* 1971 to 1977].

(c) It can be used over a very wide range of coalification, from lignite to graphite.

(d) It increases with burial depth, or metamorphism, and this helps to facilitate its use for the non-specialist.

(e) Above all, it provides greater accuracy than does carbon-ratio data for expanding our knowledge of coal deposits to fine dispersed particles found in rocks, and, consequently, for using these fine particles for the study of thermal phenomena in geological series.

3. Microscopic examination in transmitted light.

By studying microfossils and translucent organic particles, palynologists can observe the level of transformation attained due to catagenesis, such as size reduction characteristics, and, more particularly, the gradual colouration and darkness and ultimately the complete opacity of the OM.

Palynological techniques have been used to provide classification indices that increase with catagenesis, and they have been progressively correlated with the preceding parameters. Measurements are carried out by photometry [29], or by simply correlating results with a reference table: *"States of Preservation"* [15, 16]; *Thermal Alteration Index* (TAI), [68, 59]. The accuracy of this later process has proved itself sufficient in recent years, and the method has been developed widely, on account of its low cost (due to being linked to stratigraphical studies) and the fact that it complements reflectance measurements for all ranks (paragr. II.C).

Their main advantage when applied on stratigraphical fossils, is that they are characteristic of the series evolution, avoiding mistakes due to allochthonous or derived material.

4. Induced UV or blue light fluorescence.

Fluorescence under the microscope, the latest in the line of optical processes [25, 26], has shown itself capable of contributing towards an appreciation of organic catagenesis. Applying the method of Ottenjann *et al.* [52] to **sporinites,** it has been established that:

(a) Visible *light spectra* move from lower to higher wavelengths during the course of catagenesis.

(b) Recording these spectra provides calculated parameters which can be correlated with the preceding ones.

(c) The method is suitable for low rank measurements only, due to the disappearance of visible fluorescence in medium volatile bituminous coals, at about 1.0 to 1.3% R_o, depending upon the nature of maceral.

Some other fluorescent macerals can be used in such a way to provide rank scales:

(a) Algae *tasmanaceae* are intensively coloured from green to yellow, orange and dark orange between the stages "subbituminous and bituminous coals". With references to photometric surveys on typical series established by the method of Ottenjann *et al.* [52] they provide a qualitative aid for rapid determination of rank (see Alpern and Cheymol *in:* Alpern, Chapter 11).

(b) Colonial algae, *botryococcus*-type also exhibit an intense fluorescence colouration. Indications obtained from them are less accurate than those obtained from *tasmanaceae,* because of a colour variation which is more progressive and has a shorter range (green-yellow to orange); however they can be useful in series devoid of vitrinite.

(c) Many other fluorescent macerals are more or less variously-coloured, depending upon rank: *fluorinites, cutinites, resinites* (secretions of high plants) and so are the stratified organo-mineral material which constitutes the "oil shales" matrix (Teichmüller's groundmass, [73, 74]). Their contribution in determining rank is only complementary and approximate.

C. Difficulties arising from analyses in bore-holes and surface sections: conventional choice of a main parameter, vitrinite reflectance.

It is advisable to use all three usual methods of optical examination and it would be unwise to employ just one of these methods without the other two. Experience in this matter proves that even in the simplest cases, one method alone cannot guarantee sufficient reliability and accuracy.

Vitrinite reflectance R_o, was used at first; it has largely been standardized and its measuring devices have been thoroughly tried and tested. It was the only R_o to be submitted to international testing. It has been used in defining the stages of hydrocarbon generation. Always measured, never estimated, R_o is practical for graphic records and statistical surveys. It is the only R_o which covers the whole range of rank necessary in petroleum exploration.

Due to the variations of quality of samples in oil drilling and surface sections, R_o which are reliable and easy to interpret are relatively rare; poor samples are more frequent due to the following difficulties.

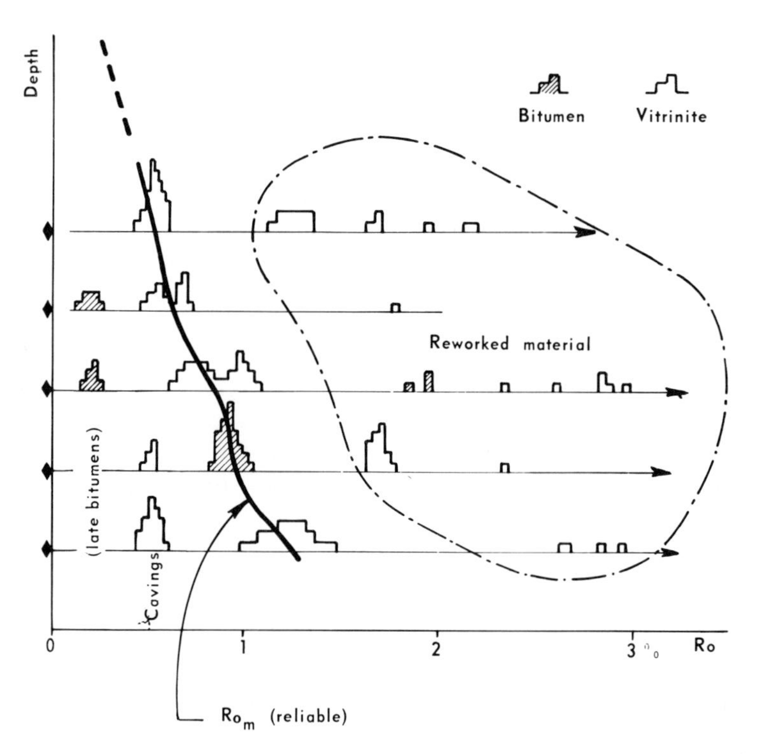

Fig. 12.1. — Schematic reflectance curve showing statistical increase with depth: kerogen populations (Robert 1973).

1. Lack of vitrinite.

This occurs frequently in series devoid of a supply of continental OM, a common feature with carbonates and almost the rule in very old rocks (Cambro-Silurian), whatever their lithology, due to the reduced development of terrestrial flora. In these cases, asphaltic particles can be used or other methods adopted.

2. Reliable vitrinite populations.

It is often difficult to pick out the *reliable* vitrinite population. This appears in Fig. 12.1, which indicates that many "anomalies" may occur [60, 11], that are liable to give wrong interpretations. These anomalies may be due to:

(a) Incorporation of cavings during the course of drilling which results in the R_o being too low.
(b) The presence of "*allochthonous*" particles that are reworked from coalified levels, and, which have thus reached higher ranks prior to burial. It is often easy to distinguish these particles of diverse origins because of their wide dispersion on the statistical reflectance diagram. It also happens that they alone constitute the whole concentrate. In general, their R_o is higher than the reliable value.

The best way to solve these difficulties is to use the TAI (paragr. II.B.3.) which, when applied to characteristic fossils at the coalification level under study, may distinguish them from derived or reworked material: thus it is often possible to choose a single reliable R_o population from among several different ones.

3. The environment of sedimentation.

This also may cause major problems in rank analyses.
For example, *sapropelic series,* —i.e. algal in reducing conditions— contain smaller or larger amounts of humic material including vitrinite:

(a) When vitrinite is present, it has been deposited in conditions of low sedimentation and long transportation and it may belong to the reworked or "allochthonous" population (Fig. 12.1).
(b) Even if they are representative, it may be that these vitrinites are fluorescent and their R_o, weakened, is not reliable.

In such cases, rank estimations or measurements from fluorescent macerals are more reliable and accurate.
Overall results show that R_o is a suitable parameter that may be presented alone in graphic recordings or comprehensive surveys. However it must always be supported by transmitted light and fluorescence surveys.
Such analyses are suitable for kerogen dispersed in rocks and aim to survey a thermal history of sediments. One could suggest that they be replaced or completed by chemical analysis of hydrocarbon extracts, whose composition could provide information on rank: due to the

uncertain origin of hydrocarbon extracts —syngenetic or migrated— the results obtained are less rigorous than those obtained from kerogen. Chemical analysis of kerogen itself gives best results [78].

D. The use of asphaltic bitumens.

Organic concentrates of sedimentary rocks frequently contain some asphaltic bitumens, insoluble in most solvents, whether associated with coaly material or not. They are located in the pores of reservoirs or are due to first generation in oil source rocks. They generally look like coaly particles, and then can be confused with homogeneous coal macerals (as dispersed in rocks).

Bitumens are sometimes abundant in the old series (Precarboniferous) where coals are rare. In carbonated series they are likely to be more frequent than coals; they are present in all permeable rocks which drained oil migrations.

The use of asphaltic bitumens is not simply a last resort, but is useful when vitrinite is lacking (paragr. II.C.1); also there is a need to recognize bitumens in the concentrates, since their R_o generally does not express the same thermal history as does that of vitrinites.

When these particles are found in association with coaly material, their R_o is generally lower than that of vitrinite. There are two reasons for this:

(a) The first is their asphaltic nature, and a chemical composition close to that of some liptinites (or exinites).

(b) Secondly, they are frequently emplaced, in the series, through oil migration, and have taken place after the main thermal events affecting the series. They can be found, for example, in certain large-size oil pools, in which the reservoir is marked by a strong earlier catagenesis, as in the Algerian Sahara [60]. Here, there is an ancient population containing high reflecting bitumen (R_o = 1.0 to 2.0%), together with a later, low reflecting bitumen R_o = 0.5%) which is in equilibrium with the oil of the pool.

In the case where they are linked to oil migrations, these bitumens may originate in several successive generations characterized by various R_o-values. Thus they may aid in understanding the termal history of the basins and provide evidence of successive stages in it.

Evaluating a comparative R_o scale, between bitumens and vitrinite is one of the most difficult problems, for the following reasons:

(a) The series which are rich in bitumens and likely to provide continuous vertical sections with bitumens are frequently poor in or devoid of vitrinite.

(b) The series which contain both coals and bitumens are rare and often contain lithological alternations, some rich in bitumens, others rich in coals.

In these cases, however, the two reflectances can be compared:

(a) In most cases, R_oB (bitumen R_o) is considerably lower than R_o (vitrinite) —in the range of "bituminous coals"— and geological results often lead to the conclusion that the bitumens are of late migration.

(b) It may happen that R_oB is very close to R_o —between values of 0.5 and 1.5%. This gives evidence that, for the same thermal history, and only in the rank-range "high and medium volatile bituminous'' coals, there is only a very slight difference of 0.1 to 0.2 R_o% between R_oB and R_o.

Teichmüller [72] considers that the bituminous substances present in coals undergo a kind of ''*coalification jump*'' (paragr. III.A.3) in the range 0.6-1.0 R_o, and that their R_oB, at first lower than R_o (vitrinite), becomes higher than R_o at the end.

In addition, some difficulties do arise in the use of bitumen particles, because of the frequent uncertainty in characterizing them and distinguishing them from coals. This is particularly so when the coals are represented solely by ''gelinites'' or homogeneous vitrinites or huminites, dispersed as fine particles in sediments.

1. Recognition of bitumens.

Bitumens can be recognized occasionally by empirical morphological criteria such as droplets, conchoid outlines, or geometric shapes from crystallized rock pores or veinlet fillings. Their shapes and reflectance are affected by radiations from radioactive minerals, and they have an earlier anisotropy than do coals [36, 39, 57, 58, 61]. They lose their fluorescence as soon as 0.3-0.4% reflectance is reached [37]. The best way to avoid confusion between coals and bitumens is to etch the polished surface with a chemical reagent such as a sulfomanganate mixture [50], which reveals the coaly or cellular structure in homogeneous vitrinite. However, this method is only a negative one because of the lack of typical etching figures for bitumens; moreover bitumens give no result when their R_o is greater than 0.5%.

2. Petrology of bitumens.

Some attemps have been made by H. Jacob [36, 37, 38], Alpern [2], to classify bitumens optically and a working group in *ICCP,* directed by H. Jacob, has undertaken to introduce bitumens into the petrological classification of coals.

H. Jacob's studies, based on the chemical composition and microscopic observations of about thirty bitumen deposits, result in a diagenetical classification using such parameters as R_o, fluorescence, solubility in various solvents. Alpern put bitumens in a group distinct from ''exinites'', with the name ''bitumites'' [2].

The main difficulty we have in classifying bitumens probably results from our lack of knowledge of their genesis. An improvement has been made thanks to studies in fluorescence by Teichmüller [33], who observed the generation of bituminous substances from liptinites (exinites) in coals. The ''*exsudatinite*'' which is produced is a kind of not reflective but fluorescent bitumen, which forms various exsudates (apophyses, veinlets, pore fillings). It is not yet known whether exsudatinite may be considered the sole origin of all insoluble bitumens present in the sediments. As far as we know, exsudatinite has been observed only in humic facies (highest plants) and has not yet been observed in algal facies which are, however, more favourable to hydrocarbon generation.

Solid bitumens are probably also generated by alteration from liquid bitumens. This occurs in the course of the oil migrations in reservoirs, due to temperature or various physical causes such as oxidation or radioactive irradiation.

E. Studying sedimentary basins.

Continuous vertical measurements and a certain amount of lateral data are required.

1. *Vertical survey.*

A good understanding of the catagenetic evolution of a sedimentary sequence and of its thermal history needs continuous vertical information. Frequent variations in the thermal history over the course of geological periods (see below paragr. IV) may be revealed only by continuous cross-sections grouping the whole sequence of a basin.

Such deduced catagenetic variations may be due to:

(a) Unconformities, tectonic repetitions or stratigraphical gaps.
(b) Lithologies with varying thermal conductivities.
(c) Thermal events.

These main factors justify some continuity in analytical processes; although sample spacing may be wider (usually 200 to 300') than for stratigraphical purposes.

Furthermore, since R_o depth measurements are sometimes equivocal (paragr. II.C), accurate results at all stratigraphical levels can only be obtained by intra- or extrapolation of values measured in a continuous vertical survey.

2. *Horizontal survey.*

Regional investigations of basins prove that the rank frequently varies rapidly when following a horizontal and/or stratigraphical level within the basin; such variations cannot usually be foreseen. Therefore, the catagenesis survey has to cover fairly large areas to help resolve the "geometry" of certain thermal phenomena which can often be related to some large-scale features concerning the paleogeography or the structural and sedimentary history of the bassin.

III. HYDROCARBON GENERATION AS SEEN BY OPTICAL ANALYSIS

A. Thermal evolution, as the main cause of hydrocarbon generation and transformation.

1. *Historical viewpoint: empirical and statistical notions of hydrocarbon generation.*

In 1915, in studies undertaken in the Appalachian region, White [84] observed, for the first time, that the regional distribution of oil and gas was related to the coal rank. Oil pools occurred in zones where the carbon-ratio ranged from 0.60 to 0.70 and gas pools where it was greater than 0.70. In 1958, microscopic methods were introduced : in studies of the Lower Saxony region, of West Germany, Teichmüller [69] showed the existence of a regional limit between oil and gas pools, coinciding with a reflectance value R_o = 0.7%, a value considered as the oil "*dead-line*". In 1961, Ammosov's [5] application of this concept to basins in the Soviet-Union led to the generalisation of an economical *dead-line* at R_o-value of 1.0%.

Finally, geochemical analyses by, among others, Philippi [55] and Tissot *et al.* [78] have shown that the generation of hydrocarbons is temperature-controlled and that there is a thermal maturation threshold or "*birth-line*" for oil.

The idea has been expressed by Vassoyevitch [80], whose generation sequence comprises:

(a) An early "diagenetic" gas phase at the low ranks (R_o lower than 0.40%).

(b) The "main phase" of oil generation, R_o-values ranging from 0.55 to 1.35%. This was later termed the "liquid window" or **oil window**, [56];

(c) The "catagenetic" gas phase at R_o-values higher than 1.35%; at first "wet" gas (with condensate), and, then "dry" gas (CH_4).

In recent years, economical and statistical boundaries for oil generation and conservation have been ascertained where reflectivity values are in the range of about 0.5 to 1.0% [4,61,6], with gas pools disappearing when reflectivity values are about 3.0% [60].

2. *The particular case of gas generation.*

a. *Catagenetic gas.*

As with oil, gas generation was at first explained as originating from sedimentary OM: generation from Carboniferous coals was proposed by Boigk and Stahl [10] from carbon isotope ($\delta^{13}C$) data obtained from North European pools (Schlochteren in the Groningen region and others).

Many other examples are now known and it is agreed that any OM (kerogen), whether coaly or not, may be the source of such gas pools (e.g. Aquitaine, France; West Sahara, Algeria) with gas originating from coal being "**dry**"(CH_4 alone) whilst gas from other sources may be "**wet**". In addition, the composition depends upon the conditions of thermal generation.

These gases are a product of *coalification* and the process of generation has been studied in several coal basins, including the Ruhr Basin [40], and the Saar Basin [43]. Fig. 12.2 shows the volumes of N_2, CO_2 and CH_4 released, in liters by kg of coal, during the course of coalification, between the rank of high volatile bituminous B (R_o # 0.7%) and anthracite (R_o # 3.0%): on the whole, 1 kg of coal releases 31 g N_2, 150 g CO_2, and 143 g CH_4, that is a total amount of 320 g of gas and 107 g carbon as CH_4.

Such figures can be used to estimate the gas potential of a particular basin and they account for the substantial amounts of gas discovered to date, for example in the Southern North Sea.

According to these data, it is obvious that anthracites with R_o = 3.0% (VM # 5.0% and C # 93.%) are very close to exhausting their gaseous production thus accounting for the disappearance of gas pools, as the source rock reaches such a high rank.

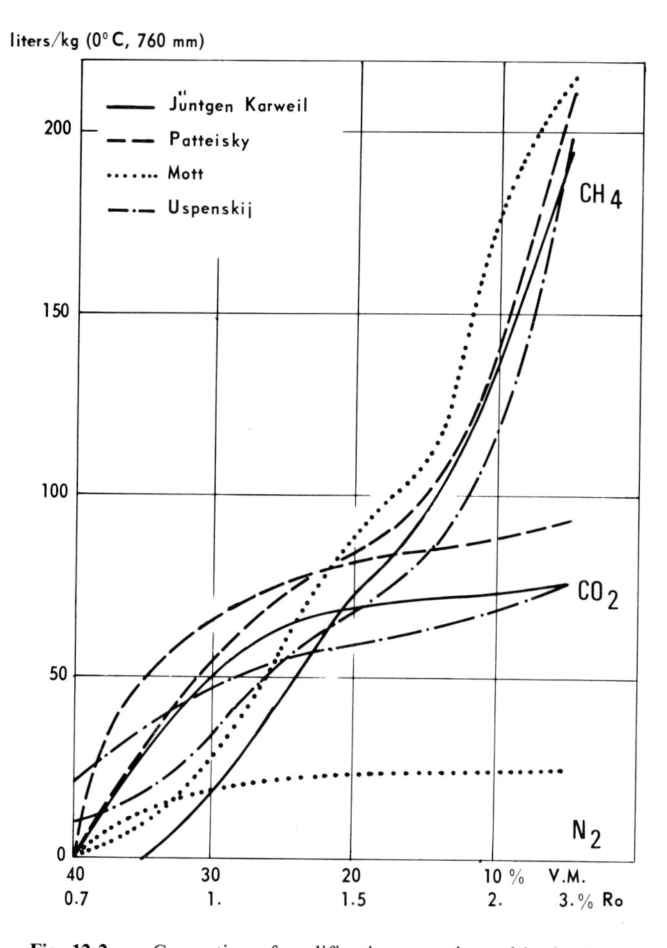

Fig. 12.2. — Generation of coalification gases in coal basins [43].

An illustration of the coalification effect is given (Fig. 12.3) by the regional coincidence between Permian gas pools, in the Southern British North Sea and the major coalification axis in the Upper Carboniferous [60].

b. Early diagenetic gas.

The generation of "biogenic" gas is a well known phenomenon, but it seems to be restricted to very shallow depths and, consequently, to the lowest ranks.

There is somewhat less known about the early stages of the formation of this "diagenetic" gas (R_o less than 0.40%). The suggestion that some gas pools are generated in this way, such as the giant pools of North West Siberia [86, 24], is founded on their carbon isotope analyses, which show that they are relatively low in ^{13}C ($\delta^{13}C$ # − 60‰ in Urengoy pool). The source rocks of such pools seem to be not so well identified as those of Northern Europe.

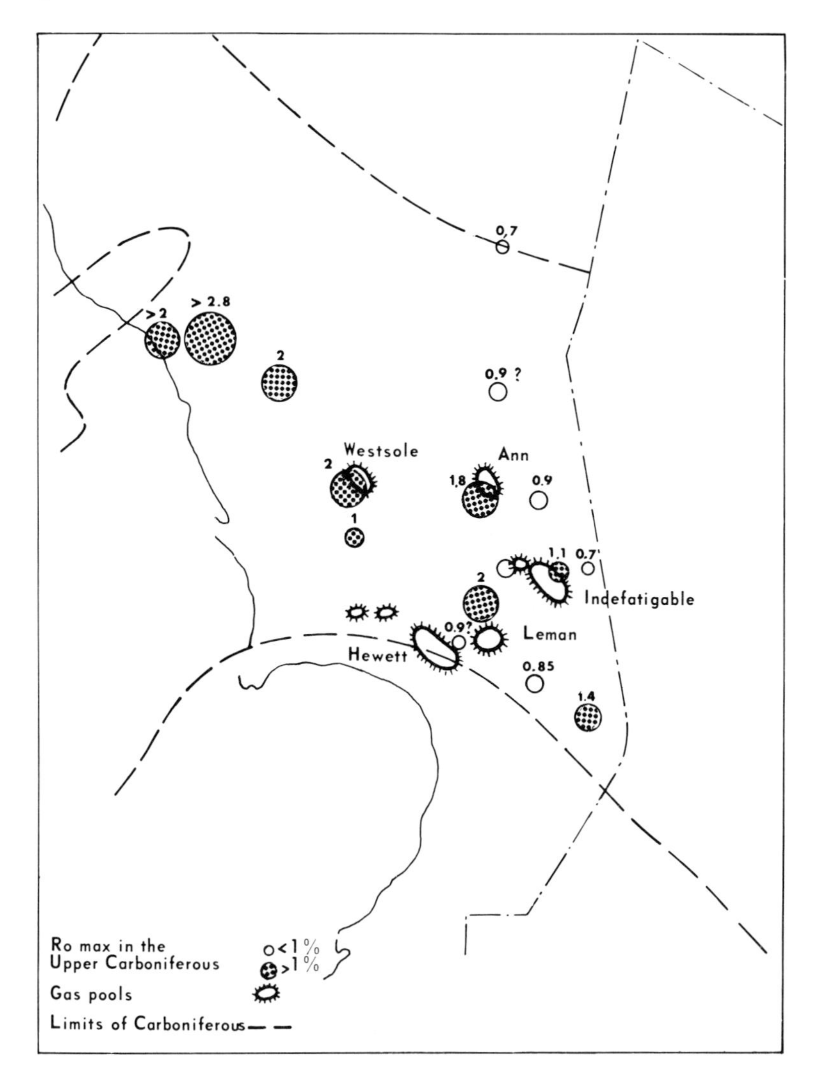

Fig. 12.3. — North Sea British Zone. Regional variation of coalification in the Upper Carboniferous [60].

One further point remains to be made, in speaking of the rank limits of hydrocarbon generation as stated above (paragr. III.A.1-2): Many authors refer to reflectance measurements in the reservoir, in default of source rocks measurements.

Source rock, frequently located lower than the producing reservoir, generated oil in an earlier period and was afterwards submitted to a catagenesis higher than the oil dead-line; thus, the location of the reservoir in the oil-window rank attests the preservation of oil but not its generation.

Our conception [61], for the boundaries of the " oil window ", at R_o-values of about 0.50 and 1.0%, refers to main generation *in the source rock,* with an economical-statistical meaning, but does not exclude some oil or condensate generation after an R_o-value of 1.0% has been reached.

PLATE 12.1

Well Fregate, 2 700 m – R_o = 0.7 %. Fluorescent and reflected white light. Gelification of a fluorescent tissue, losing its structure.

N.B. : The addition of blue light is artificially induced by high reflecting pyrite and the need to retain the lowest fluorescent colours.

Left : fluorescence. Right : white reflected light. (White = pyrite. Clear grey = gelinite = homogeneous vitrinite).

Photographs by Photomicroscope Zeiss.

100μ
10μ
100μ

3. *Microscopic observation of hydrocarbon generation.*

While statistical views on oil generation have become well established, very few sample-scale observations were formerly able to support them. It is only in recent years that these views have been borne out by more data available from physical and chemical investigation as well as by improvements in optical methods. Thus fluorescence microscopy now provides a means of showing the main transformation undergone by OM: oil formation can be seen in coals under the microscope [72], in the form of smear films, oil droplets, exsudates, fluorescent vitrinites, and by a fluorescent maceral, "exsudatinite", that appears during coalification in the "oil window" rank. Same observations can be made in "inorganic" rocks, on organic concentrates.

The catagenetic transformations observed under the microscope give evidence that certain stages of coalification, narrow in range, are marked by quicker alterations, similar to the boundaries of the "oil window"; these stages are the "**coalification** jumps".

One *coalification jump* that is linked to a major discontinuity in the properties of coals has been known for a long time [66], and occurs at the rank of medium volatile bituminous coals (R_o # 1.2%) equivalent to the gas to fat coal limit on the DIN scale (Table 12.1). From normal light and fluorescence investigations, Teichmüller [72] considers that *another jump* takes place at the rank of B/C high volatile bituminous coals (brown coal/hard coal limit DIN scale; with R_o-value about 0.60%). Thus both jumps coincide with two important stages of oil generation within coals (and to a certain degree, by extension, in source rocks): the "lower" or *first jump* R_o # 0.6 – 0.65% corresponds to the maximum rate in oil generation, whilst the "higher" or *second jump* R_o # 1.2% corresponds to the end of oil generation.

These coalification jumps, as seen in coal seams, are located at R_o-values higher than the limits of the "oil window" in source rocks. However the two phenomena may be similar; indeed, Teichmüller [72] observed that oil generation in coals progresses later than in "inorganic series" because of the physico-chemical properties of coal in coal seams.

Numerous other phenomena arise from the generation of bituminous substances in coals:

(a) **Micrinite,** a maceral classified in the inertinites due to having a reflectance level higher than that of vitrinite, also forms in the "oil window" rank and appears [71], to be the residue remaining after the release of lighter hydrocarbons from liptinic macerals (initially lower in reflectance than vitrinite). Thus, the presence of micrinite indicates previous oil (or gas) generation.

(b) Some other properties of coals, such as "**diagenetic gelification**" are associated with oil generation within the coal. They contribute to the ability of coals in the "oil window" range to undergo coking because of the formation of a "solubilizing" bituminous phase [72].

(c) Generally, **fluorescence disappears** gradually as the rank increases to R_o-values of about 1.2-1.3%. The reflectance increasing at the same time on previously fluorescent macerals is also related to the transformations described above; this is due to (fluid or volatile) hydrocarbon release. What happens is that a heavier part of the kerogen "gelifies" or "condenses" on release from the mobile phase: Plate 12.1 illustrates this process and shows a fluorescent tissue losing its structure and fluorescence while gelifying.

B. Contribution of optical studies in research of hydrocarbon source rocks.

Optical methods provide a characterization of the biological components of the kerogen. Thus, they give a qualitative and semi-quantitative appraisal of the petroligenous quality of the kerogen.

1. Examination in transmitted light.

The description of "palynofacies" [13, 15, 59], enables us to distinguish:

(a) Distinctive substances, microfossils, remains of tissues, organs, secretions, etc.
(b) Amorphous matter forming the main part of oil shale concentrates. This matter is frequently called colloidal, algal, sapropelic by various authors.

The palynofacies, often composed of a mixture of these evolves between a **"ligneous"** pole and an **amorphous** pole and the proportion of both components is an indication of the petroligenous potential.

2. Examination in reflected light and fluorescence.

Coal petrology in reflected light which affords a diagnosis of the humic material is completed by fluorescence. Fluorescence provides a visualisation of highly hydrogenated matter, invisible in white light and favourable to hydrocarbon generation. On the basis of this examination, the constituents of kerogen can be classified into two extreme groups:

(a) Humic, originating in high vegetals, their secretions and spores.
(b) Algal and sapropelic (= consisting of algal but not distinctive matter).

The intensity and distribution of fluorescence are closely related with the petroligenous character.

C. Theoretical approach to the thermal generation of hydrocarbons by chemical kinetics.

It is now well known that the major cause of coal evolution and hydrocarbon generation is temperature change. This has been accepted to such an extent that certain parameters have a direct "thermal" attribution, for instance, the "thermal alteration index" TAI of Staplin [68] and Raynaud [59]. It has been shown that the effect of present geothermy varies in the same direction as hydrocarbon generation: increasing geothermal gradients result in a vertical rise in oil generation zones and vice-versa [48].

The burial effect is considered as fundamental by all authors (*cf.* for example, Bostick [9], Alpern [4]), and one of the major preoccupations is to evaluate the maximum *"paleotemperature"* — due to burial — reached by the geological formation. This was one of the aims of the *ICCP* symposium held in Paris in 1973, directed by Alpern [4].

Specialists in coal chemistry have carried out a quantitative evaluation of the coalification reaction, that involves *time and temperature*. The time factor is taken into account with the hypothesis that the chemical reaction is one of the first order and thus simply an exponential function of time:

$$V = \frac{dn}{n_0} = -k\,dt$$

The variation of temperature is taken into account in the coefficient k by means of the Arrhenius relationship.

$$k = a \exp\left(-\frac{E}{RT}\right)$$

with:

V = reaction rate
t = time
n = amount of transformable kerogen
n_0 = initial amount of transformable kerogen
E = activation energy
R = perfect gases constant
T = absolute temperature ($\theta°$ C + 273)
a = constant

Karweil's chart, [41] which is based on these calculations, provides the rank of a coal from its geologic age and its maximum temperature. This latter is obtained from the present geothermal gradient and the maximum depth of burial.

A variant of this rule was proposed by Lopatin [49]. In principle it consists in applying the same law, but to account for temperature variations during the burial history of the series, this time-span is divided into a variable number of constant temperature time-steps. Some simplification is obtained, by assuming that the Arrhenius relationship approximately doubles the reaction rate for a temperature increase of 10° C.

Lopatin's calculations do not take in account the time-variations of the geothermal gradient and consider that the temperature variations are due to burial-effect alone, admitting as an approximation that the geothermal variations vs. geologic time only slightly affect the final result.

A determinant improvement in applying chemical kinetics to hydrocarbon generation from kerogen has been made by means of a *mathematical simulation* [77, 79]. This calculation is based on extensive knowledge of kerogen, which has been classified by element analysis into three types corresponding to three (catagenetic) ways of evolution [22]: type I = higher hydrogen content (more algal and lipidic) to type III = higher oxygen content (i.e. humic and mainly consisting of terrestrial high vegetals); type II is intermediate in chemical composition.

The degradation of kerogen is simulated, not as a whole, but by type (I, II, III), each one split up into parts distinguished by their activation-energy (in classes covering steps of 10 or 20 kcal/mole). The basis of the cinetic calculation remains the same: first order reaction, thermal control by the Arrhenius law.

Operations are carried out by a computer; the verification of the programme is performed at the end of the operation by comparing results with laboratory data (chemical, optical), thus adjusting the main selected parameters by successive cycles.

This calculation takes into account both sediment burial and variations of geothermal flow during geologic time.

This simulation, by means of an equivalence table between R_o and the rate of the hydrocarbon generation, is able to calculate a relation between alteration of vitrinite and its thermal history: thus it proves that coalification is not simply an effect of a maximal temperature, but is probably similar to hydrocarbon generation in that it is an integral fonction of two variables, time and temperature. This is the true meaning of the word *"paleothermometer"*.

IV. ANALYSIS OF THE THERMAL HISTORIES OF GEOLOGICAL FORMATIONS BY MEANS OF REFLECTANCE CURVES

A. Variations in the deep thermal flow.

Present-day observations show that some regions, considered as "cold" or "hot", have geothermal gradients that vary between values of less than 20° C/km and 80 to 100° C/km — the latter for example in the Rhine Graben [19, 20]. It is therefore obvious that the basins may have been affected by similar variations in the past.

Indeed, the vertical R_o curves, in boreholes, exhibit individually or regionally such high variations as to prove the occurrence of similar fluctuations in the thermal histories of sedimentary basins.

In order to interpret such variations, it is, at first, convenient to look for and to characterize an eventual normal or standard law of R_o increase vs. depth. This should give a better understanding and calibration for diverse measured curves.

1. Vertical statistical survey: Evidence of a "normal gradient".

This was performed by means of a computer analysis of reflectance data covering 10 years from about 400 wells in 25 different basins — as carried out by Y. Dagens, see [61, 62] —: this analysis proves the *statistical* existence of a "standard" or more frequent R_o increase vs. depth.

The figures used are for *vitrinite* only; they exclude both the unrepresentative data (eliminated after vertical curve interpretation and well-basin information), and bitumens (which do not necessarily indicate the entire thermal history); data are grouped in successive levels by means of histograms which measure the number of points against R_o-value (Fig. 12.4). At each level, they provide one particularly predominant mode (distribution maximum) and 2 or 3 secondary modes; the first corresponds to the lower R_o-value, and the latter ones to R_o-values from 1.0 to 4.0%.

The R_o of this dominant mode has values of about 0.50% at 3 000 m and 0.65% at 4 000 m (sea level depths) and its medium vertical gradient is about 0.15%/km. It was verified that the secondary modes correspond to hyperthermal anomalies located in certain basins; moreover they do not show a progressive change with depth as does the main mode.

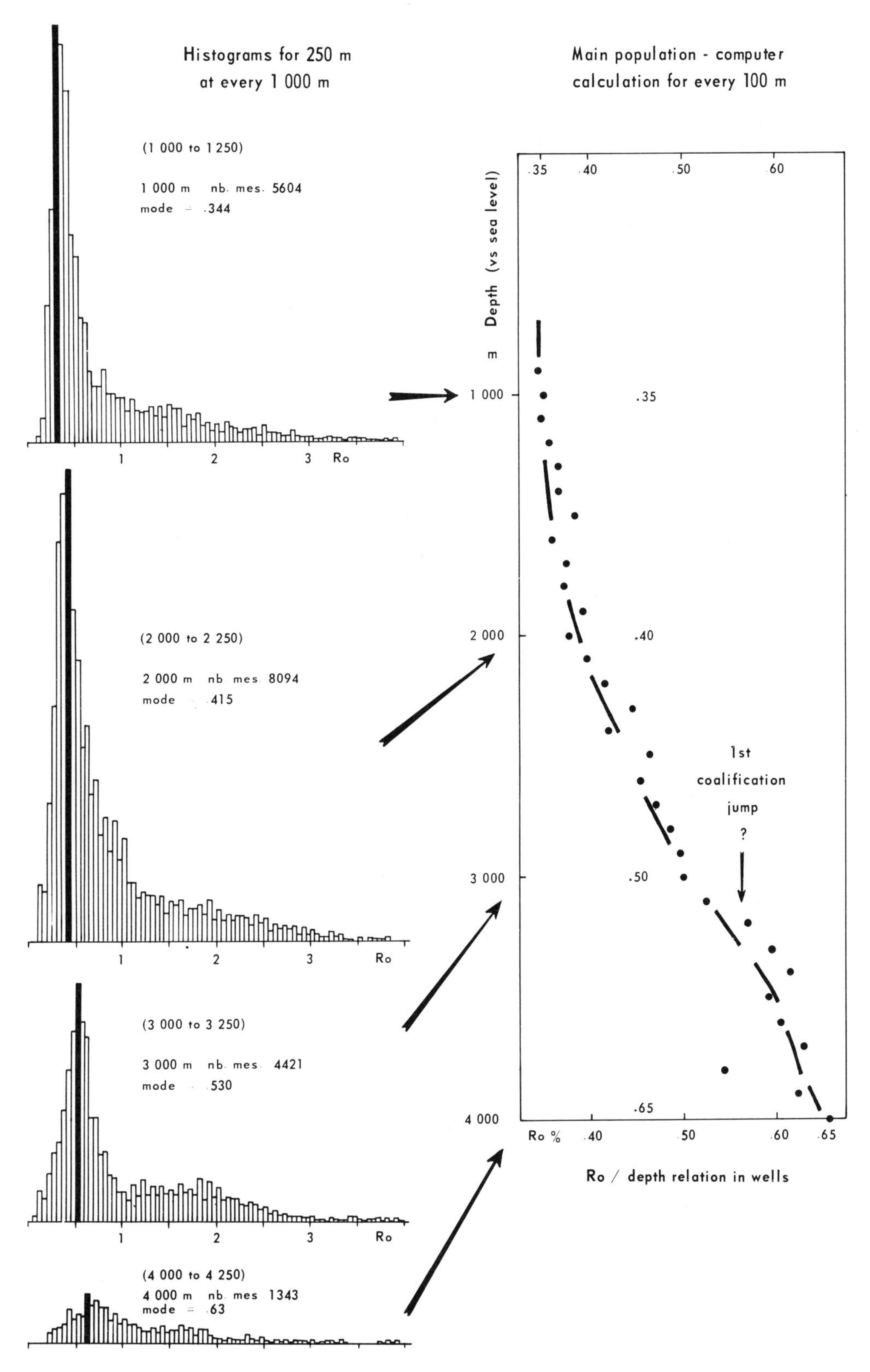

Fig. 12.4. — Statistical increase R_o/depths. Analysis of a computer file.

It should be noted that the gradient of this vertical "normal" progression is half as much as that of coal basins [3]. This is not surprising, since coal Carboniferous Basins were formed in warm parts (orogenic) of the earth's crust, while our statistics cover all kinds of basins.

However, we must recognize that this is only an approximation, since the data providing this "normal" progression originate more from the Mesozoic and Cenozoic than the Paleozoic. But the results obtained have been confirmed for the whole Mesozoic period, down the Permian, that is for about 200-250 million years, by separate processing. Separate processing for *lower* Paleozoic only gives higher results (about 0.8% R_o at 3 000 m for the dominant population).

This *"standard"* progression (Fig. 12.4 on the right) is not linear. Our interpretation of a 0.15%/km gradient is only presented for reference purposes. Between 0.55 and 0.60% R_o the curve shows an inflexion which could reflect the first *"coalification jump"* [72].

Consequences.

It is known that the present geothermal flow has a relatively constant mean-value at the surface of the earth. This value is estimated to be about 1.2-1.4 μcal/cm^2/s [27], but may be affected locally by hyperthermal anomalies. Such an analogy between both phenomena —present geothermy and paleothermal catagenesis— suggests the following:

In the case of an "eventless" geothermal history, in which the gradient has maintained a normal value (about 25° C/km) the coalification of MOD seem to "stabilize", as far as R_o is concerned, at the end of a (geologically speaking) relatively short period, that may be perhaps only a few million years. The statistical progression would otherwise be monitored by the age of the series and not by its depth. This opinion has already been expressed by other authors [14], and, according to Heling and Teichmüller [31], under the same temperature conditions "stabilization" of kerogen is reached earlier than that of clay minerals. This can also be deduced from the kinetic equation of the paragraph III. C, which shows that temperature (i.e. generally depth) has an exponential effect on the reaction rate, while time has only a linear effect.

In practice, this statistical progression constitutes a reference, which is useful for comparing basins with one another and with the present geothermal gradients.

2. *Main shapes of the R_o/depth curves.*

Sublinear curves are the most frequent, in agreement with the statistical increase, but there are frequently other shapes of curves, as shown in Fig. 12.5. They include:

(a) One with two different slopes, the deeper one having a lower slope $\frac{d_{Depth}}{dR_o}$, which indicates a higher reflectance gradient for the oldest part of the series.

(b) A curve showing, at a given level, a large inflexion which separates two normal sloping branches by a zone of rapidly increasing R_o; this increase may be as high as 1.0 to 2.0 % R_o over a depth of only 200 to 300 m, whilst above and below, the gradient may remain around 0.1 or 0.2 % R_o/km.

(c) Finally, a curve whose shape falls between these two types, with a low deep slope, a higher shallow slope and an intermediate long curved portion, the whole curve suggesting a rather progressive transition between a hot early state and more common recent rate.

The extreme case (case b) indicates a strong thermal perturbation; when it is repeated in several neighbouring boreholes at about the same stratigraphical level, in a continous uninterrupted series, it suggests a higher than normal deep thermal flow, during a limited period, followed by a relatively quick return to normal thermal conditions.

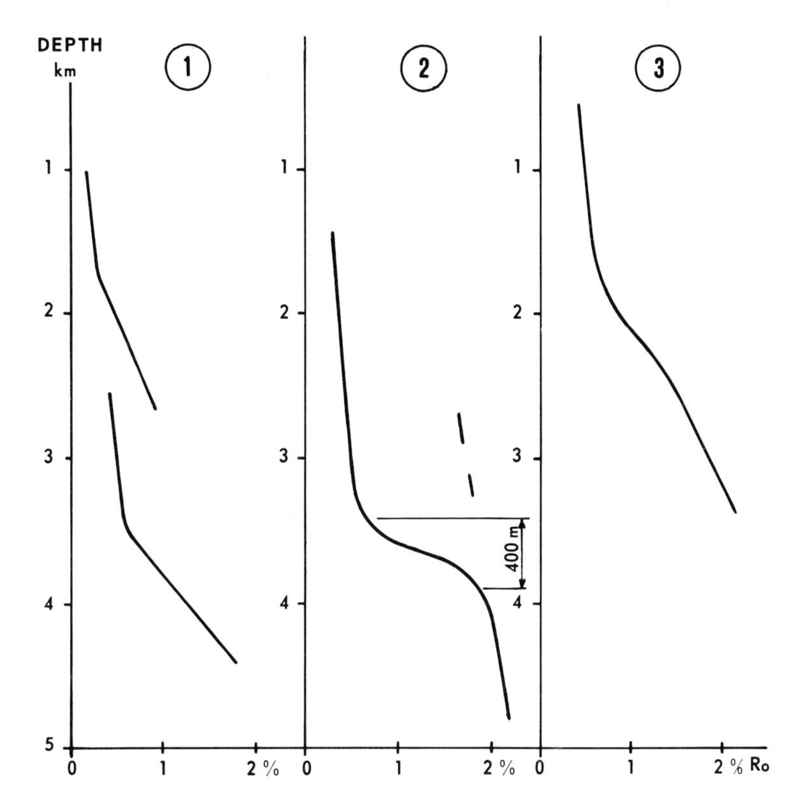

Fig. 12.5. — Common schematic types of R_o/depth curves, with high paleogeothermies. Depths indicated here are reference only and may vary largely. In the same way, R_o may reach 7% and more.

1. 2 slopes. **2.** 1 gap/R_o jump ≥ 1%. **3.** Intermediate.

Successive thermal states can also be recorded otherwise than by the R_o/depth curve, that is when successive bitumen populations are present, where the more recent one has the lowest reflectance, or a lower reflectance than that of vitrinite (the latter only assumed of the same age as the sediment).

With such observations within a region or a basin some "thermal anomalies" can be checked.

3. Main types of thermal anomalies.

Thermal anomalies have been recognized by many authors including Koch [44], Teichmüller [76], Damberger [18] and may be classified into three groups:

a. Magmatic.

A good example is the "Bramscher Massif" in Lower Saxony, West Germany [7], where a basic laccolite, intruded during the Middle Cretaceous, induced a metamorphic halo at its periphery (pyrophyllite, anthracite).

In a similar manner, the Reinisches Schiefergebirge's front (Sauerland; Lippstadt's high) is affected by deep anomaliės related to granitic intrusions [83, 85, 35].

b. Tectonic.

The Rhine Graben is an example that has already been mentioned, with its strong geo- and paleogeothermy [21].

In the Lacq Basin, Aquitaine, France (Fig. 12.6) the continuous sedimentary series is divided into a highly coalified deep portion and a shallow part (Cretaceous 3-4 km) showing normal catagenesis with R_o-values between 0.5-0.6% at 3 000 m.

The same has been observed in the British North Sea along the high coalification axis (Fig. 12.3): a rapid R_o increase at the top of Upper Carboniferous suggests a strong thermal phenomenon at the end the Hercynian period.

c. Related to the variable thermal conductivity of the sedimentary section.

Such anomalies do not originate in the deep crust and they are of moderate amplitude. However, their effect, as regards the "oil window" rank can be so great as to modify the maturation conditions and to give rise to commercial oil accumulations. This is so for the 2 to 4 km high salt colomns (Gulf of Mexico, West Germany, Gabon, etc.) which more or less "short-circuit" the vertical thermal balance in some regions; note for instance, how some oil pools are "coupled" with salt masses, in Lower Saxony, West Germany.

More generally, the conductivity of rocks affects and transforms the deep thermal flows:

(a) Formations that are badly conductive or not conductive can be considered as thermal barriers; the concentration of the thermal effect in them emphasizes coalification and sometimes induces lateral refraction of deep calories.

(b) Very conductive horizons, like salt, involve higher R_o in overlaying sediments, for example in the top of salt domes, which focus upward thermal flows; on the other hand, they involve lower values than normal in the underlaying strata.

In the same way, provinces with vertical, continuous carbonated lithology have smaller geothermal gradients and a smaller R_o increase than the "standard" one; maturation attained at deeper levels in such provinces may result in increased oil generation and decreased gas generation.

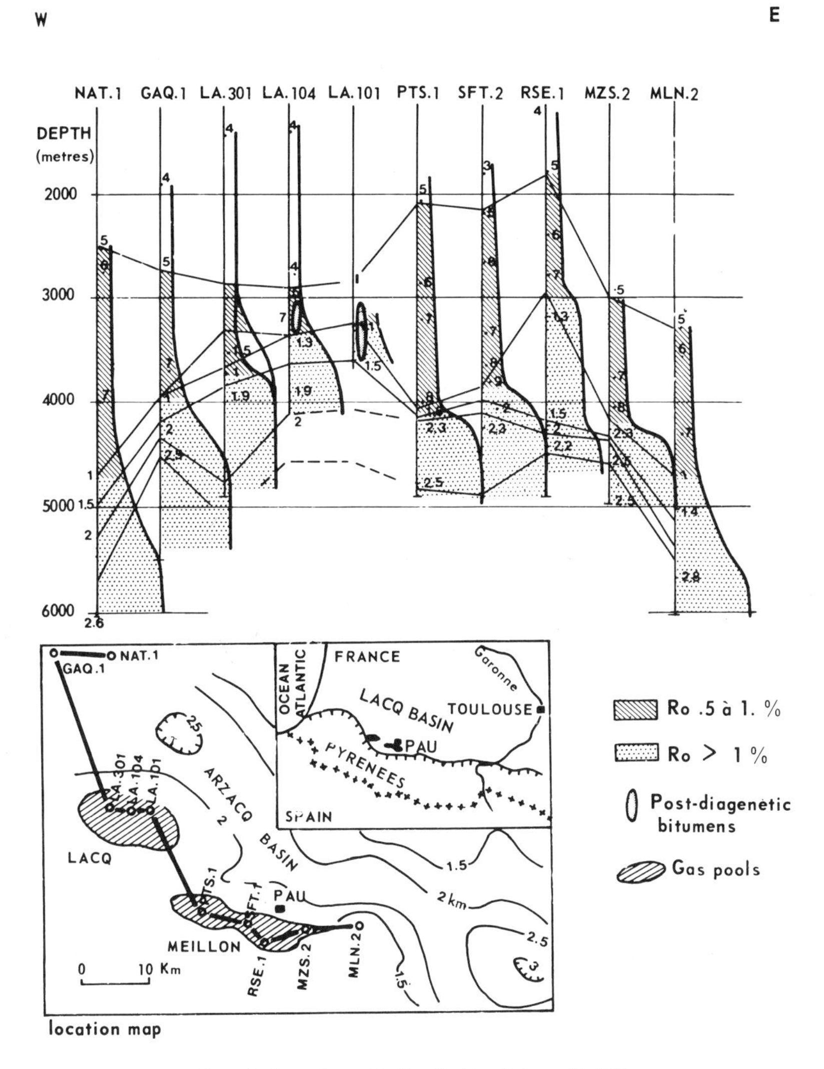

Fig. 12.6. — Lacq Basin. R_o/depth in wells [62].

B. Geological factors controlling geothermal histories.

As fluctuations in thermal histories are the main obstacle to thermodynamic interpretation of coalification (paragr. IV.A), and hence the resulting hydrocarbon generation, an attempt can be made to anticipate them by seeking those geological factors which control the deep thermal flows. In order to do this, we shall try to classify the basins or geothermal provinces into two main groups, with reference to a number of known examples. This is an attempt, based on the author's experience and world references and using simplifying hypothesis, at grouping the deep thermal flows in various kinds of basins.

1. Basins with a "normal" geothermy.

The foregoing statistics provide evidence that there exists a general and more frequent case where the sublinear increase of catagenesis suggests a medium geothermal gradient during all the geological history of the basins.

a. Oceanic margins.

These conditions are found in oceanic margins, Mesozoic to Cenezoic in age, on the edge of stable platforms. This is so for the West Africain margin (Angola, Congo, Gabon, Nigeria) where the oil maturation of source rocks arises normally at depths of about 3 000 m (0.50% R_o according to the statistic analysis, Fig. 12.4). Some zones in the Gulf of Mexico appear to exhibit the same conditions: for example in Terrebone Parish Louisiana, R_o attains a value of 0.50% at 3 000 m [31] and would reach 1% below a depth of 6 000 m.

The major disturbances in these basins can be related to the conductive effect of salt domes, but the amplitudes of thermal variations generally remain moderate.

b. Stable platforms of any age.

Such conditions are also found in various platforms. For example, the Lower Paleozoïc of the Baltic region [8], where Cambro-Ordovician rocks generate oil and have as low a rank as 0.7% R_o.

In the Middle to Lower Paleozoïc of the Illizi Basin in the Algerian Sahara some oil-producing structures in the Devonian with a maximum burial (in paleozoic times) of about 2 500 m [60], do not exceed 0.60% R_o.

The namurian brown coals in the Moscow basin have R_o-values of 0.4 to 0.45% and therefore have not reached an oil producing rank; but their depth of burial seems to have remained relatively low.

2. Hyperthermal zones and their geological framework.

Hyperthermal zones generally coincide with the mobile parts of the crust and for this reason are often paleogeographic regions which encourage the accumulation of large amounts of OM. They are highly estimated in petroleum exploration, their favourable conditions being due, to their thermal history [42], but also, and perhaps primarily, to the relatively high amount of source rocks.

We suggested [62], that the main thermal anomalies are more or less directly related with the phenomena of crust distension:

(a) For example, the *Rhine Graben* is affected by high geothermies, the maximum ones coinciding fairly well with the higher tertiary catagenesis ([21] and the author's survey of Alsace [62]). Oil maturity is sometimes reached (0.50% R_o) at depths as shallow as 800 m (in Oligo-Miocene sections).

(b) *The Red Sea* exhibits all of the transitional terms between and old and a recent graben: the warm periods coincide, in every location, with periods of high tectonic activity and subsidence, and disappear when they pass, as in the case of the Suez Gulf.

(c) The *"Bramscher Massif"* in Lower Saxony exhibits, in addition to the magmatic aspect that has already been explained (paragr. IV.A.3.a), a comprehensive development of a subsident trough during the first part of its history, that was subsequently lifted in a true **"tectonic inversion"** [67]; this inversion is linked with a lateral compression which makes it overlap its borders (mostly in the prolonging Harz Massif). A high heat flow developed at the end of the subsidence phase, linked to the intrusion of a basic laccolite; the heat flow ended before the beginning of compression. This is the typical scheme of an **"Aulacogen"** [63], which comprises a cycle covering distension —subsidence— and then, compression.

Such tectonic systems are not rare, in the geological history of various world basins (even though the cycle is not always complete) e.g. the Naragansett Basin, Eastern U.S.A. [18], the Lacq-Arzacq Basin, Aquitaine, France (paragr. IV.A.3.b and Fig. 12.6).

Many hot distension basins can be cited: the Viking Graben in the North Sea with a present thermal anomaly [30], the West African margin, with series affected by high paleothermal anomalies, during the rift period preceding ocean formation [62], the Douala Basin with a tertiary thermal anomaly [23], are examples.

These hyperthermal zones generally have an elongated form along their respective grabens; their transverse development is frequently reduced to a few dozen kilometres. Their thermal anomalies appear to be vertically limited in geological sections, with a top corresponding to the disappearance of strong heat flows over time.

When such thermal anomalies are linked to a geosynclinal system, they are often localized on its margins or in molassic intratroughs, and, in many cases, it has been observed that the short-lived thermal event ends before the compression phases which complete their history: as, for example, in the Ruhr coal Basin [76], and the anticlines of the Rheinisches Schiefergebirge [85, 8], where coalification is assumed to be preorogenic. A similar observation has been made recently for the Pyrenean "Aulacogen" by Souquet *et al.* [64], where an Albian distension phase with high temperature metamorphism is followed by a long compression period, from Cenomanian to Eocene.

These anomalies, linked to the major breaks in the Lithosphere, seem to be linked to a thinning of the Lithosphere and to the related lifting of the Asthenosphere and mantle material. In addition, it is probable that the heat flows are further reinforced by the earth's convection currents, considered by physicists as being more or less responsible for these diastrophic phenomena [17, 27]. In contrast to this, certain parts of the subduction zones, where the lithosphere is thickened, are abnormally cold, as in Bavaria [75], the Tejon Basin in California [12], or in the present Western Pacific Ocean [82].

It is important to distinguish these anomalies from circumscribed (or batholitic) ones, arising from cooling granitic instrusions. Within the folded orogenes both kinds of anomaly may be present, but intrusions generally develop later, sometimes contemporaneous with folding [35].

Finally, it is probable that the thermal history of a sedimentary basin belongs to its general evolution, affecting certain stages or phases of this evolution in time; these phases are likely to be related with the evolution of the underlaying deep crust [53, 54].

V. CONCLUSION

The petroleum industry has benefited greatly from the development of optical means of studying catagenesis and identifying source rocks. When applied statistically, these methods have contributed to expanding our overall knowledge in the following different fields:

(a) Hydrocarbon generation.
(b) Relationship between maturation of coals and petroleum formation.
(c) Variations in the thermal history of sedimentary basins.

Thermal histories of sediments can now be analyzed and are better understood, forming one of the most direct observation linkages between sediments and the deep crust in the past.

(Manuscript received december 1977)

Acknowledgments. — Thanks are due to *Elf Aquitaine Oil Co* for permission, to publish.

REFERENCES

1. Alpern, B., (1969), *Ann. Soc. Géol. Nord,* **89,** 143.
2. Alpern B., (1970), *Rev. Inst. Franç. du Pétrole,* **XXV,** 1233.
3. Alpern, B., (1971), *C.R. 7e Congrès Géologie Stratigraphie du Carbonifère, Krefeld,* **1,** 91.
4. Alpern, B., (1975), *Coll. Internat. Pétrographie Matière Organique Sédiments, Relation avec Paléotempérature et Potentiel Pétrolier, 1973,* CNRS, Paris, 278 p.
5. Ammosov, I.I., (1961), *Soviet. Geol.,* **4,** 7.
6. Ammosov, I.I., (1975), *in: C.R. 8e Congrès Internat. Géologie Stratigraphie du Carbonifère,* **2.** 85.
7. Bartenstein, H. and Teichmüller, M. and R., (1974), *Fort. Geol. Reinld. Westf.,* **18,** 501.
8. Bartenstein, H. and Teichmüller, R., (1974), *Fort. Geol. Reinld. Westf.,* **24,** 129.
9. Bostick, N.H., (1971), *7e Congrès Internat. Géologie Stratigraphie du Carbonifère, Krefeld,* **2,** 183.
10. Boigk, H. and Stahl, W., (1970), *Erdöl und Kohle,* **6,** 325.
11. Castano, J.R., (1975), *in: Coll. Internat. Pétrographie Matière Organique Sédiments, Relation avec Paléotempérature et Potentiel Pétrolier, 1973,* ed. Alpern B., CNRS, Paris, 123.
12. Castano, J.R. and Sparks, D.M., (1974), *Geol. Soc. Amer. Sp.,* **153,** 31.
13. Combaz, A., (1964), *Rev. Micropaléontol.,* **7,** 4, 205.
14. Cornelius, C.D., (1975), *N.G.U. Bergen,* **316,** 29.
15. Correia, M., (1967), *Rev. Inst. Franç. du Pétrole,* **XXII,** 9, 1285.
16. Correia, M., (1969), *Rev. Inst. Franç. du Pétrole,* **XXIV,** 12. 1417.
17. Coulomb, J., (1969), "La Science Vivante", P.U.F., Paris.
18. Damberger, H.H., (1974), *Geol. Soc. Amer. Sp.,* **153,** 53.
19. Delattre, I.N., Heutinger, R. and Lauer, J.P., (1970), *in:* "Graben Problems", ed. Illies J.H. and Müller S., Schweizbt., Stuttgart.
20. Doebl, F., (1970), *in:* "Graben Problems", ed. Illies J.H. and Müller S., Schweizbt., Stuttgart, 110.
21. Doebl, F., Heling, D., Homann, W., Karweil, J., Teichmüller, M. and Welte, D., (1974), in "Approaches to Taphrogenesis", ed. Illies, J.H. and Fuchs, Schweizbt., Stuttgart, 192.
22. Durand, B., and Espitalié, J., (1973), *C.R. Acad. Sci. Paris,* **276,** 2253.
23. Durand, B., Dunoyer de Segonzac, G., Albrecht, P. and Vandenbroucke, M., (1975), *20e Congrès Internat. Sédimentologie, Nice,* 39.

24. Galimov, E.M., (1974), *in: Advances in Organic Geochemistry 1973,* ed. Tissot B. and Bienner F., Editions Technip, Paris, 439.

25. Gijzel, P. van, (1971), *in:* "Sporopollenin", ed. Brooks *et al.,* Academic Press, London.

26. Gijzel, P. van, (1975), *in: Coll. Internat. Pétrographie Matière Organique Sédiments, Relation avec Paléotempérature et Potentiel Pétrolier, 1973,* ed. Alpern, B., CNRS, Paris, 67.

27. Goguel, J.L., (1975), "La Géothermie", Doin, Paris, 171 p.

28. Gransch, J.A., and Eisma, E., (1966), *in: Advances in Organic Geochemistry, 1965,* ed. Hobson and Speers, Pergamon Press, 407.

29. Gutjahr, C.C.M., (1966), *Leidse Geol. Meded.,* **38.** 1.

30. Harper, H.L., (1971), *Nature,* **230,** 235.

31. Heling, D., and Teichmüller, M., (1974), *Fort. Geol. Rheinld. Westf.,* **24,** 113.

32. Hilt, C., (1873), *Ann. Assoc. Ingén. Liège,* 387.

33. "Houillification et Pétrole", (1974), *Symposium Fort. Geol, Rheild. Westf.,* 24 French Translation, ref. 31, 35, 52, 71, 72; 220 p.

34. Hood, A., Gutjahr, C.C.M., and Heacock, R.L., (1975), *AAPG Bull.,* **59,** 6, 986.

35. Hoyer, P., Clausen, C.D., Leuteritz, K., Teichmüller, R., and Thome, K.N., (1974), *Fort. Geol. Rheinld Westf.,* **24,** 161.

36. Jacob, H., (1967), *Erdöl und Kohle,* **20,** 393.

37. Jacob, H., (1975), *in: Coll. Intern. Pétrographie Matière Organique Sédiments, Relation avec Paléotempérature et Potentiel Pétrolier, 1973,* ed. Alpern B., CNRS, Paris, 103.

38. Jacob, H., (1976), *Compendium 76/77, Add. Erdöl und Kohle,* 36.

39. Jedwab, J., (1964), *in: Advances in Organic Geochemistry,* ed. Hobson and Louis M., Pergamon Press.

40. Jüntgen, H., and Karweil, J., (1966), *Erdöl und Kohle,* **19,** 251, 339.

41. Karweil, J., (1955), *Z. Dtsch. Geol. Ges.,* **107,** 132.

42. Klemme, H.D., (1972), *Oil and Gas J.,* **17,** 140; **24,** 76.

43. Kneuper, G.K., and Hückel, B.A., (1972), *in: Advances in Organic Geochemistry, 1971,* ed. von Gaertner H.W.R. and Wehner H., Pergamon Press, Oxford, 93.

44. Koch, J., (1974), *Erdöl und Kohle,* **27, 3,** 121.

45. Kouznetsova, A.A., Golytsin, M.V., and Krylova, N.M. (1976), "Atlas of Paleozoic Coals in Karaganda", Nauka, Moscow, 103 p.

46. Krevelen, D.W. van, (1953), *Brennst. Chem. Essen,* **34,** 167.

47. Le Tran Khan, B., and Weide, B.M. van der, (1969), *Bull. Centre Rech. Pau, SNPA.*

48. Landes, K.K., (1967), *AAPG Bull.,* **51,** 828.

49. Lopatin, N.V., (1971), *Izv. Akad. Nauk SSSR, Ser. Geol.,* **3,** 95.

50. Mackowsky, M.T., (1971), *C.R. 7e Congrès Internat. Géologie Stratigraphie du Carbonifère, Krefeld,* **3,** 375.

51. Noël, R., (1976), *Bull. Centre Rech. Pau, SNPA,* **10,** 301.

52. Ottenjann, K., Teichmüller, M., and Wolf, M., (1974), *Fort. Rheinld. Westf.,* **24,** 1.

53. Perrodon, A., (1976), *C.R. Acad. Sci. Paris,* 1265.

54. Perrodon, A. (1977), *Bull. Cent. Rech. Explor.-Prod. Elf-Aquitaine,* **1,** 111.

55. Philippi, G.F., (1965), *Geochim. et Cosmochim. Acta,* **29,** 1021.

56. Pusey, W.C., III, (1973), *World Oil,* **4,** 71.

57. Ragot, J.P., (1976), *Bull. Centre Rech. Pau, SNPA,* **10,** 1, 221.

58. Ragot, J.P., (1977), Thèse Toulouse, 150 p.

59. Raynaud, J.F., and Robert, P., (1976), *Bull. Centre Rech. Pau, SNPA,* **10,** 1, 109.

60. Robert, P., (1971), *Rev. Inst. Franç. du Pétrole,* **XXVI,** 2, 106.

61. Robert, P., (1974), *in: advances in Organic Geochemistry, 1973,* ed. Tissot B., and Bienner F., Editions Technip Paris, 549.

62. Robert, P., (1976), *Bull. Centre Rech. Pau, SNPA,* **10,** 1, 271.

63. Schatzki, N.S., (1961), *Fortsch. Soviet. Geol.,* **4,** 220 p.

64. Souquet, P., Peybernes, B., Billotte, M., and Debroas, E.J., (1977), Géologie Alpine.

65. Stach, E., Taylor, G.H., Mackowsky, M.T., Chandra, D. and Teichmüller, M. and R., (1975), "Stach's Textbook of Coal Petrology", Borntraeger, Berlin, 428 p.

66. Stach, E., (1953), *Brennst. Chem.,* **34,** 353.

67. Stadler, G., and Teichmüller, R., (1971), *Fort. Geol. Rheinld. Westf.,* **18,** 547;

68. Staplin, F., (1969), *Bull. Canad. Petrol. Geol.,* **17,** 1, 47.

69. Teichmüller, M., (1958), *Rev. Ind. Min. Spec. Paris,* 99.

70. Teichmüller, M., (1971), *Erdöl und Kohle,* **24,** 69.

71. Teichmüller, M., (1974), *Fort. Geol. Rheinld. Westf.,* **24a,** 37.

72. Teichmüller, M., (1974), *Fort. Geol. Rheinld. Westf.,* **24b,** 65.

73. Teichmüller, M., and Wolf, M., (1977), *J. Microscop.,* **109,** 49.

74. Teichmüller, M., and Ottenjann, K., (1977), *Erdöl und Kohle,* **9,** 49.

75. Teichmüller, M., and Teichmüller, R., (1975), *Geol. Bavaria,* **73,** 123.

76. Teichmüller, R., (1973), *Z. Dtsch. Geol. Ges.,* **124,** 149.

77. Tissot, B., (1969), *Rev. Inst. Franç. du Pétrole,* **XXIV,** 2, 470.

78. Tissot, B., Califet-Debyser, Y., Deroo, G., and Oudin, J.L., (1971), *AAPG Bull.,* **55,** 12, 2177.

79. Tissot, B., and Espitalié, J., (1975), *Rev. Inst. Franç. du Pétrole,* **XXX,** 5, 743.

80. Vassoyevitch, N.B., Korchagina, J.I., Lopatin, N.V., and Tchernitchev, V.V., (1969), *Z. Angew. Geol.,* **15,** 612.

81. Vries, H. de, Habets, P., and Bokhoven, G., (1968), *Brennst. Chem.,* **49.**

82. Watanabe, T., (1976), *Ewing Symp.,* **1.**

83. Weber, K., (1972), *Neu. Jb. Geol. Pal.,* **141,** 333.

84. White, D., (1915), *J. Washington. Acad. Sci.,* **5,** 189.

85. Wolf, M., (1972), *Neu, Jb. Geol. Pal.,* **141,** 222.

86. Yermakov, V.I., Lebedev, V.S., Nemchenko, N.N., Rovenskaya, A.S. and Grachev, A.V., (1970), *Izv. Akad, Nauk SSSR,* **190,** 3, 683.

13

Structure of kerogens as seen by investigations on soluble extracts

M. VANDENBROUCKE*

I. INTRODUCTION

The structure of kerogen can be approached by global characterization methods: elemental analysis (EA), infrared (IR) spectroscopy, etc. Such methods make up what can be called a mineralogical approach, in that they essentially provide information on the relative amounts of aromatic and aliphatic forms of carbon. Because there are no methods giving detailed insight into the molecular structure of solid materials, structural information can be obtained only by analyzing degradation products, either artificial or natural ones.

Artificial degradation products may be obtained by chemical or thermal processes. The evolution of kerogens in their geological environment (mainly as the result of burial) leads to the formation of natural degradation products, including oil and gas, and this may also provide information on the structure of kerogens.

Natural degradation products have a better chance of preserving structural information because they are produced at low temperatures and without any additionnal chemical reactants. However, many natural products are less easily recovered under geological conditions than as the result of artificial degradation; for instance light hydrocarbons, carbon dioxide and water are mostly lost during sampling of natural samples and analytical procedure. Fortunately for structural analyses, since their structures are simple, they do not retain much information about molecular arrangement inside kerogen. On the contrary, the heaviest compounds, such as asphaltenes, are difficult to analyze within natural evolution products as well as in artificial ones. To some extent, analytical problems resemble those of the structural determination of kerogens, althouth their solubility in solvents enables specific methods such as NMR (nuclear magnetic resonance) or Raman spectroscopy to be applied. Actually, aside from some studies on light hydrocarbons [26, 27, 14], most research on soluble compounds has been done on hydrocarbons in the C_{12}-C_{35} range.

---

*Institut Français du Pétrole. Rueil-Malmaison, France.

II. EXTRACTS: ANALYTICAL PROCEDURES AND CHEMICAL COMPOSITION

A. Extraction and separation processes.

Soluble organic matter (OM) in a natural sample must be extracted from the rock before being subjected to separation and identification procedures. Generally, organic solvents such as hexane, chloroform or methanol-benzene mixtures are added to the crushed rock in a Soxhlet extractor or in a beaker with a stirring device such as a magnetic rod, or ultrasonic waves. Gentle heating is often used to improve extraction [35, 36]. The solution is then filtered and the solvent removed by a rotary evaporator.

Separation methods are mainly based on chromatographic techniques, some specific chemical reactions (such as saponification [18], silylation [42], etc.) or liquid-liquid extractions. Adsorption chromatography on column [36], paper or thin-layer [25, 40] is performed on an analytical or preparative scale depending on the weight of the sample. Adsorbents such as silica gel or alumina, and eluants such as hexane, benzene, methanol, are commonly used. High-pressure liquid chromatography, a method now being used in many laboratories, results in a best resolution and shortest time span for such separations [39]. Other types of chromatography such as ion exchange or gel filtration may also be effective in some particular cases [2]. Fractions obtained by thin-layer chromatography are detected by direct visualization under normal or UV light or by spraying (berberine sulfate for instance). The different zones are scraped and extracted separately. In the case of column chromatography, the effluent can be checked by refractometry for saturated hydrocarbons or by UV absorption for aromatic hydrocarbons.

Identifications by powerful analytical methods have been used extensively during the last decade. Among such methods, mass spectrometry (MS) [7], gas chromatography (GC) with different detectors, which may be selective or not, and the coupling of this technique with MS can be used to identify a great many molecules [22, 21]. Mass spectra of the sample and of reference substances may be compared by means of a computerized spectrum library [28].

B. Main types of constituents of extracts and their analysis.

This section will be restricted to rock extracts which have an evolution state corresponding to the generation of oil and gas, i.e. mainly composed of hydrocarbons.

1. Saturated hydrocarbons (Fig. 13.1).

These compounds include normal alkanes (straight carbon chains, I), isoalkanes (branched chains, II) among which isoprenoids (regular branching of a methyl group on every fourth carbon, III, IV), and cycloalkanes (V). Their elution from adsorbents such as alumina or silica gel is generally performed in the C_{12}-C_{35} range by solvents such as heptane or cyclohexane. Normal alkanes in the C_{12}-C_{35} range are easily identified in the total saturated fraction by gas-liquid chromatography (GLC) on silicon-coated columns [39]. Their distribution (percentage of each *n*-alkane in a given carbon range) can be determined from the planimetry of the peaks. Their absolute amount can be determined either by direct weighing, after insertion of iso-cyclo alkanes into 5 Å molecular sieves [45], or by GLC using planimetry with a known weight of internal standard [17]. Isoprenoids, particularly pristane (C_{19}, III) and phytane (C_{20}, IV) can also be measured by the planimetry of the corresponding peaks. The distribution of total alkanes according to the number of their rings can be determined by MS under high ionization voltage (70 eV). Intensities corresponding to well-defined fragment peaks are used to calculate the initial distribution of normal plus iso, one ring... six rings, according to Hood and O'Neal [23].

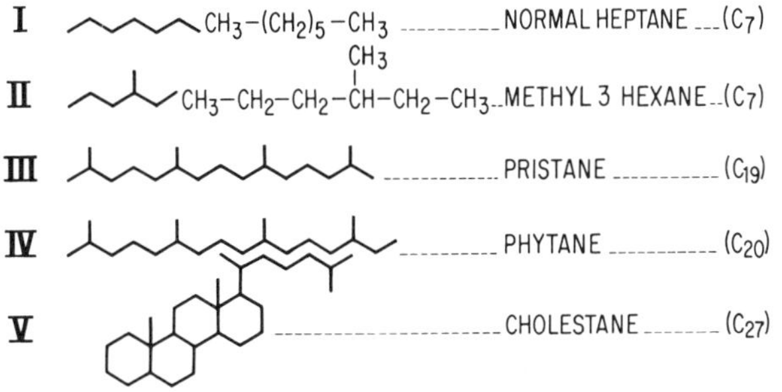

I $CH_3-(CH_2)_5-CH_3$ NORMAL HEPTANE (C_7)

II $CH_3-CH_2-CH_2-CH(CH_3)-CH_2-CH_3$ METHYL 3 HEXANE (C_7)

III PRISTANE (C_{19})

IV PHYTANE (C_{20})

V CHOLESTANE (C_{27})

Fig. 13.1. — Types of saturated hydrocarbons.

The coupling of GC and MS can be used either to scan the entire mass range for each interesting peak of the chromatogram or to follow a chosen mass which is characteristic of a specific fragment during the entire chromatogram. This technique is often used to obtain "mass fragmentograms", for instance at m/e = 191 (hopanes) or m/e = 217 (steranes) [12].

2. *Unsaturated hydrocarbons* (Fig. 13.2)

These compounds may be linear (*n*-alkenes I, branched alkenes II) or cyclic (hopenes III, sterenes). Since they are less stable than the corresponding saturates or aromatics, they can be observed only if the samples have not undergone too high levels of maturation, particularly for linear alkenes. These compounds cannot be separated from saturates with the standard

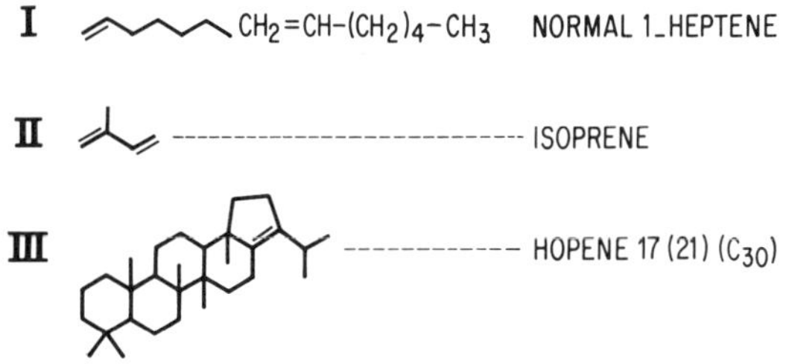

I $CH_2=CH-(CH_2)_4-CH_3$ NORMAL 1_HEPTENE

II ISOPRENE

III HOPENE 17 (21) (C_{30})

Fig. 13.2. — Types of unsaturated hydrocarbons.

methods since they are eluted very close to them, and a further separation of the saturated + unsaturated fraction must be performed, generally by thin layer chromatography (TLC) on silica/silver nitrate plates [38]. GLC analyses can be made on unsaturates in the same way as on saturates.

3. *Aromatic hydrocarbons* (Fig. 13.3).

Besides true mono-, di- and polyaromatics with aliphatic side chains, the aromatic fraction contains naphtheno-aromatic hydrocarbons, i.e. partly aromatized polycyclic structures. Thiophenes and other sulfur-containing molecules are also generally eluted with these hydrocarbons.

This fraction can be examined by GC on silicon-coated capillary columns. A flame photometry detector (FPD) with a suitable filter can be used together with a flame ionization detector (FID) to record the trace of sulfur-containing molecules [8, 10]. Individual peaks cannot be identified in the same way as for *n*-alkanes, but some general features (such as characteristic groups of peaks, baseline deformation, gross carbon range) can be deduced from FID and FPD chromatograms. The main use of this analytical technique is for making a comparison between samples so as to help the selection for further analyses [11].

Individual molecules are identified by MS under low ionization voltage (around 10 eV) from molecular ion peaks [30, 31]. This method requires prior separation of aromatic molecules according to the number of their aromatic rings [10] so as to avoid interference between mass families (Fig. 13.3). Peak heights are processed by a computer program which, for each mass family *p,* gives the weighted distribution by carbon number *n* of the naphtheno-aromatic molecules (for which the general formula is C_nH_{2n-p}). Mass fragmentograms for specific m/e can also be obtained from GC-MS coupling.

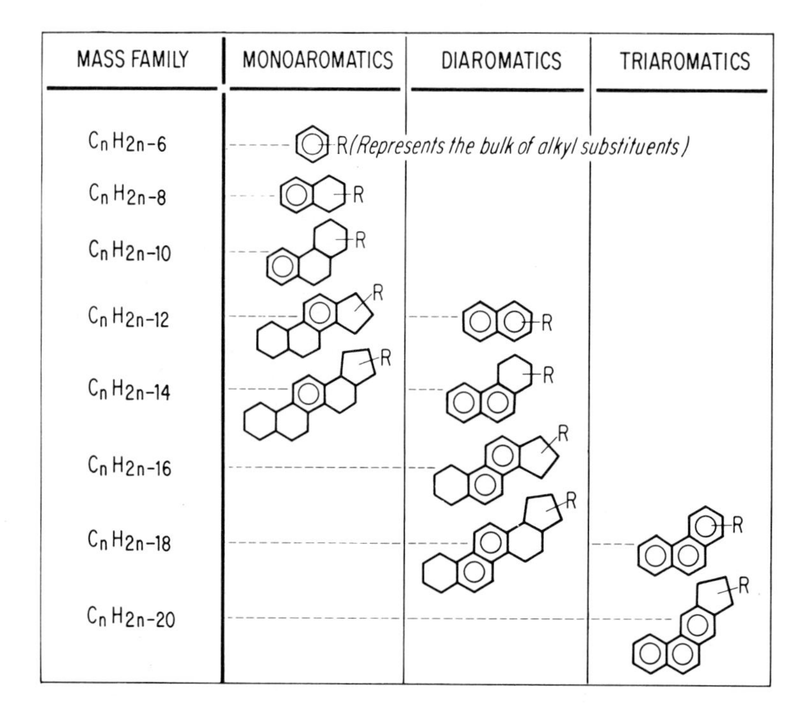

Fig. 13.3. — Principal structural types of aromatic hydrocarbons.

4. Heteroatomic compounds (Fig. 13.4).

They make up the non-hydrocarbon fraction of oils andc rock extracts. Their main heteroelement is oxygen. Two sorts of compounds, for which analytical problems are very different, can be found amongst the heteroatomic products. On the one hand, one can find "simple" molecules, the general structure of which is fairly well known, as for instance waxes or porphyrins. Provided sufficiently efficient isolation procedures might be applied, their molecular structure can be analyzed with a great detail. On the other hand, one can find in heavy ends of crude oil distillation and in fractions of extracts precipitated by hexane or petroleum ether, heteroatomic compounds like resins, asphaltenes, carbenes, which can only be analyzed by global methods, very similar to those used for kerogen studies.

Isolation procedure and analysis of fatty acids and porphyrins, which belong to the first category of heteroatomic compounds defined above, will be briefly examined.

Fatty acids (I), whether free or bound in the form of esters, are isolated from sediments or extracts by treatment with a potash-methanol mixture [19]. The acids are then precipitated by calcium chloride and transformed into methylic esters for GC [34]. The neutral fraction is then transformed into saponifiable and unsaponifiable fractions by alkali hydrolysis [32]. The acids of the saponifiable fraction are analyzed as above. The unsaponifiable fraction is separated by its solubility in water or hydrocarbon solvents.

I $CH_3-(CH_2)_7-CH=CH-(CH_2)_7-C(=O)-OH$ **OLEIC ACID** ($C_{18:1}$)

II **VANADYL DPEP**

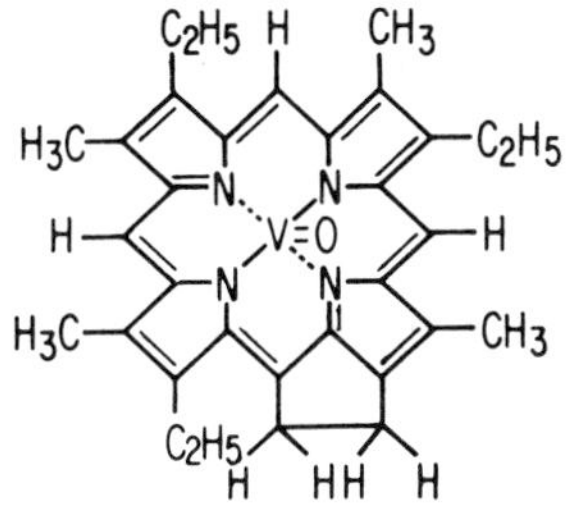

Fig. 13.4. — Types of heteroatomic compounds.

Porphyrins (II) are organometallic compounds, and the most frequent metals are originally Mg and Fe which are replaced in the petroporphyrins isolated from extracts by Ni and V. The most usual process for isolating them is acid extraction with a hydrobromic-acetic acid mixture at 50° [51]. This process has the disadvantage of losing the distinction between porphyrins bound to Ni or V. It is possible to concentrate porphyrins in their organometallic form by liquid chromatography on alumina columns, but it is difficult to obtain fractions containing more than 10% porphyrins. Another concentration method is liquid-liquid extraction with a nitrogenous solvent (pyridine, aniline, dimethylformamide). Likewise, it is possible to obtain a fraction enriched with porphyrins by means of gel permeation chromatography. The separation of the enriched fractions by classes of compounds is generally carried out by column, thin-layer or paper chromatography. The spectra are then studied with visible or UV light and are characterized by MS [3].

A review on structure of heavy ends and general methods for their study can be found in papers of Yen [56]. These methods are mainly spectroscopic ones, like X-rays, UV, IR, NMR (nuclear magnetic resonance) etc., besides elemental analysis and chemical characterization of functional groups. Results obtained by these methods on Lagunillas asphaltenes are mentioned later in this Chapter.

III. PRINCIPAL MARKERS OF THE ORIGIN OF ORGANIC MATTER AND THEIR IDENTIFICATION IN OILS AND ROCK EXTRACTS

A. Definition of markers.

Markers are compounds which characterize the origin of OM because of their special chemical structure. For this reason they are found essentially at states of evolution which are not too advanced, where the chemical skeleton has been largely preserved. The quantitative significance of these compounds should not be overlooked, for they are often found only as traces or in very small quantities compared with the bulk of kerogen. As the isolation and detection methods which are applied to them are often complex and very specific, it is in most cases impossible to establish a weight balance.

Amongst the markers can be distinguished those which in general characterize the biological origin of the carbonaceous matter of the sediment (in opposition to the inorganic theory) and those which are specific to a type of OM in accordance with the distinction made in the present volume between the various types of kerogens. In the first case, for example, are found the porphyrins derived from pigments, which exist throughout the animal and plant kingdoms except for fungi and certain protozoa and bacteria. In the second case, can be classed, for example, *n*-alkanes with a high predominance of the odd numbered homologues in the C_{25}-C_{35} range, which are derived from cuticular waxes of terrestrial plants and serve to protect them against evaporation.

B. Main biological compounds from which markers originate.

1. Proteins and carbohydrates (Fig. 13.5).

a. Proteins.

These are fundamental components of living matter as the constituent tissues of cells as well as catalysts for cellular synthesis or regulation (enzymes, certain hormones). They are formed by the condensation of amino acids and may consist of 100 to 10 000 of these constituent blocks (I). They are relatively easily hydrolyzed under the influence of enzymes (autolytic enzymes after the death of organisms, or bacterial enzymes) and are transformed into free amino acids (II). The latter are either used as metabolites in order to form other proteins, or undergo deamination and/or decarboxylation, or may also be recombined, for example, with carbohydrates to form Maillard polymers (pseudo humic acids), and also with phenols. In this case they participate in the formation of kerogen in recent sediments (as is shown by the high proportion of amino nitrogen in the latter). Except in this latter case, or when the proteins are protected by mineral substances (bone or shell), these compounds are destroyed very rapidly after the death of the organisms and do not participate as such in kerogen formation.

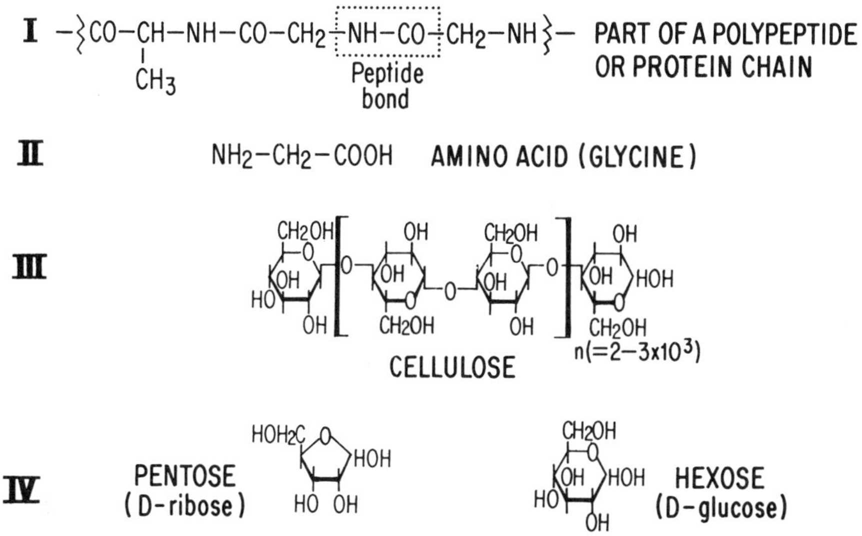

I $-\{CO-CH(CH_3)-NH-CO-CH_2-NH-CO-CH_2-NH\}-$ Peptide bond — PART OF A POLYPEPTIDE OR PROTEIN CHAIN

II NH_2-CH_2-COOH AMINO ACID (GLYCINE)

III CELLULOSE $n(=2-3\times10^3)$

IV PENTOSE (D-ribose) HEXOSE (D-glucose)

Fig. 13.5. — Chemical structure of some biological compounds: Proteins and carbohydrates.

b. Carbohydrates.

These are like proteins major constituent elements of cellular tissues (cellulose (III), chitin, mucilage, nucleic acids); they are also used in energy exchanges. Like proteins they are generally rapidly destroyed and metabolized by bacteria after the death of the organisms. Therefore they can only be incorporated into kerogen by polycondensation (for example of the Maillard reaction type). For this reason, except for recent sediments from which free monosaccharides (IV) can be extracted with water or alcohol, these compounds can be obtained only by acid hydrolysis of ancient sediments. Only small quantities, of the order of some tens of *ppm* [43], can generally be set free in this way.

2. Lipids (Fig. 13.6).

In the broad sense, lipids include a large number of compounds which possess the characteristic of being soluble in "fatty" solvents (hydrocarbons, chloroform, ether, etc.). Although these compounds in general represent less than 10% of living OM, their quantitative role in kerogen is far above this proportion. They are used in organisms either for the construction

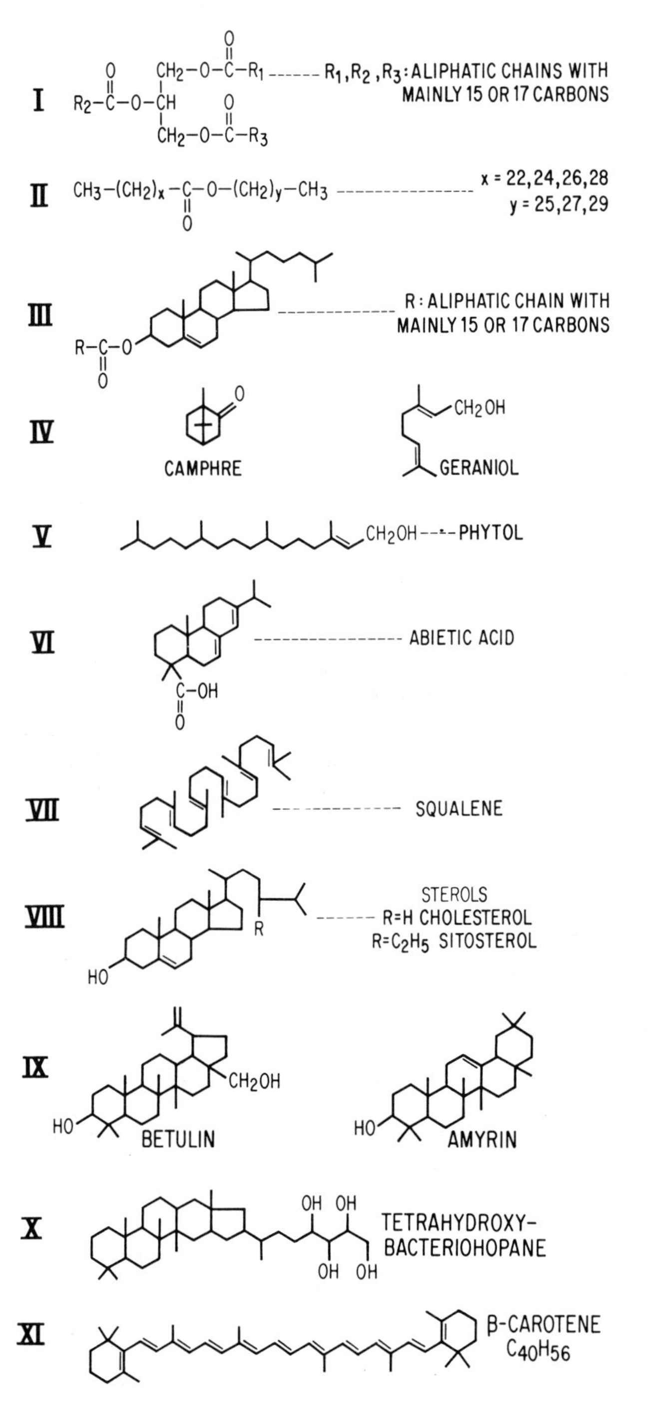

Fig. 13.6. — Chemical structure of some biological compounds: Lipids and terpenoids.

of membranes, or as energy reserves, or as means for external protection (shells of certain algae or bacteria, cuticular waxes of higher plants). Because of their relatively high chemical inertia, they undergo less oxidation or hydrolysis by bacterial enzymes than proteins and carbohydrates, and this explains why they are preferentially incorporated in sedimentary OM. Therefore the products of thermal degradation of kerogen contain a large number of markers belonging to this class of compounds.

a. *Classification of lipids.*

Lipids in the specific sense are esters of fatty acids and alcohols. Depending on the nature of the alcohol, a distinction is made between fats, which are mainly esters of glycerol and fatty acids (I), and waxes, subdivided into cerides (II) and sterides (III) depending on whether the alcohols are respectively fatty alcohols or sterols. The chemical inertia of fats is less than that of waxes: the acids found in them most frequently contain 16 and 18 carbon atoms and may have unsaturated bonds (Fig. 13.4, I). It is for this reason that they have biological functions (energy reserves in the case of triglycerides, constitution of cellular walls in the case of phospholipids). Waxes contain carbon chains which are longer and are generally completely saturated. They have a protective role against chemical aggressions because of their inertia, and against evaporation because of their hydrophobic character. They also can be used as energy reserves in some marine organisms.

Along with true lipids, lipidic fractions recovered from biological organisms contain compounds with similar chemical properties such as free fatty acids, saturated long-chain hydrocarbons, free fatty alcohols and ketones [33]. The common characteristic for fatty esters, acids and alcohols is that they contain an even number of carbon atoms, which is the result of their biosynthesis by means of acetyl-S-coenzyme A which operates by successive additions of two carbon atoms.

b. *Other compounds biochemically linked to lipids: terpenoids.*

These compounds are extremely widespread in all organisms. Their basic skeleton results from the combination of isoprene type units (Fig. 13.2, II) with five carbon atoms, biosynthesized by means of isopentenylpyrophosphate. This basic skeleton may be cyclic or acyclic and may contain functional groups and have a total number of carbon atoms slightly above or below a multiple of five.

The following classes of terpenes and terpenoids are distinguished:

(a) Monoterpenes (C_{10}): these are very widespread in the higher plants in the form of essential oils (camphor, geraniol (IV), etc.);

(b) Sesquiterpenes (C_{15}): these are also found in the essential oils of higher plants, especially of conifers.

The compounds belonging to these two classes of terpenes are generally quite volatile, so that they are probably not preserved under the usual sedimentation conditions.

(c) Diterpenes (C_{20}): amongst the acyclic compounds of this family is to be found phytol (V), which by esterification supplies the long side chain of chlorophyll a. Amongst the cyclic compounds in this class, there are a large number of conifer resins with di- and tricyclic structures (abietic acid (VI), agathic acid, etc.).

(d) Triterpenes (C_{30}): the most important of the acyclic compounds of this family is squalene (VII) which is found in large quantities, for example in the liver of certain fish. However, it is mainly the four and five cycle compounds derived from the cyclization of squalene which are very widespread in all living organisms. Among them steroids (VIII), with three hexagonal cycles and one pentagonal cycle are often found, the most frequent of them being cholesterol (C_{27}) in animals and β-sitosterol (C_{29}) in the higher plants [24]. Five cycle triterpenes are also found frequently in the higher plants (betulin, β-amyrin (IX)). Many bacteria synthesize five cycle triterpenes of the hopane family (X).

(e) Tetraterpenes (C_{40}): amongst these are found all the photosynthetic pigments of the carotenoid type, whose structure is linear except possibly one or two hexagonal cycles at the ends of the chain (XI). They are not very stable and are observed only in slightly evolved sediments.

c. *Distinctive properties of lipids according to the type of organic matter.*

Around years 1970, some researchers hoped it would be possible, from the existence of specific geochemical fossils in sedimentary environments, to establish precisely from what organisms they were issued. It seems now that this paleochemotaxonomic approach is generally not feasible. The number of living species in a given sedimentary medium is generally too large, except in special cases (for example bogheads), although it can be supposed that an environment favorable to species with the same needs would lead to the synthesis of similar metabolites. However the main obstacle is that the chemical composition of many species varies considerably as a function of the environmental conditions. Nevertheless, in the case of lipids it is possible to indicate some very general specific characteristics of the plant biomass, according to its aquatic or terrestrial origin, and also of the bacterial biomass.

For example, it can be observed that protective waxes with long carbon chains (from C_{25} to C_{35} per chain) characterize almost exclusively the higher plants [52]. They also contain fats with 16 and 18 C atoms, but since their resistance to oxidation is much lower than that of waxes, it is the long chains generally with odd numbers (decarboxylation) and sometimes with even numbers (highly reducing medium) which are found in the normal alkanes of the corresponding extracts [55]. On the other hand, in marine material where these waxes are absent [54], the normal alkanes derived from the evolution of kerogen generally have a maximum of C_{15} or $_{17}$ [5]. The contribution of the bacterial biomass which reworks the sediment at the time of deposition is indicated by the presence of waxes with very long chains ($> C_{30}$) without parity characteristics [49]. Isoalkanes containing a methyl group at the first or second carbon of the linear chain (iso and anteiso alkanes) are also characteristic of bacteria [29].

Among cyclic compounds, di- or tricyclic diterpenes are found essentially in the resins of the higher plants [41]. Tetracyclic steroids are abundant in the aquatic biomass [20]. Pentacyclic triterpenes are present in continental OM and in the bacterial biomass in the form of hopenes [15]. All these compounds are not very stable in the unsaturated form and are transformed rapidly by saturation or aromatization of one or several cycles. They also quickly lose their oxygenated functions, either to reach a more stable form as free molecule, or to be linked by this point to kerogen. The carbon skeletons of these compounds are therefore found in the extracts both in saturated and aromatic hydrocarbons.

3. *Natural pigments* (Fig. 13.7).

In addition to the already mentioned carotenes (tetraterpenes), the most widespread pigments are nitrogenous heterocyclic compounds of the porphyrin type. The basic structure is a tetrapyrrolic nucleus stabilized by a metal which is magnesium for chlorophyll and iron for hemin. Different forms of chlorophyll exist in all terrestrial and aquatic plants as well as in most bacteria. The initial structure of chlorophyll a (I) chelated with magnesium is unstable [4] and is rapidly replaced during the sedimentation of OM by a complex of nickel or vanadium (Fig. 13.4, II). In this form porphyrins are extremely stable. They appear in the heteroatomic fraction (resins and asphaltenes) of oils and rock extracts, and probably also participate in the formation of kerogen.

There is another series of natural pigments which is very widespread in the higher plants: these are the flavones and anthocyanins, which are oxygenated heterocyclic compounds. It is difficult to form an idea as to the possibilities of their incorporation into kerogen, for they are water soluble compounds, but they could be the source of some phenols and aromatic acids in the extracts.

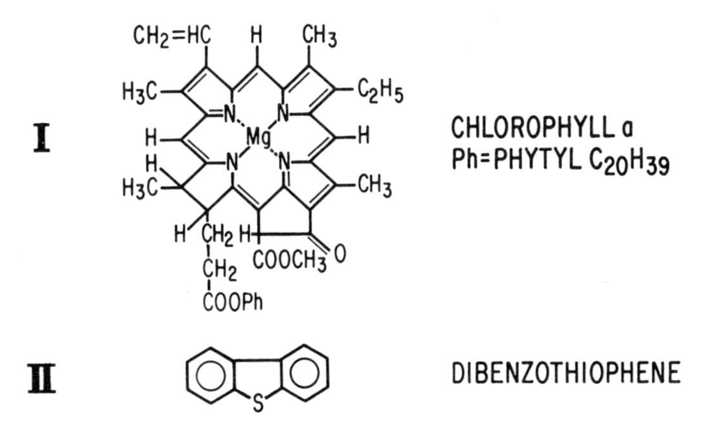

Fig. 13.7. — Chemical structure of some biological compounds: Natural pigments and sulfur compounds.

4. *Sulfur compounds* (Fig. 13.7).

Although these compounds cannot be considered as genetic markers because they do not exist in living OM (except in some amino acids) they are sometimes abundant in oils and rock extracts (especially in the case of carbonate source rocks). They are found in the aromatic fraction, and it is very hard to separate them from hydrocarbons. The most abundant belong to the thiophene families (II), but thiols and organic sulfides can also be found. Although it is relatively easy to detect them with GC and high-resolution MS [10], they are used essentially as fingerprints in the comparison of oils and source rocks.

Sulfur is incorporated in living OM essentially by means of amino acids, although some vitamins and enzymes contain heterocyclic sulfur compounds. It is possible that the formation of thiophenes results from abiogenic processes which redistribute sedimentary sulfur. This latter is derived from the metabolism of sulfate-reducing bacteria living in an anaerobic medium on sedimentary OM and drawing oxygen from sulfates dissolved in seawater. Sulfur released in this way (in the form of sulfide ion) by bacteria recombines with the available iron in clays to form pyrite, and probably forms organic sulfur compounds in carbonates where iron is usually not very abundant.

IV. FORMATION OF EXTRACTS THROUGH KEROGEN EVOLUTION

A. Types of kerogen, their thermal evolution and their relation to the formation of petroleum products.

In previous Chapters we have seen that sedimentary OM is the residue of the chemical and biochemical alteration of an initial biological material. On the basis of the elemental composition of kerogens (Chapter 4), three main types of sedimentary OM and their evolution paths in a H/C, O/C-diagram have been defined Fig. 13.8. To a certain extent, it is possible to relate the source of each type of organic material with the environment of deposition, for instance by sedimentological criteria, nature of associated minerals, etc:

(a) **Type I** has an initial elemental composition such that H/C is slightly under 2, which theoretically indicates a carbon skeleton very rich in saturated structures, and mainly linear ones. Material of this kind is not very widespread. The reference series in the Uinta Green River Shales, which consists in a lacustrine deposit which has been so completely reworked by microorganisms that all that remains is the most resistant OM, i.e. their own constituent lipids and some cuticular vegetal waxes. A nearly similar chemical composition can also be found in sediments which are generally very rich in organic carbon and consist almost entirely of remains of unicellular algae of the family of *Botryococcacea,* rich in linear lipids (Autun boghead coal, torbanite of Scotland and Australia, coorongite).

(b) **Type II** has an initial elemental composition such that H/C is slightly under 1.5, corresponding to a highly saturated polycyclic (naphthenic) carbon skeleton. OM of this type are derived from planktonic material (zoo- and mainly phytoplankton) but they can also be found in sediments which are very rich in organic carbon and are made up of peculiar algae (*Tasmanite* of Australia, for instance). Amongst the constituents of these organisms which resist the reworking of the sediment by benthic flora and fauna are lipids, a large fraction of which is in the form of sterides [20]. They are frequently found in more or less carbonated clays, deposited in a shallow and calm marine medium, where anoxic conditions are frequent.

(c) **Type III** has an initial elemental composition such that H/C is slightly under 1, and is particularly rich in oxygen. The corresponding carbon skeleton is theoretically of the heterocyclic or aromatic polycyclic type. Remains of higher plants rich in lignin are preponderant; lipids are present in lesser quantity and consist mainly of cuticular waxes. Amongst the constituents which resist oxidation can also be found terpenic resins (abietic acid of conifers, amber, etc.). Microorganism action generally seems not to be very important in this type of deposit. Such OM is frequently found in detrital series where subsidence is rapid and derives essentially from continental material issued from vegetal cover of soils by erosion (leaching, transport of solid particles). It is associated with detrital sedimentation consisting of clays and silts. Coals with very similar organic progenitors can also be associated with such type III material.

These various organic materials evolve with the burial of the sediments and the resulting thermal action [48]. They progressively lose their functional groups (principal variation in

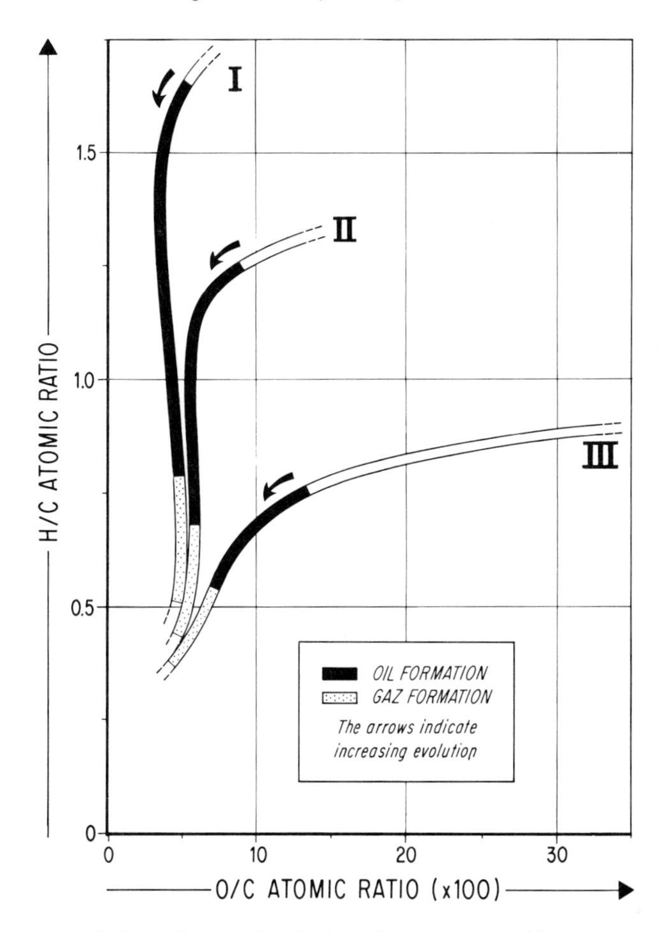

Fig. 13.8. — Structural evolution of various types of kerogen.

O/C, Fig. 13.8), then their saturated chains (principal variation in H/C, Fig. 13.8). The insoluble residue (kerogen) is correspondingly enriched with aromatic carbon so that it moves towards the state of thermodynamic equilibrium represented by the structure of graphite (stacking of layers of aromatic rings).

The representative points of elemental composition of kerogens in the H/C, O/C-diagram thus describe an evolution path, whose position is different according to the type to which they belong. Oil and gas are formed during this evolution, during the final phase of O/C variation and the phase of H/C variation.

B. Quantitative evaluation of extract formation.

The mathematical representation, in terms of chemical kinetics, of the transformation of kerogen into petroleum products has been the subject of many publications. We will make frequent use of the most recent study [47]. Without going into computation details, we will mention only that the equations for the formation of oil and gas from kerogen, which are used in this model, are of the following type:

$$\text{kerogen} \begin{cases} \nearrow CO_2,\ H_2O,\ \text{etc.} \\ \rightarrow \text{petroleum products} \begin{cases} \nearrow CO_2,\ H_2O,\ \text{etc.} \\ \rightarrow \text{gas} \\ \searrow \text{residue} \end{cases} \\ \searrow \text{residue} \end{cases}$$

They are expressed mathematically by equations similar to those for first-order kinetic reactions. The reaction constants depend on the temperature by an exponential relation of the Arrhenius-law type. For each equation this law introduces an activation energy which characterizes the reaction under consideration. The higher the activation energy, the higher the temperature required to set up the corresponding reaction. The various constants for each equation are adjusted on the basis of measurements on different samples from each type, as well on quantities of extract already formed, as on the final rate of transformation of kerogen into hydrocarbons, measured in the laboratory.

The results of these adjustements for the three types of kerogen used as references here are shown in Fig. 13.9. Thermal evolution is considered to have occurred under identical geological conditions (constant rate of burial up to 6 000 m within 100 millions years, with an uniform geothermal gradient of 35° C/km). The distribution of activation energies, and the percentage of kerogen entering into the corresponding reactions, is shown in Fig. 13.10. Although in view of the chemical complexity of kerogen it is not possible to write a definite chemical reaction equation for each activation energy, the effects of these latter can be roughly classified as follows in terms of increasing values:

Phase 1: breakdown of weak bonds (adsorption, hydrogen bridge, etc.).
Phase 2: breakdown of unconjugated heteroatomic bonds (esters, ketones, etc.).
Phase 3: breakdown of carbon-carbon bonds in saturated cycles or in lateral chains attached to cycles.
Phase 4: breakdown of carbon-carbon bonds at any point of the aliphatic chains.

The main oil formation occurs essentially in phase 3 with possibly a small contribution in phases 1 and 2. The main gas formation occurs essentially in phase 4. The breakdown of aromatic bonds in the carbon or heteroatomic cycles requires activation energies which are much higher than those reached in sedimentary rocks. They might however be reached at the stage of metamorphism.

If Figs. 13.9 and 13.10 are examined for each type of kerogen it is already possible, from these data alone, to form a certain idea of the corresponding chemical structure (characterized by the proportion of kerogen entering into each reaction with a given energy), and also of the characteristics of the compounds resulting from thermal evolution (by means of the activation energy values):

(a) **Type I:** the proportion of kerogen which is capable of being transformed into hydrocarbons is very high. On the other hand, the activation energy spectrum is concentrated around high values. The result is that hydrocarbon formation occurs late and these hydrocarbons will have a marked aliphatic character since the principal reactions will be breakdowns of C – C bonds.

(b) **Type II:** the amount of kerogen which can be transformed into oil is little less. On the other hand the activation energy spectrum is more extended. Hydrocarbon formation begins

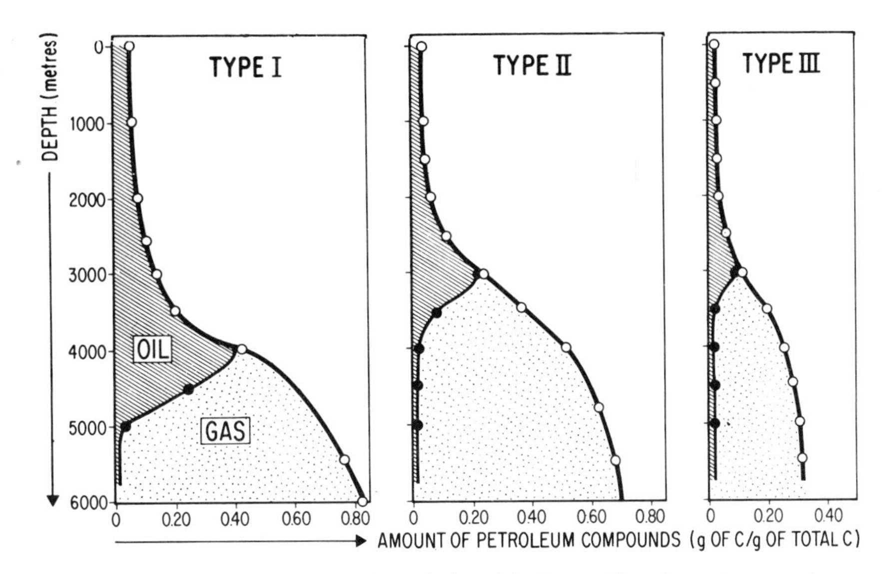

Fig. 13.9. — Computation by mathematical model of quantities of petroleum products formed by thermal evolution of the three main types of kerogen (according to Tissot and Espitalié, [47]).

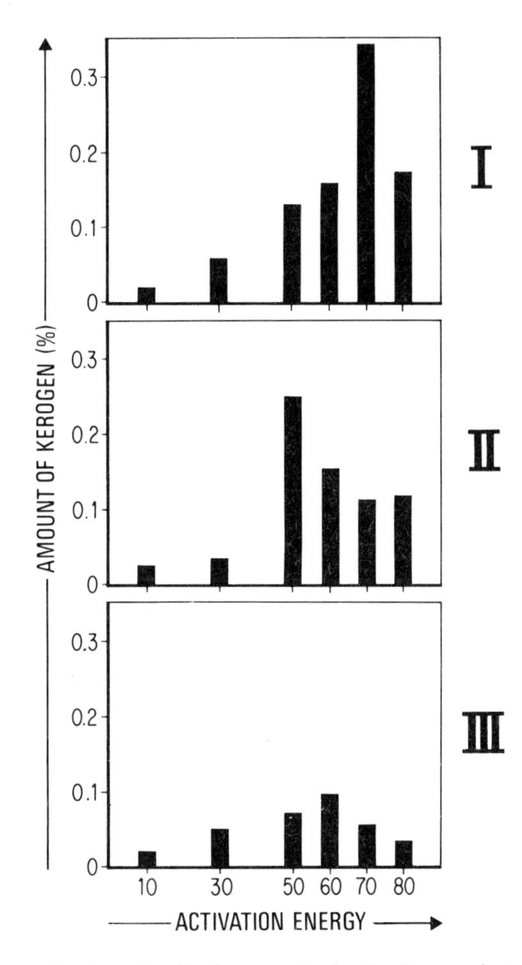

Fig. 13.10. — Distribution of activation energies in the three main types of kerogen (according to Tissot and Espitalié, [47]).

earlier, and these hydrocarbons will have a more marked cyclic character since the energies will be lower than in the preceding case.

(c) **Type III:** the amount of kerogen which can be transformed into oil is small. Low activation energies are important, and this means that kerogen degradation begins soon, and that at this stage will be formed many labile heteroatomic compounds (CO_2, H_2O) and hydrocarbons with characters of geochemical fossils, since there will be little change in their carbon skeleton.

C. Composition of petroleum products formed by the various types of kerogen: Data obtained from the study of extracts.

Chemical structures of kerogens have been set up above from considerations on timing and characters of petroleum formation, on the basis of activation energy spectra. More precise data can be obtained by considering the chemical composition of the extracts in the vicinity of maximum oil production. Although the corresponding evolution state of the OM is not defined very precisely in this way, this process enables to make comparisons between kerogen degradation products, at a time when they are representative because of their abundance while their original properties are not too altered by cracking. A danger in using this criterion is that the soluble products may not be representative of all the OM in place, because of migration phenomena. Series of samples will therefore always be used, so that statistical study will be possible.

It is also difficult to find evolution criteria which are valid for all the various types of OM. Moreover, burial and geothermal conditions differ for each particular basin. Besides comparisons on amounts of oil extracted, which have the disadvantage of requiring a relatively large number of samples, it is also possible to use the pyrolysis peak temperature of the kerogen [16], or the vitrinite reflectance [37, 44] if the latter is representative of the OM (type III) or if it is relatively frequent in the deposition medium.

We will present below a comparison of extracts formed by organic matters of type I, II and III in the vicinity of maximum oil production, based upon the series studied in the following publications:

(a) Type I: Green River Shales [50]
(b) Type II: Toarcian of the Paris Basin [46, 53]
(c) Type III: Logbaba Series [53, 1]

1. Distribution of hydrocarbons.

The results are shown in Figs. 13.11 to 13.15. It can be observed (Fig. 13.11) that the proportions of OM extractable with chloroform decrease from type I sample to type III sample. This is also true for the hydrocarbon fraction alone. In type I as well as type III there is a high proportion of saturated hydrocarbons in the extract, and the saturates/aromatics ratio is high. On the other hand, in type II the heteroatomic products are predominant in the extract, and the saturates/aromatics ratio is close to 1.

In type I the proportion of normal alkanes in saturated hydrocarbons is very high. In addition, the amount given is less than the real quantity since, for analytic reasons, it takes into

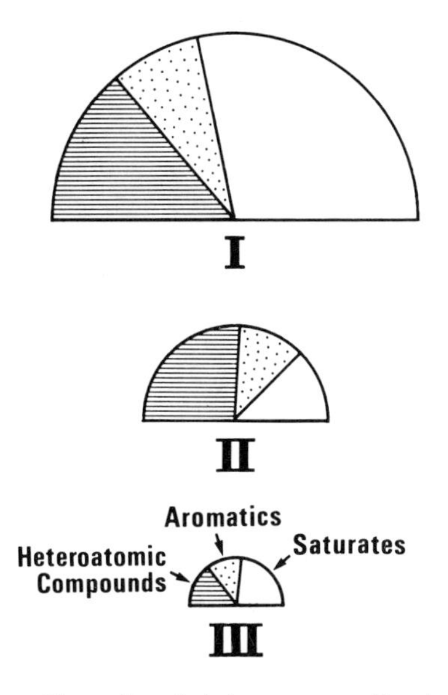

Fig. 13.11. — Extracts: quantities and compositions for the three examples mentioned.

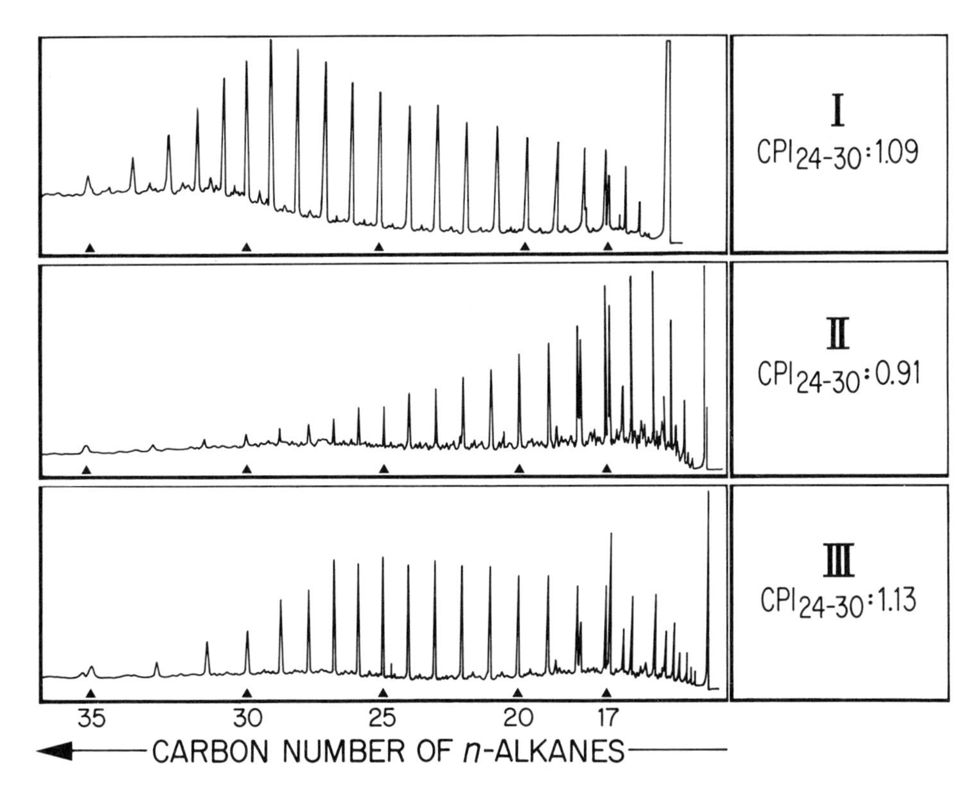

Fig. 13.12. — Gas chromatograms of saturated compounds for the three examples mentioned (25 m × 0.5 mm column, OV 17 phase, temperature programming 2° C/min. from 100-270° C).

account only the n-alkanes from C_{14} to C_{30}. Examination of the chromatogram in Fig. 13.12 shows that the fraction above C_{25} is preponderant; above C_{30}, the n-alkanes are still very abundant. Moreover, there is no predominance of odd or even numbered homologues. The n-alkanes are still well represented in type III, although in a lesser proportion than in type I. The chromatogram of saturated hydrocarbons between C_{15} and C_{30} shows that n-alkanes are abundant principally between C_{20} and C_{30}, and shows a slight predominance of odd numbered homologues.

In type II the n-alkanes are less abundant and their distribution between C_{15} and C_{30} drops regularly as the number or carbons decreases. There are few of them beyond C_{25}.

As concerns cyclanes, it can be seen (Fig. 13.13) that there are very few of them in type I. Their distribution according to the number of cycles shows no marked characteristics. There are slightly more of them in type III, and the di- and tricyclic molecules are best represented. Saturated cyclic hydrocarbons are more abundant in type II than linear hydrocarbons, and tetracyclic molecules predominate.

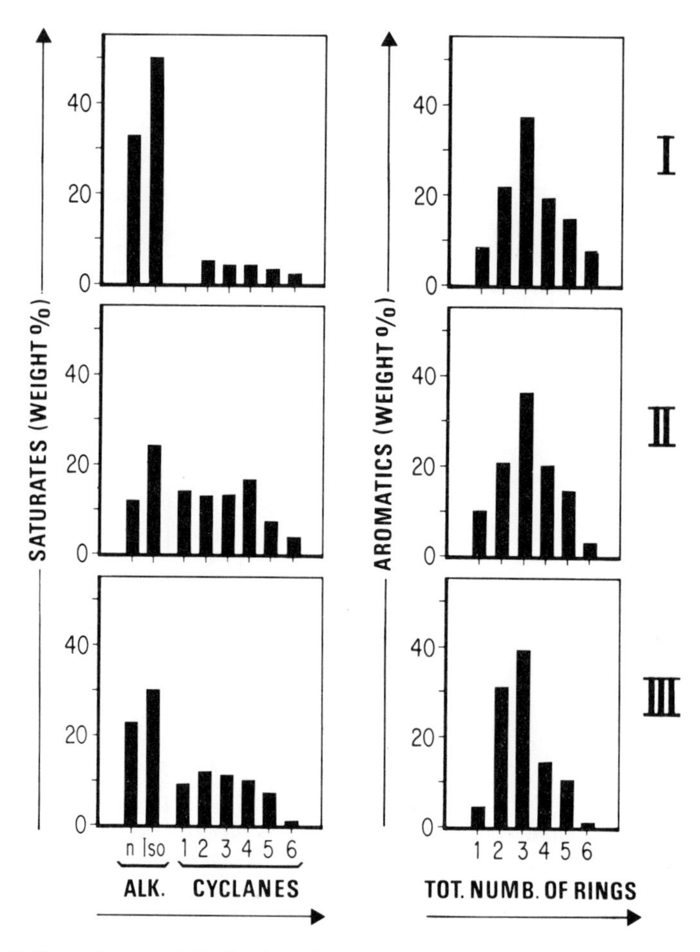

Fig. 13.13. — Structural distribution of saturated and aromatic compounds for the three examples mentioned.

To sum up, it can be stated that saturated hydrocarbons derived from the thermal evolution of kerogen have characteristics which are quite similar in type I and type III. They are rich in linear carbon structures with long chains. On the other hand they have a cyclic nature in type II. A remark about type III saturates must be made. It can be observed, for an evolution state less advanced than that of the samples described above, before the main oil producing zone, a small hydrocarbon production (about 1% of the carbon) consisting essentially of heavy *n*-alkanes (C_{25} to C_{35}) with a marked odd-number predominance.

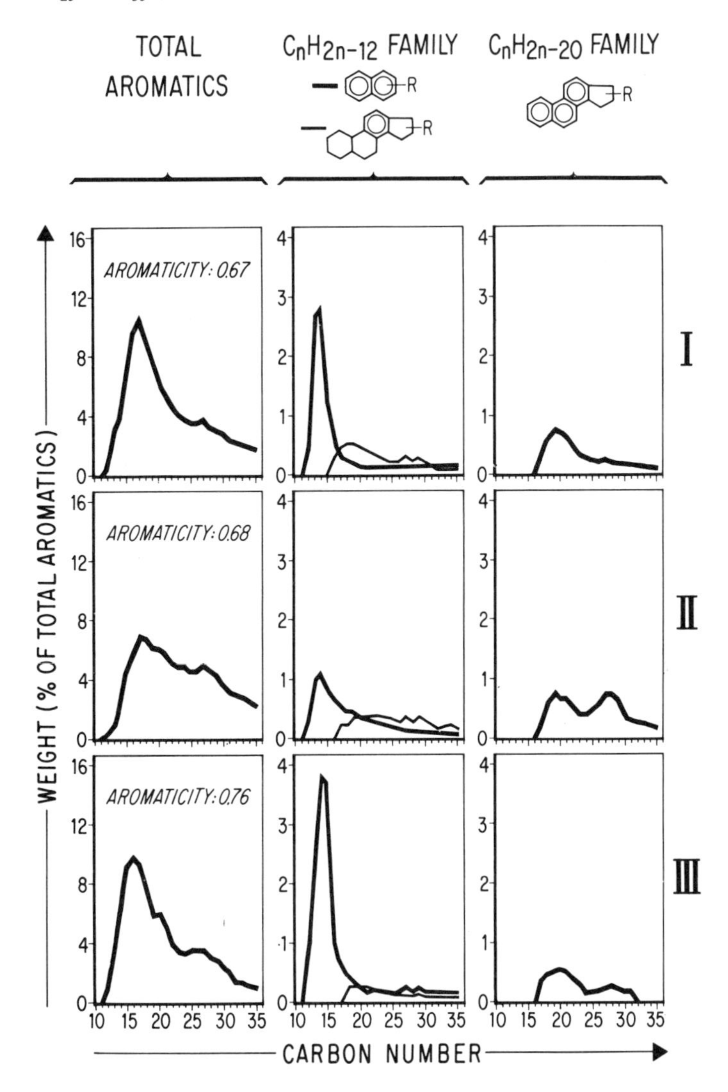

Fig. 13.14. — Distribution by carbon atoms of aromatic hydrocarbons for the three examples mentioned (data by low-voltage ionisation MS — 15 eV).

The characteristics of aromatic hydrocarbons are presented in Fig. 13.14. They account for nearly half of the total hydrocarbons of type II, and their distribution by carbon numbers, obtained by low-tension MS, shows the presence of a noticeable proportion of heavy molecules > C_{25}. On the contrary, light molecules between C_{15} and C_{20} are best represented in type I and in type III. In the same way as for saturated hydrocarbons, tetra- and pentacyclic aromatic molecules are frequent in type II. The distribution by carbon numbers makes it possible to determine that these are mainly molecules between C_{25} and C_{30}, as can be seen in Fig. 13.14. They are found in fraction 1 in the $C_n H_{2n-12}$ family (mainly monoaromatics with three saturated cycles) and in the $C_n H_{2n-20}$ family (essentially triaromatics with one saturated cycle). In type I and even more in type III, the small molecules predominate, as is shown in Fig. 13.14 where the $C_n H_{2n-12}$ family is represented mainly by C_{13} to C_{15} molecules of fraction 2 (true diaromatics with short side chains). Abundant two-cycle molecules can be noted in the aromatics of type III.

For each sample studied here it is possible to calculate a coefficient of aromaticity ([1]), equal to the sum of the proportions, in each family and each fraction, of the masses which can be attributed to the aromatic cycles in the specific sense in the one to five cycle hydrocarbons

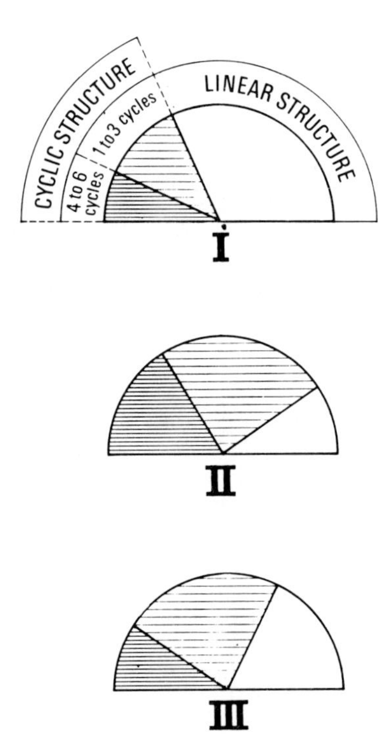

Fig. 13.15 — Structural distribution of all hydrocarbons for the three examples mentioned.

([1]) This calculation is not equivalent to that for the factor of aromaticity F_a [6], which results from use of parameters obtained by NMR.

(for example the monoaromatics of the $C_n H_{2n-12}$ family, with one aromatic cycle and three saturated cycles, have a coefficient of aromaticity of 25%). The results of Fig. 13.14 show that this aromaticity is higher in type III than in types I and II where it is nearly the same.

To sum up, the aromatic hydrocarbons in the C_{12}-C_{35} range, which are formed during the thermal evolution of kerogen, are not very abundant in type I, and their distribution is centered around low numbers of carbons, which is not the case for saturated hydrocarbons. Their aromaticity is low. On the other hand, they are abundant in type II, their average number of carbons is higher and their aromaticity is low. They include many partly aromatized polycyclic molecules. Aromatics are quite abundant in type III and they are essentially made up of molecules of the families of naphtalene and phenanthrene, with short side chains.

The structural distribution of total hydrocarbons extracted from the three samples examined above is shown in Fig. 13.15.

2. *Heteroatomic compounds.*

a. *Sulfur compounds in the aromatic fraction (essentially thiophenes).*

The sulfur distribution was not studied quantitatively for the three examples presented above. However it is possible to mention the qualitative results obtained, for the aromatic fractions of other samples from these three reference series, by chromatography with a selective sulfur detector (FPD). It was observed that the aromatic fractions from samples of type II are richer in thiophene compounds than those of types I and III, which in some cases do not even contain any at all. The cause, wether it is the nature of the OM or wether there is a sedimentation medium which does not allow biochemical reworking by sulfate-reducing bacteria, is actually unknown. The distribution of thiophenes by structural types and by number of carbons, as can be deduced from the chromatograms [10], changes with the evolution state of the OM. In slightly evolved samples many compounds are found giving well differentiated peaks around the zone of C_{30}. With evolution, these peaks disappear and give way to high peaks corresponding to dibenzothiophenes with short side chains.

b. *Resins and asphaltenes.*

Although there are few publications on these compounds, it is of interest to mention them since it can be supposed that heteroatomic products, aside from their molecular mass, have a structure which is quite similar to that of kerogen.

The most complete structural study on the subject concerns asphaltenes obtained by precipitation with pentane from Lagunillas crude oil (Venezuela) [56]. The author used a group of physical analysis methods and, for the example studied, deduced a model formula which is given in Fig. 13.16. According to this study, asphaltenes are complex polymers which can be characterized by a microstructure and a macrostructure. The microstructure (A) consists of substituted polycondensed aromatic layers assembled in piles of four to six by means of the π electrons of the aromatic rings. The dimensions of this structure, as determined by X-rays, are less than 40 Å. These particles can agglomerate in a macrostructure (D) consisting of micelles of up to about 2 000 Å.

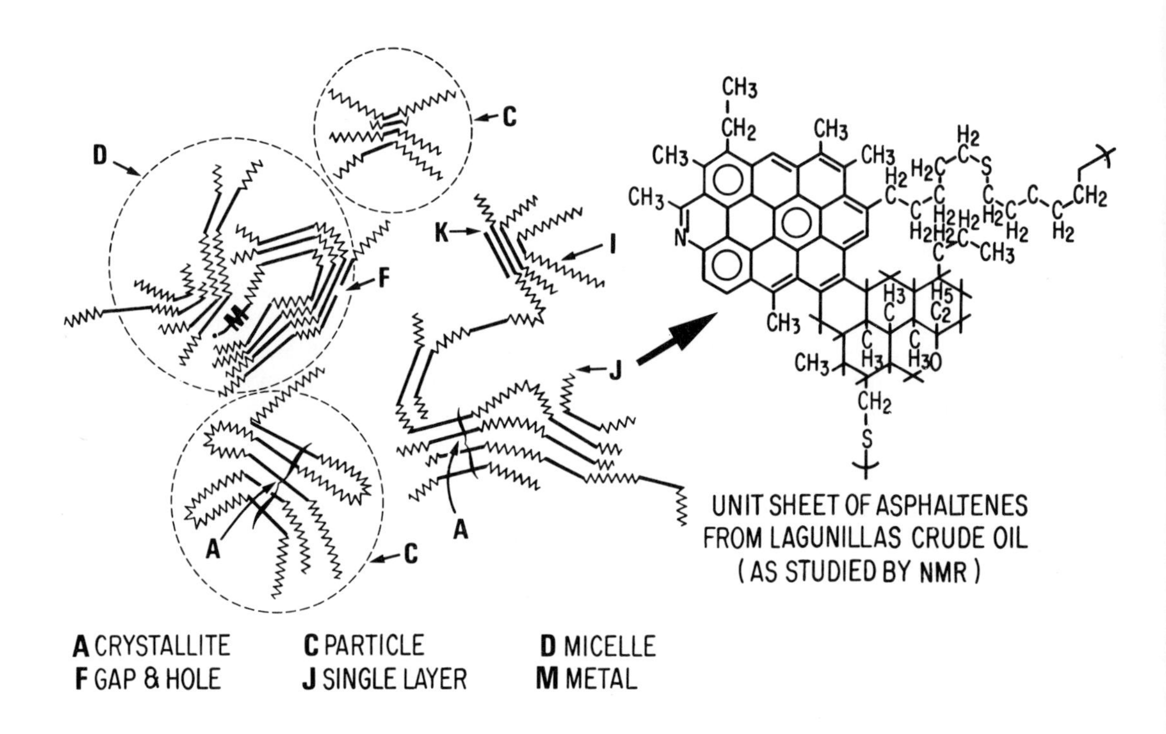

Fig. 13.16. — Model of structure of asphaltenes from Lagunillas crude (according to Yen, [56]).

In the example studied by NMR, the aromatic rings forming the microstructure are 50-70% substituted by naphthenic cycles and alkyl groups, mainly methyl ones. There is a high proportion of heteroatoms, with sulfur occurring mainly in the form of thiophene, nitrogen in the form of quinoline, and oxygen in the form of ether. The assemblage of layers in the macrostructure is due to bridges either by π bonds or by coordination of heteroatoms with metals, especially vanadium and nickel. There metals may occur in a structure of the porphyrin type but may also participate in other structures.

At the present time we have no data on the type of OM which is at the origin of this oil, but it can be supposed that several of the conclusions of this study concerning the macro- and microstructures of asphaltenes will be applicable to heavy products isolated from rock extracts, whatever the original OM may have been.

TABLE 13.1
STRUCTURAL PARAMETERS OF RESINS AND ASPHALTENES COMING FROM OM OF DIFFERENT TYPES IN THE MAXIMUM FORMATION ZONE OF THE C_{15+} EXTRACT [9]

Parameter	Type I		Type II		Type III
	Resins	Asphaltenes*	Resins	Asphaltenes	Resins
Molecular mass (in $CHCl_3$)	370	—	470	800	360
Elemental analysis (weight %)					
C	83.7	78.2	81.7	84.8	74.1
H	9.2	8.6	8.7	7.5	7:8
O	5.0	5.5	5.0	4.1	7.0
S	1.8	—	2.8	1.2	4.3
N	0.3	2.8	1.5	1.7	0.7
Atomic ratios					
H/C	1.32	1.32	1.28	1.06	1.26
O/C	0.045	0.053	0.046	0.036	0.071
NMR data (%)					
F_a	38-45	32-44	41-44	48-63	46-48
σ	48-53	71-75	56-64	55-67	50-64
C_N	16.5	23.3	31.6	26.4	26
L_C	4.5-6	6-8	4.1-5.2	2.6-4.1	3.5-4

F_a Aromaticity factor (proportion of carbon in an aromatic form).
σ Degree of substitution (proportion of aromatic carbon in which the proton is replaced by a substituent).
C_N Proportion of carbon in a naphtenic form.
L_C Average length (in number of carbon atoms) of aliphatic chains.
* The resins and asphaltenes analyzed here do not come from the same sample (as opposed to those in type II).

A study of resins and asphaltenes from rock extracts derived from well-defined types of OM, and concerning elemental analyses, osmometry and NMR measurements, is now being published [9]. Table 13.1, although incomplete, brings out several structural characteristics which better differentiate the types of OM, at the same evolution stage (maximum formation of C_{15+} extract) as the reference samples used above (for which the hydrocarbons were previously compared). Although the H/C atomic ratio differs only slightly in all cases, it can be noted that, in passing from type I to type III, the aromaticity increases and the average length of the aliphatic chains decreases. The proportion of carbon in naphthenic form is at a maximum in type II.

V. USE OF KNOWLEDGE OBTAINED BY STUDY OF EXTRACTS IN THE DETERMINATION OF KEROGEN STRUCTURES

A. Representativity of extracts in relation to kerogens.

The use of extract study, as described above, for the elucidation of kerogen structure requires some precautions. It is necessary to be sure that the extract is reasonably representative of the total OM, and it is immediatly clear that this characteristic depends on the type of OM under consideration and its degree of evolution. With type I OM, in which the kerogen, as shown by its elemental analysis, is transformable almost exclusively into hydrocarbons, and where the rate of this transformation is very high, it is clear that the study of these hydrocarbons will give more data than with type III OM, where the amounts of extract formed are always small (1)

In the same way, the extractable quantities depend on the degree of evolution of the OM. For this reason, it was decided to describe the characteristics of products obtained in the zone of maximum extractability. It should also be noted that we know how to make detailed analyses of only a certain fraction of these extractable products, i.e. mainly hydrocarbons in the C_{15}-C_{35} range. Volatile or unrecoverable products such as water, carbon dioxide, volatile sulfur compounds and hydrocarbons from C_1 to C_{15} are generally not taken into consideration. Heteroatomic products are often known only by their weight and sometimes by their elemental analysis.

However in spite of its fragmentary nature, the study of extracts is of great interest for the structural determination of kerogens. Analyses concerning the total OM are necessarily global ones and supply information essentially on the degree of aromaticity. On the other hand, analysis of hydrocarbons make it possible to study separately and quantitatively the various structural types and to give a distribution by carbon atoms within each type.

Another point of basic interest is the possibility, on the basis of geochemical markers identified in these extracts, of going back to some of the biochemical precursors of the OM, and so of at least partly reconstructing the initial composition of the OM and its condensation possibilities due to functional groups, thus determining the causes of kerogen insolubility, etc.

B. Determination of kerogen structure on the basis of extract study.

On the basis of the characteristics of extracts from the three main types of OM, defined on the above-mentioned examples (I: Eocene of the Green River Shales; II: Toarcian of the Paris

---

(1) It should be noted that from the standpoint of extract characteristics, a mixture of distinct OM, for example an authochtonous type II and a detrital type III, will after extraction have characteristics close to those of a type II alone, since the latter produces proportionally much more extract. This case may occur when there is little organic material in the sediment.

Basin; III: Cretaceous of Logbaba), an attempt will now be made to give a structural description of each type of kerogen at the beginning of the oil formation stage. We will successively describe the average structural unit (equivalent to a monomer or a microstructure) and the possible modes of association of these units (equivalent to a polymer or a macrostructure). This description naturally is largely theoretical. When necessary we will refer to the representative schemes which were given in the Chapter on electron microscopy (Chapter 7).

1. Type I: Example of Green River Shales (Eocene, Uinta Basin).

The extract may represent a large part of the evolution products of kerogen. We have seen that it consists mainly of long-chain saturated hydrocarbons with little or no branching. Since it is known that this kerogen, even at the beginning of its evolution, contains little oxygen (less than 5% in atomic O/C), it can be supposed that it is built essentially by linear carbon chains, between C_{30} and C_{40}, linked by oxygenated functions of the ester or ether type (an ester with two C_{30} chains linked by a carboxyl group would have an O/C atomic ratio of 3.3%, and for a triglyceride with three C_{30} chains the ratio would be 3.7%). The average structural unit of this type of kerogen would therefore have approximately 80 carbon atoms. The straight carbon chains would have a tendancy to fold back on themselves, and the association between the units would be mainly by steric crosslinking. This entangling would also make it possible to trap some cyclic molecules with short side chains. The fact that the cyclic and long straight-chain molecules do not have chemical bonds would thus explain that there is no relation, either quantitative or in the average number of carbon atoms, between these two structural types: they would come from different biological precursors.

2. Type II: Example of Toarcian of Paris Basin.

The extract still represents a large share of the products of kerogen evolution, but heteroatomic compounds are abundant, which is not true in the preceding type. The hydrocarbon structure has a naphthenoaromatic tendency with a predominance of four and five cycle molecules. It can be supposed that the corresponding kerogen also consists of polycyclic molecules with an average of four to five cycles with frequent heteroatomic substitutions, especially of sulfur, in the cycles. The initial aromaticity must be low (of the order of a single aromatic ring per molecule, and this corresponds to the structure of the possible precursors such as the steroids which have a single unsaturation in the A ring). The straight carbon chains have essentially less than 20 carbon atoms, probably with a large number of branchings of the methyl or ethyl type. These chains can either be carried directly by the rings (isoprenoid chain of steroids) or be attached to the cycles by functional groups. Thus for the average structural unit there would be about 40 carbons atoms, i.e. two times less than in type I. The mode of association of the naphthenoaromatic molecules should be relatively close to that described for asphaltenes by Yen [56] i.e. interaction between π electrons of aromatics rings, coordination links between metals and heteroatoms, and compensation between deficit and excess of negative charges due to free electronic doublets of heteroelements. The bonding energy of these interactions must be relatively low, since solvents which are not polar or are very slightly so, like benzene or chloroform, are able to extract a high proportion of the heteroatomic compounds.

3. *Type III: Example of Logbaba Series (Upper Cretaceous, Cameroon).*

The extract never constitutes a large fraction of the evolution products of kerogen. Data concerning the kerogen structure of type III OM are therefore very fragmentary. The proof of this is that elemental analysis shows an aromatic structure for kerogen, while the hydrocarbons have a well represented linear chain fraction. It is also understandable that a kerogen with an aromatic nature is not very extractable since, even if the aromatic domains are relatively small, the interactions between π electrons make possible very stable associations which are resistant to solvolysis.

In this type of kerogen there is probably a small fraction of esters with long carbon chains (C_{25}-C_{35}) from plant cuticular waxes, which are simply juxtaposed to aromatic structures derived from lignin, but without any chemical bond between them as in the case of type I. The existence of a high proportion of oxygen in slightly evolved kerogen of type III, aside with the slight representation of heavy products in the extract, leads to the supposition that oxygen can be eliminated as labile products (CO_2, H_2O). It is therefore concluded that, in slightly evolved kerogen, oxygen falls into groups of the alcohol or phenol type, rather than into heterocyclic structures.

VI. CONCLUSIONS

A. Specific contributions of extract data to the structural determination of kerogen.

Throughout this Chapter we have attempted to reconstruct, in as detailed as possible a manner, the chemical structure of the various types of kerogen on the basis of analytical data obtained from extracts. It is however necessary not to make excessive generalization, and it should not be forgotten that the three main types of kerogen studied throughout this work are determined on the basis of samples from very special geological series. The concept of the frequency of the various types of OM thus defined should not be overlooked, since it is only really possible to speak of types if certain characteristics are found constantly.

As concerns this point, it should be noted that type I is quite rare and covers biochemical entities which are theoretically very different, such as a lacustrine deposit in which the bacterial reworking becomes preponderant (Green River Shales), as well as a concentrated deposit (we could almost say cemetery) of algae of the *Botryococcus* type (while algae with other origins would rather be classified in type II). Under these conditions, an extrapolation for properties other than aliphaticity could be venturesome. As we have seen for the Green River Shales, the study of extracts makes it possible to determine the average number of carbon atoms in the aliphatic chains, but such a method is by no means a general one and must be performed again in each case of type I organic material.

This is probably not true for types II and III, as these illustrate the biochemical difference which exists between aquatic and terrestrial OM. In order to colonize the land, the plants (which make up most of the biomass) needed supporting tissues (cellulose, lignin, etc.) and agents for protection against evaporation (cuticular waxes). These special needs, together with the chemical resistance of the constituents created for these purposes, are at the origin of the aromatic characteristics of type III kerogen and of the paraffinic characteristics of the corresponding extracts.

In the case of these two types of OM, aquatic and terrestrial, the study of extracts is of great interest since it leads to the concept of markers for the origin of OM. It thus makes it possible not only to give structural description of part of the carbon skeleton of the kerogen but, most important of all, to arrive at the biological precursors of the latter and so to extend the field of geochemistry to biochemistry. Knowledge of the chemical composition, functional groups and the stability of the molecules making up living matter, thus make it possible to form an idea of the way in which the latter may have been incorporated into kerogen, and then were expelled during catagenetic evolution.

B. Other components whose study could contribute to understanding of kerogen structure.

1. Gaseous compounds.

In the introduction to this Chapter, it was stated that the study of such compounds would not supply structural data on kerogen because of their simple formula. However this is not altogether the case. While their structure does not make it possible to reach a chemical formula, the timing of their genesis or, in other terms, the bonding energies which link them to kerogen, is significant. In type III, in particular, where only a small quantity of extract but apparently large quantities of H_2O, CO_2 and CH_4 can be obtained, the study of the degree of catagenesis necessary for obtaining these products, and their respective proportions, might perhaps make it possible to form a better idea of their mode of incorporation into kerogen. The case of methane, whose migration is facilitated by its small mass, high coefficient of diffusion and relative insolubility in water (unlike carbon dioxide), is significant. When it is found in shallow deposits, geochemists often hesitate between an early origin (bacterial degradation of slightly evolved kerogen before the main oil formation zone) and a deep origin (cracking of already formed liquid hydrocarbons). Knowledge of its mode of incorporation into kerogen, possibly by an indirect method such as that of the study of the isotopic distribution of carbon (Chapter 9), might lead to an answer since it is clear that the chemical bonds linking the precursors of methane to kerogen are not the same in the two hypotheses.

2. Heteroatomic compounds.

To date, few studies have taken into account all the heteroatomic products extracted from samples belonging to well defined types of OM. We have seen, however, that these compounds can represent half of the C_{15+} extract at maximum extract formation, and much more at a lower evolution stage. There should be further studies of this type of products, because their solubility gives them an advantage compared to kerogen in certain physical analysis methods, and much structural data could probably be collected in that way.

REFERENCES

1. Albrecht, P., Vandenbroucke, M. and Mandengue, M., (1976), *Geochim. et Cosmochim. Acta,* **40**-7, 791.
2. Altgelt, K.H. and Segal, L., (1979), "Gel Permeation Chromatography", Marcel Dekker, New York.
3. Baker, E.W., (1969), *in:* "Organic Geochemistry. Methods and Results", ed. G. Eglinton and M.T.J. Murphy, Springer Verlag, Berlin, 464.
4. Baker, E.W. and Smith, G.D., (1974), *in: Advances in Organic Geochemistry 1973,* ed. B. Tissot and F. Bienner, Editions Technip, Paris, 649.
5. Blumer, M., Guillard R.R.L. and Chase, T., (1971), *Mar. Biol.* **8**, 183.
6. Brown, J.K., Ladner, W.R. and Sheppard, N., (1960), *Fuel,* **39**, 79.
7. Burlingame, A.L. and Schnoes, H.K., (1969), *in:* "Organic Geochemistry. Methods and Results", ed. G. Eglinton and M.T.J. Murphy, Springer Verlag, Berlin, 89.
8. Castex, H., (1972), *Rev. Inst. Franç. du Pétrole,* **XXVII**, 219..
9. Castex, H., to be published, *in: Advances in Organic Geochemistry 1977,* Moscow.
10. Castex, H., Roucaché, J. and Boulet, R., (1974), *Rev. Inst. Franç. du Pétrole,* **XXIX**, 3.
11. Deroo, G., Powell, T.G., Tissot, B. and Mc Crossan, R.G., (1977), *Geological Survey of Canada Bull.,* **262**.
12. Dorsselaer, A. van, (1975), Thesis, Univ. of Strasbourg, France.
13. Douglas, A.G., (1969), *in:* "Organic Geochemistry. Methods and Results", ed. G. Eglinton and M.T.J. Murphy, Springer Verlag, Berlin, 161.
14. Durand B. and Espitalié J., (1972) *in: Advances in Organic Geochemistry 1971,* ed. H.R.v. Gaertner and H. Wehner, 455, Pergamon Press, Oxford, 455.
15. Ensminger, A., Thesis (1977), Louis Pasteur University, Strasbourg France.
16. Espitalié, J., Madec, M., Tissot, B. and Leplat, P., (1977), *OTC Pub. 2935,* Offshore Technology Conference, 439.
17. Fabre, M., Leblond, C. and Roucaché, J., (1972), *Rev. Inst. Franç. du Pétrole,* **XXVII**, 469.
18. Farrington, J.W. and Quinn, J.G., (1971), *Geochim. et Cosmochim. Acta, 35,* 735.
19. Forsman, J.P., (1963), *in:* "Organic Geochemistry", ed I.A. Breger, Pergamon Press, New York, 148.
20. Goodwin, T.W., (1973), *in:* "Lipids and Biomembranes of Eukaryotic Microorganisms", ed. J.A. Erwin, Academic Press, 1.
21. Heller, S.R., (1972), *Anal. Chem.,* **44**, 1941.
22. Hertz, H.S., Hites, R.A. and Biemann, K., (1971), *Anal. Chem.,* **43**, 681.
23. Hood, A. and O'Neal, M.J., (1959), *Anal. Chem.,* **31**, 1.
24. Huang, Wen Yen and Meinschein, W.G., (1976), *Geochim. et Cosmochim. Acta,* **40**, 323.
25. Huc, A.Y., Roucaché, J., Bernon, M., Caillet, G. and Da Silva, M., (1976), *Rev. Inst. Franç. du Pétrole,* **XXXI**, 67.
26. Hunt, J.M., (1974), *in: Advances in Organic Geochemistry 1973,* ed. B. Tissot and F. Bienner, Editions, Technip, Paris, 593.
27. Jonathan, D., L'Hote G. and Du Rouchet, J., (1975), *Rev. Inst. Franç. du Pétrole,* **XXX**, 65.
28. Kwok, K.S., Venkataragharan, R. and McLafferty, F.W., *J. Amer. Chem. Soc.,* **95**, 4185.
29. Leo, R.F. and Parker, P.L., (1966), *Science,* **152**, 649.
30. Lumpkin, H.E., (1958), *Anal. Chem.,* **30**, 321.
31. Lumpkin, H.E. and Aczel, T., (1964), *Anal. Chem.,* **36**, 181.
32. McCarthy, R.O. and Duthie, A.H., (1962), *J. Lipid Res.,* **3**, 117.
33. Mead, J.F., Howton, D.R. and Vevenzel, J.C., (1965), *in:* "Comprehensive Biochemistry", ed. M. Florkin and E.H. Stotz, Elsevier, Amsterdam, **6**, 1.
34. Metcalfe, L.D. and Schmitz, L.M., (1966), *Anal. Chem.,* **38**, 514.
35. Monin, J.C., Pelet, R. and Février, A., (1978), *Rev. Inst. Franç. du Pétrole,* **XXXIII**, 223.
36. Oudin, J.L., (1970), *Rev. Inst. Franç. du Pétrole,* **XXV**, 3.
37. Robert, P., (1971), *Rev. Inst. Franç. du Pétrole,* **XXVI**, 105.

38. Roucaché, J., Boulet, R., Da Silva, M. and Fabre, M., (1977), *Rev. Inst. Franç. du Pétrole,* **XXXII,** 981.

39. Snyder, L.R., (1971), *Proceedings Internat. Symposium Gas Chromatography,* **8,** 81.

40. Stahl, E., (1962), "Thin Layer Chromatography", Springer, New York.

41. Streibl, M. and Herout, V., (1969), *in:* "Organic Geochemistry. Methods and Results", ed. G. Eglinton and M.T.J. Murphy, Springer Verlag, Berlin, 401.

42. Supina, W., Kruppa, R. and Henly, R., (1967), *J. Amer. Oil Chem. Soc.* **44,** 74.

43. Swain, F.M., (1969) *in:* "Organic Geochemistry. Methods and Results", ed. G. Eglinton and M.T.J. Murphy, Springer Verlag, Berlin, 374.

44. Teichmüller, M., (1971), *Erdöl u. Kohle,* **24,** 2.

45. Thomas, T. and Mays, R., (1961), *in:* "Physical Methods in Chemical Analysis", ed. W. Berl, Academic Press, New York, **4,** 45.

46. Tissot, B., Califet-Debysser, Y., Deroo, G. and Oudin, J.L., (1971), *AAPG Bull.,* **55,** 2177.

47. Tissot, B. and Espitalié, J., (1975), *Rev. Inst. Franç. du Pétrole,* **XXX,** 743.

48. Tissot B., Durand, B., Espitalié, J. and Combaz, A., (1974), *AAPG Bull.,* **48,** 499.

49. Tissot, B., Pelet, R., Roucaché, J. and Combaz A., (1977), *in: Advances in Organic Geochemistry 1975,* ed. R. Campos et J. Goni, Enadimsa, Madrid, 117.

50. Tissot, B., Deroo, G. and Hood, A., to be published, *in: Geochim. et Cosmochim. Acta,* 1978.

51. Treibs, A., (1934), *Ann. Chem.,* **509,** 103.

52. Tulloch, A.P., (1976), *in:* "Chemistry and Biochemistry of Natural Waxes", ed. P.E. Kolattukudy, Elsevier, Amsterdam, 235.

53. Vandenbroucke, M., Albrecht, P. and Durand, B., (1976), *Geochim. et Cosmochim. Acta,* **40,** 1241.

54. Weete, J.D., (1976), *in:* "Chemistry and Biochemistry of Natural Waxes", ed. P.E. Kolattukudy, Elsevier, Amsterdam, 349.

55. Welte, D.H. and Ebhardt, G., (1968), *Geochim. et Cosmochim. Acta,* **32,** 465.

56. Yen, T.F., (1974), *Energy Sources,* **1,** 447.

14

Origin and formation of organic matter in recent sediments and its relation to kerogen

A.Y. HUC*

I. INTRODUCTION

In old rock formations, kerogen, which is defined as the fraction that cannot be extracted by organic solvents, comes close to the chemical preparation: «kerogen», obtained by the complete destruction of the mineral phase. Successive acid hydrolyses to eliminate carbonates and silicates and to concentrate the organic matter (OM) have hardly any affect on the latter and even serve to restore it under good conditions without appreciably modifying its overall structural properties [32], (Chapter 2).

For recent sediments, the concept of kerogen thus defined and chemical preparations overlap only in exceptional cases. Whereas it is always possible to define a kerogen entity that is «non-extractible by organic solvents», such kerogen can seldom be isolated from these sediments without considerably altering it. Acid treatments, which can so successfully be applied to older formations, result in a great loss of OM, for recent sediments (30 to 70%) [23, 25, 26, 47].

Under the circumstances, the usual definition of kerogen as **sedimentary OM insoluble in current organic solvents,** could best be adhered to.

This definition has indeed the advantage of implying a genetic link between the bulk of the insoluble OM deposited in a basin and the fossilized insoluble OM set in rock formations. Hence an understanding of kerogen in recent sediments cannot be based on a single, direct and overall analysis but must involve a variety of techniques such as:

(a) Optical analysis, electron microscope investigation and electronic microdiffraction, which yield information both on shaped organic constituents (algal debris, fecal pellets, lignin-bearing remnants, spores, etc.) and on organic fragments already having gone through one or more sedimentary cycles [20].

---

*Institut Français du Pétrole. Rueil-Malmaison, France.

(b) Chemical analyses based on degrading techniques (acid or alkaline hydrolyses, etc.) or on fractionating methods by extraction (hydrosoluble, alkali-soluble, acid-soluble).

The great diversity in the nature and concentration of OM in sedimentary formations has been known for a very long time. It serves as a basis for quite a number of classifications, both petrographical [54, 1] and physico-chemical [31, 106]. This diversity can be explained in terms of deposition environment. It presupposes a solid understanding of the processes controlling the amount and quality of the OM deposited in different sedimentary contexts. Recent environmental models should certainly be considered because their parameters may be evaluated advantageously. The data thus acquired should then be completed and discussed in the light of the alterations the OM may have undergone in the course of fossilization.

Beyond its immediate paleo-ecologic and sedimentological signifiance, this approach can also be expected to help in understanding problems raised during oil prospection on the scale of an entire basin. It should lead, in the end, to predicting, for a given sedimentary environment, the location, nature, amount and prospective interest of potential source rocks.

Unfortunately, based on our present knowledge, it seems premature to formulate a real synthesis concerning the genesis of source rocks. The data available are still too dispersed and their representativeness is poorly defined.

This Chapter therefore pretends only to make an inventory of observed facts related to questions raised by the organic geochemistry of recent sediments:

(a) What is the origin of the OM?
(b) What are its properties and how have they been acquired?
(c) Where and why does it accumulate?
(d) How does it evolve during diagenesis?

The study is based partly on a review of references on biological, oceanographic and ecologic phenomena, etc. which may explain the qualitative and quantitative diversity of organic deposits on earth, and partly on more specific studies into the nature of the diagenesis of recent sedimentary OM.

II. ORIGINS OF SEDIMENTARY ORGANIC MATTER

The primary source of sedimentary OM is living organisms; they contribute to it by their excretions and secretions, and later by their dead bodies.

The most abundant weighted components of this material of biologic origin are carbohydrates, proteins, lipids and lignin (Fig. 14.1) which vary greatly for different organisms (Fig.14.2).

Fluctuations within a single species exist in addition to the diversity between species; the former are sometimes of importance mainly due to environmental influences and the physiological state of the organisms.

Zenkevich, cited by Bordovskiy [10], Lee *et al.* [57] have pointed out, for instance, that the lipid content of numerous planktonic species tends to increase when going from the equatorial towards the temperate and cold regions. Cane *et al.* [15] found that *Botryococcus braunii* (a fresh water alga) has a lipidic composition which varies according to its physiological state: most of the lipid material in the resistant form consists of isoprenoid unsaturated hydrocarbons (Botryococcene), while under active development it consists essentially of linear alkadienes (which are probably at the origin of coorongite).

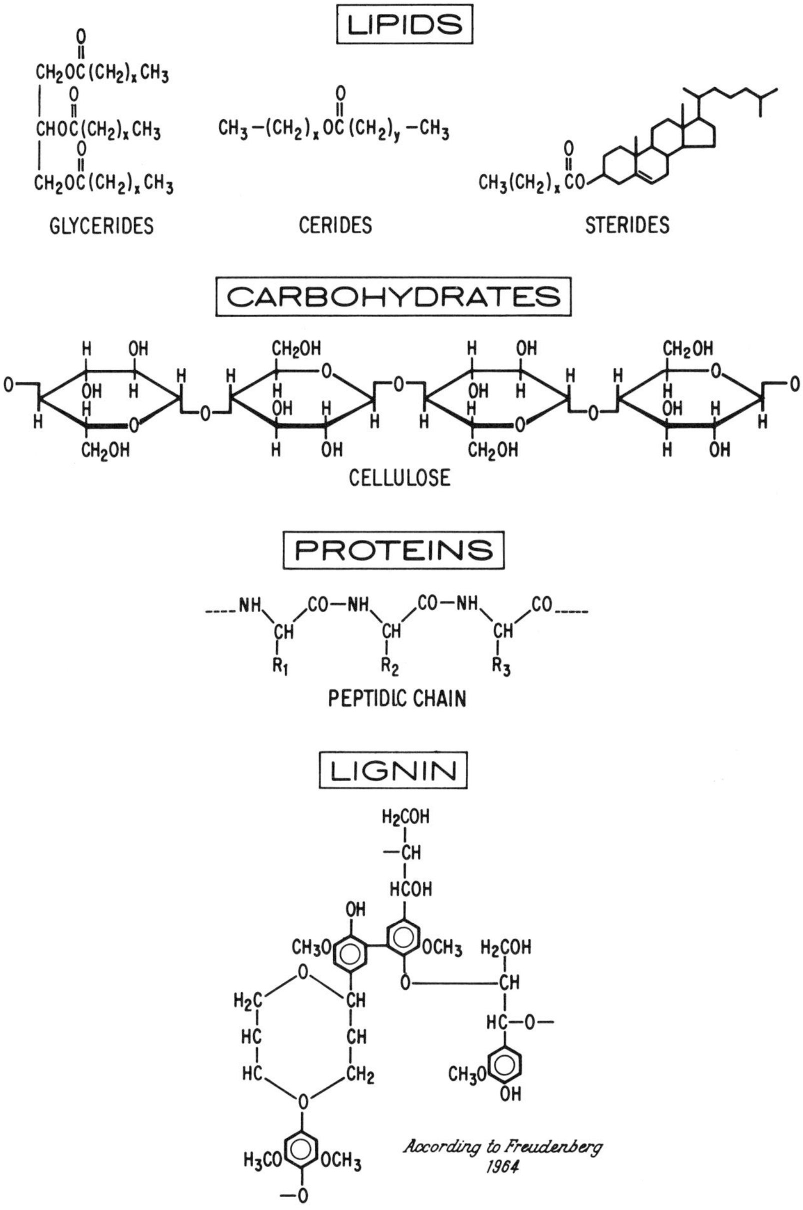

LIPIDS

GLYCERIDES CERIDES STERIDES

CARBOHYDRATES

CELLULOSE

PROTEINS

PEPTIDIC CHAIN

LIGNIN

According to Freudenberg 1964

Fig. 14.1. — Structures of some biological molecules.

However, as pointed out by Debyser *et al.* [24] the species differences might not be reflected as much in the chemical properties of the sedimentary OM as the bulk chemical properties of biocoenosis, i.e. organisms gathering together as the result of various environmental factors.

On the continent, it is mainly climate which determines the major bio-geographical regions (temperate forests, steppes, etc.), while superimposed local factors govern the growth of natural associations more or less limited in space (e.g. peat bogs) [22, 69]. In water, the essential factors which control life are first light, limiting the primary (photosynthetic) productivity in the oceans to the euphotic zone (0-200 m), and secondly available nutrients [24, 22, 69].

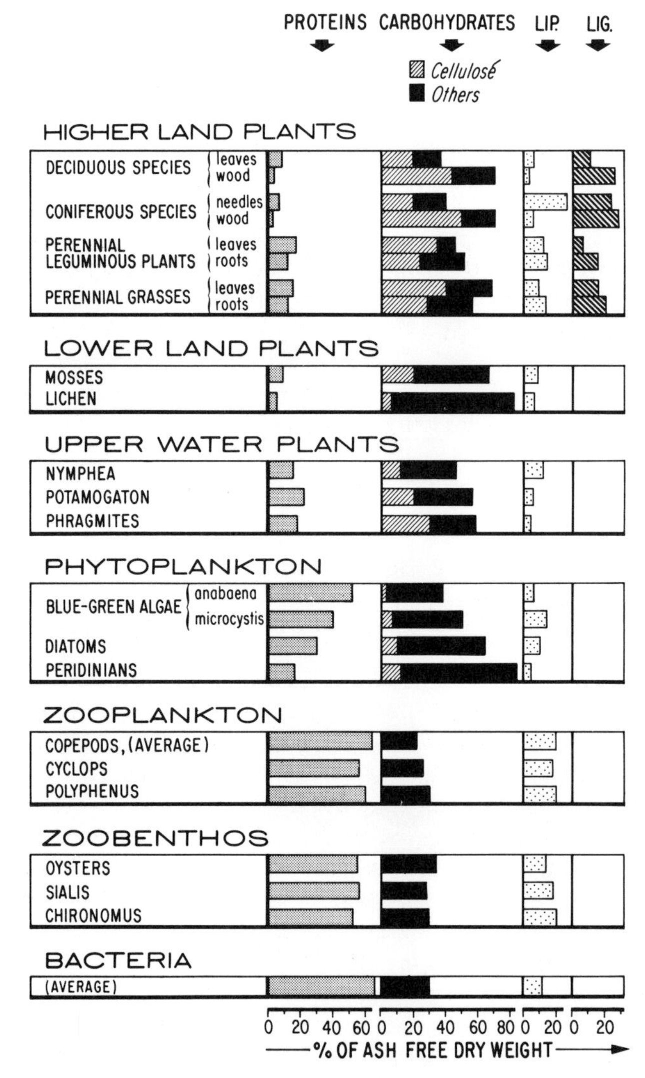

Fig. 14.2. — Comparative chemical composition of some organisms. (from data cited by Kononova [55], Manskaya *et al.* [62], Bordovskiy [9]).

In general, a classic distinction is made between the terrestrial and the marine domains. The former is characterized by the presence of higher plants, rich in lignin, the latter by the presence of organisms which do not contain lignin (Fig. 14.2).

This difference, however valid in general, should be looked at more closely. Although a high lignin content is the rule under tropical and temperate climates, it changes with higher latitudes where swamps and peat bogs, the great primary producers of OM on continents with low temperatures, contain plants with less and less lignin the colder the climate gets (moss peats).

Besides the chemical constituents which are liberated after the organisms die, biological activity produces organic substances (excretion and secretion) which could play an important role in some cases. Sieburth *et al.* [87] made a study of some littoral algae which produce great quantities of exudates of a polyphenol, protein and carbohydrate nature. Thomas [104] believes that phytoplankton excretes 5 to 10% of its photosynthetic production (excretions including simple components such as amino acids and sugars), and Baraskov, mentioned in Bordovskiy [10], thinks that diatoms discharge, during their life, more than 10% of their dry weight in the form of lipids. Fecal pellets may contribute an important even major part of the organic supply; this is the case in some tropical lakes [11], for the Autun bogheads in France which contain admirably preserved fishes and reptiles coprolites [6], or for some coastal areas around the Gulf of Mexico where they are the source of carbon-rich biogenic sediments [74].

Quantitative data concerning masses of living organisms and their biogenic productivities are essential for understanding and assessing the extent of organic sedimentation processes.

The difficulties inherent in assessing the biomass and the small number of measurements now available prevent a general reliable worldwide picture form being given, as shown by the scattering of published figures which range from 30 to 300 $\times$ 10^{10} t of organic carbon [48, 75, 107, 111, 42, 18, 115].

However, data on the primary productivity are more coherent notwithstanding the diversity of bibliographic references:

(a) 1.5 to 7.0 $\times$ 10^{10} t of organic carbon/year in the oceans [66, 91, 92, 113], (Winberg cited by Whittle [113]) [82], (Bogorov, [8] cited by Whittle, [113]).

(b) 1.4 to 7.8 $\times$ 10^{10} t of organic carbon/year on the continents [3, 8, 22, 91], (Whittaker *et al.* cited by Dajoz [22]).

The photosynthetic production of OM is very irregularly distributed on the surface of the earth; it depends on different ecosystems (Figs. 14.3 and 14.4). On the continents, organic production is by far the more important in tropical zones. In the oceans, the greater part of OM is in general produced in humid temperate zones and in the higher latitude regions. The main characteristic of oceanic productivity is that it is restricted to the euphotic zone, i.e. the layer of water in which sunlight penetrates in sufficient quantity for the development of photosynthesis. The thickness of this privileged zone varies substantially. It amounts to 200 m in the open oceans, to some 50 m in coastal waters and to only a few meters in lagoons.

These depths are primarily a function of the concentration both of organic particulates (living or inert) and of mineral products in suspension. The important role which the geological settling plays (orogeny, weathering, etc.) thus becomes immediately apparent. It determines the supply of detritus (dissolved salts, mineral load) hence the turbity of the water in a basin and, in part, its fertility or, in other words, the sum of fundamental components making up its organic productivity.

On a scale as vast as the general history of the earth, the horizontal extension of the euphotic zone, where life is to be found, has undergone enormous variations. Periods of major transgressions, when open seas invaded the continental shelf, stand out most favourably in respect to organic production (Middle Cretaceous). General regressive periods (Triassic), on the contrary, with few epicontinental seas and with oceans restricted to deep troughs, correspond to minima in organic production [24, 105].

The supply of nutrient salts (nitrates, phosphates) is essential to life in the oceans. A good correlation exists between maps showing, on the one hand, the distribution of phosphates and nitrates in superficial water layers and, on the other hand, the fertility of ocean water.

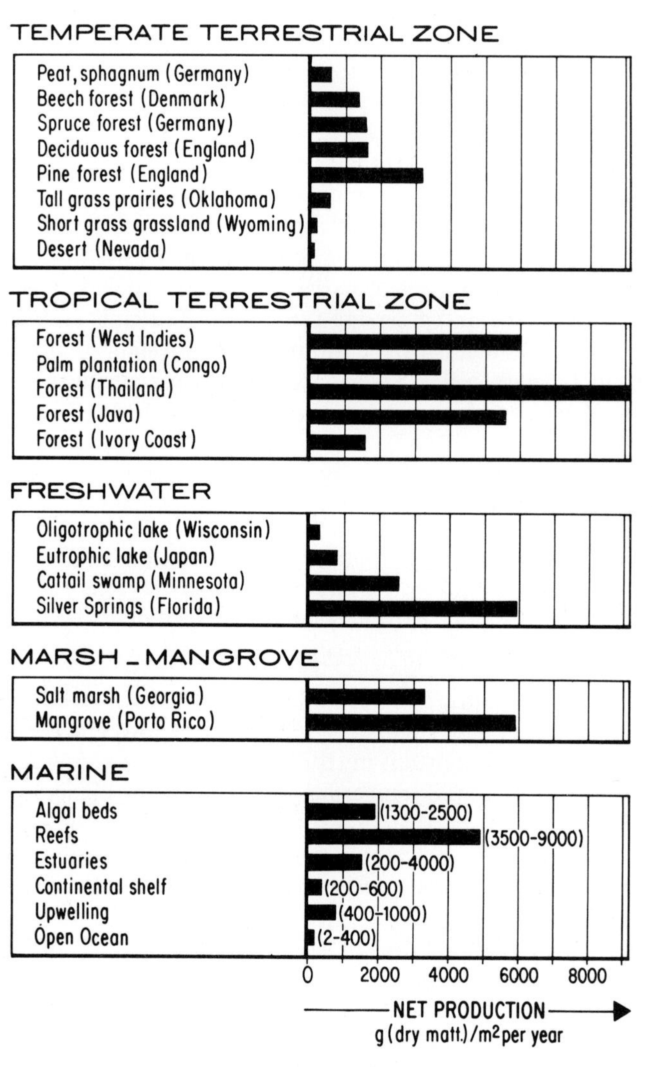

Fig. 14.3. — Primary production of some natural ecosystems (from data from Odum [69], Dajoz [22], Whittle [113]).

This clearly demonstrates that the development of planktonic life goes hand in hand with the presence of nutrient salts (Hentschel, cited by Debyser *et al.* [24]). Because organisms consume these nutrient salts the concentration of the latter in superficial waters is generally poor; it increases considerably with depth where the waters benefit from a supply due to the mineralization of organic particles settling down from the surface. At higher latitudes, superficial waters possess temperatures, and thus densities, which differ little from water at depth, and mixing is therefore easier. In tropical and equatorial areas, on the other hand, the higher temperatures of superficial waters provoke appreciable differences in density, and this favours a more distinctive layering of the waters and prevents the fertilization of the euphotic zone by bottom water. Locally, however, ascending currents of deeper waters, rich in nutrient salts, exist and this results in a notable increase in fertility at these latitudes, to which should be added propitious sunlight conditions for photosynthesis. These upwelling phenomena correspond to divergent zones between two currents or between one current and the coast where predominant or even permanent trade winds move the superficial waters towards the open ocean, with this being compensated for by deeper waters moving upward. These upwelling zones are a frequent occurrence close to the west coasts of continents (Africa, California, Peru, etc.). Such regions are known for a sedimentation which is exceptionally rich in organic

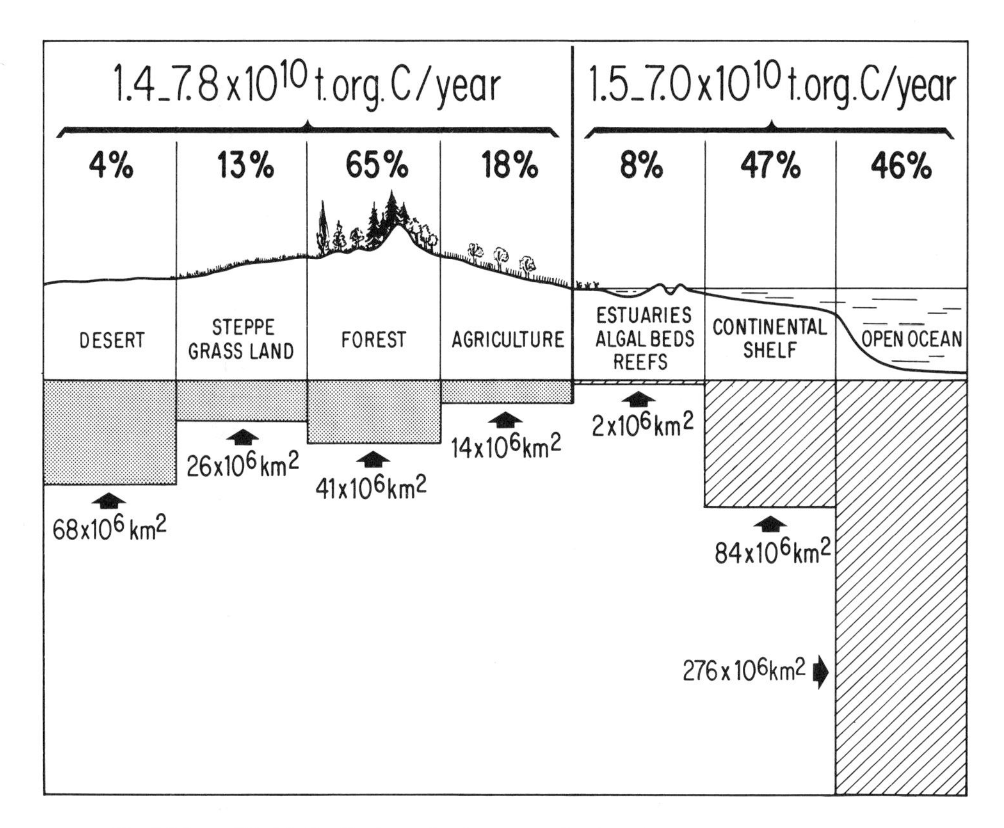

Fig. 14.4. — World distribution of primary production (from data from Odum [69], Dajoz [22], Whittle [113]).

carbon; in the deposits offshore from Angola (Walvis Bay) for instance, the organic carbon content may reach up to 30% of the sediment [14].

III. TRANSITION TO ORGANIC MATTER IN SUPERFICIAL FORMATIONS

Between the release of biological products (deads organisms, excretions) into the surrounding environment and their incorporation in the sediment, a great many physico-chemical and biological factors intervene which will affect their chemical structure and also determine their spatial distribution within the deposits.

A. Transformation processes of organic matter.

Biological molecules undergo modifications of which the resulting products could be present in the form of various components, in particular:

(a) **Products of decomposition:**
- . Simple molecules, resulting from the complete decomposition of OM: CO_2, CH_4, NH_3, H_2S, H_2O, etc.
- . Or molecules resulting from the selective decomposition of biological precursors the original structure of which they will preserve, fully or in part (heritage): amino acids, peptids, simple sugars, polycarbohydrates, fatty acids, lipids, phenols, acid phenols, lignin, etc.

(b) **Products of neogenesis:** Complex components, synthesized «*de novo*» from the former ones, at various stages of their decomposition.

1. Decomposition.

Agents responsible for this decomposition are of a physical nature (laceration of the tissues by rain, hail, wind, hydrodynamic movements, etc.), a chemical nature (oxygen, pH, light, etc.) or biochemical nature (enzymes, bacteria, fungi, burrowing organisms). This destructive action will depend on the specific resistances of the different biochemical entities, largely determining their relative accumulation within the natural environment.

Compounds most prone to decomposition are first proteins which are destroyed by enzymes (not very specific ones and therefore quasi ubiquitous) and transformed into amino acids destined, in turn, to be mineralized, and, second, carbohydrates which are equally accessible to numerous enzymatic reactions that decompose them into simple sugars. The enzymatic hydrolysis of carbohydrates resembles the proteolytic decomposition in that the transformation of the polymers into individual units or into shorter chains requires a minimum amount of energy. Because of their fragility, these structures are almost totally eliminated under geochemical conditions. Amino acids and sugars are generally not abundant in old rocks and only exist as a result of exceptional trapping mechanisms and conservation circumstances [101, 28].

Lipids and lignin, on the other hand, are generally more resistant to decomposition. Indeed, lipids, the ester links excepted, possess only C – C bonds the cleavage of which necessitates considerable energy. Lignin, consisting of polymers and copolymers of phenyl-propenyl alcohols (Fig. 14.1), decomposes only very slowly under attack by specific organisms (lignolytic fungi in particular).

In addition to its specific resistance, OM may be protected from decomposition by a number of factors inherent to the environment :

(a) The lack of oxygen limits the oxidation reactions and prevents aerobic organisms from developing; more specifically, lignolytic fungi cannot grow and lignin is thus preserved [108]. As an example, we can mention the perfect conservation of lignin structures of leaves in a sedimentary formation, sampled offshore from the mouth of the Amazon, in 3 000 m of water [25].

Observations made long ago already stressed the point that the most favourable sedimentary environments for the conservation of OM are waters with a restriction in oxygenation. The classic example is the Black Sea where waters below a depth of 200 m do not contain oxygen anymore. Anaerobic processes come into play; in particular seawater sulphates are reduced, the concentration in H_2S increases and the environment becomes strictly anaerobic.

This type of mechanism develops in embayments having great water depths (Kaoe Bay), in oceanic trenches (Cariaco Trench), in closed or semi-closed basins (Black Sea, Abidjan lagoon); it is controlled by the physiography of the sea floor in conjunction with the fact that the water masses possess a distinct layering. Sediments associated with those environments in general contain a higher than average organic carbon content:

Kaoe Bay	3- 4%
Black Sea	1- 5%
Abidjan lagoon	4-12%
Cariaco Trench	3- 4%

In lacustrine environments, where such layering of waters frequently occurs because they are stagnant, the concentration of organic carbon may reach very high values, frequently in excess of 5% [24].

(b) The lack of nitrogen compounds is a limiting nutrient factor for many microorganisms. Absence of nitrogen-containing substrates is reflected, amongst others, by a slowing down of the decomposition of cellulose [30]. Conversely, too great an abundance of nitrogen compounds and of nutrients salts in general could trigger the eutrophisation process in closed basins (lakes, closed seas), with particularly favourable consequences for the preservation of OM (proliferation of surface plankton and hence an important supply of OM to deeper waters, resulting in the consumption of dissolved oxygen and the development of anaerobic processes with production of H_2S).

(c) The presence of toxic products in the form of metabolites, discharged into the environment by numerous organisms, could inhibit some bacterial developments [2]; in isolated environments in particular, bacteria intoxicate themselves. In reducing environments, the presence of H_2S adds its toxic effet to the absence of free oxygen. A great concentration of humic substances (peat bogs, some lacustrine water) could also play an antibiotic role [100]. This is an important characteristic from a geological point of view because it helps in preserving unstable organic compounds associated with this humus. One of the most astonishing

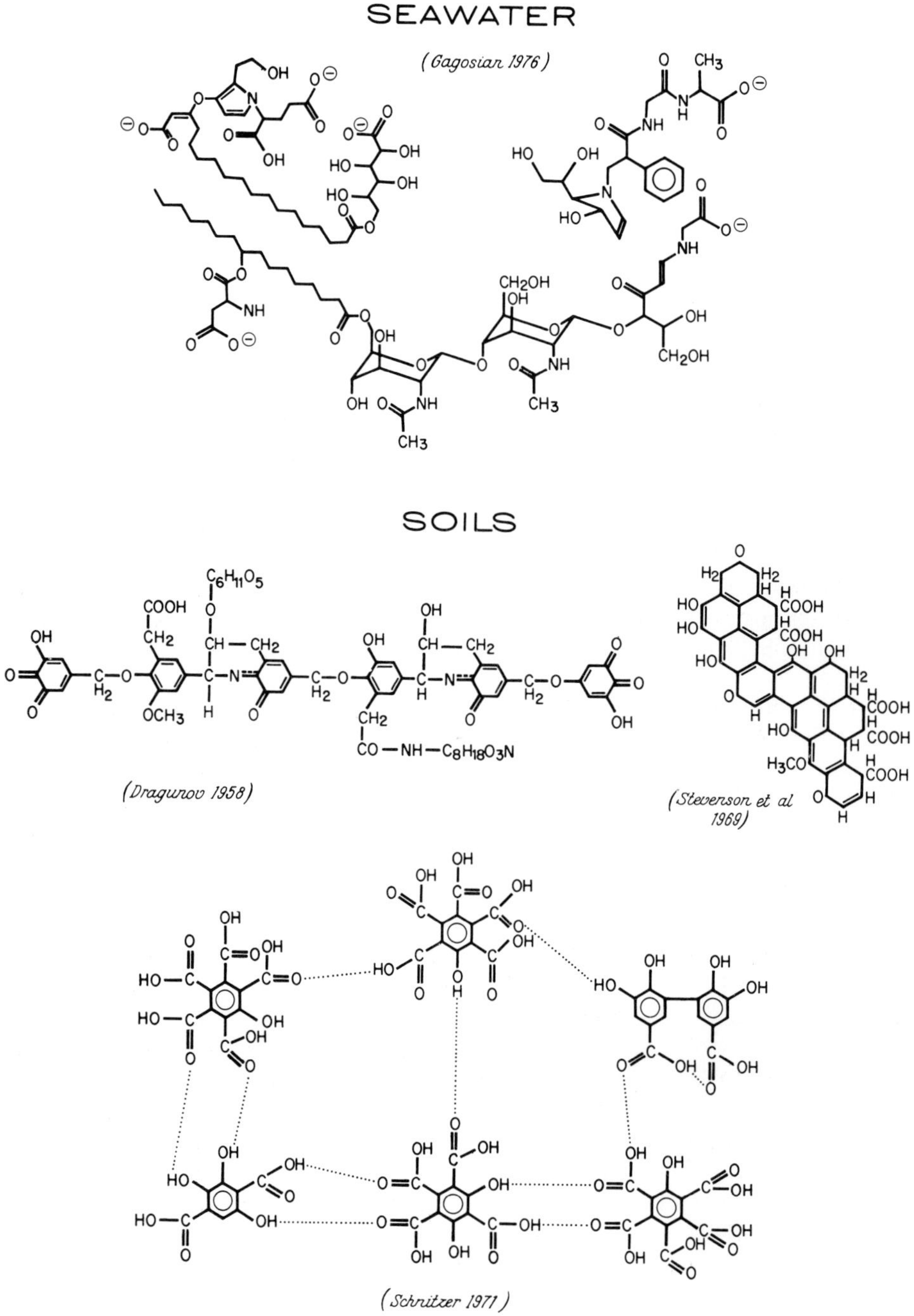

SEAWATER

(Gagosian 1976)

SOILS

(Dragunov 1958)

(Stevenson et al 1969)

(Schnitzer 1971)

Fig. 14.5. — Hypothetical structures of humic substances [36, 27, 83, 95].

examples in this respect is the organic remains (human bodies, clothing, etc.) preserved intact for several thousand years in peat bogs in Denmark [37].

(d) Physical conditions, notably temperature and pressure, are known to be responsible for the decrease in bacterial activity in deeper waters [51]. Low winter temperatures also cause a slowing down of microbial activity in many soils.

2. *Neogenesis.*

The alteration of biogenic OM leads to the release of more or less decomposed compounds with a lower molecular weight than their precursors, even to monomers (phenols, amino acids, simple sugars, fatty acids, etc.). All these products will interreact and form complex structures, such as humic substances, consisting of polycondensed nuclei, supporting carbonaceous chains and functional groups (COOH, OCH_3, NH_2, OH, etc.), joined together by hetero-atomic bonds (carbonyl, carboxyl, sulfur, ether, peptidic linkage, etc.) or by C – C bonds [83, 27, 95, 36] (Fig. 14.5) The processes leading up to the creation of these structures have been the subject of numerous soil investigations, known under the general term of humification. It is, in fact, a polycondensation, depending on the reciprocal reactivity of the different available organic units (the composition of the parent material plays a role) in association with the physicochemistry and biology of the environment which will favour or, conversely, retard these reactions, either directly by influencing their kinetics (temperature, pH, humectance, presence of enzymes) or indirectly by providing a medium for the reaction (presence of a liquid phase) or else by ensuring the preservation (or the destruction) of intermediate neogenetic products.

The main reactions which have been put forward to account for humification phenomena make use of condensations of the phenol-phenol type [35], phenol-nitrogen compounds[38], nitrogen compounds-sugars [60] and, more recently, phenols-fatty acids [84] (Figs. 14.6 to 14.9). Because there are so many phenol compounds (lignin decomposition [35], flavenols [21], microbial synthesis [63, 39]) and because they are so resistant, pedologists believe that they play an essential role in the formation of humic substances. Phenol-phenol reactions are probably caused by oxidative condensations, notably ones involving quinones as intermediate steps [35] (Fig. 14.6).

Oses-amino acids polycondensation [60, 44] (Fig. 14.8) is considered to be a subordinate phenomenon in soils [98]. However, this humification process, which is also called the Maillard reaction, could exist in a subaquatic environment when autochthonous OM is being supplied, which is thus devoid of lignin but rich in proteins and carbohydrates, especially since these reactions develop in a liquid medium and are favoured by the pH conditions occurring in seawater and sediments (pH 8 approximately) [41, 51]. The way in which hydrophobic organic compounds of the lipid type participate during the formation of humic substances is not very well understood. Ester links could exist between phenol compounds and fatty acids (Fig. 14.9), in conjunction with adsorptions in which hydrogen bonds and Van der Waal's forces play a role [84].

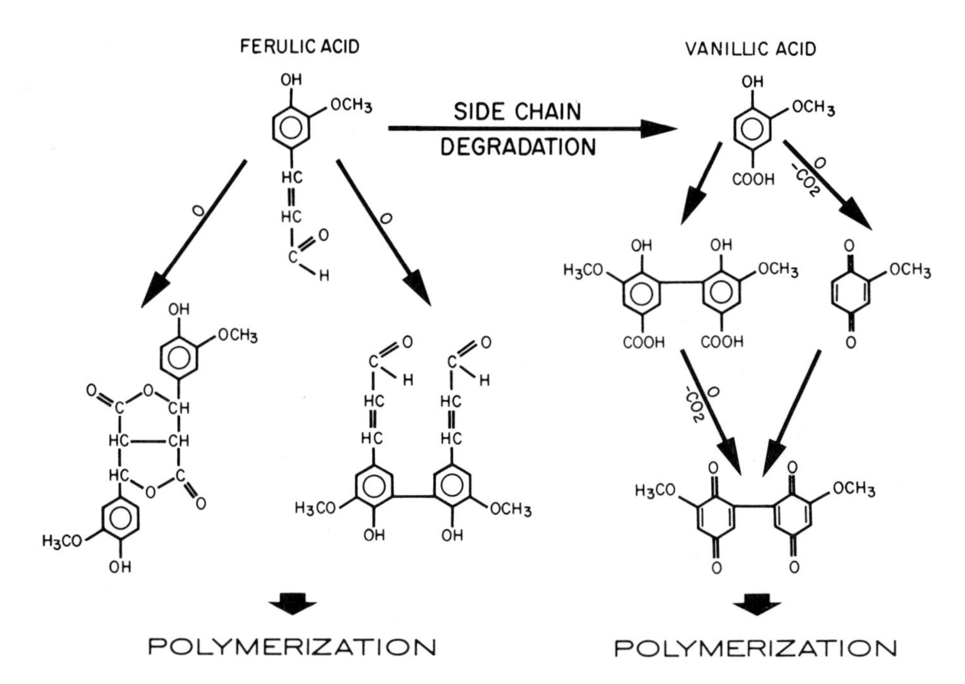

Fig. 14.6. — Formation of polymers by phenol-phenol reactions [35].

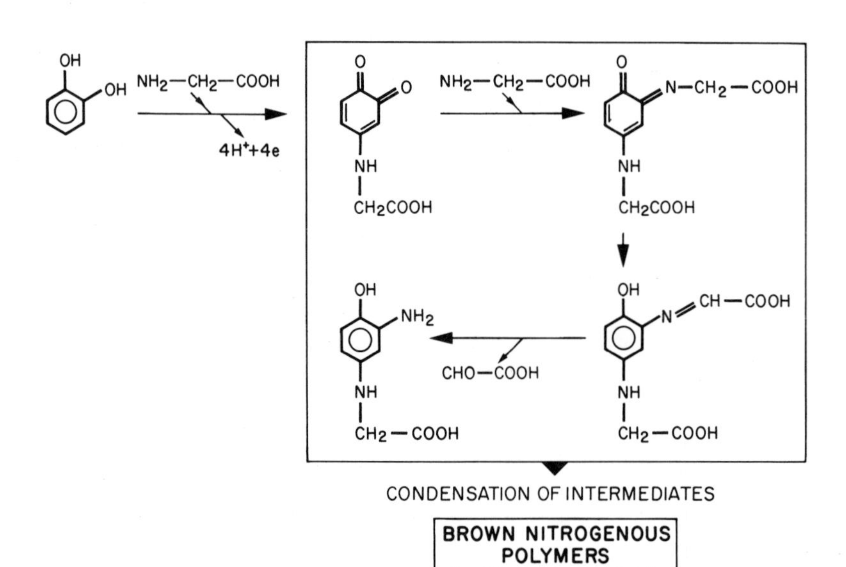

Fig. 14.7. — Formation of polymers by condensation of amino acids and phenols (Swaby *et al.* [99] cited by Stevenson *et al.* [95]).

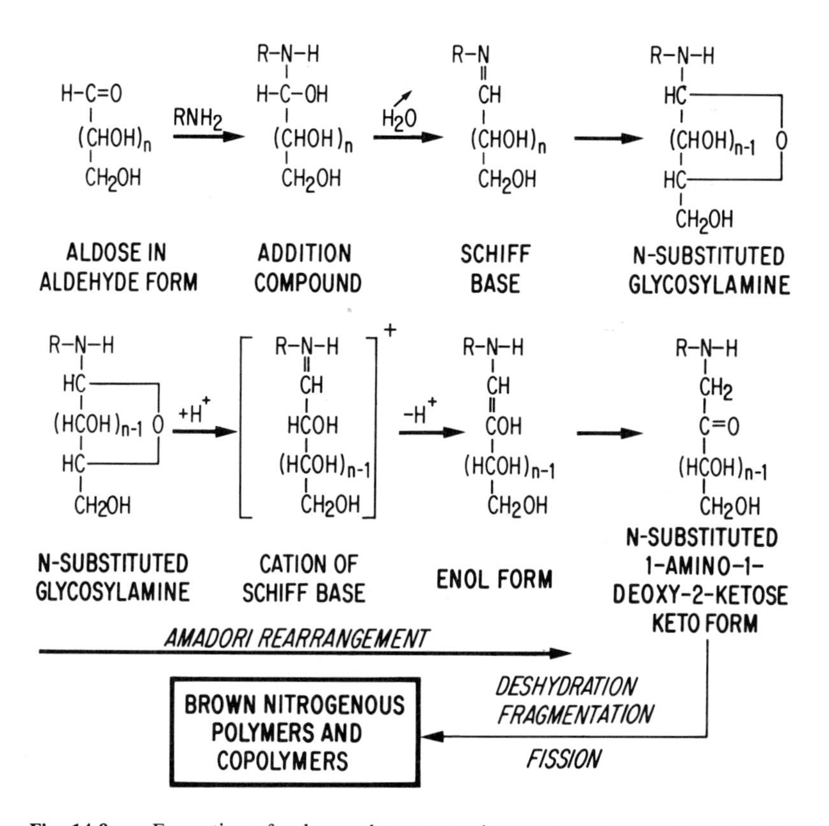

Fig. 14.8. — Formation of polymers by sugar-amine condensation reaction [60, 44].

Fig. 14.9. — Phenol-fatty acid ester in humic substances [84].

B. Role of environment.

The influence both of the nature of biologic precursors and of the conditions which exist in the medium, on the structure of OM in superficial formations may be determined by studying the physico-chemical properties of humic substances which form a major fraction of it.

An elemental analysis of humic acids [62, 55, 46, 23] (Fig. 14.10), for instance, distinguishes among OM from peat, chernozem ([1]), podzolic soils ([2]) and subaquatic sediments (lacustrine or marine, in which the origin of the organic constituents has been recognized as being mainly autochthonous) [23, 46, 50].

The differences noticed may be explained, taking into account :

(a) The quality of the parent OM:
- The absence or presence of lignin has, a priori, an influence on the relative importance of the aromatic elements (a high H/C ratio for precursors devoid of lignin, i.e. subaquatic sediments).
- The content of protein compounds, characterizing humic acids stemming from aquatic organisms by their high N/C ratio.

(b) The habitat: seasonal humectance and desiccative alternations of soils favour the condensation of structures in the case of chernozems (low H/C ratio), while low temperatures and acid conditions slow down the humification processes in podzols (high H/C ratio).

Infrared (IR) spectroscopy provides supplementary information by defining the nature of the functional groups associated with humic acids. The saturated CH groups (2 900 cm^{-1}) are more abundant in subaquatic sediments [93], slightly less so in podzols, while they are particularly scarce in peat and chernozem (Fig. 14.11).

The peptid bonds responsible for the 1 540 cm^{-1} band (Amide II) and taking part in the 1 600 cm^{-1} massif (Amide I) are clearly discernible in subaquatic samples for which the N/C ratios proved to be highest (Fig. 14.11).

The oxidation method in air, developed by Mazumdar [65] on coals and applied by Wright *et al.* [116] and Ishiwatari [49] to humic acids can be used to make an evaluation, though non precise, of the relative proportion of aliphatic and aromatic structures in humic molecules, i.e. and overall picture of their carbonaceous skeleton. The results obtained (Fig. 14.12) confirm the idea which the elemental analysis and IR spectroscopy had suggested, namely the clearcut aliphatic nature of the humic acids of subaquatic sediments and the clearcut aromatic nature of humic acids in chernozem, with the humic acids in podzols occupying an intermediate position.

Finally recent investigations on nuclear magnetic resonance ^{13}C by Stuermer *et al* [97] and Dereppe (personal communication) reveal a greater abundance of resonances due to aliphatic carbon in the case of humic substances of marine origin, while conversely, for humic substances of terrestrial soils, resonances from aromatic and olefinic carbons predominate.

([1]) Chernozem : neutral soils, occurring in continental climates, characterized by pronounced alternations of dry and humid seasons; they develop underneath a vegetation of the steppe or forest-steppe type [30].

([2]) Podzolic soils : soils whose evolution depends on the presence of a very acid humus with slow decomposition, either because of climatic conditions (cold regions) or because of particular environmental factors, notably the presence of an acidifying vegetation (heather, conifers) releasing great quantities of organic acids [30].

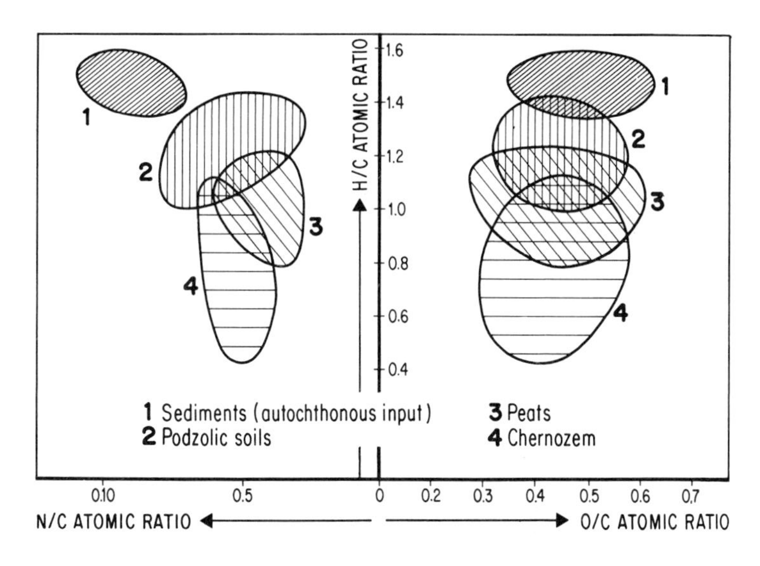

Fig. 14.10. — Humic acids from various origins in H/C vs. O/C and H/C vs. N/C diagrams (from data from Manskaya *et al.* [62], Kononova [55], Huc *et al.* [46], Debyser *et al.* [23]).

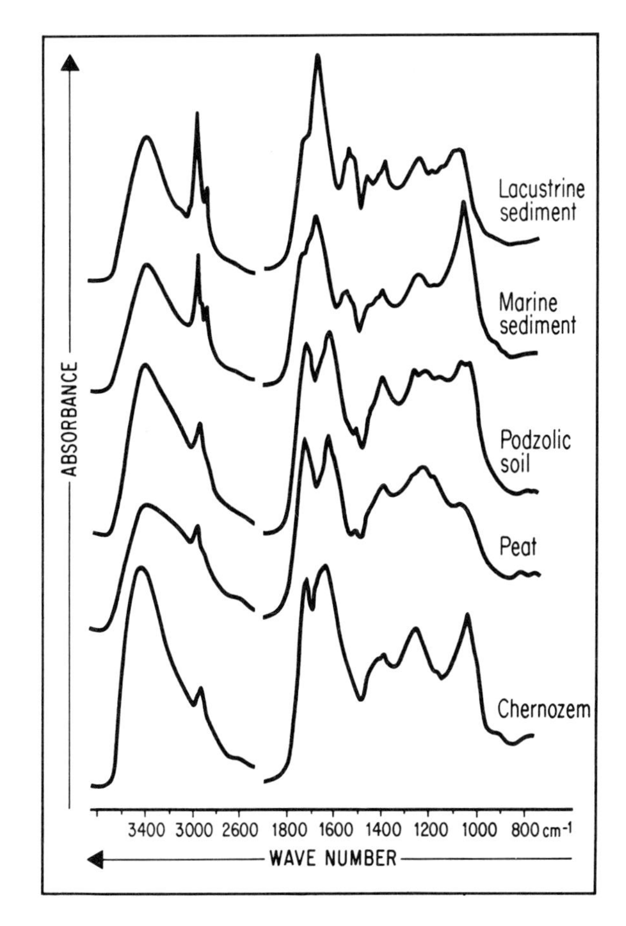

Fig. 14.11. — IR spectra of humic acids from various origins (from data from Huc *et al.* [46], Debyser *et al.* [23] and unpublished data).

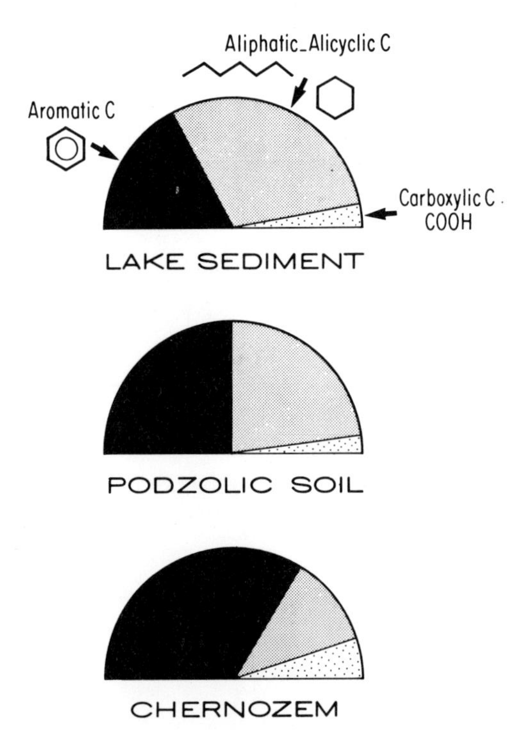

Fig. 14.12. — Comparative distribution of carbon in humic substances (from data from Wright *et al.* [116], Ishiwatari [49]).

C. Transportation of organic matter.

OM, either originating directly from the biomass or having already undergone various modifications (humification, catagenesis, etc.), is taken up by transporting agents and carried towards sedimentation centers where it is distributed. Sedimentary OM may, *a priori,* be considered as the result of a mixture (in various proportions) of a fraction of the biomass existing in the basin itself, and of an allochthonous fraction coming from the emerged surroundings.

The supply of allochthonous OM to a basin results from weathering, erosion and transportation phenomena, i.e. it depends on the physiographic and climatic conditions of the drainage area. The main supply agents of OM of terrestrial origin are glaciers, atmospheric phenomena (wind, rain), and essentially rivers:

(a) Glaciers, by abrading, wear away detrital material (mineral and organic) which is later discharged during melting. However, the organic supply role wich ice plays is generally considered to be very restricted and localized [87].

(b) The effects of eolian transportation are very poorly understood but probably affect arid or semi-arid regions. Studies on atmospheric particles collected over Atlantic Ocean far offshore have revealed the existence of OM (in particular shaped constituents from higher plants, e.g. cuticles of leaves) whose origin would have to be sought in African lakes which periodically dry up and from which plant debris are possibly torn away and then transported by trade winds [87]. However, the exact way such material penetrates into sedimentary basins and the amounts of OM involved are far from having been elucidated. Mention should nevertheless be made of the estimates made by Sellers [85], Duce [29], Blanchard [7] and Romankevitch [80] and which give figures varying from 3 to 15 $\times$ 10^8 t of organic carbon/year. This is a considerable mass which is certainly greatly overestimated according to the authors themselves.

(c) River transportation has been the subject of innumerable studies determining the annual amount of OM discharged into sedimentary basins at from 1 to 5 $\times$ 10^8 t of organic carbon [115, 66, 110, 29, 80].

The great rivers of the earth such as the Amazon, the Hudson, the Mississipi, the McKenzie and the Danube carry approximately 2 to 15 mg/l of organic carbon [61, 80], consisting primarily (70 to 90%) of matter with a high molecular weight of the type of humic substances [40, 4].

Because the qualitative and quantitative properties of river transported OM, and the geological and climatic conditions of drainage basins are interdependent, it would seem justified to treat these phenomena in a much larger context such as proposed, for instance, by Erhart [34] in his biorhexistasy theory.

During a rhexistasy phase which corresponds to an orogenic or climatic imbalance of the environment, intense erosion occurs and the transportation of superficial material (mineral and organic), essentially in particulate form, takes place without any great qualitative selection. OM tends to be torn away as a whole without any preference being given to any given fraction on account of its physico-chemical properties.

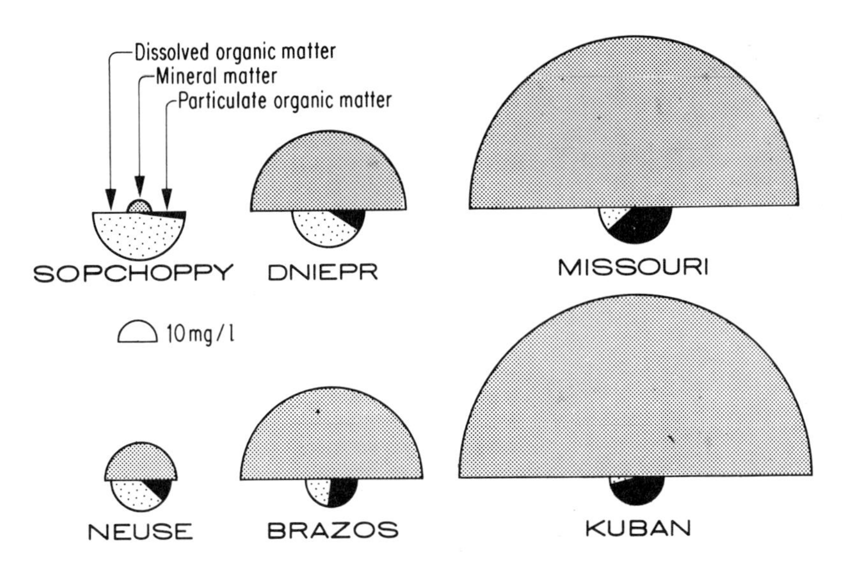

Fig. 14.13. — Organic and mineral loads of some rivers (from data from Lopatin [59], Strakhov [96], Shimkus *et al.* [88], Skopintsev [90], Malcom *et al.* [61]).

On the other hand, during the biostasy phase the continent is in equilibrium and the erosion products feeding the rivers are mainly restricted to the most soluble fractions (organic and mineral) of the superficial formations. This phenomenon is best illustrated by comparing the organic content of some rivers, e.g. the Dniepr (USSR), the Brazos (Texas), the Neuse (North Carolina) and the Sopchoppy (Florida), which discharge into the sea after having streamed through extensive lowland areas (even only coastal swamps in the case of the Sopchoppy). Their organic load is characterized by the importance of the dissolved phase, as opposed ro rivers like the Missouri (USA) and the Kuban (USSR) which run through mountainous regions and in which particulate-form OM by far predominates [59, 96, 88, 90, 61] (Fig. 14.13).

Under biostatic conditions, a selective wearing away of OM can be imagined on the basis of its nature, with the fractions carried away being made up of the compounds that are the least condensed and the richest in hydrophilic groups. It is evident therefore that the outgoing supply of OM from emerged land areas should not be considered as a global phenomenon but that each specific geological setting of the continent bordering a sedimentary basin and the prevailing climate there will be reflected in an original, qualitative and quantitative nature of the OM transported into a sedimentary basin.

D. Sedimentation of organic matter.

The existence of depositional models where OM is in the process of being synthetized in great abundance, right at the place where it has come to rest, such as peat bogs, various lagoons in which real algal mats are being formed (Mormona Lagoon in Baja California, [73]; Khor el Bazam in Abu Dhabi, [17]), some lakes in which phytoplankton proliferates (Lake Coorong in Australia, [19]; Lake Baloe in USSR, [17, 19]), explains the formation of organic-rich rocks such as coals, kukersites, tasmanites and torbanites [19, 107] as an *in situ* accumulation whose dynamics is quite well understood.

Except for these particular cases which represent only a few percent of the bulk of all sedimentary OM (Chapter 1), the sedimentation of the OM present in a dispersed form in deposits (with a content of less than 10%) is a complex problem that is far from being resolved. Romankevitch [80] has published a compilation map showing the concentration of organic carbon in superficial worldwide oceanic-bottom (Fig. 14.14). This map emphasizes the extremely low average organic carbon content of the majority of the seabeds but also the existence of some privileged zones with high concentrations, notably the rim of continents, boreal zones and interior seas. For deposits rich in organic carbon to be formed, the biological material, or at least its decomposition or neogenesis products, should both be able to reach the bottom in sedimentary basins and to do so under conditions preventing it from being dispersed over large areas; on the contrary, it should be concentrated even further.

To reach the seabed, organic substances should evidently possess sufficient hydrodynamic properties (size and specific gravity) to overcome those forces preventing their settling (buoyancy, water turbulence, etc.). This appears to be a simple remark but it is enough to exclude all dissolved OM from the sedimentation process, as long as it remains in such form, as well as a siezable proportion of OM even in the form of particulates (a current with a velocity of only 0.0002 cm/s is enough to keep a particle of 2 μ in suspension).

Except for those regions with very shallow waters and/or being highly anoxic, organisms after death likewise have little chance of participating directly in the enrichment of sediments

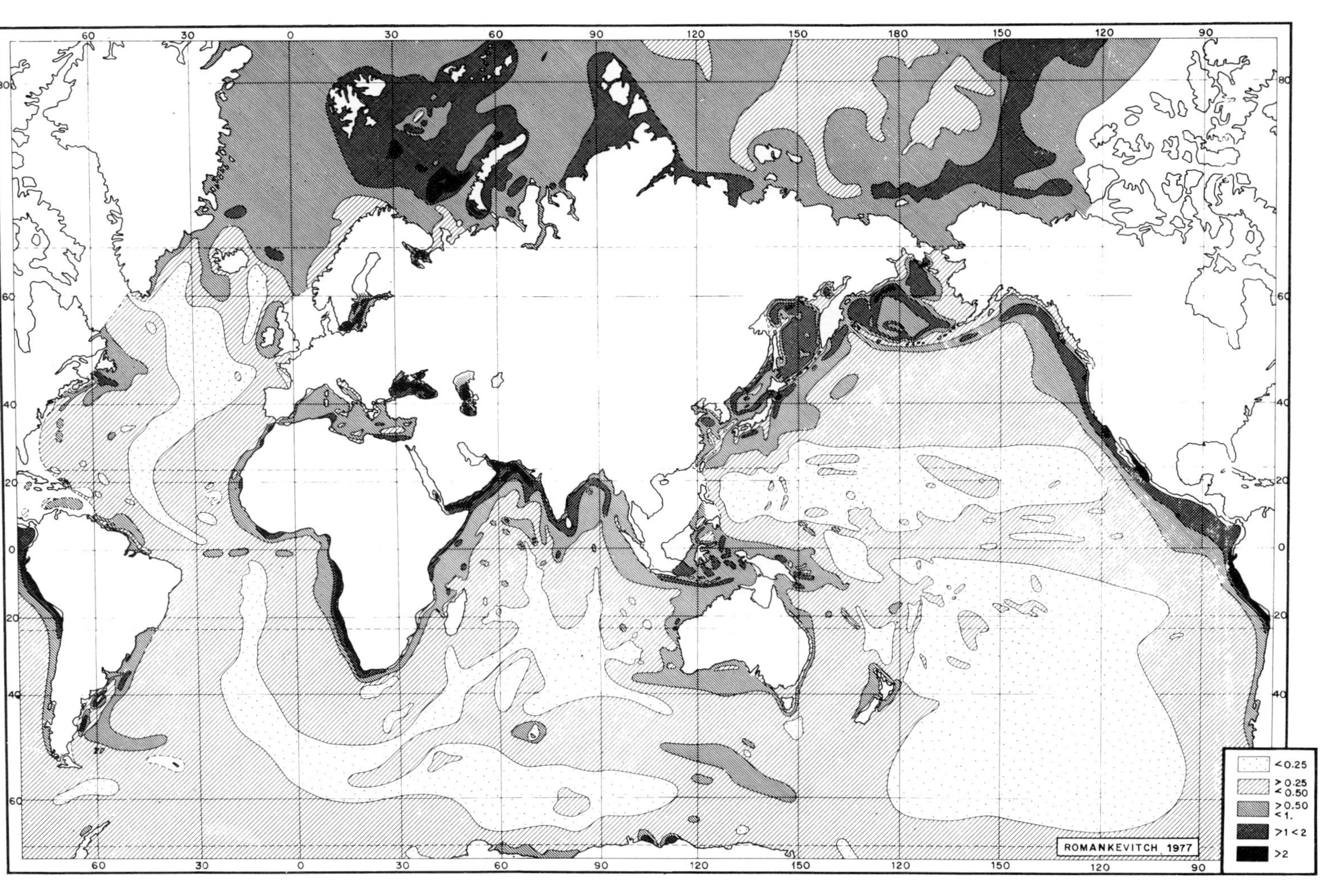

Fig. 14.14. — World distribution of modern sedimentary carbon [80].

in OM. Indeed, during their fall through the water, they are attacked by heterotrophic organisms which feed on organic fragments and by dissolution and oxidation phenomena. Menzel *et al* [67] estimate that almost the entire primary production is recycled within the first 200 to 300 m. Studies on the sedimentation of foraminifers (Belyayeva in Bordovskiy [10]) show that the number of tests devoid of OM increases rapidly with depth: 2% from 0 to 200 m, 10 to 13% from 200 to 500 m, 20 to 26% from 500 to 1 000 m, and most tests are empty below 1 000 m. Consequently, in most cases for OM to reach the sediment it must exist in a form other than the bodies of organisms or as the dissolved or particulate phase in forms too tiny to settle.

In fact, OM is rapidly transformed into material with a high molecular weight of the humic compound type [64]. It may coalesce with mineral products and thus constitute a particulate phase which in turn could assemble into aggregates, thereby combining an increase of their settling velocity (1-2 m/day for particles $> 30\ \mu$) [43, 78], with a greater resistance to degrading agents [70].

This flocculation into aggregates is controlled by the mineralogical environment (clays in suspension), the concentration of salts in the medium, the Van der Waals forces, and the frequency of collisions between particles [76].

Flocculation in deltaic or estuarine environments is certainly one of the most important means for OM to settle. The organic load in rivers, consisting essentially of humic substances [40, 4] (in colloidal or particulate form) is in equilibrium with the chemism of river water. The moment this water mixes with sea-water, this equilibrium is broken, essentially because of the increase in salinity, causing flocculation and the formation of organo-mineral aggregates with a tendency to settle [102, 89, 87]. Such organo-mineral aggregates have indeed been described in the superficial parts of coastal sediments [52].

In the open sea, in general, the formation of humic substances with colloidal properties favouring flocculation is certainly an appreciable factor for aggregates to take shape in a medium where all other elements are in equilibrium. Some authors believe that foam and air bubbles are privileged centres at the air-water interface for such aggregates to form [78].

In addition to these physico-chemical phenomena, OM can be entrained down through the water by means of the trophic cycle. Superficial photosynthetic production serves indeed as a nutrient for primary consumers (zooplankton) and during the nychthemeral migrations of the latter they can be entrained down to depths of several hundred metres where they serve in turn as a source of energy for secondary consumers, with a production of fecal pellets rich in OM (1.5 to 10% organic carbon) [74] at each stage in the food chain.

This organic supply by means of fecal pellets is far from negligible. In waters suitable for production it has been evaluated that from 20 to 30% of phytoplankton material is redistributed in this way after ingestion by zooplankton (Steele in Whittle, [113], Butler *et al.* [13]).

Research by Pryor on fecal pellets from *Callianassa* and *Onuphis* stresses the resistance of these structures during transportation. Moreover, from a hydrodynamic point of view, their size and weight resemble that of fine to medium-grained sand, thus perpetuating rapid settling through the water [114].

When in transit towards the seabed, OM faces a number of biological (trophic cycles) or physical factors such as differences in density of water (temperature, salinity) and turbulences (permanent currents, tidal currents, wave effects, etc.), from which it may be deduced that the enrichment of OM in a sediment will only exceptionally reflect the euphotic productivity in a vertical zone directly above it (Fig. 14.14). In most cases, the logic of organic depo-

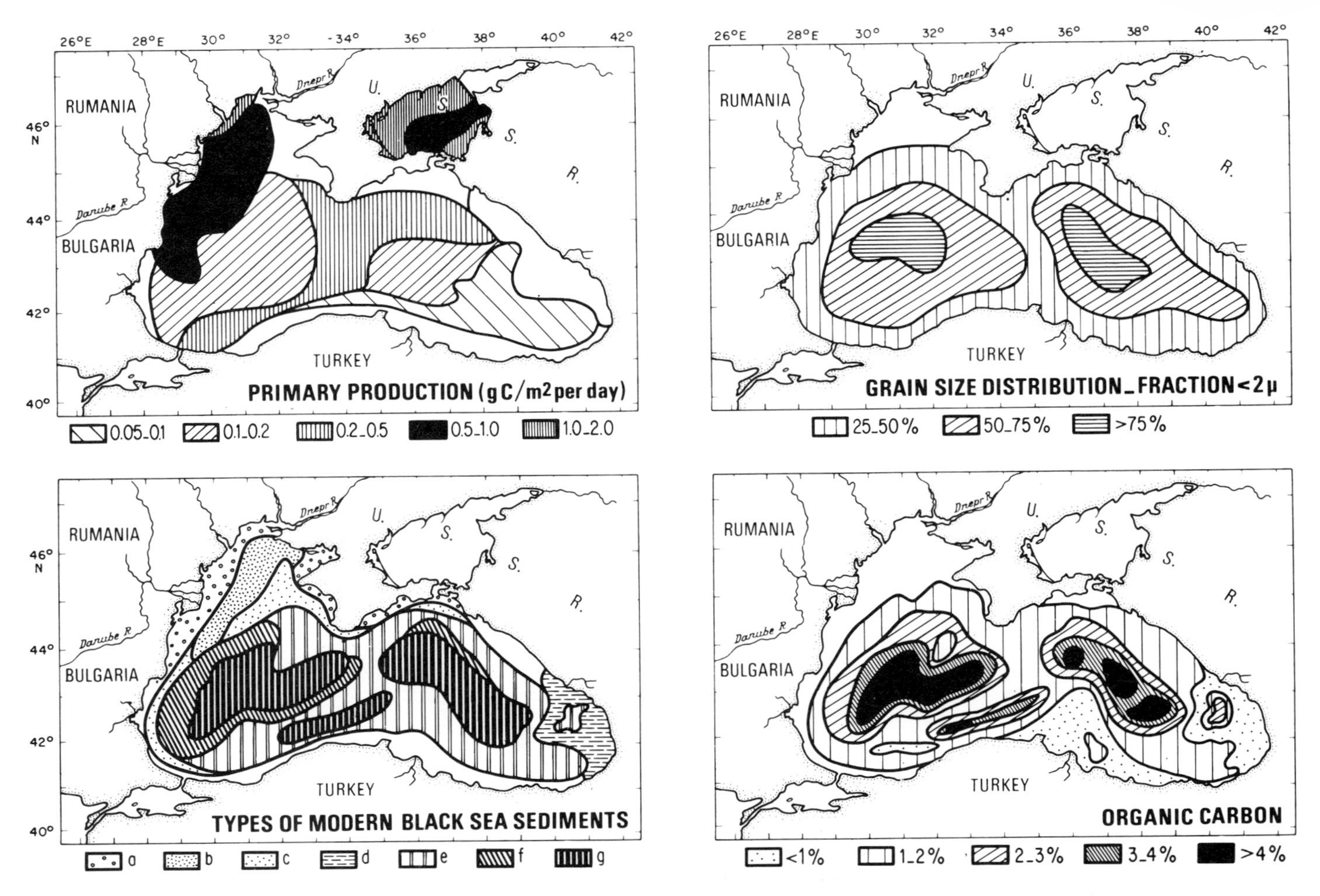

Fig. 14.15. — Maps of primary production, grain size distribution, types of modern sediments and organic carbon distribution in the Black Sea [88, 68]. Types of modern sediments:

a. Organogenic-clastic, very coarse-grained and coarse-grained sediments. **b.** Carbonate-rich shelly sediments ($CaCO_3 > 50\%$). **c.** Carbonate-poor and carbonate-bearing organogenic-terrigenous mytilid muds and phaseolina muds ($CaCO_3 = 10\text{-}50\%$). **d.** Carbonate-free terrigenous sediments ($CaCO_3 < 10\%$). **e.** Carbonate-poor organogenic-terrigenous muds ($CaCO_3 = 10\text{-}30\%$). **f.** Carbonate-bearing organogenic-terrigenous, finely dispersed coccolith muds ($CaCO_3 = 30\text{-}50\%$). **g.** Carbonate-rich, finely dispersed coccolith muds ($CaCO_3 > 50\%$).

sits will be the result of the inherent physico-chemical properties of the OM or its transporting agents with respect to the sedimentary environment. A look at the maps showing the concentration on the sea-bottom of both organic carbon and mineral deposits in a basin, such as the Black Sea, stresses the close links that exist between the organic carbon content and the nature (grain size, mineralogy) of the sediments (Fig. 14.15) [88, 68].

The relation between organic carbon content and finest grain-size sediments is well known, for both old rocks [48, 81, 16] and recent sediments [10, 103, 33].

This behaviour may be explained on the one hand by the colloidal nature of OM which is identical to that of clays and, on the other hand, by the links which are preferentially made with small-size minerals (notably clays) and which closely link the sedimentary destiny of OM to that of the smallest fractions of the mineral matter.

The following low-energy sedimentary environments are thus privileged depocenters of OM:

(a) The deepest parts of closed basins: Lake Ontario [103]; the Black Sea [88]; Lake Maracaïbo [77]; the Caspian Sea [9].

(b) Depressions in the continental shelf: Walvis Bay [70]; the Bering Sea [58]; California Basin [33].

(c) The base of the continental slope on the edge of the oceans [58] (Bezrukov *et al. in:* Bordovskiy [10]).

(d) Deltaic beds in shales separating more-sandy higher-energy channels (Mahakam Delta, Indonesia) where the clayed lentils indeed contain up to 4 times more organic carbon than the sandy channels (Deroo, personal communication).

This general rule of OM being associated with fine-grained sediments is nervertheless not absolute. We have seen that fecal pellets possess hydrodynamic properties which resemble that of more coarse-grained particles, and their sedimentation is thus associated with sandy deposits. Pryor [74] has even observed a segregation into two types of fecal pellets produced by animals living in the same ecological habitat offshore from Florida (*Callianassa* and *Onuphis*). The fecal pellets released by *Callianassa Major* were transported and deposited within medium-grained sands (0.5-0.125 mm) on a high-energy beach, while the fecal pellets released by *Onuphis Microcephala* were carried and deposited farther away in a lower-energy zone with fine-grained sands (0.21-0.125 mm).

IV. INFLUENCE OF EARLY DIAGENESIS ON THE PROPERTIES OF SEDIMENTARY ORGANIC MATTER

Once incorporated in muddy sediments, OM will gradually be buried further under the effect of continued sedimentation. With increasing distance from the seawater-sediment interface, the probability diminishes that it will again be brought into suspension and redistributed by movements of the water mass (wave action, tidal currents, longshore currents, turbidity currents, etc.). Within the sediment the physico-chemical and biologic environment will then gradually be modified (compaction, decrease in water content, disappearance of bacterial activity, qualitative transformation of the mineral phase, increase in temperature). During this diagenesis, OM will undergo transformations and will acquire properties such as

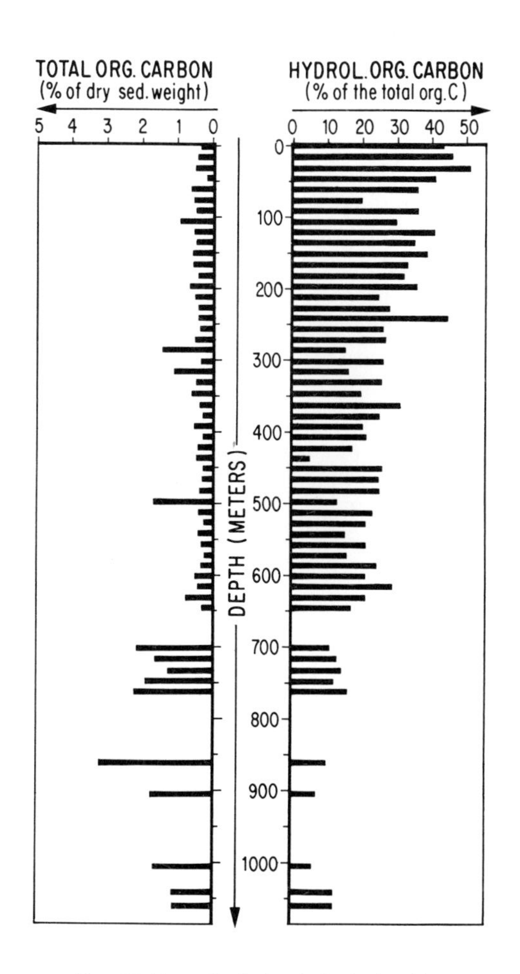

Fig. 14.16. — Variation in hydrolyzable organic carbon during diagenesis in Black Sea cores (Deep sea drilling project [DSDP] Leg 42 B).

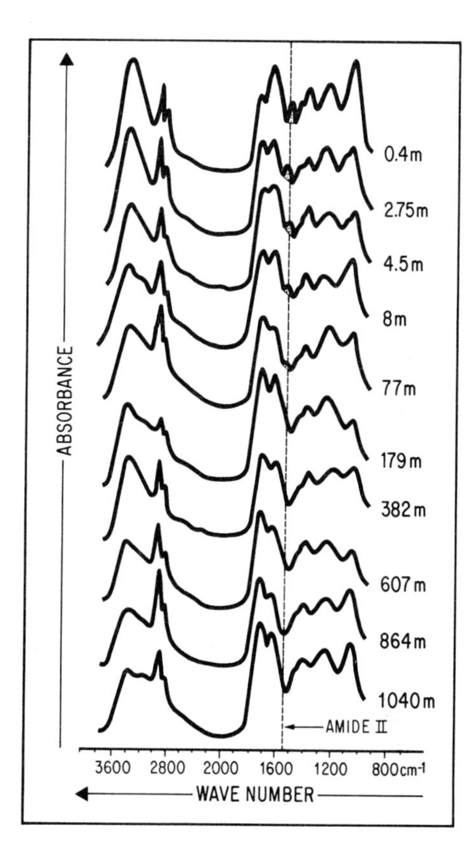

Fig. 14.17. — Variations in IR spectra of humic acids during diagenesis from Black Sea Cores (from data from Pelet *et al.* [72], Huc *et al.* [47]).

OM is known to possess in lithified old rocks. We have already stressed that OM from recent sediments and sedimentary rocks behaves differently under acid hydrolysis (paragr. I).

	Hydrolyzable carbon (%)	Humic carbon (%)
Recent sediments (10 samples)	35 – 75	25 – 40
Toarcian sediments (8 samples)	0 – 2	0.4 – 2.2

This difference is illustrated in the above table in which the percentage of hydrolyzable carbon is compared in samples from subrecent sediments and from the Toarcian in the Paris Basin (immature) with regard to supposed equivalent sediments as far as the nature of the OM is concerned (autochthonous marine origin). This increase in resistance to hydrolysis during natural evolution has also been followed by studying several cores from recent sediments in the Bering Sea [10], from offshore Brazil [25] and from offshore Mauritania [26] and even from long core samples taken in the Black Sea during Leg 42 B of the JOIDES (*Joint Oceanographic Institutions for Deep Earth Sampling*) project [47] (Fig. 14.16).

Naturally, this general observation may be related on the one hand to the decrease in simple organic compounds, i.e. amino acids, sugars, etc. during diagenesis [53, 12, 79, 94, 112], and on the other hand to the destiny of the peptid bonds in humic acids in recent sediments whose alteration and complete disappearance may be followed, for instance, by the behaviour of the 1 540 cm^{-1} band (Amide II) in IR spectroscopy [45, 47, 72] (Fig. 14. 17).

This phenomenon apparently does not correspond to any orderly modification of the organic carbon content and would seem to indicate that the evolution does not progress from a relative concentration of the most resistant material through a destruction of the most unstable matter, but instead from a restructuration of the OM during which the initially hydrolyzable fraction probably establishes resistant links with the rest of the organic phase. Modifications of OM by diagenesis have been carefully studied in coals which are perfectly suited to direct examination because of their low mineral products content [54, 32]. One of the outstanding features of the geological evolution of coals is the progressive decomposition of the hydrophilic functions, i.e. hydroxyl ($-OH$), carboxyl ($-COOH$), methoxyl ($-OCH_3$) and carbonyl ($C=O$) groups, and the disappearance of humic acids whose solubility in alkalis is in part associated with the existence of these hydrophilic groups.

When the OM is dispersed within a mineral matrix, it would seem that these phenomena are fundamentally the same, although an overview of the organic content is lacking. In particular, the fossilization of OM is accompanied by a decrease in the humic fraction [45, 46, 56] (Fig. 14.18). These phenomena could be interpreted here as being the result of an increase in organo-mineral interactions and/or a relative decrease in the functional hydrophilic groups of the OM. The decrease in the O/C atomic ratio, observed when going from humic acids in recent sediments to "fossil" humic acids, or when following their evolution in transition sediments (Fig. 14.19), may indeed be related to such an impoverishment in functional oxygenated groups.

Apart from these characteristics, some properties of OM pass the diagenetic stage without notable modifications. This is the case for the relative richness in aliphatic and alicyclic structures which on the whole seem to be preserved. For coals, for instance, this richness, estimated by means of the H/C ratio and the importance of the infrared 2 920 cm^{-1} band) (satura-

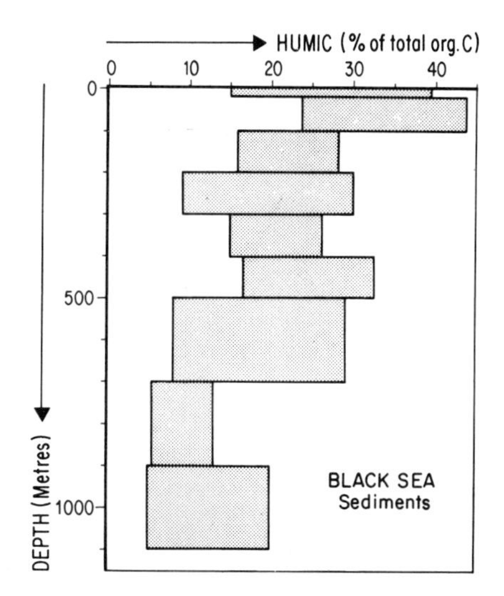

Fig. 14.18. — Variations in humic compounds content during diagenesis in Black Sea cores (from data from Pelet *et al.* [72], Huc *et al.* [47]).

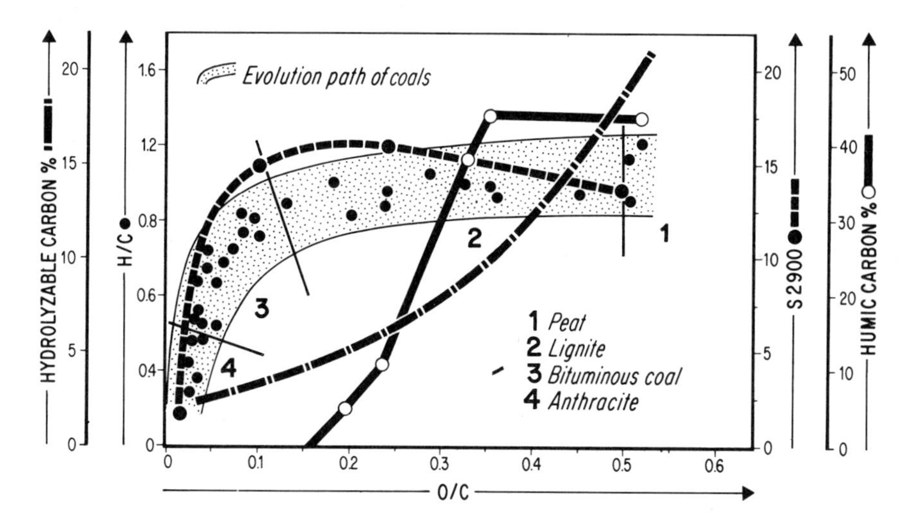

Fig. 14.19. — Inflence of diagenesis on elemental composition of humic acids from various origins, and variation of elemental composition of humic acids during diagenesis in Black Sea cores (from data from Pelet *et al.* [72], Huc *et al.* [47]).

ted CH groups), remains almost at its initial level until the stage of bituminous coal (Fig. 14.20) [54, 32]. Similarly, sedimentary humic acids, either recent or old, the moment they have an equivalent origin, show remarkable analogies for these parameters (H/C, IR 2 920^{-1} band) (Figs. 14.19 and 14.21) [46]. This remanant property of the saturated CH groups during the first stages of evolution is not contradicted by what has been observed on the long core samples collected in the Black Sea. Occasional fluctuations in the humic acid aliphatic properties are evidently due to differences in nature of the organic matter at the time of deposition (Fig. 14.17).

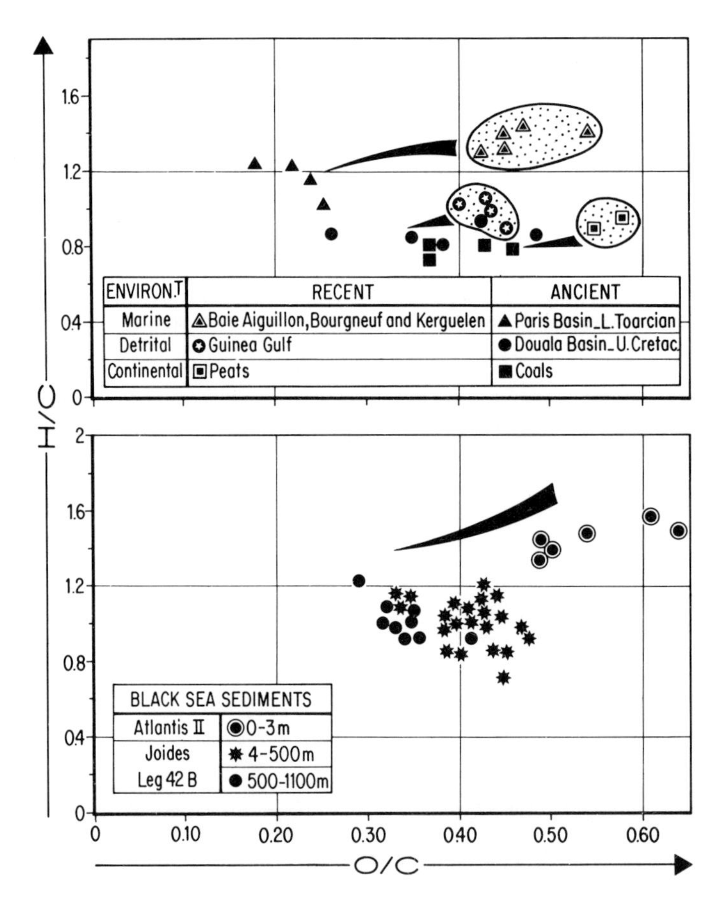

Fig. 14.20. — Variation of some properties of coals during natural evolution (from data from Durand *et al.* [32], Huc *et al.* [46]).

The foregoing remarks are an additional justification for the use of the term kerogen as defined earlier. They add a notion of continuity to the intuitive idea of a genetic relation between insoluble OM of ancient and recent sediments. Except for the most superficial sediments which still belong to the biosphere [71], transformations of organic substances apparently take place with neither an appreciable influx nor loss of matter, and as a first approxi-

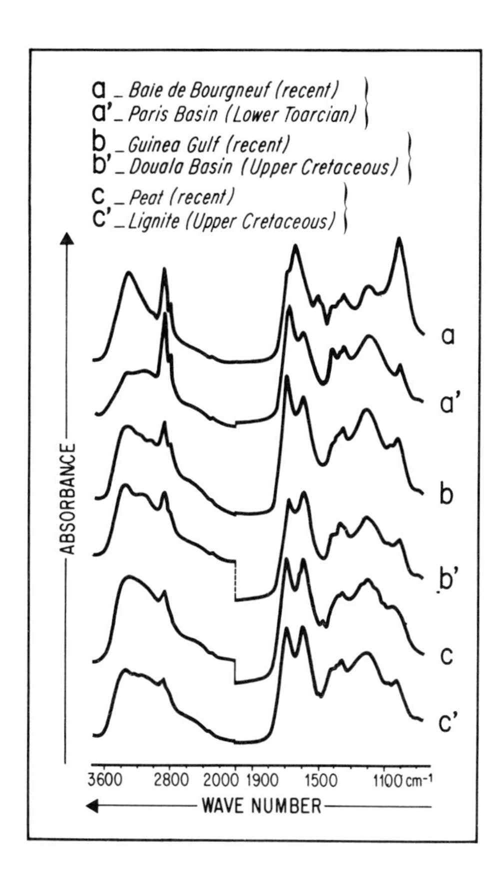

Fig. 14.21. — Influence of diagenesis on IR spectra of humic acids from various origins.

mation it can thus be assumed that kerogen is preserved quantitatively during diagenesis. From a qualitative point of view on the other hand it would seem that early natural evolution (as opposed to catagenesis) hardly modifies various important structural characteristics acquired at the moment sedimentation takes place (notably the relative importance of the aliphatic and alicyclic structures and of the aromatic structures). Diagenetic phenomena essentially affect the functional nitrogen and oxygen groups which are progressively decomposed (reducing by just as much the share of the soluble alkaline fractions) and the most unstable organic fractions (simple compounds and hydrolyzable fractions) which disappear and are certainly recombined to form more resistant structures.

REFERENCES

1. Alpern, B., (1970), *Rev. Inst, Franç. du Petrole,* **XXII,** 1233.
2. Aubert, M. and Gauthier, M.J., (1976), *Symposium on Concepts in Marine Organic Chemistry,* Edinburgh.
3. Basilevitch, N.I., Rodin, L.E. and Ronov, N.N., (1971), *Priroda,* **1.**
4. Beck, K.C., Reuter, J.H. and Perdue, E.M., (1974), *Geochim. et Cosmochim. Acta,* **38,** 341.
5. Becking, L.G.M., Kaplan I.R. and Moore, D., (1960), *J. of Geol.,* **68,** 243.
6. Bertrand, C.E., (1893), *Bull. Soc. Belge Geol.,* **7,** 45.
7. Blanchard, D.C., (1975), *in: Advances in Chemistry Series,* American Chemical Society, **145,** 360.
8. Borogov, V.G., (1968), "Atlas of Living Ressources of the Sea", FAO Dept. Fish. Rome.
9. Bordovskiy, O.K., (1969), *Oceanology,* **9,** 799.
10. Bordovskiy, O.K., (1965), *Marine Geol.,* **3,** 3.
11. Bradley, W.H., (1966), *Geol. Soc. of Am. Bull.,* **77,** 1333.
12. Brown, F.S., Baedecker, M.J., Nissenbaum, A. and Kaplan, I.R., (1972), *Geochim. et Cosmochim. Acta,* **36,** 1185.
13. Butler, E.T., Corner, E.D.S. and Marshall, S.M., (1970), *J. Mar. Biol. Assoc. UK 50,* 525.
14. Calvert, S.E. and Price, N.B., (1971), "The Geology of the East Atlantic Continental Margin", Delany ed., London 4.
15. Cane, R.F. and Albion, P.R., (1973), *Geochim. et Cosmochim. Acta,* **37,** 1543.
16. Cazes, P. and Reyre, Y., (1976), *Bull. BRGM.* **2,** 2.
17. Christopher, G., Kendall, C., Patrick, A. and Skipwith, D.E., (1968), *J. of Sed. Petrol.* **38,** 1040.
18. Combaz, A., (1975), *in:* "Pétrographie de la Matière Organique des Sédiments. Relation avec la Paléotempérature et le Potentiel Pétrolier", éd. Alpern, B., CNRS Paris, 93.
19. Combaz, A., (1974), *in: Advances in Organic Geochemistry,* 1973, éd. Tissot, B. and Bienner, F. Editions Technip, Paris, 423.
20. Combaz, A., Bellet, J., Poulain, D., Caratini, C. and Tissot, C., (1977), *in: Géochimie Organique des sédiments marins profonds. Orgon I - Mer de Norvège,* CNRS, Paris, 139.
21. Coulson, C.B., Davies, R.I. and Lewis, D.A., (1960), *J. Soil. Sci.* **11,** 20.
22. Dajoz, R., (1970), "Précis d'écologie", Dunod, Paris.
23. Debyser, Y., Dastillung, M. and Gadel, F., *in: Advances in Organic Geochemistry,* 1977, Moscow (in press).
24. Debyser, J. and Deroo, G., (1969), *Rev. Inst. Franç. du Pétrole,* **XXIV,** 21 and **XXIV,** 151.
25. Debyser, Y., Gadel, F., Leblond, C. and Martinez, M.J., (1978), *in: Géochimie Organique des sédiments marins profonds. Orgon II. Atlantique. N-E Brésil,* CNRS, Paris, 339.
26. Debyser, Y. and Gadel, F., *in: Géochimie Organique des sédiments marins profonds. Orgon III. Côtes de Mauritanie et Iles du cap vert,* CNRS, Paris (to be published).
27. Dragunov, S.S., (1958), *Trudi. Mosk. Torf.* **8,** 244.
28. Drozdova, T.V., (1974), *in: Advances in Organic Geochemistry,* 1973, éd. Tissot, B. and Bienner, F., Editions Technip, Paris, 285.
29. Duce, R.A., (1976), *Symposium on Concepts in Marine Organic Chemistry,* Edinburgh.
30. Duchaufour, P., (1970), "Précis de Pédologie", Masson, Paris.
31. Durand, B. and Espitalié, J., (1973), *C. R. Acad. Sci. Paris* **276,** 2253.
32. Durand, B., Nicaise, G., Roucaché, J., Vandenbroucke, M. and Hagemann, H.W., (1977), *in: Advances in Organic Geochemistry,* 1975, ed. Campos, R. and Goni, J., Enadimsa, Madrid, 601.
33. Emery, K.O., (1958), *in: Habitat of Oil,* ed. L.G. Weeks. AAPG Publication, Tulsa, 955.
34. Erhart, H., (1935), "Traité de Pédologie", Strasbourg **1.**
35. Flaig, W., (1966), *in:* "The Use of Isotopes in Soil Organic Matter Studies", Pergamon Press, 103.

36. Gagosian, R.B., (1976), *Symposium on Concepts in Marine Organic Chemistry,* Edinburgh.
37. Glob, P.V., (1951), *Illust. Lond. News,* **219,** 862.
38. Haider, K., Frederick, L.R. and Flaig, W., (1965) *Plant and Soil,* **22**-1, 49.
39. Haider, K. and Martin, J.P., (1967), *Soil Sci. Soc. Amer. Proc.,* **31,** 766.
40. Hair, M.E and Bassett, C.R., (1973), *Estuarine and Coastal Marine Science,* **1,** 107.
41. Hedges, J., (1975), *Carnegie Institution of Washington Year Book,* 792.
42. Himus, G.W., (1951), "Oil Shale and Cannel Coal", ed. Institute of Petroleum, London, vol. **2.**
43. Hobson, L.A., (1967), *Limnol. and Oceanogr.* **12,** 642.
44. Hodges, J.E., (1953), *Agr. Food. Chem.* **1,** 928.
45. Huc, A.Y. and Durand, B., (1974), *in: Advances in Organic Geochemistry, 1973,* ed. Tissot, B. and Bienner F., Editions Technip, Paris, 53.
46. Huc, A.Y. and Durand, B., (1977), *Fuel,* **56,** 73.
47. Huc, A.Y., Durand, B. and Monin J.C., *Initial Reports of the Deep Sea Drilling Project,* 42 B (in press).
48. Hunt, J.M., (1972), *AAPG Bull.,* **56,** 2273.
49. Ishiwatari, R., (1969), *Soil Science,* **107,** 53.
50. Ishiwatari, R., (1970), *in: Advances in Organic Geochemistry, 1966,* ed. Hobson G.D. and Speers G.C., Pergamon Press, 285.
51. Jannasch, H.W., (1971), *Science,* **171,** 672.
52. Johnson, R.G., (1974), *J. Mar. Res.* **33,** 313.
53. Kemp, A.L.W., (1973), *Geochim. et Cosmochim. Acta,* **37,** 2191.
54. Krevelen, D.W. van, (1961), "Coal", Elsevier, Amsterdam.
55. Kononova, M.M., (1966), "Soil Organic Matter", Pergamon Press.
56. Kuprin, P.N., *in: Advances in Organic Geochemistry, 1977,* Moscow, in press.
57. Lee, R.F. and Hirota, J., (1973), *Limnol. and Oceanogr.,* **18,** 227.
58. Lisitsyn, A.P., (1955), *Dokl. Akad. Nauk SSSR,* **103,** 2.
59. Lopatin, G.V., (1950), *Priroda,* **7,** 19.
60. Maillard, L.C., (1913), *C. R. Acad. Sci. Paris,* **156,** 1159.
61. Malcom, R.L. and Durum, W.H., (1976), *US Geol. Survey Water Supply.* Paper 1817-F.
62. Manskaya, S.M. and Drozdova, T.V., (1968), *Geochemistry of Organic Substances,* ed. Shapiro and Breger, Pergamon Press.
63. Martin, J.P. and Haider, K., (1971), *Soil Sci.,* **111,** 53.
64. Maurer, L.G., (1976), *Deep sea Research,* **23,** 1059.
65. Mazumdar, B.K., Chakrabartty, S.K. and Lahiri, A., (1957), *Sci. Ind. Res,* **16 B,** 275.
66. Menzel, D.W., (1974), "The sea", ed. Goldberg, Wiley, New York, **5,** 659.
67. Menzel, D.W. and Ryther, J.H., (1970), *Inst. Mar. Publ.,* University of Alaska, **1,** 31.
68. Müller, G. and Stoffers, P. (1974), *in:* "The Black Sea Geology, Chemistry and Biology", ed. Degens and Ross, AAPG Publ. **20,** 200.
69. Odum, E.P., (1959), "Fundamentals of Ecology", Saunders, Philadelphia.
70. Ogura, N., (1976), *Symposium on Concepts in Marine Organic Chemistry,* Edinburgh.
71. Pelet, R., (1976), *Symposium on Concepts in Marine Organic Chemistry,* Edinburgh.
72. Pelet, R. and Debyser, Y., (1977), *Geochim. et Cosmochim. Acta,* **41,** 1575.
73. Philp, R.P. and Calvin, M., (1975), *in:* "Environmental Biogeochemistry", ed. Nriagu., **1,** 131.
74. Pryor, W.A., (1975), *Geol. Soc. of Am. Bull.,* **86,** 1244.
75. Putman, P., (1953), "Energy in the Future", Van Nostrand, New York.
76. Rashid, M.A, Buckley, D.E., and Robertson, K.R., (1972), *Geoderma,* **8,** 11.
77. Redfield, A.C., (1958), *in: Habitat of Oil,* Symposium AAPG, ed. Weeks L.G., 968.
78. Riley, G.A., (1970), *Advances Mar. Biol.* **8,** 1.
79. Rittenberg, S.C., Emery, K.O., Hülsemann, J., Degens, E.T., Fay, R.C., Reuter, J.H., Grady, J.R., Richardson, S.H. and Bray, E.E., (1963), *J. Sed. Petrol.,* **33,** 140.
80. Romankevitch, E.A., (1977), *in:* "Geochemistry of Organic Matter in the Oceans" (in Russian), Akad. Nauk SSSR.
81. Ronov, A.B., (1958), *Geochemistry,* **5,** 510.
82. Ryther, J.H., (1969), *Science,* **166,** 72.
83. Schnitzer, M., (1971), *Agron. Abstr.,* 77.
84. Schnitzer, M. and Neyroud, J.A., (1975), *Fuel,* **54,** 17.
85. Sellers, W.D., (1965), "Physical Climatology", Univ. of Chicago Press.

86. Sieburth, J. and Jensen, A., (1968), *J. Exp. Mar. Biol. Ecol.,* **2,** 174.

87. Simoneit B.R.T., (1975), Thesis, Bristol University.

88. Shimkus, K.M. and Trimonis, E.S., (1974), *in:* "The Black Sea Geology, Chemistry and Biology", ed. Degens and Ross, AAPG Publ. **20,** 249.

89. Skolkovitz, E.R., (1976), *Geochim. et Cosmochim. Acta,* **40,** 831.

90. Skopintsev, B.A. (1955), *Dokl. Akad. Nauk SSSR,* **105,** 770.

91. Skopintsev, B.A., (1961), *in:* "Recent Sediments of Seas and Oceans" (in Russian), Moscow, 285.

92. Steeman Nielsen, E. and Jensen, J.A., (1957), *Galathea Reports,* **1,** 49.

93. Stevenson, F.J. and Goh, K.M., (1971), *Geochim. et Cosmochim. Acta,* **35,** 471.

94. Stevenson, F.J. and Tilo, S.N., (1970), *in: Advances in Organic Geochemistry, 1966,* ed. Hobson G.D. and Speers G.C., Pergamon Press, 237.

95. Stevenson, F.J. and Butler, J.H.A., (1969), *Organic Geochemistry,* ed. Eglinton and Murphy, Springer-Verlag.

96. Strakhov, N.M., (1961), *Akad. Nauk Moscow SSSR* (in Russian), **5,** 27.

97. Stuermer,, D.H. and Payne, J.R., (1976), *Geochim. et Cosmochim. Acta,* **40,** 1109.

98. Swaby, R.J. and Ladd, J.N., (1966), "The Use of Isotopes in Soil Organic Matter Studies", Pergamon Press.

99. Swaby, R.J. and Ladd, J.N., (1966), *Trans. Intern. Soc. Soil Sci. Comm.* **4,** 3.

100. Swain, F.M., (1956), *AAPG Bull.,* **40,** 600.

101. Swain, F.M., (1969), *Organic Geochemistry,* ed. Eglinton and Murphy, Springer-Verlag.

102. Swanson, V.E. and Palacas, J.G., (1965), *US Geol. Survey Bull.,* **1214 B.**

103. Thomas, R.L., Kemp, A.L.W. and Lewis, C.F.M., (1972), *J. Sed. Petrol.* **42,** 66.

104. Thomas, J.P., (1971), *Marine Biol.,* **11,** 311.

105. Tissot, B., (1977), *La Recherche,* **8,** 327.

106. Tissot, B., Durand, B., Espitalié, J. and Combaz, A., (1974), *AAPG Bull.* **58,** 499.

107. Trager, E.A., (1924), *AAPG. Bull.* **8,** 301.

108. Varossieau, W.W. and Breger, I.A., (1951), *C.R. Congrès Internat. Stratigraphie Geol. Carbonifère,* Heerlen 3, 637.

109. Wall, D., (1962), *Geol. Mag.* **94,** 353.

110. Wangersky, P.J., (1976), *Marine Ecology,* **4.**

111. Welte, D.H., (1970), *Naturwissenshaften,* **57,** 17.

112. Whittaker, K. and Vallentyne, J.R., (1957), *Limnol. and Oceanogr.* **2,** 98.

113. Whittle, K.J., (1976), *Symposium on Concepts in Marine Organic Chemistry,* Edinburgh.

114. Wiebe, P.H., Boyd, S.H. and Winget, C., (1976), *J. Mar. Res.* **34,** 341.

115. Williams, P.M., (1975), "Chemical Oceanography", ed. Riley and Skinow, **2,** 301.

116. Wright, J.R. and Schnitzer, M., (1961), *Nature,* **190,** 703.

15

Évolution géochimique de la matière organique

R. PELET *

I. INTRODUCTION

« Conformément à la loi et selon toutes les règles, j'aurais dû parler le dernier. Mais il y a des cas où les lois et les règlements se retournent contre ceux qui les respectent, et il faut alors les ignorer. Je prendrai le premier la parole, parce que je ne puis me taire plus longtemps. Je parlerai le premier parce que je n'ai pas envie d'attendre et je ne souffrirai aucune objection. »

A. et B. Strougatsky (La Troïka)

A. La matière organique ne se confond pas avec la matière vivante.

On ne reviendra pas sur les arguments qui montrent qu'à peine née, la chimie organique a cessé d'être la chimie des organismes vivants. Dès 1828, la synthèse de l'urée par Wöhler démontrait, par une voie un peu oblique, l'unité du monde matériel évidente aux yeux des savants antiques — et spécialement d'Epicure dont le XIXe siècle a vu la réhabilitation éclatante. A l'heure actuelle, la chimie organique est sûrement la science des composés du carbone — mais, traditionnellement, certains, par exemple les carbonates métalliques, en sont exclus — et tout aussi sûrement ne se sent plus que des rapports distants avec le monde vivant.

Il faut consentir un effort identique de révision des définitions pour la matière organique sédimentaire et, par voie de conséquence, pour la matière organique (MO) en général. Il y a déjà un hiatus entre les concepts du géochimiste pour qui, bien sûr, cette MO est un produit du monde vivant — âgé, usé, changé certes, mais cependant encore reconnaissable bien souvent à telle porphyrine ou tel isoprénoïde — concepts presque vitalistes, et la pratique quotidienne du laboratoire où toute forme réduite du carbone est, par un consensus général mais jamais explicité, dosée comme carbone organique provenant d'une MO — pratique qui permettrait aux escaliers du métro, roches silicatées riches en carbures de silicium, de s'inventer

*Institut Français du Pétrole. Rueil-Malmaison, France.

des parents animés. Il faut donc franchir le pas et reconnaître en la MO, simplement, l'objet de la chimie organique.

Cet effort amène des récompenses immédiates. Rien ne nous empêche plus d'étudier la MO du Précambrien le plus ancien et le plus métamorphique : les inférences sur son origine viendront après. Et pourquoi ne diraient-elles pas que cette origine n'est pas la matière vivante? Voilà pour le temps.

Pour l'espace, l'univers entier s'ouvre à nous. A ne considérer que la vie, même les plus audacieux, quittée la Terre, ne se permettent guère que des probabilités. Ce sont des certitudes sur la présence de molécules, ions ou radicaux organiques, au sens où nous l'entendons, que nous apporte la lumière sidérale. 43 de ces entités avaient été recensées par spectrométrie en 1976, de nouvelles le sont chaque jour — certaines trop simples et rébarbatives, mais que dire de l'éthanol? — par-delà les gouffres cosmiques, plus loin que les lointains inimaginables où se fatigue la lumière, le bon vieux Silène nous adresse un signe rassurant. La MO est un composant banal de l'univers.

Si banal que cela? Il n'existe pas d'autre élément dont la chimie soit aussi riche et complexe que celle du carbone. Imprévisible aussi : l'abondance et la stabilité des composés organiques ne se laissent pas appréhender par un simple calcul thermodynamique qui ne permet, quelles que soient les conditions envisagées, que des teneurs évanescentes en molécules organiques complexes [9]. Mais qui ne voit que cette tendance à la complexité, que cet acharnement à nier la thermodynamique sont la promesse de l'émergence de la vie? Et, incidemment, qui, ne comprend que le message du vieux Silène est que la vie est une dans tout le cosmos, que tous les êtres vivants sont constitués de matière qui est de la MO au sens terrestre du terme?

Et ainsi le cercle se referme : la MO n'est pas la matière vivante; c'est la matière qui est, a été, ou sera vivante.

B. Il faut connaître l'histoire de l'évolution de la matière organique sédimentaire.

La thermodynamique est l'apparence positiviste du destin. Que le chemin soit long ou court, simple ou tortueux, elle connaît l'arrivée. Sous sa forme vivante, la MO semble libérée de cette obligation; on reconnaît qu'elle est morte à cela justement qu'elle s'abandonne, que, malgré quelques lenteurs, elle parcourt bien un chemin qui l'amènera à l'équilibre. Qui l'amènera sûrement, mais ne l'amène pas toujours, faute du temps nécessaire.

Sur la Terre, la MO sédimentaire est, à l'instant premier de son dépôt, d'une complexité tout à fait hors d'équilibre. Coupée des régions favorables à la vie, elle va évoluer de plus en plus rapidement à mesure de son enfouissement, vers sa composition d'équilibre. Cette accélération, due à l'accroissement de la température dans les profondeurs terrestres —c'est-à-dire à l'augmentation de l'agitation interne des molécules, génératrice de ruptures de liaisons— amènera, à la limite des zones métamorphiques, la MO à une composition quasi équilibrée : un solide carboné —qui, dernier effort, conserve quelques hétéroatomes et refuse de prendre la forme requise du graphite— des gaz : eau, méthane, azote, gaz carbonique. Nous voici revenus aux temps archéens.

Le rythme de cette évolution dépend de l'horloge de qui l'observe. Pour l'homme elle est infiniment lente, mais il y a eu cependant un commencement, comme il y aura une fin; si cette dernière nous importe peu parce qu'elle ne conditionne pas le présent, il n'en est pas de même du passé. Il convient donc de remonter aux origines.

II. ORIGINE DE LA MATIÈRE ORGANIQUE SÉDIMENTAIRE

« Quelqu'une des voix
— Est-elle angélique —
Il s'agit de moi
Vertement s'explique :

Ces mille questions
Qui se ramifient
N'entraînent, au fond,
Qu'ivresse et folie.

Reconnais ce tour,
Si gai, si facile
C'est toute Onde et Flore
Et c'est ta famille,... etc.

A. Rimbaud (Poèmes, l'Age d'Or)

A. Il existe un kérogène abiotique.

Les remarques qui précèdent comme les développements qui suivent sont évidemment fondés sur le principe d'actualisme : toutes choses égales par ailleurs, les mêmes causes reproduisent les mêmes effets, et le temps ne fait rien à l'affaire. Les cartes peuvent changer —et elles changent effectivement— les règles du jeu sont immuables. Ce principe n'est jamais tant évident qu'en astrophysique, où les informations que véhicule la lumière sont rigoureusement datées et montrent un passé d'autant plus reculé que les distances sont plus lointaines. De ce tableau disparate, l'astrophysicien tire sans aucune gêne une description synchronique de l'univers, d'ailleurs essentiellement équivalente à son apparence, diachronique par sa nature même. Comme l'astrophysicien, nous dirons donc que l'abondance « actuelle » de la MO cosmique montre que cette abondance est contemporaine de l'univers lui-même. Il va de soi que cette abondance, au sens trivial du mot, découle premièrement de la grande abondance cosmique des éléments fondamentaux de la MO : carbone, oxygène, hydrogène, azote (cf. par exemple [7]). Point encore plus intéressant, si l'on suit Hoyle [19, 39] et Knacke [22], l'étude du rayonnement IR intergalactique apporte non seulement la preuve de l'existence des composés simples dont on a parlé dans l'introduction, mais aussi celle de la présence à la surface des grains de poussière interstellaire d'un revêtement organique de nature polymérique et plus spécialement polyosidique — de kérogène (1) pour parler brièvement. Forts maintenant de ces enseignements de la lumière cosmique, qui nous parlent de distance et d'époques incroyablement reculées, nous pouvons alors attribuer sans crainte la plus grande généralité aux informations plus locales que nous apportent les météorites.

---

(1) Dans un sens évidemment plus large que sa définition (*cf.* Durand, même ouvrage). On entend par là un ensemble de macromolécules, où qu'elles se trouvent et quelle que soit leur origine, dont toutes les propriétés tant physiques que chimiques sont celles du (des) kérogène (s).

Les météorites ont le grand avantage d'être des corps matériels, et donc de nous permettre de dresser un tableau exhaustif de leur nature. Quoique originaires du système solaire, elles sont bien plus représentatives des corps planétaires que la seule Terre. Leur contenu en MO, nul dans les blocs de fer nickelifère, croît à mesure que leur densité diminue, et atteint son maximum dans les chondrites carbonées, fragments de croûtes planétaires. Il est naturel en effet que la ségrégation gravifique à l'origine des planètes concentre à leur surface les éléments les plus légers —dont C, H, O, N. Ce qui a une conséquence importante : la MO est l'hôte normal de l'interface entre le corps planétaire solide et l'espace dans le cas général, l'atmosphère, voire l'hydrosphère quand la planète en possède. Or cette interface est le lieu du mouvement, où déplacements et réactions chimiques sont facilités. Devant l'espace nu, comme sur la surface lunaire, ces réactions sont stérilisantes et figent pour l'éternité une MO peu différenciée [13]. En présence de l'océan, lieu obligatoire des naissances, la complexification de la MO va se poursuivre d'un mouvement sans cesse accéléré.

La MO des météorites ne porte pas l'empreinte de la vie. On y trouve — en traces — des acides aminés, mais pas seulement ceux d'origine biologique; on y trouve des hydrocarbures, mais pas isoprénoïdes; on y trouve une foule de composés, mais en mélanges racémiques ou optiquement inactifs. Enfin, en quantités importantes —jusqu'à 3-4 % en poids de carbone [3, 40], on y trouve un matériau de nature polymérique, donnant par pyrolyse essentiellement des hydrocarbures [24, 1, 35] — autant dire du kérogène. Il est intéressant de remarquer que ce matériau est d'autant plus « évolué », d'autant plus proche du carbone pur (amorphe, le graphite est rare dans les chondrites) — et même d'autant moins abondant que la météorite a subi des températures plus élevées au cours de ses périples intersidéraux.

Ce kérogène primordial n'est pas une spécialité extra-terrestre. Les savants qui tentent la simulation de la naissance de la vie sur la Terre en soumettant à des flux énergétiques des modèles d'atmosphère et d'océan primitifs, rapportent avec exactitude les analyses soigneuses qu'ils font du bouillon de leur soupe démiurgique. Ils laissent pour compte les grumeaux brunâtres de matériel insoluble qui apparaissent de-ci, de-là dans leurs appareils (Ponnamperuma, Buvet, communications personnelles) et dont le désordre moléculaire est une négation de la vie.

Ainsi, dès l'origine des temps, apparaît la dualité de la matière vivante et du kérogène. Tous deux comportent les mêmes groupements constitutifs — briques issues des liaisons covalentes carbone-carbone — mais différemment assemblées. Un ordre architectural magnifique signale la matière vivante, et permet aux différentes parties de l'édifice d'assumer des fonctions spécifiques; au kérogène appartient le désordre, générateur d'une stabilité qui est surtout inertie, mais qui lui permet d'être, sur Terre et dans l'Univers entier, la forme la plus probable de la MO.

Il ne fait donc aucun doute que du kérogène s'est formé et accumulé dans les sédiments avant la naissance de la vie. En quelles quantités? Les teneurs en carbone des météorites sont extrêment variables; s'il existe des chondrites carbonées extrêmement riches comme Ivuna ou Orgueil (3 à 4 %), les valeurs moyennes les plus récentes [23] sont de $3,4.10^{-4}$ pour les météorites pierreuses et de 11.10^{-4} pour les métalliques — mais ces dernières, témoins des intérieurs planétaires, ne contiennent que du carbone cristallin et des carbures; on tiendra ces teneurs pour des ordres de grandeur, les conditions régnant sur la Terre archéenne étant évidemment très favorables à l'épanouissement de la chimie du carbone. D'un autre côté, les compilations de teneurs en carbone organique des roches sédimentaires [32] semblent montrer, sous des variations importantes, une sorte de tendance à la diminution avec l'âge des roches. Dans ces conditions, il faudrait penser, et cela ne paraît pas déraisonnable, que ce kérogène primordial n'a jamais été très abondant. D'autres interprétations sont d'ailleurs possibles.

A cette faible abondance originelle s'ajoute l'usure par le temps. Sans doute la plupart, sinon tous les sédiments archéens ont été repris par le métamorphisme et leur kérogène a pris le vêtement anonyme des carbones plus ou moins graphitisés, qui n'évoque plus rien de son origine. Nous laisserons au lecteur le soin de décider si la recherche du kérogène abiotique est envisageable, et à partir de quels traits diagnostiques. La question n'a pas d'intérêt pratique, et pour la théorie on doit la considérer comme résolue.

B. Mais l'apparition de la vie assure la prééminence quantitative du kérogène.

Vint un moment dans l'histoire de la Terre où la vie cessa d'être créée par les orages dans les nuées; par la force des choses, cette fin de la création devait être précédée de l'apparition d'êtres autotrophes; peut-être d'ailleurs cette apparition est-elle une des causes du déclin de la biogenèse, par les changements de composition, dont la production d'oxygène libre, amenés à l'atmosphère. Quoi qu'il en soit, au Précambrien la vie n'est plus issue que de la vie et atmosphère et hydrosphère ont des compositions voisines de l'actuelle. Tous ces changements se sont faits lentement et progressivement, et il n'y a donc pas de raisons de penser que la MO sédimentaire précambrienne soit radicalement différente de ses prédécesseurs.

Cependant l'existence d'êtres autonomes implique qu'ils se différencient clairement du milieu aqueux homogène qui les entoure, c'est-à-dire qu'ils s'équipent de membranes séparatrices hydrophobes. Cette apparition progressive de l'hydrophobie dans un milieu initialement à l'état dissous a été considérée [18] comme le moteur de l'évolution; quoiqu'il en soit, elle implique la naissance, puis l'accroissement quantitatif, des chaînes aliphatiques longues — et effectivement, dans un premier temps, les chaînes en C_{16} et C_{18} vont s'imposer. Le kérogène météoritique est spécialement aromatique [16] : les molécules qui lui donnent naissance ont des chaînes aliphatiques courtes facilement cyclisables. B et L. Nagy [26] étudiant la série précambrienne du Lesotho remarquent que le kérogène de la formation d'*Onverwacht* (3,5 milliards d'années) est plus aromatique que celui de la formation plus récente de *Fig-Tree* (3,2 milliards d'années); ce résultat — d'ailleurs discutable sur le plan analytique — peut être dû à des conditions locales; il n'est pas interdit d'imaginer que nous saisissons là une étape de l'évolution cosmique du kérogène.

Si donc globalement la MO paraît peu évoluer, par contre, dans le détail, le changement est complet. Les structures complexes nécessaires à la vie autotrophe sont toutes là : sucres [29], isoprénoïdes [21], plus maintenant les acides aminés biologiques [15] — on trouve même des stéranes et triterpanes [5], squelettes de métabolites d'une architecture pourtant compliquée. Ces trouvailles sont d'ailleurs de moins en moins étonnantes, car à mesure que le temps passe, se multiplient dans les roches les témoignages directs, les fossiles proprement dits, d'une organisation déjà si complexe que, comme on l'a dit plus haut, il faut faire remonter l'origine de la vie sur Terre à un passé très reculé.

L'émergence de la vie précambrienne fait faire à la MO des progrès quantitatifs nets. Des teneurs en carbone de l'ordre de 10^{-4} envisagées pour les roches archéennes, on passe dans les roches précambriennes à des teneurs de l'ordre de 10^{-3}. Les accumulations de MO deviennent fréquentes — des « charbons », des pétroles apparaissent (*cf.* la synthèse de Sidorenko et Sidorenko, [34]). Vers la fin du Précambrien, un monde sédimentaire fort semblable à l'actuel s'est constitué — simplement plus monotone, moins diversifié; le règne vivant y voit le triomphe des « algues » bleu-vert [8] autotrophes et fixatrices de l'azote, ancêtres communs de l'ensemble des formes de vie actuelles, à la morphologie très primitive — mais qui montrent déjà pratiquement tous les types moléculaires biochimiques de leur descendance multiforme.

C. Les végétaux terrestres apportent des traits originaux.

L'étape suivante va être la conquête des terres continentales, la sortie de l'océan matriciel. Le milieu émergé présente des caractéristiques bien tentantes pour les autotrophes : une luminosité incomparable, une disponibilité permanente en éléments minéraux — et, pour l'azote, sous des formes faciles à assimiler qui rendront inutile l'attirail enzymatique nécessaire à la fixation de l'azote moléculaire, permettant par là même l'acquisition de potentialités nouvelles. En contrepartie, un approvisionnement en eau — le fluide vital — limité et irrégulier et la contrainte de la gravité, que ne masque plus la poussée d'Archimède. De nouveaux composés encore plus hydrophobes — les cires cuticulaires — vont permettre le contrôle des échanges d'eau avec le milieu extérieur, l'apparition de vaisseaux dans le corps du végétal en permettant le contrôle interne. Dès que la trop grande taille de la plante rendra impossible l'érection vers la lumière par simple turgescence, les tissus vont se rigidifier par imprégnation de lignine, deuxième invention nécessaire de la conquête des terres émergées.

La lignine appartient en propre à la végétation terrestre ([1]) et spécialement à celle de grande taille. C'est un composé de nature ambiguë; son rôle de soutien pérenne, non engagé dans les relations métaboliques quotidiennes, demande obligatoirement l'inertie chimique; cette stabilité est obtenue par un désordre contrôlé dans la polymérisation qui crée la macromolécule; l'enchaînement des monomères aromatiques polyfonctionnels se fait « n'importe comment », conduisant à une réticulation capricieuse, cause à la fois de la rigidité physique et de l'aspect décourageant de la lignine pour les attaques enzymatiques. D'où l'ambiguïté : la lignine, dès sa naissance, participe déjà du kérogène. Ce fait a une conséquence très importante pour le monde sédimentaire : quelles que soient les conditions présidant à l'érosion et au transport des matériaux continentaux — à l'exception peut-être du transport en solution — la lignine ne pourra pas ne pas constituer une fraction importante de la MO, soit sous un aspect amorphe si elle a été (un peu) modifiée chimiquement, soit sous des formes où se reconnaîtront des tissus végétaux si elle a plus ou moins conservé sa nature. Il va donc exister une MO continentale quantitativement marquée par la lignine, et spécialement sa structure aromatique.

Les chaînes aliphatiques très longues des cires cuticulaires sont peut-être une solution moins originale au contrôle des échanges extérieurs. Il ne s'agit finalement que du renforcement de l'hydrophobie des membranes par un procédé éprouvé, et qui existe sous des formes un peu différentes chez les algues unicellulaires et surtout chez les bactéries : aucun céride végétal ne bat les acides mycoliques qui présentent deux fonctions alcool et une fonction

([1]) Il y a, pour les mêmes raisons, des précurseurs chez les algues érigées de grande taille : fucales, laminariales [25].

acide pour 88 atomes de carbone. Il est impossible, dans l'état actuel de nos connaissances, de savoir si l'acquisition de ces mécanismes chez les protistes a précédé ou non l'ascension au sec des plantes. Si les témoignages fossiles directs de plantes vasculaires terrestres ne remontent guère en deçà du début du Dévonien, ces premières formes sont déjà si perfectionnées qu'on est obligé de reculer bien plus la date de l'émersion première. Sur des arguments indirects (pédogenèses fossiles) Ehrart [11] fait même remonter cet événement à la fin du Précambrien. S'il n'y a pas eu contamination ultérieure, les distributions de *n*-alcanes d'extraits de roches précambriennes données par Rodionova et Sidorenko [30] ou Sidorenko et Sidorenko [34] et qui montrent des quantités non négligeables de molécules lourdes jusqu'en C_{33} au moins, avec une légère prédominance des nombres de carbones impairs au-delà de C_{25}, sans être plus déterminantes, vont dans le même sens.

Quoi qu'il en soit, l'importance quantitative des cires cuticulaires a eu un début, et ne connaîtra son importance actuelle qu'à partir de l'explosion des phanérogames à l'aurore du Crétacé. Cependant, ces cires ne constituent jamais qu'une fraction extrêmement restreinte de la biomasse végétale (de l'ordre du millième en poids sec, à comparer à l'ordre du dixième pour la lignine) et, quoique spécialement stables, elles n'ont pas vocation à être généralement un constituant quantitativement important de la MO continentale. Par contre, leur mise en évidence analytiquement très facile va constituer un moyen privilégié (et dont on a peut-être abusé) de reconnaissance des apports continentaux.

La sédimentation, due au fait qu'à côté de l'océan la Terre a très tôt présenté des continents émergés soumis à l'action des météores, est un phénomène d'âge immémorial. Aux produits minéraux issus du démantèlement du continent et, éventuellement, de réactions chimiques en milieu aqueux, se mêle dès le début des temps une MO de type kérogène. D'abord très rare et d'origine côtière — parce que les réactions de synthèse abiotique demandent des concentrations possibles seulement dans des corps aquatiques de faible profondeur et de faible superficie — elle devient de plus en plus fréquente et océanique à mesure que la soupe prébiotique épaissit, tout en restant de caractère plutôt aromatique. Le Précambrien, avec le foisonnement de la vie, voit le triomphe absolu de la MO océanique, dont les teneurs ont décuplé, et qui a un caractère aliphatique marqué. Mais déjà les algues colonisatrices rampent vers le continent. Au Crétacé, sinon bien avant, les biomasses marine et continentale, et donc en première approximation leur contribution sédimentaire, sont du même ordre de grandeur. La MO continentale se distingue nettement de sa sœur marine par une aromaticité marquée. A l'unicité d'origine de la matière sédimentaire minérale va donc maintenant s'opposer la dualité d'origine de la matière sédimentaire organique. Un des problèmes favoris de la géochimie organique est posé.

III. LA SÉDIMENTATION ET LA DIAGENÈSE PRÉCOCE

« De même, tout ce qui est sorti de la terre s'en retourne à la terre... ainsi la mort ne détruit pas les corps au point d'anéantir leurs éléments : elle ne fait que dissoudre leur union. Puis elle en forme d'autres combinaisons, et fait en sorte que toutes choses modifient leurs formes et changent leurs couleurs, acquièrent la sensibilité ou la rendent en un instant — ce qui te montre de quelle importance sont, pour les mêmes atomes, les combinaisons, les positions, les mouvements qu'ils s'impriment entre eux. »

Lucrèce (De la nature, livre II, traduction Ernout)

A. La sédimentation ajoute une différenciation propre à celle donnée par l'origine de la matière organique.

A la diversité originelle de la MO, les processus sédimentaires vont ajouter une différenciation propre. Mais on voit tout de suite que cette différenciation ne va pas jouer avec une intensité comparable pour les MO marine et continentale; cette dernière aura en effet l'occasion et le temps de connaître toutes les étapes de la sédimentation : érosion, éventuellement précédée d'un stade d'altération continental — pédogénétique —, transport fluviatile, à l'état solide, dissous ou colloïdal, avec toutes les altérations chimiques et biochimiques correspondantes — avant l'arrivée au bassin de sédimentation, où son destin rejoindra celui de la MO marine. Cependant, dans les deux cas, les principes restent les mêmes, et c'est la géographie qui contrôle étroitement la suite des événements.

1. Le milieu marin ne se révèle que loin des continents ou en bordure de régions désertiques dont l'apport sédimentaire est nul. On constate alors que le régime des courants marins et la topographie du fond jouent un rôle essentiel. Il est d'ailleurs bon de remarquer que la topographie du fond est un des éléments déterminant, à grande échelle, le régime des courants marins. En contrôlant, en surface, la disponibilité des sels nutritifs nitrates et phosphates, les courants marins dirigent la productivité océanique. En ralentissant, accélérant ou déviant les transits vers le fond, ils imposent les zones d'accumulation. La topographie intervient à l'échelle générale par l'épaisseur de la tranche d'eau : plus celle-ci est grande, moins les débris organiques ont de chance d'arriver au fond, minéralisés par les dégradations, essentiellement biologiques, qu'ils auront subies pendant leur longue descente. Il convient en effet de remarquer que cette MO n'ayant pas quitté son milieu d'origine, sa dégradation par des processus purement physicochimiques avec des vitesses appréciables à l'échelle de temps de la sédimentation est peu probable; cela reviendrait en effet à postuler un déséquilibre thermodynamique tel entre les êtres marins et leur milieu vital que justement toute vie deviendrait impossible. Bien entendu, ceci ne s'applique pas à l'interface océan-atmosphère, qui n'est l'habitat que d'un nombre infime d'espèces. Les dégradations biologiques successives effectuent un recyclage efficace de la MO, et seules les fuites de ce cycle, alimentent la sédimentation. A chaque étape du cycle, la fraction de la MO qui a échappé à la minéralisation ou à l'incorporation à la biomasse va évoluer vers le kérogène, et d'autant plus complètement que son temps de séjour dans l'eau sera grand — c'est-à-dire que les profondeurs d'eau seront considérables —

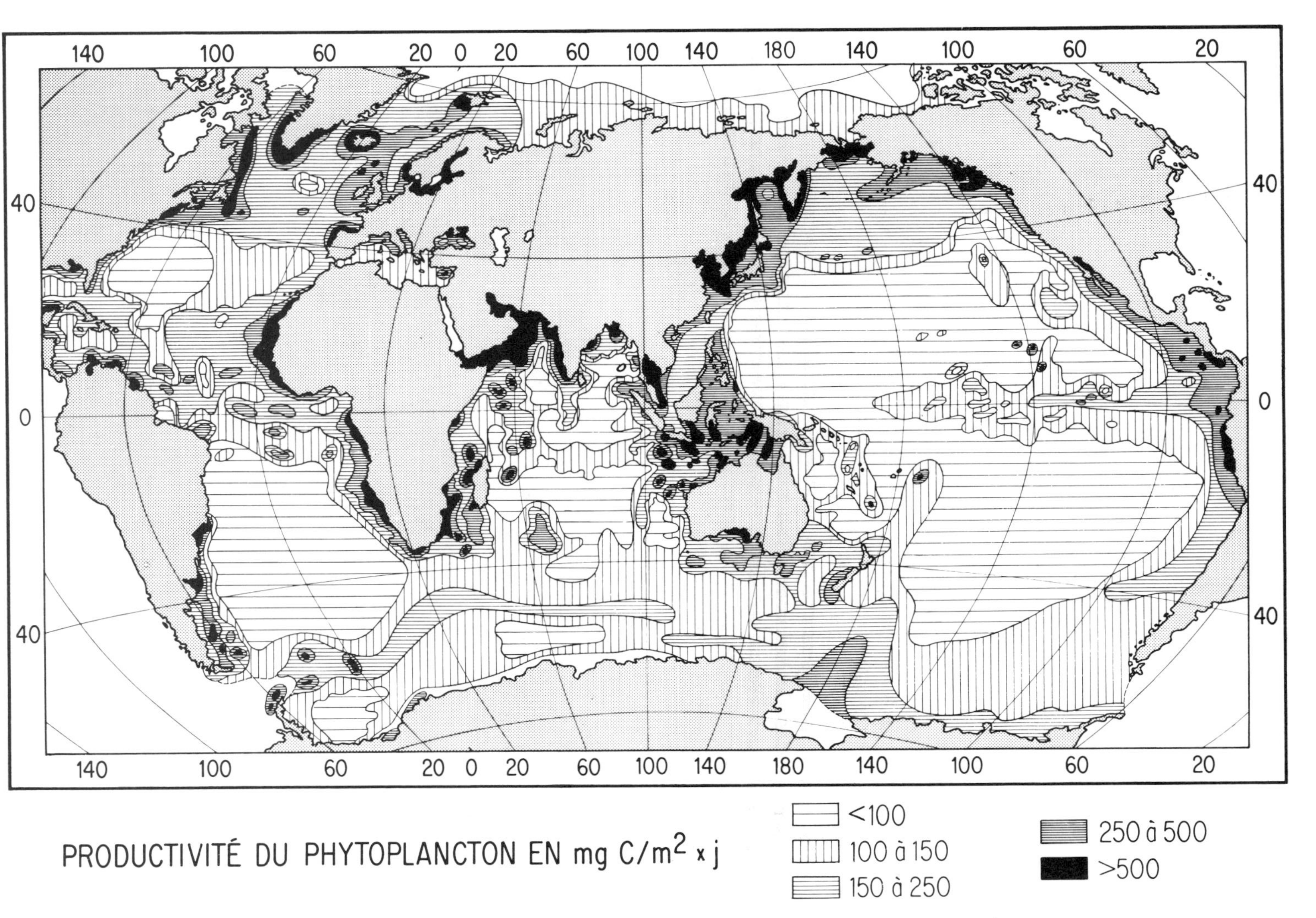

Fig. 15.1. — Carte des productivités primaires phytoplanctoniques de la surface océanique (d'après Romankevic [31]).

et que la concurrence sera plus âpre entre les besoins biologiques et les réactions de polycondensation — c'est-à-dire que le milieu sera plus pauvre et plus inhospitalier. Seules donc les zones peu profondes ([1]) et riches en MO verront sédimenter éventuellement une MO marine comprenant autre chose que du kérogène — mis à part les fragments non digérés des pelotes fécales, providence du benthos.

Au total, pour les zones dégagées de l'influence continentale, l'ensemble de ces processus aboutit à une sédimentation organique généralement négligeable dans le monde actuel. Tout d'abord, les zones à productivité marine importante sont habituellement en bordure des continents (*cf.* par exemple [31]). Pour les zones au large, on se convaincra à l'examen des figures 15.1 et 15.2 que, à des régions de productivité moyenne comme l'Atlantique central ou le Pacifique Est équatorial, correspondent des sédiments à faible teneur en carbone organique. De façon plus générale, il n'existe actuellement des sédiments riches en carbone qu'au voisinage des continents, quelle que soit la productivité marine en surface : comme on l'a vu plus haut, l'explication de ces observations est dans la dégradation biologique, le « recyclage », de la MO lors de sa lente descente vers le fond. Cependant, en quelques zones de faible étendue, peu perceptibles à l'échelle de la figure 15.2, comme la mer Noire et le bassin de Cariaco (ORGON 2), la conjugaison de facteurs locaux (essentiellement une topographie de cuvette fermée jointe à des apports organiques importants) permettent l'établissement de l'anoxie dans la colonne d'eau sus-jacente au sédiment; on observe alors un ralentissement extrêmement important des processus de biodégradation, qui ne peuvent plus être le fait que d'une microflore anaérobie, avec, comme corollaire, un accroissement très net des teneurs en carbone organique sédimentaire.

Il n'existe donc pas à l'heure actuelle de grands bassins marins anoxiques. L'explication en est l'existence d'une circulation océanique générale autour de la Terre qui permet, par la plongée des eaux polaires froides, donc saturées en oxygène et denses, le renouvellement de l'oxygène des fonds marins; d'où le profil vertical bien connu (fig. 15.3) des teneurs en oxygène dissous, où un minimum apparaît aux profondeurs intermédiaires.

Il n'en a pas été toujours ainsi au cours de l'histoire géologique. Pour l'Océan Atlantique Sud, dont le développement est maintenant assez bien connu, il a existé au Crétacé Inférieur et Moyen des périodes où des bassins plus vastes que la Méditerranée actuelle étaient coupés des eaux polaires (fig. 15.4). L'établissement de conditions anoxiques dans de tels bassins est attesté justement par les formations (« black shales ») très riches en MO qu'on y retrouve aux époques correspondantes [10]. Toutes ces conditions témoignent d'un monde fort différent du monde actuel, mais cependant intelligible. Il faut être actualiste avec discernement : qu'en tout temps les mêmes causes reproduisent les mêmes effets ne signifie pas que les mêmes causes sont en tout temps présentes. Si les clés sont toujours les mêmes, et donc que le monde actuel nous les livre, les serrures et les portes anciennes nous apparaissent souvent inconnues, et le trésor lui-même peut quelquefois être unique.

2. Avant d'arriver à la mer — où il subira, au débouché des fleuves, un changement complet de milieu physicochimique — le matériel organique continental a subi bien des vicissitudes. Le cas le moins fréquent est celui où les organismes, puis leurs fragments, sont pris en charge par les météores dès leur mort ou leur libération et sont transportés en l'état jusqu'au

---

([1]) Ou anoxiques, c'est-à-dire où la colonne d'eau dépourvue d'oxygène ne permet que l'existence d'une microflore anaérobie peu dégradante. A l'heure actuelle, il n'existe pas de grands bassins marins anoxiques. Ce point va être repris plus loin.

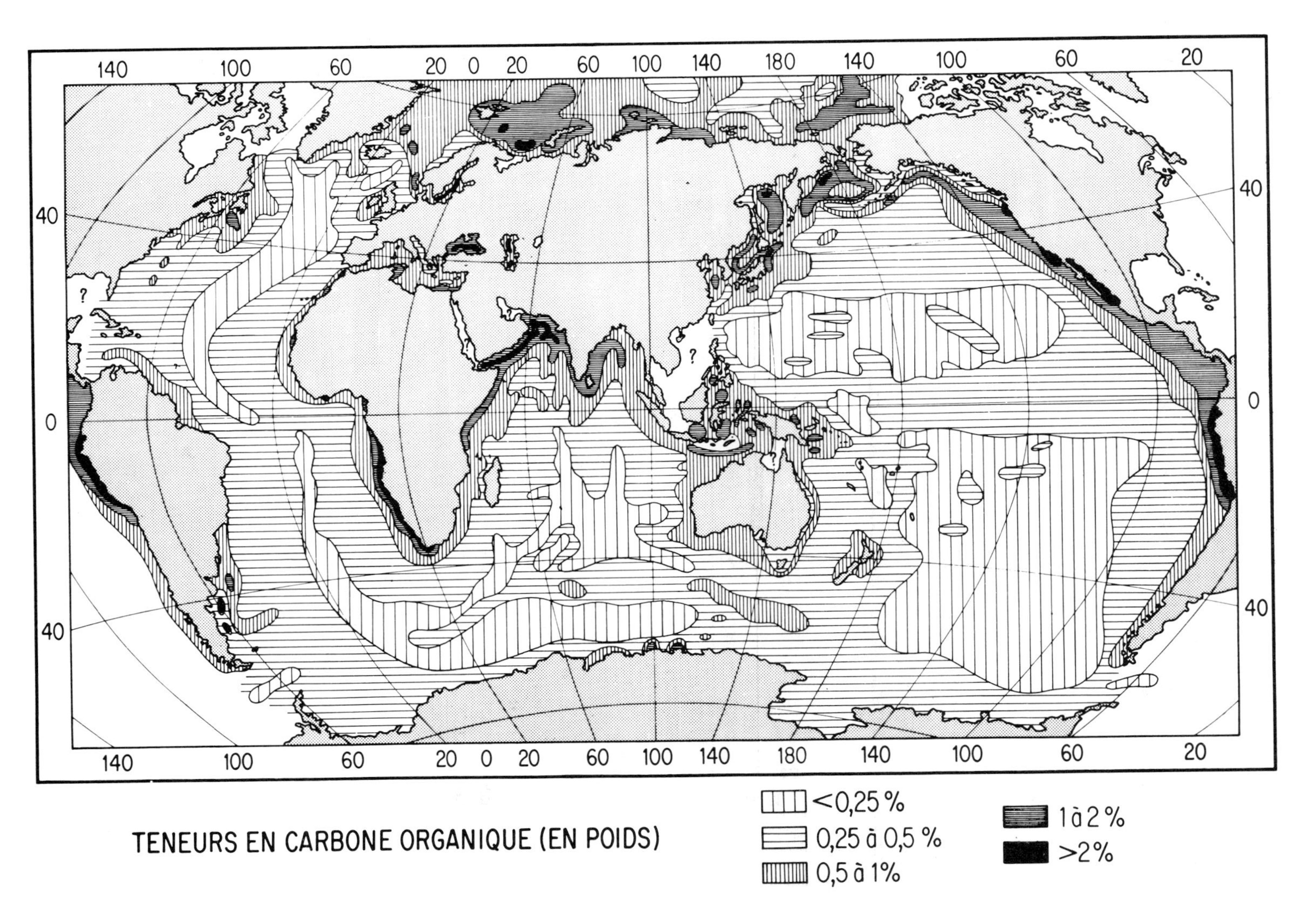

Fig. 15.2. — Carte des teneurs en carbone organique des 5 premiers centimètres du fond océanique (d'après Romankevic [31] et Orgon).

bassin de sédimentation; si ce mode d'alimentation est très important pour les petits bassins lagunolacustres intracontinentaux ou paraliques, il l'est moins pour la grande sédimentation détritique marine. Le plus souvent, le matériel organique continental est engagé dans le cycle pédologique, pendant terrestre du recyclage biologique de la MO marine, en plus complexe parce que la diversité des conditions climatiques et le rôle nécessaire de la surface minérale de la Terre n'ont aucun correspondant en mer. Au fil des dégradations biologiques successives, la MO jonchant le sol et qui n'est pas minéralisée va donner plus ou moins directement du kérogène (humines, acides humiques, acides fulviques *pro parte,* dans la terminologie des pédologues) immobilisé sur place ou à courte distance, et des produits solubles exportés par les cours d'eau qui représentent la fuite hors du cycle. Les proportions relatives de ces produits dépendent du climat et, dans les limites qu'il permet, de la nature de la végétation et de la topographie. Par exemple, la grande forêt équatoriale primitive, qui représente une biomasse imposante, connaît un recyclage particulièrement efficace avec une accumulation minimale de kérogène et une exportation de produits solubles faible : contrairement aux apparences, elle est donc une source de MO sédimentaire peu prolifique.

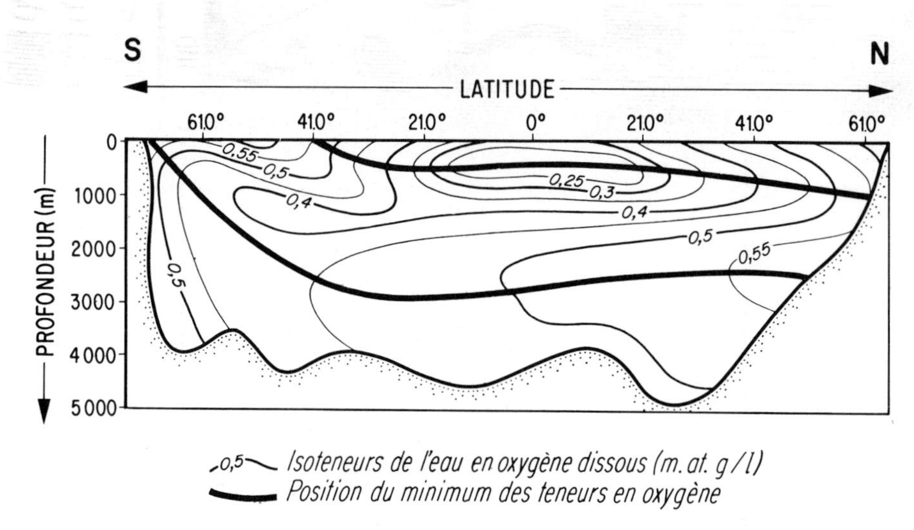

Fig. 15.3. — Teneurs en oxygène d'une coupe méridienne de l'Océan Atlantique (très simplifié, d'après Bubnov et Krivelevic [4]).

Vienne une variation du niveau des mers ou de la topographie, et tout change. L'érosion s'attaque à ce qui était jusque-là sol de végétation et démantèle les accumulations de kérogène pédologique, dont les débris sont évacués pêle-mêle avec les minéraux. Si cette source de MO est certainement moins importante que la précédente, elle est aussi certainement très constante : les destructions d'anciens sols et leur pendant, la colonisation d'affleurements azoïques, se voient en effet autour de nous tous les jours, ne serait-ce que dans la divagation des méandres des rivières.

Ces trois formes d'apports continentaux — par ordre d'importance : en solution, sous forme de débris de kérogène, en fragments biologiques non transformés — vont passer du

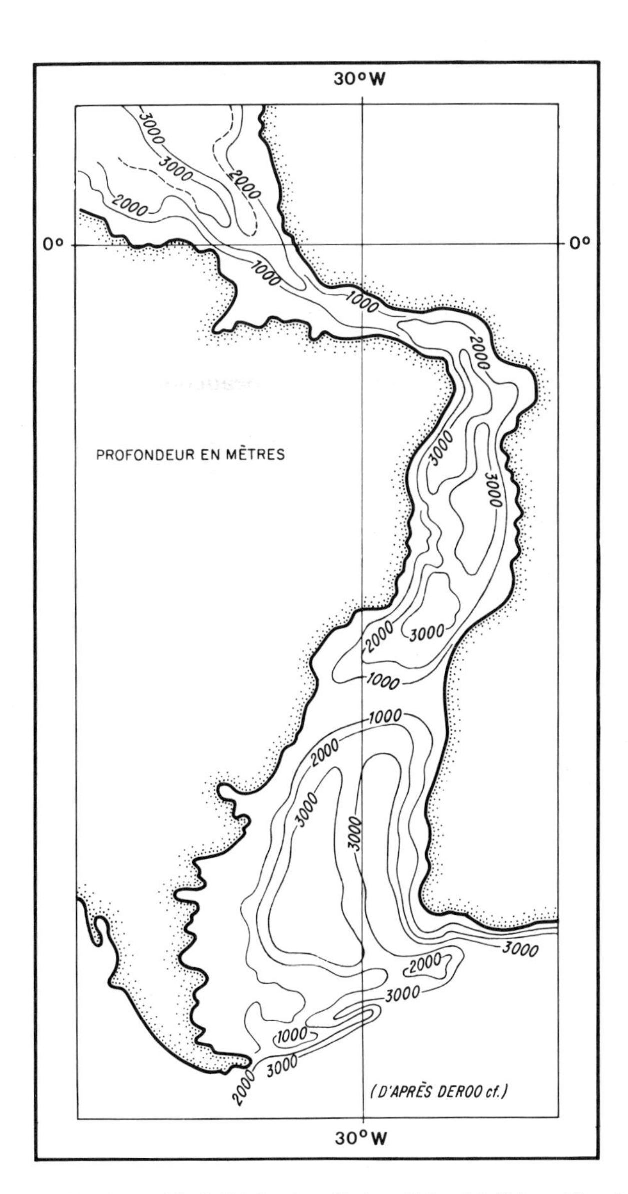

Fig. 15.4. — Paléogéographie de l'Atlantique Sud au Crétacé Inférieur (d'après Deroo *et al.* [10]).

fleuve à la mer; si ce passage n'a guère d'effets sur le kérogène ou la matière biologique en l'état, il n'en est certainement pas de même pour la fraction soluble [1]. Encore que les rares études faites ne soient pas complètement éclairantes, il semble qu'une partie au moins de ce matériel soluble — qui est de la nature du kérogène — va floculer, accroissant ainsi la quantité totale de matériel susceptible de sédimentation. A partir de ce moment, MO continentale et marine se rejoignent, ou plutôt la MO marine va rejoindre les apports détritiques continentaux, tant minéraux qu'organiques. On a vu plus haut que dans la nature actuelle, au total, la MO continentale domine. On comprend que cela signifie aussi que, toujours actuellement, les processus de sédimentation détritique dominent largement les processus chimiques. Là encore, il n'en a pas toujours été ainsi.

Au total, dès sa sédimentation, la matière MO sédimentaire apparaît comme essentiellement formée de kérogène. S'y ajoutent, en quantités faibles, des composés plus simples mais inertes (hydrocarbures, autres lipides) et en quantités encore plus faibles des métabolites (sucres, acides aminés, lipides insaturés). Cependant, il ne faut pas oublier que ce kérogène présente des différences de composition fondamentales suivant son origine. Beaucoup plus apliphatique — par manque de composés aromatiques — quand il est d'origine marine, aromatique — encore qu'avec des nuances — quand il est un ancien kérogène terrestre, il présentera une composition intermédiaire lorsqu'il est issu de la floculation de produits solubles, d'origine surtout continentale [27]. Le kérogène est simplement la forme adéquate pour retirer la MO du cycle biologique.

B. La diagenèse précoce est l'époque de l'azote.

Intuitivement et pratiquement, il n'est guère difficile de distinguer la mer de son fond. On peut cependant imaginer le cas d'un passage continu, la densité du matériel d'abord en suspension augmentant sans cesse. Même dans ce cas apparaîtra en fait une discontinuité : lorsque les éléments constitutifs de la suspension sont suffisamment proches les uns des autres, ils constituent une barrière au mouvement du fluide qui les baigne et, à un moment donné, aux ions en solution : il y a apparition d'une interface entre le milieu marin et ce qui, par définition, sera le milieu sédimentaire. Une fois acquis, ce caractère se conserve en général : le fluide des loupes de glissement ou des courants de turbidité est distinct de l'eau de mer. Ceci dit, pendant un certain temps, des échanges vont encore avoir lieu entre le sédiment fraîchement déposé et l'océan qui le baigne. Cette période sera pas définition celle de la diagenèse précoce. Elle voit la fin de l'évolution biologique de la MO, et donc le triomphe du kérogène. Mais déjà s'amorcent les processus de défonctionnalisation caractéristiques de la phase suivante — la diagenèse — de l'évolution, abiotique et proprement souterraine.

Dans un premier temps, le sédiment nouvellement formé contient encore des êtres vivants. Si l'on laisse de côté les zones côtières à tranches d'eau suffisamment minces pour laisser pénétrer la lumière — zones où la sédimentation est généralement très réduite — on a donc affaire à un milieu obscur même à l'interface eau-sédiment, froid, à forte pression et où les substrats organiques, essentiellement composés de kérogène, sont de très faible valeur nutritive. Ces caractéristiques sont rébarbatives pour la plupart des êtres vivants, et effectivement

---

(1) On emploiera ce terme par commodité : on entend par là tout ce qui se comporte comme formant avec l'eau une phase unique au point de vue hydrodynamique, qu'il s'agisse de solutions vraies ou de colloïdes, et de matériel organique pur ou des combinaisons organominérales.

les populations qu'on rencontre ne sont ni abondantes, ni métaboliquement très actives (ORGON). Leur action va toutefois contribuer à réduire encore plus dans la composition de la matière organique sédimentaire la part des métabolites et à augmenter la part du kérogène — avec bien sûr minéralisation d'une fraction relativement importante. Que ces processus ne soient pas quantitativement significatifs se voit directement au fait qu'il n'y a pas d'abaissement systématique de la teneur en MO quand on passe des niveaux encore biologiquement actifs aux niveaux sous-jacents azoïques. Au total, un dernier effort, mais discret, pour assurer la prépondérance du kérogène.

Ce triomphe quantitatif du kérogène, qui nécessite le concours des organismes vivants, s'accompagne d'un triomphe qualitatif, fruit de réactions abiotiques : le kérogène va perdre petit à petit les groupes fonctionnels qui lui donnent une certaine réactivité, et devenir de plus en plus inerte. Au stade de la diagenèse précoce, cet effet porte principalement sur l'azote des liaisons peptidiques — CONH — qui va être éliminé essentiellement sous forme d'ion ammonium (ORGON 1, 2). Cette perte sera bien entendu en raison directe des teneurs initiales, et sera donc différente pour les quatre (1) types que nous avons distingués plus haut : faible pour les débris végétaux en l'état, faible à moyenne pour les kérogènes terrestres, faible à forte pour les produits de floculation suivant qu'ils sont d'origine continentale ou marine. La teneur en azote qui, au moment de la sédimentation, est un bon critère d'origine, va se trouver rapidement uniformisée par ce processus, et perdra donc son caractère diagnostique.

La fraction lipide (2), quantitativement peu importante, mais si intéressante pour l'analyste, voit elle aussi intervenir des réactions de défonctionnalisation (décarboxylation des acides, déshydratation des alcools) qui amènent la néogenèse — en quantités infimes à ce stade — d'hydrocarbures portant les traits spécifiques de leur origine.

A la fin de la diagenèse précoce, la MO sédimentaire, qui ne sera plus désormais soumise à l'action des organismes vivants, est composée quasi totalement d'un matériel polycondensé en désordre — le kérogène (encore soluble en partie dans les liqueurs alcalines s'il a suffisamment de fonctions acides disponibles); des lipides — à cause de leur relative inertie chimique — sont aussi présents, mais en quantité très limitée; les métabolites ont presque totalement disparu. Les débris végétaux ont perdu métabolites et cellulose (ou équivalents); leur squelette de lignine, composé à peine modifié dans les conditions sédimentaires (3) puisque c'est en quelque sorte un protokérogène, leur permet cependant de retenir leurs formes primitives; ils évoluent lentement vers le lignite. Ce dernier cas est celui qui permet la plus grande préservation dans la sédimentation de la MO considérée; mais on a vu qu'en moyenne seule une faible fraction de celle-ci échappait au recyclage biologique. L'autre extrême peut être représenté par des milieux laguno-lacustres à haute productivité mais à recyclage si efficace que même le kérogène sera en partie consommé et où ne pourront s'accumuler que les lipides de poids moléculaire élevé, dont personne ne veut, et spécialement les lipides des micro-organismes, dernier maillon de la chaîne trophique.

---

(1) Cet embryon de classification, purement génétique, n'a guère d'intérêt que didactique. Le lecteur intéressé par ces problèmes de classification pourra se reporter, je n'ose pas dire utilement, au travail de Subbota *et al.* [36].

(2) C'est-à-dire soluble dans les solvants organiques.

(3) La destruction de la lignine s'effectue normalement sur le continent et en conditions aérobies par action de champignons spécialisés. Elle n'a pas lieu dans les corps aquatiques permanents et stagnants (marécages, étangs, lagunes) même aérés, milieux impropres au développement des lignivores.

Les teneurs correspondantes en matière (carbone) organique sont finalement étroitement dépendantes des facteurs d'origine et de conditions de sédimentation qu'on vient d'exposer. Le matériel soluble ne semble pas pouvoir donner des teneurs supérieures à 1 % (ORGON 2); il est présent à des teneurs de l'ordre de 0,3-0,5 %, dans les plaines abyssales au pied de la pente continentale (ORGON 1, 2, 3). Il est masqué par les produits détritiques continentaux dès que les taux de sédimentation augmentent; la teneur totale dans les grands épandages détritiques peut atteindre 2-3 %; signalons au passage que, pour des raisons qui ne sont pas toutes claires, il y a accroissement de la teneur avec l'augmentation des taux de sédimentation (ORGON 1, 2, 3 et [14]) — dans certaines limites évidemment — ce qui fait des grands édifices détritiques les dépositaires de quantités gigantesques de MO. Au large et loin des côtes, il n'y a pratiquement plus rien (teneurs < 0,01 % dans certains échantillons du Pacifique central). Par contre, les forts taux en carbone, qu'ils soient le fait de matériel continental (lignite) ou aquatique, demandent nécessairement dans la nature actuelle des bassins de faible dimension et des conditions locales exceptionnelles. La quantité totale de MO mise en réserve par ces milieux riches est — encore une fois dans la nature actuelle — bien inférieure à celle emmagasinée dans les séries détritiques.

IV. LA DIAGENÈSE

> « L'empereur Claude était lettré. Il citait les poètes. Il fit ajouter trois lettres à l'alphabet, mais elle ne furent jugées utiles que pendant la durée de son règne. »
>
> L. Magre (La vie amoureuse de Messaline)

La limite entre diagenèse précoce et diagenèse proprement dite n'est pas définie en termes d'une propriété intrinsèque de la MO, ou d'une de ses fractions déterminable sans ambiguïté. Cet état de fait reflète le peu de connaissances actuellement acquises sur le stade diagénétique, période la plus mal connue de l'évolution de la MO sédimentaire. Par contre, pour des raisons inverses, la limite entre diagenèse et catagénèse est aujourd'hui à peu près universellement repérée par une réflectance de la vitrinite de 0,5 %. La figure 15.5, établie d'après la compilation de Vassoyevitch [38], reprenant elle-même les travaux de Vassoyevitch, Kalinko, Karcev, Kontorovic, Lopatin, Nerucev, Hood, Gutjahr, Teichmüller, Alpern, Stach et Van Krevelen, est l'objet aujourd'hui d'un large consensus pour ce qui est des valeurs de la réflectance de la vitrinite qui repèrent les limites de stades. Il n'en est malheureusement pas de même de la terminologie. Celle qu'on propose de retenir n'a d'autres mérites que d'être généralement bien comprise, et de ne pas être trop difficile à prononcer. On ne discutera pas plus avant ici de ces questions, qui sont en fait inextricables : pour s'en convaincre, qu'on compare les différentes classifications nationales des charbons ([37], et fig. 15.5 pour les systèmes allemand, américain et français), toutes raccordées cependant à des propriétés objectivement mesurables des charbons, mais avec des nomenclatures souvent différentes et des limites subtilement décalées lorsque par hasard une classe porte le même nom dans deux systèmes.

D'un point de vue qualitatif, la diagenèse ne récèle pas de mystères. A mesure qu'elle suit son cours, le kérogène devient de plus en plus inerte, perdant aussi bien sa susceptibilité à

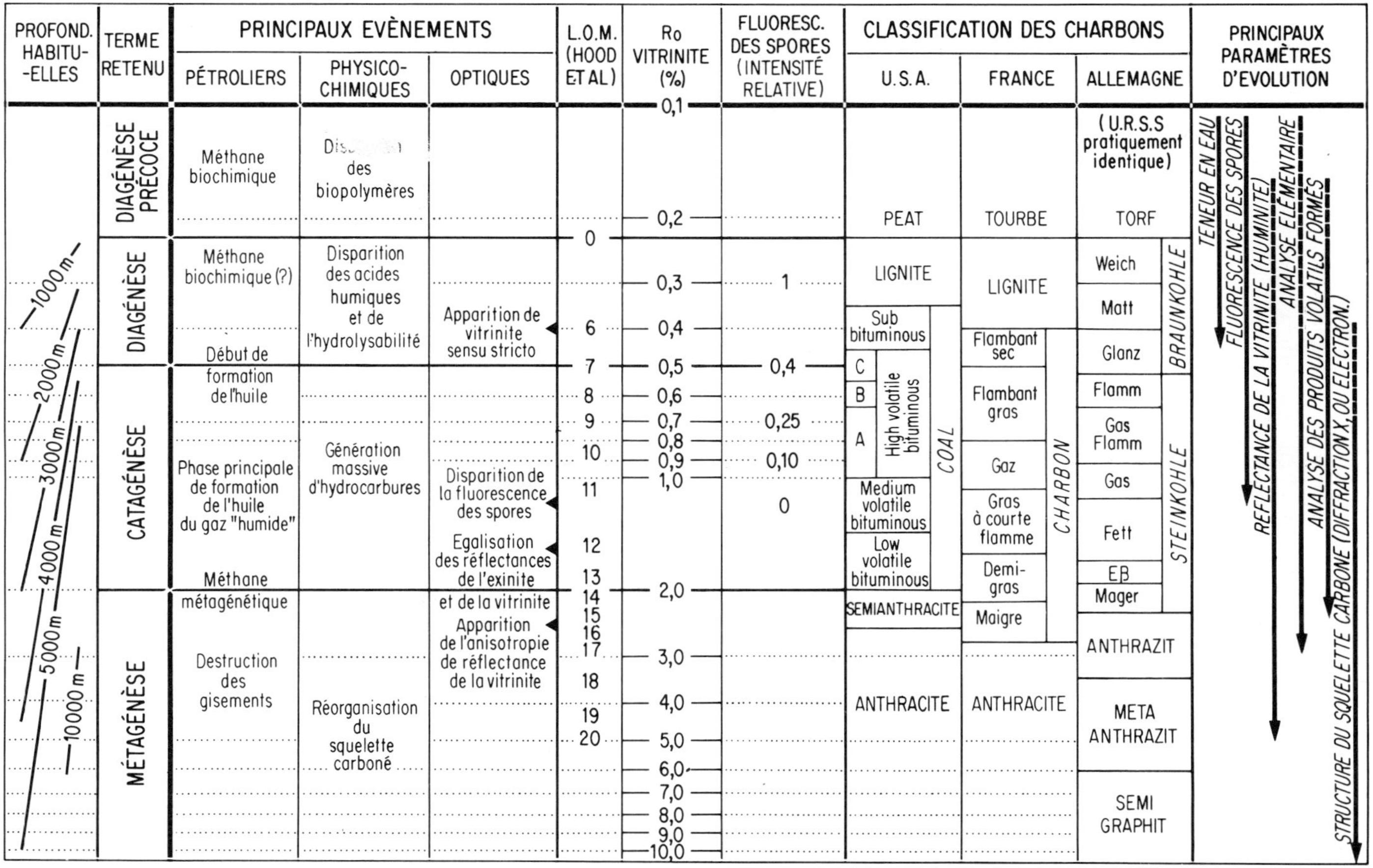

PROFOND. HABITU-ELLES: 1000 m; 2000 m; 3000 m; 4000 m; 5000 m; 10000 m

TERME RETENU: DIAGÉNÈSE PRÉCOCE; DIAGÉNÈSE; CATAGÉNÈSE; MÉTAGÉNÈSE

PRINCIPAUX EVÈNEMENTS

- PÉTROLIERS: Méthane biochimique; Méthane biochimique (?); Début de formation de l'huile; Phase principale de formation de l'huile du gaz "humide"; Méthane métagénétique; Destruction des gisements
- PHYSICO-CHIMIQUES: Dis… des biopolymères; Disparition des acides humiques et de l'hydrolysabilité; Génération massive d'hydrocarbures; Réorganisation du squelette carboné
- OPTIQUES: Apparition de vitrinite sensu stricto; Disparition de la fluorescence des spores; Egalisation des réflectances de l'exinite et de la vitrinite; Apparition de l'anisotropie de réflectance de la vitrinite

L.O.M. (HOOD ET AL): 0; 6; 7; 8; 9; 10; 11; 12; 13; 14; 15; 16; 17; 18; 19; 20

Ro VITRINITE (%): 0,1; 0,2; 0,3; 0,4; 0,5; 0,6; 0,7; 0,8; 0,9; 1,0; 2,0; 3,0; 4,0; 5,0; 6,0; 7,0; 8,0; 9,0; 10,0

FLUORESC. DES SPORES (INTENSITÉ RELATIVE): 1; 0,4; 0,25; 0,10; 0

CLASSIFICATION DES CHARBONS

- U.S.A.: PEAT; LIGNITE; COAL: Sub bituminous; High volatile bituminous C, B, A; Medium volatile bituminous; Low volatile bituminous; SEMIANTHRACITE; ANTHRACITE
- FRANCE: TOURBE; LIGNITE; CHARBON: Flambant sec; Flambant gras; Gaz; Gras à courte flamme; Demi-gras; Maigre; ANTHRACITE
- ALLEMAGNE (U.R.S.S pratiquement identique): TORF; BRAUNKOHLE: Weich, Matt, Glanz; STEINKOHLE: Flamm, Gas Flamm, Gas, Fett, EB, Mager; ANTHRAZIT; META ANTHRAZIT; SEMI GRAPHIT

PRINCIPAUX PARAMÈTRES D'EVOLUTION: TENEUR EN EAU; FLUORESCENCE DES SPORES; REFLECTANCE DE LA VITRINITE (HUMINITE); ANALYSE ELEMENTAIRE; ANALYSE DES PRODUITS VOLATILS FORMÉS; STRUCTURE DU SQUELETTE CARBONE (DIFFRACTION X, OU ELECTRON.)

Fig. 15.5. — Différents paramètres d'évolution de la matière organique sédimentaire (inspiré de Vassoyevitch [38]).

l'hydrolyse acide que son extractibilité par les solvants alcalins; la disparition quasi totale de ces deux propriétés coïncide avec la fin de la diagenèse. Cette évolution du kérogène, qui est une défonctionnalisation et une condensation, s'accompagne du rejet dans le milieu extérieur d'eau, de gaz carbonique, de méthane, moteur de cette évolution. Tout est simple : la machine entropique va paisiblement.

Si l'on veut maintenant aborder le calcul quantitatif ou, ce qui revient au même, le détail des mécanismes réactionnels, on demeure à peu près impuissant. Un point négatif est acquis : les simulations (chauffages) de laboratoire habituelles reproduisent imparfaitement l'évolution naturelle. Les résultats des pyrolyses expérimentales ne nous sont donc pas d'un grand secours. Il n'existe aucune observation d'ensemble d'une série géologique en cours de diagenèse et, à notre connaissance, une seule partielle [20]. Aucune description ne nous aidera donc. Deux points sont pourtant essentiels, aussi bien du point de vue pratique que de la connaissance pure. Le premier est de savoir, à partir d'une masse donnée de MO en début de diagenèse, ce qu'il en subsistera à la fin : les variations de composition élémentaire (les chemins d'évolution dans le diagramme de Van Krevelen) ne disent évidemment rien là-dessus. Le deuxième point est de savoir quelles quantités d'eau, de gaz carbonique — surtout de méthane et, pourquoi pas, d'autres hydrocarbures — ont été corrélativement produites. Il semble en effet à peu près certain maintenant, sur des indices très indirects, que la production de méthane n'est pas négligeable — quoique la controverse existe toujours; la chose est beaucoup plus douteuse pour les hydrocarbures de rang supérieur mais, sur une échelle modeste, apparaît comme plausible (par transformation directe des lipides fonctionnels : alcools, acides, esters). Il est d'ailleurs probable que cette aptitude à la génération précoce d'hydrocarbures ne doit pas être équivalente suivant l'origine du matériel organique.

Un autre problème intéressant, de solution immédiate si l'on connaît les aspects quantitatifs de l'évolution, mais qui peut être abordé plus simplement, est celui de la correspondance entre origine de matériel organique et « lignée » du kérogène telle que définie par Durand dans cet ouvrage à partir de l'analyse élémentaire sur le diagramme de Van Krevelen. Si, en effet, un certain nombre de correspondances typiques sont communes (matériel autochtone marin et lignée II, matériel continental charbonneux et lignée III par exemple), il n'est pas certain que l'évolution diagénétique, suivant les conditions qui la régiront, ne puisse pas, à partir d'un matériel identique, donner deux kérogènes de lignée différente — ou, réciproquement, à partir de matériaux d'origine différente, produire un seul et même type de kérogène. La question reste ouverte.

L'élément clé de toutes ces transformations est l'oxygène : la diagenèse est, avant tout, l'étape de la perte des fonctions oxygénées. On conçoit donc que l'évolution diagénétique sera d'autant plus importante que le matériel de départ sera plus oxygéné : tout ce qui a été dit plus haut s'appliquera donc spécialement à la lignée III, et n'aura par contre qu'une influence secondaire sur la lignée I. La manière dont se fait ce départ d'oxygène est capitale pour le destin ultérieur du kérogène : s'il y a perte de gaz carbonique, la quantité de kérogène décroît fortement, mais le rapport H/C s'élève, et l'aptitude à la genèse d'hydrocarbures; au contraire, s'il y a perte d'eau, la quantité de kérogène décroît moins, mais le rapport H/C s'abaisse très fortement, comme l'aptitude à la genèse d'hydrocarbures. L'ensemble de ces points est précisé à la figure 15.6.

A la fin de la diagenèse, le kérogène a acquis son inertie chimique caractéristique. Pratiquement dépouvu de groupements fonctionnels, il n'est plus ni hydrolysable, ni extractible aux liqueurs alcalines; les hétéroéléments qu'il contient encore sont dans des positions stables,

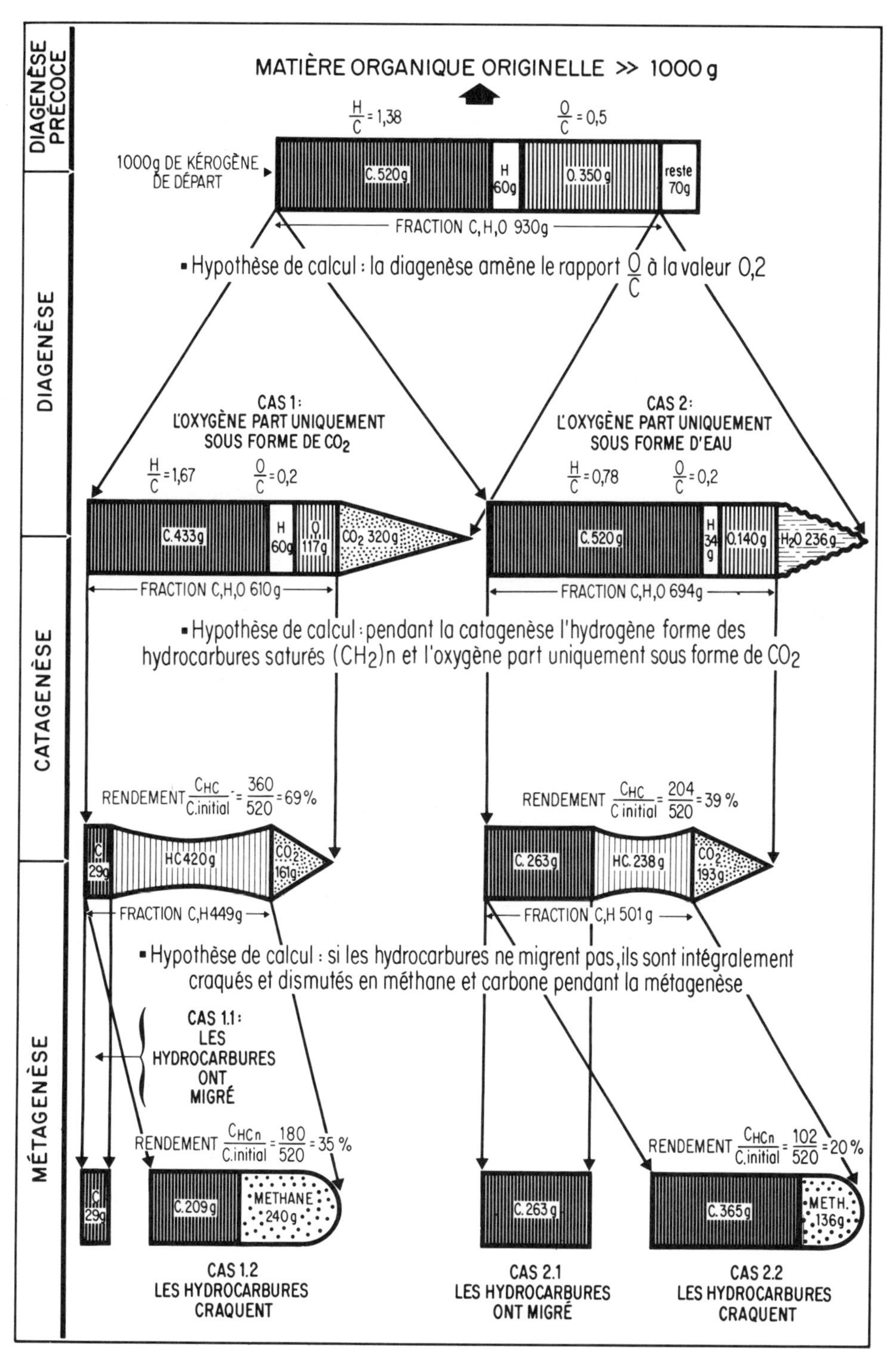

Fig. 15.6. — Evolution quantitative d'un kérogène de lignée II à partir de deux hypothèses extrêmes sur la diagenèse.

non fonctionnelles; leurs liaisons avec les atomes de carbone sont fortes et demandent pour leur rupture des énergies d'activation du même ordre de grandeur que celles nécessaires pour rompre les liaisons carbone-carbone : seul le craquage thermique peut désormais faire évoluer cet édifice.

V. LA CATAGENÈSE

« Il faut chercher sur l'objet de notre étude, non pas ce qu'en ont pensé les autres, ni ce que nous soupçonnons nous-mêmes, mais ce que nous pouvons voir clairement et avec évidence, ou déduire de manière certaine. C'est là le seul moyen d'arriver à la Science. »

R. Descartes (Règles pour la direction de l'esprit,
règle III, trad. S. de Sacy

La catagenèse est l'étape la plus étudiée et la mieux connue de l'évolution géochimique. Les raisons en sont multiples : pratiques, les plus déterminantes — alors apparaissent charbon et pétrole — mais aussi opératoires — les pyrolyses de laboratoire deviennent des simulations convenables, et analytiques — l'isolement du kérogène peut se faire sans pertes, l'étude — relativement facile — de l'extrait lipidique apporte de précieux renseignements.

Si la diagenèse précoce est l'époque de l'azote et la diagenèse celle de l'oxygène, la catagenèse apparaît comme celle de l'hydrogène. L'évolution du kérogène va en effet fournir pendant cette période essentiellement des hydrocarbures, avec un abaissement concomitant de la teneur en hydrogène, d'autant plus important que le kérogène était initialement riche en cet élément. Ce phénomène est étudié sous tous ses aspects et avec le plus grand détail par les autres auteurs de cet ouvrage : on n'y reviendra donc pas. Qu'il nous suffise de dire ici que la perte de substance du kérogène est, durant cette étape, considérable et que le résidu carboné qui subsistera en fin de catagenèse doit représenter, en masse de carbone, entre la moitié et le 1/10e du carbone initialement sédimenté (*cf.* fig. 15.6), ce qui, en masse de MO, et encore plus en volume, donne des chiffres encore plus faibles. Le kérogène résiduel, très carboné, va se réorganiser autour des îlots aromatiques qu'il contenait déjà : la structure hexagonale du graphite est en effet la structure thermodynamiquement stable dans les conditions de la colonne sédimentaire. Le kérogène va donc « s'aromatiser » : mais on voit bien que ce n'est qu'une apparence, le kérogène s'est surtout volatilisé, transformé en produits fluides, liquides ou gazeux; le kérogène d'aujourd'hui n'a plus que de lointains rapports avec celui d'hier. A mesure de l'évolution, la perte de substance s'accompagne d'une perte d'identité; à la diversité fourmillante des faciès organiques des sédiments actuels s'oppose l'uniformité anonyme des carbones pseudographitiques, où seule la répartition des défauts cristallins rappelle encore vaguement l'ordonnance classique des trois lignées de l'âge mûr.

Les trois lignées principales de kérogènes (*cf.* chap. 4) ont été en effet définies à partir d'échantillons au stade catagénétique et doivent être considérées comme caractéristiques de ce stade. Les observations comme les expériences de laboratoire ont en effet montré que les kérogènes d'une seule lignée donnée suivaient en cours d'évolution un et un seul chemin (par exemple tracé sur le diagramme de Van Krevelen). Il n'en est pas du tout de même au stade de la diagenèse précoce, il pourrait bien ne pas en être de même au stade de la diagenèse. Cette propriété a d'ailleurs reçu une application avant même d'être connue sous son aspect général, c'est le principe de la classification des charbons : ceux-ci constituant une lignée, un seul paramètre suffit à les classer de façon biunivoque.

Avec la catagenèse s'achève l'ère des pertes importantes de substance. Désormais, l'évolution du kérogène va surtout être interne; à l'évolution chimique va succéder l'évolution physique-minéralogique.

VI. LA MÉTAGENÈSE

« Le dyamant... encore a il une dureté invincible laquelle ne résiste pas seulement à la lime, ny aux métaulx, mais qui plus est, elle ne peult estre vaincue des flammes... de sorte que les philosophes ont expérimenté qu'il peult durer neuf jours assidus dans des brasiers ardens sans en estre offencé. »
J. Boaistuau, dit Launay (Histoires prodigieuses)

Les mêmes raisons pratiques qui causent l'intérêt des pétroliers pour la catagenèse motivent leur désintérêt pour la métagenèse : le stade de la formation principale de l'huile est passé, il ne se forme plus que du gaz, et de moins en moins; on peut d'ailleurs se demander jusqu'à quel point ce gaz provient de l'évolution du kérogène lui-même, ou bien du craquage des hydrocarbures libres déjà présents — avec formation correspondante de « kérogène » résiduel. Les courbes comme celles de la figure 15.7 où, apparemment, la formation de gaz représente de nouvelles quantités d'hydrocarbures, ne doivent en effet pas faire illusion : elles représentent des quantités rapportées à l'unité de masse de carbone organique; or cette dernière quantité ne représente pas du tout une quantité identique de matériel organique initial. Comme on l'a vu plus haut (*cf.* fig. 15.6) dans un volume (une masse) de roche donné, le volume (la masse) de kérogène présent en tant que tel décroît fortement au cours de l'évolution, encore que quantitativement cette décroissance soit mal connue, en particulier pour ce qui se passe durant la diagenèse. Enfin, autant il est raisonnable d'espérer récupérer quantitativement, dans les roches mères, les hydrocarbures extractibles au chloroforme ayant échappé à la migration, autant il est vain de vouloir en faire autant pour les composés les plus volatils qui ont tout le temps de diffuser lors des opérations de forage et de carottage.

Au total, on n'a donc que peu de données sûres sur la production de nouveaux hydrocarbures (gazeux) lors de la métagénèse. Il se pourrait même que la « phase principale de formation de gaz » créée à l'imitation de celle — bien assise — de formation de l'huile, n'ait pas une importance quantitative plus grande que la formation précoce, diagénétique. Peut-être les simulations de laboratoire nous apporteront-elles une réponse dans l'avenir; de toute manière, la genèse de nouveaux produits mobiles à partir du kérogène apparaît comme très limitée au cours de la métagenèse; il est d'ailleurs aisé de s'en convaincre *a priori* à partir des analyses élémentaires : le kérogène tend de plus en plus vers le carbone pur; la métagenèse est l'étape du carbone.

On a pensé pendant longtemps que la réorganisation du réseau cristallin de ce carbone qui s'épure irait, à mesure de son achèvement, jusqu'à la perfection du graphite minéralogique. Les études structurales du type de celles décrites dans cet ouvrage par Oberlin ont montré qu'il n'en était rien. Dans les conditions habituelles de la colonne sédimentaire, les îlots prégraphitiques laissés par le départ des atomes étrangers n'arrivent pas à entrer en coalescence :

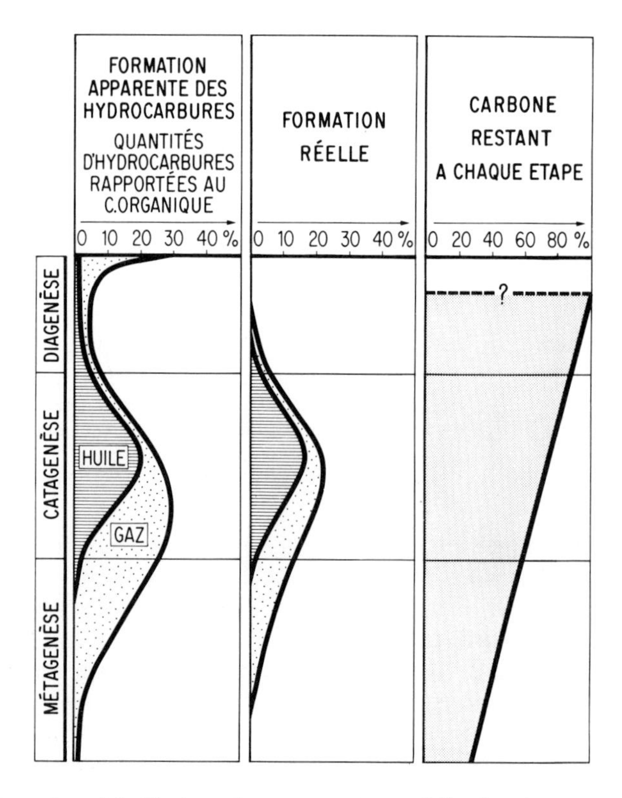

Fig. 15.7. — Quantités d'hydrocarbures apparentes et réelles formées au cours de l'évolution géochimique ([38] et Fig. 15.6).

quelques-uns de ces atomes restant en position bloquée, des imbrications ou des arboutements de feuillets empêchent les réorientations et ne permettent pas l'apparition de cristaux macroscopiques. Le plus intéressant est que l'existence et l'arrangement de ces défauts ne se font pas au hasard, mais dépendent de l'histoire passée du kérogène — et donc, d'une certaine façon, de l'origine de la MO qui lui a donné naissance.

On voit ainsi que, même parvenu à un degré d'évolution si avancé que la chimie ne peut plus percer son anonymat, le kérogène, dans son arrangement le plus intime, se souvient encore vaguement de son origine organique et peut nous permettre d'en avoir quelque idée.

Il nous reste maintenant, en changeant d'échelle, à évoquer le destin ultime du kérogène lors du métamorphisme général et, plus loin encore, dans les processus magmatiques profonds; il n'y a en effet pas de raison pour que, lors de la subduction d'une plaque continentale sous une autre, de la MO sédimentaire ne soit transportée dans les tréfonds de la planète. Apparemment, dans les deux cas, cette MO va alors accéder au repos anonyme de la perfection minérale, sous l'humble aspect du graphite dans les zones métamorphiques, sous la forme éblouissante du diamant indomptable dans l'obscurité chaude et pesante où se meuvent mollement les magmas.

Et cependant... au stade du graphite, alors même que les paramètres cristallins ne changent plus, parfaits qu'ils sont, le rapport isotopique du carbone varie encore, et de telle façon qu'il peut permettre une appréciation du degré de métamorphisme plus détaillée que les indicateurs minéraux habituels [33]. Quant au diamant, sujet d'innombrables études de par sa valeur symbolique — et marchande — on a reconnu (*cf.* par exemple [2]) qu'il comprenait deux types fondamentaux, distingués par la présence ou l'absence d'azote. Le réseau du diamant paraît être accueillant pour l'azote : les diamants synthétisés en présence d'azote en contiennent (10^{-6} à 10^{-8} atome/atome); bien entendu, ceux synthétisés en l'absence d'azote n'en contiennent pas. Or la plupart (mais pas tous) des diamants naturels en contiennent. D'où vient cet azote? On peut évidemment, aux teneurs envisagées, le faire venir des minéraux environnants. N'est-il pas plus naturel de penser à un ultime reste d'azote organique dans le kérogène qui recristallise? Comme on l'a vu plus haut, les dernières traces d'hétéroatomes, et spécialement d'azote, ne s'éliminent plus au stade de la métagenèse.

VII. CONCLUSION

« Tant plus haut je m'élevais ainsi
Et tant moins je comprenais.
C'est là ce nuage ténébreux
Qui rend la nuit toute claire
Or, cependant, qui vient à le connaître
Demeure toujours sans rien voir
Transcendant toute science. »

Saint Jean de la Croix (Poèmes, Couplets faits sur une extase de haute contemplation, trad. ancienne)

1. Avant de conclure vraiment, il faut souligner deux points qui ressortent insuffisamment, en son ordonnance, de l'exposé qui précède. Le premier est la prodigieuse sensibilité de la MO à l'évolution géologique. Alors que les minéraux perdent ou acquièrent quelques molécules d'eau, réparent tant bien que mal ou modifient très peu leur réseau cristallin, donnant lieu au total à des modifications si mineures que les spécialistes en disputent, la MO subit les importantes métamorphoses qu'on vient de décrire. Etudier l'histoire des conditions physicochimiques de la colonne sédimentaire indépendamment de son contenu organique est comme chronométrer une épreuve de ski avec un sablier.

Le second point, qui ne peut être encore précisé, et qui devra dans le proche avenir faire l'objet de grandes investigations, est le rôle de l'eau. La vie naît des eaux; le grand mérite des recherches actuelles est de dire pourquoi précisément (*cf.* par exemple [6, 12, 13]). Il en résulte que toute évolution biochimique de la MO nécessite la présence de l'eau : la momification le démontre *a contrario.* Or la dégradation biochimique antésédimentaire ou pendant la diagenèse précoce est nécessaire à la suite des événements. Pas d'eau, pas d'évolution de la MO sédimentaire. Mais, outre ce rôle initiateur, il semble que l'eau joue ensuite, au moins au stade de la diagenèse, un rôle orientateur sur la suite des réactions de réorganisation; on a vu que la diagenèse était l'époque du départ de l'oxygène; ce départ peut se concevoir soit sous forme d'eau, soit sous forme de gaz carbonique. Ces deux processus ne sont nullement équivalents pour la suite de l'évolution, en particulier catagénétique; dans le premier cas, il y a départ d'hydrogène, donc diminution de la quantité totale d'hydrocarbures productibles (et/ou de leur teneur en hydrogène); dans le deuxième cas, il y a départ de carbone seulement, donc maintien de la quantité d'hydrocarbures productibles, le carbone étant toujours en excès (et/ou augmentation de leur teneur en hydrogène). En milieu ouvert, il est impossible de prévoir lequel des deux processus sera favorisé; si, par contre, on impose la présence d'eau liquide, alors on peut prévoir de façon globale que, toutes choses égales par ailleurs, la production d'eau sera défavorisée. Ce champ de recherches, très prometteur, est presque vierge.

2. La MO, au sens large qu'on lui a donné ici, est un constituant normal de l'univers. Le kérogène en est la forme stable, ou plutôt métastable, dans une large gamme de températures et de pressions. Après le bref épisode ordinateur de la vie, l'évolution sédimentaire verra le triomphe du kérogène, puis son déclin et le retour de la MO au carbone élémentaire.

Elle y revient peu à peu. Au début, musée des souvenirs de la vie, elle voit se briser petit à petit les objets signifiants. Pendant la diagenèse précoce, époque de l'azote, disparaissent les témoins des protéines, maîtresses des biosynthèses. Pendant la diagenèse, époque de l'oxygène, sont évacuées les parties fonctionnelles des glucides et des lipides, matériaux constitutifs des édifices vivants. Pendant la catagenèse, époque de l'hydrogène, l'expulsion des hydrocarbures enlève jusqu'aux trames fondamentales. Pendant la métagenèse, époque du carbone, s'achève le nettoyage des hétéroatomes. Là s'arrête l'histoire sédimentaire.

Elle peut reprendre sans fin à toutes ces étapes; que survienne en effet une surrection tectonique et kérogène comme produits de son évolution seront repris dans le cycle de la vie, d'autant plus facilement qu'ils seront encore proches des structures vivantes, ou qu'ils auront été transformés en gaz carbonique. Il y a une espèce de grande respiration du cycle sédimentaire qui assure la prééminence du kérogène, mais maintient la vie. Mais il ne faut pas sortir de ce cycle.

Plus loin, dans les abysses hypogées aux températures et pressions inimaginables, le kérogène cèdera la place à l'élément carbone, sous les espèces modestes du graphite d'abord puis, toujours plus bas, apparaîtra la perfection du diamant.

La boucle étant bouclée, le carbone à nouveau au degré zéro de son histoire attendra l'heure du recommencement : inerte et mort dans les conditions terrestres, seule une explosion cosmique lui rendra le mouvement.

BIBLIOGRAPHIE

1. Bandurski, E.L. et Nagy, B., (1976), *Geochim. et Cosmochim.* Acta **40**, 1396.
2. Bardet, M.G., (1972), « Géologie du Diamant », éd. BRGM, Paris.
3. Boato, G., (1954), Geochim. et Cosmochim. Acta, **6**, 209.
4. Bubnov, V.A. et Krivelevic L.M., (1973), *Trudy Instituta Okeanologii,* **95**, 60, Kaliningrad.
5. Burlingame, A.L., Haug, P., Belskyi et Calvin M., (1965), *Proc. Natl. Acad. Sci.* **54**, 1406.
6. Buvet, R., (1974), *in : Ecole de Roscoff 1974,* CNRS, Paris, 7.
7. Cameron, A.G.W., (1968), *in : Origin and Distribution of the Elements,* ed. by L.H. Ahrens, 125. Pergamon Press.
8. Danjusevskaja, A.J., Vojcekhovskaja, A.G., Kolotova, L.F. et Krasil'scikov A.A., (1970), *Geologija nefti i Gaza,* **3.**
9. Dayhoff, M.O, Lippincot, E.R., Eck, R.V. et Nagarajan, G., (1967), « Thermodynamic Equilibrium in Prebiological Atmospheres », NASA publ. SP 3040.
10. Deroo, G., Herbin, J.P., Roucache, J. et Tissot B., (1977), Initial Reports, DSDP, **41**, 865,

11. Ehrart, H., (1973), « Itinéraires Géochimiques et Cycle Géologique de l'Aluminium » — éd. Doin. Paris, 215.

12. Fox, S.W., (1973), *in : « Chemistry in volution and Systematics »,* Buttenworths, London, 641.

13. Fox, S.W., Hare, P.E., Harada, K. et Windsor C.R., (1973), *in : Proceedings of Symposium on Hydrogeochemistry and Biogeochemistry,* The Clarke Company, Washington DC, **11**, 65.

14. Gersanovic, D.E., Gorskova, T.I. et Konjukhov, A.I., (1974), *in : La Matière Organique des Sédiments Actuels et Fossiles et les Méthodes de son Etude,* éd. Nauka, Moscou, 63.

15. Harington, J.S. et Cilliers, J.J., (1963), *Geochim. et Cosmochim. Acta,* **27**, 411.

16. Hayatsu, R., Matsuoka, S., Scott, R.G., Studier, M.M. et Anders E., (1977), *Geochim. et Cosmochim. Acta,* **41**, 1325.

17. Hayatsu, R., Studier, M.M., Moore, L.P. et Anders, E., (1975), *Geochim. et Cosmochim. Acta,* **39**, 471.

18. Horne, R.A., (1970), *Proceedings of Symposium on Hydrogeochemistry and Biogeochemistry,* Clarke and Co, Washington DC, 114.

19. Hoyle, F., Wickramasinghe, N.C., (1977), *Nature,* **268**, 610.

20. Huc, A.Y., Durand, B. et Monin J.C., *Initial Reports, DSDP,* **42B XLII, 2,** 737, Washington (U.S. Government Printing Office).

21. Johns, R.B., Belsky, T., McCarthy, E.D., Burlingame, A.L., Haug, P., Schnoes, H.K., Richter, W. et Calvin, M., (1966), *Geochim. et Cosmochim. Acta* **30**, 1191.

22. Knacke, R.F., (1977), *Nature,* **269**, 132.

23. Levin, B.J., Kozlovskaja, S.V., et Starkova, A.G., (1956), *Meteoritika,* **14**, 38.

24. Levy, R.M., Grayson, M.A. et Wolf, C.J., (1973), *Geochim. et Cosmochim. Acta,* **37**, 467.

25. Manskaja, S.M. et Kodina, L.A., (1975), « Géochimie de la Lignite », éd. Nauka, Moscou.

26. Nagy, B. et Nagy, L.A., (1969), *in : Advances in Organic Geochemistry 1968,* Pergamon Press, 209.

27. Pelet, R., (1978), *« Géochimie Organique des Sédiments Marins Profonds, Mission Orgon 2, 1975.* A. Combaz et R. Pelet, éd. CNRS, Paris 1978, 371.

28. Pollock, G.E., Cheng, C.N., Cronin, S.E. et Kvenvolden K.A., (1975), *Geochim. et Cosmochim. Acta,* **39**, 1571.

29. Prashnowsky, A.A. et Schidlowsky, M., (1967), *Nature,* **216**, 560.

30. Rodionova, K.F. et Sidorenko, S.A., (1973), *Trudy VNIGRI, vypusk,* **138.**

31. Romankevic, E.A., (1977), « Géochimie de la Matière Organique Océanique », éd. Nauka, Moscou.

32. Ronov, A.B., (1958), *Geochemistry,* **5**, 510.

33. Rumble, D., Hoering, T.C. et Grew, E.S., (1977), « Annual Report, Geophysical Laboratory », Carnegie Institution, Whasington DC, 623.

34. Sidorenko, S.A. et Sidorenko, A.V., (1975), « La Matière Organique des Roches Sédimentaires et Métamorphiques du Précambrien », éd. Nauka, Moscou.

35. Studier, M.H., Hayatsu, R. et Anders, E., (1972), *Geochim. et Cosmochim. Acta,* **36**, 189.

36. Subbota, M.I., Khodzkukliev, J.A. et Romanjuk, A.F., (1976), « Classifications de la Matière Organique Dispersée des Roches Sédimentaires in *Etude de la Matière Organique des Sédiments Récents et Fossiles,* éd. Nauka, Moscou.

37. Van Krevelen, (1961), « Coal », Elsevier.

38. Vassoyevitch, N.B., (1975), *Vestnik Moskovskogo Universiteta,* **5**, 3.

39. Wickramasinghe, N.C., Hoyle, F., Brooks, J. et Shaw, G., (1977), *Nature,* **269**, 674.

40. Wilk, H.B., (1956), *Geochim. et Cosmochim. Acta,* **9**, 279.

Subject Index

Index des matières (des chapitres 3, 11 et 15 en langue française)

ACHEVÉ D'IMPRIMER
EN FÉVRIER 1980
PAR L'IMPRIMERIE BAYEUSAINE
14401 BAYEUX
N° d'impression : 2974
N° d'éditeur : 496
Dépôt légal : n° 4001 - 1er trimestre 1980

IMPRIMÉ EN FRANCE